AF869962

Advanced Concurrent Engineering

For further volumes:
http://www.springer.com/series/8083

Daniel D. Frey · Shuichi Fukuda
Georg Rock
Editors

# Improving Complex Systems Today

## Proceedings of the 18th ISPE International Conference on Concurrent Engineering

Springer

Assoc. Prof. Daniel D. Frey
Massachusetts Institute of Technology
Mass Ave 77, Room 3-449D
Cambridge, MA 02139
USA
e-mail: danfrey@mit.edu

Prof. Georg Rock
University of Applied Science Trier
Schneidershof
54293 Trier
Germany
e-mail: g.rock@inf.fh-trier.de

Consult. Prof. Shuichi Fukuda
Stanford University
228 Hamilton Ave
Palo Alto, CA 94301
USA
e-mail: shufukuda@cdr.stanford.edu

ISSN 1865-5440
ISBN 978-0-85729-798-3
e-ISBN 978-0-85729-799-0
DOI 10.1007/978-0-85729-799-0
Springer London Dordrecht Heidelberg New York

British Library Cataloguing in Publication Data.
A catalogue record for this book is available from the British Library.

© Springer-Verlag London Limited 2011

Post-it is a registered trademark of 3M Company, 3M Center, 2501 Hudson Road, St Paul, Minnesota 55144-1000, United States of America.

Apart from any fair dealing for the purposes of research or private study, or criticism or review, as permitted under the Copyright, Designs and Patents Act 1988, this publication may only be reproduced, stored or transmitted, in any form or by any means, with the prior permission in writing of the publishers, or in the case of reprographic reproduction in accordance with the terms of licenses issued by the Copyright Licensing Agency. Enquiries concerning reproduction outside those terms should be sent to the publishers.

The use of registered names, trademarks, etc., in this publication does not imply, even in the absence of a specific statement, that such names are exempt from the relevant laws and regulations and therefore free for general use.

The publisher makes no representation, express or implied, with regard to the accuracy of the information contained in this book and cannot accept any legal responsibility or liability for any errors or omissions that may be made.

*Cover design:* eStudio Calamar, Berlin/Figueres

Printed on acid-free paper

Springer is part of Springer Science+Business Media (www.springer.com)

## Preface

Engineering has transformed the world so thoroughly that it is difficult to imagine our lives without its benefits. Most every action we take – meeting our basic needs for food and shelter, moving from one location to another, communicating with others, carrying out or work – all these have changed significantly over the past century and continue to change at a rapid pace today.

In many cases, the transformation of human activity by engineering has come about through creation of complex systems. For example, the infrastructures that provide clean water, electric power, and communication bandwidth are amazing due to their scale and complexity. Though we may pay these systems little attention in our daily lives, it is worthwhile to pause and take note of the chain of technical innovations and investments needed to bring them about and keep them in service.

Concurrent Engineering (CE) seeks to augment the successes of engineering by making our professional activities even more effective. Building on the engineering practices of the past, we seek to add insights from a diverse set of scholarly disciplines. The research practices of the social sciences help us to understand the engineering process more deeply as a human activity. Leveraging information technology ensures that information flows effectively, takes ever more useful forms, and can be visualized and shared. Today, CE concentrates on enterprise collaboration and its many different elements, from integrating people and processes to very specific complete multi/inter-disciplinary solutions. Each sub-discipline of engineering, science, and management has informed engineering practice in the past. CE seeks to ensure that record of successful integration will continue and intensify.

The conference CE2011 is a showcase for all the ways that research, development, and scholarship can advance the engineering profession in its broadest view. The theme of "Improving Complex Systems Today" is suggestive of the scale and ambition of the field as it is currently practiced. It is reflected in the papers presented here covering contemporary system design challenges such as sustainability, international development, and information technology. We hope readers will be inspired by the ideas in this proceedings and find many building blocks for further work.

--------------------------------

Daniel Frey
General Chair, CE 2011
Massachusetts Institute of Technology, USA

Shuichi Fukuda
General Chair, CE 2011
Stanford University, USA

Georg Rock
Program Chair, CE 2011
University of Applied Science Trier, Germany

Program Committee CE 2011

*General Chair:*
Daniel Frey, Massachusetts Institute of Technology, USA

*General Chair:*
Shuichi Fukuda, Stanford University, USA

*Local Chair:*
Sanjay Sarma, Massachusetts Institute of Technology, USA

ISPE Technical Committee CE 2011

Ahmed Al-Ashaab, Cranfield University, UK
Jianzhong Cha, Beijin Jiaotong University, China
Richard Curran, Delft University of Technology, Netherlands
Shuichi Fukuda, Stanford University, USA
Parisa Ghodous, University Lyon , France
Ricardo Goncalves, UNINOVA, Portugal
Geilson Loureiro, INPE, Brazil
Jerzy Pokojski, Warsaw University of Technology, Poland
Rajkumar Roy, Cranfield University, UK
Eric Simmon, NIST, USA
Mike Soboleswki, Texas Tech University, USA
Amy Trappey, National Taipei University of Technology, Taiwan
M.J.L.Tooren, Delft University of Technology, Netherlands
P.M.Wognum, Wageningen University, Netherlands

| | | | |
|---|---|---|---|
| Carlos | Agostinho | Univ. Nova de Lisboa | Portugal |
| Ahmed | Al-Ashaab | Cranfield University | UK |
| Kazuhiro | Aoyama | The University of Tokyo | Japan |
| Otavio | Bogossian | INPE | Brazil |
| Monica | Bordegoni | Politecnico di Milano | Italy |
| Adam | Borkowski | IFTR, Polish Academy of Sciences | Poland |
| Marcio | Branco | INPE | Brazil |
| Diogo | Branquinho | INPE | Brazil |
| Tadeusz | Burczyński | Silesian University of Technology | Poland |
| Osiris | Canciglieri Jr | PUCPR | Brazil |
| Wei-Lun | Chang | Tamkang University | Taiwan |
| Edward | Chlebus | Wrocław University of Technology | Poland |
| Chen | Chun-Hsien | Nanyang Technological University | Singapore |
| Shane | Colton | MIT | USA |
| Carlos | Costa | UCS | Brazil |
| Richard | Curran | TU Delft | Netherlands |
| Walter Abrahão | dos Santos | INPE | Brazil |
| Jan | Duda | Cracow University of Technology | Poland |
| Alex | Duffy | University of Strathclyde, Glasgow | UK |
| Joze | Duhovnik | University of Ljubljana | Slovenia |
| Marcos | Esterman | Rochester Institute of Technology | USA |
| Fatima | Farinha | Algarve University | Portugal |
| Daniel | Frey | MIT | USA |
| Ernst | Fricke | BMW | Germany |
| Shuichi | Fukuda | Stanford University | USA |
| Shinichi | Fukushige | Osaka University | Japan |
| Yoshiyuki | Furukawa | National Institute of Advanced Industrial Science and Technology | Japan |
| John | Gabrieli | MIT | USA |
| Rajit | Gadh | University of California, Los Angeles | USA |
| Parisa | Ghodous | LIRIS CNRS | France |
| Maciej | Gil | Warsaw University of Technology | Poland |
| Edison | Goncalves | University of Sao Paulo | Brazil |
| Ricardo | Goncalves | Univ. Nova de Lisboa | Portugal |

| | | | |
|---|---|---|---|
| Antonio | Grilo | Universidade Nova de Lisboa | Portugal |
| Balan | Gurumoorthy | Indian Institute of Science | India |
| Arturo Molina | Gutiérrez | ITESM Campus Mexico City | Mexico |
| Timothy G. | Gutowski | MIT | USA |
| Kazuo | Hatakeyama | Federal University of Technology, Parana | Brazil |
| Kazuo | Hiekata | The University of Tokyo | Japan |
| Masaru | Hirakata | National Maritime Research Institute | Japan |
| Wolfgang | Huhnt | Berlin Univ. of Techn. | Germany |
| Masato | Inoue | University of Electro-Communications | Japan |
| Haruo | Ishikawa | University of Electrocommunications | Japan |
| Pingyu | Jiang | Xian Jiao Tong University | China |
| John | Jianzhong Cha | Beijing Jiaotong University | China |
| Germano | Kienbaum | INPE | Brazil |
| Yoshinobu | Kitamura | Osaka University | Japan |
| Zbigniew | Kłos | Poznan University of Technology | Poland |
| Tsuyoshi | Koga | Yamaguchi University | Japan |
| Roman | Konieczny | Poznan University of Technology | Poland |
| Vlaho | Kostov | Panasonic | Germany |
| Andrew | Kusiak | Iowa University | USA |
| Jay | Lee | University of Cincinnati | USA |
| Maria | Lígia | ITA | Brazil |
| Celson Pantoja | Lima | Univ. Nova de Lisboa | Portugal |
| Geilson | Loureiro | INPE/ITA | Brazil |
| Kohei | Matsuo | National Maritime Research Institute | Japan |
| Joao Pedro | Mendonca | Universidade do Minho | Portugal |
| John | Miles | Cardiff University | UK |
| Nozomu | Mishima | National Institute of Advanced Industrial Science and Technology | Japan |
| Wojciech | Moczulski | Silesian University of Technology | Poland |
| Davy | Monticolo | University of Technology UTBM | France |
| Sara | Mountney | Sheffield Hallam University | UK |

| | | | |
|---|---|---|---|
| Jorge | Muniz | UNESP | Brazil |
| Flores | Myrna | EPFL | Switzerland |
| Masashi | Okubo | Doshisha University | Japan |
| Egon | Ostrosi | Universite de Technologiede Belfort-Montbeliard | France |
| Michael | Pecht | University of Maryland | USA |
| Marcus Vinicius Pereira | Pessoa, | Brazilian Air Force | Brazil |
| Madhav S. | Phadke | Phadke Associates Inc | USA |
| Jerzy | Pokojski | Warsaw University of Technology | Poland |
| Yaqub | Rafiq | Plymouth University | UK |
| Ravi | Rangan | Centric Software | USA |
| Łukasz | Rauch | AGH University of Science and Technology | Poland |
| Rajkumar | Roy | Cranfield | UK |
| Andrew | Sage | George Mason University | USA |
| Kurt | Sandkuhl | School of Engineering at Jönköping University | Sweden |
| Marcus Venicius | Santos | Monsanto | Brazil |
| Joao | Sarraipa | Univ. Nova de Lisboa | Portugal |
| Troy | Savoie | MIT | USA |
| Andy | Schürr | Darmstadt University | Germany |
| Vladimir | Sedenkov | | Byelorussia |
| Wojciech | Skarka | Silesian University of Technology | Poland |
| Ian | Smith | EPFL | Switzerland |
| Michael | Sobolewski | TTU US | USA |
| Ram | Sriram | NIST | USA |
| Dragan | Stokic | ATB - Institute for Applied Systems Technology Bremen GmbH | Germany |
| Tatsuhiko | Sugiyama | UNICUS Co., Ltd. | Japan |
| Karol | Szustakiewicz, | Warsaw University of Technology | Poland |
| Shun | Takai | Missouri University of Science and Technology | USA |
| Kenji | Tanaka | University of Tokyo | Japan |

| | | | |
|---|---|---|---|
| Toshitake | Tateno | Advanced Institute of Industrial Technology | Japan |
| Luís Gonzaga | Trabasso | ITA | Brazil |
| Amy | Trappey | National Taipei University of Technology | Taiwan |
| Mitchell M. | Tseng | Hong Kong University of Science and Technology | China |
| Ricardo | Valerdi | MIT | USA |
| Wilfred | van der Vegte | TU Delft | Netherlands |
| M.J.L. | Van Tooren | TU Delft | Netherlands |
| Emilia | Villani | ITA | Brazil |
| Yue | Wang | Hong Kong University of Science and Technology | Hong Kong |
| Dan | Whitney | MIT | USA |
| Jerzy | Wróbel | Warsaw University of Technology | Poland |
| Hiroyuki | Yamato | University of Tokyo | Japan |
| R.I. | Young | Loughborough University, | UK |
| Kenji | Yura | University of Electro-Communications | Japan |

**Part I Systems Engineering**

## Part II Sustainable Environment



## Part III New ICT for Concurrent Engineering



## Part IV Knowledge Engineering

## Organizers

**International**
**Society**
**for**
**Productivity**
**Enhancement**

**150 Anniversary**
**of**
**Massachusetts**
**Institute**
**of**
**Technology**

MIT
+150
INVENTIONAL WISDOM

# Part I
# Systems Engineering

# Underestimation in the "When It Gets Worse Before it Gets Better" Phenomenon in Process Improvement

Ricardo Valerdi[a,1] and Braulio Fernandes[b]

[a]MIT, Lean Advancement Initiative, Cambridge, MA, USA.
[b]National Institute for Space Research, São José dos Campos, SP, Brazil.

**Abstract.** When people make interventions to a system they expect the effects to be nearly instantaneous. Unfortunately, in most of the cases the intervention intended to improve the process actually causes outcomes to get worse before they get better, if they get better at all. The challenge in these types of situations is being able to readjust expectations that there is a delay in the improvement. This is not simply a case of learning curve where people get better by performing repetitive tasks over time. What we are describing is a delay in the improvement and, in some cases, a degradation of performance.

In this paper we discuss why humans tend to underestimate such delays in process improvement across a variety of circumstances. To illustrate this, we compare data collected from a survey with three well-documented scenarios of process improvement: the implementation of a preventative maintenance program at DuPont, the modification of Tiger Woods' golf swing, and the implementation of a platform engineering initiative for the embedded software product line at Rolls-Royce. We discuss potential reasons for the chronic underestimation of these types of improvements and recommend mechanisms for making these estimates more realistic.

**Keywords.** Process delays, improvement paradox, planning fallacy, hyperbolic discounting, system dynamics.

## 1 Introduction

When people make interventions to a system they expect the effects to be nearly instantaneous. Traffic engineers, for example, expect the addition of a new lane of highway to immediately decrease traffic congestion. Similarly, organizations look for immediate productivity improvement when new processes or tools are put in place. Unfortunately, in most of these cases the intervention intended to improve the process actually causes outcomes to get worse before they get better, if they get better at all.

[1] Massachusetts Institute of Technology – Lean Advancement Initiative, 77 Massachusetts Avenue, Building 41-205, Cambridge, MA 02139. United States of America. Phone: (617) 253-8583. Fax: (617) 258-7845. rvalerdi@mit.edu.

D. D. Frey et al. (eds.), *Improving Complex Systems Today*,
Advanced Concurrent Engineering, DOI: 10.1007/ 978-0-85729-799-0_1,
© Springer-Verlag London Limited 2011

The challenge in these types of situations is being able to readjust expectations that processes will not perform better immediately after an intervention, but that there is a delay in the improvement. The causes of delay may be due to the chaos caused by the intervention itself. The unlearning-relearning process has been shown to slow people down – lowering their productivity – for a period of time before they recover their previous productivity levels. This is not simply a case of learning curve where people get better by performing repetitive tasks over time. What we are describing is a delay in the improvement and, in some cases, a degradation of performance.

In this paper we discuss why humans tend to underestimate such delays in process improvement across a variety of circumstances. The estimated delay times vary significantly across people based on their perception of the speed in which certain changes can affect the system. To illustrate this we compare three well-documented scenarios of process improvement: the implementation of a preventative maintenance program at DuPont, the modification of Tiger Woods' golf swing and the implementation of a platform engineering initiative for the embedded software product line at Rolls-Royce.

Optimizing engineering design cycles based on performing tasks concurrently is the overall goal of concurrent engineering. A key concept for concurrent engineering is the idea that all processes of a systems life-cycle should be taken into consideration in early design efforts. Since these are relevant interventions in processes, it is important to understand them and estimate accurately the impact of the performance dip (improvement delay) on change management strategy.

We discuss potential reasons for the chronic underestimation of these types of improvements and recommend potential mechanisms for making these estimates more realistic. The contribution of this research is twofold. First, it helps calibrate future planning by leveling expectations for process improvements. Second, it facilitates discussion of dynamic behavior of systems and enables visualization of complex behaviors so that decision makers can make more informed decisions.

## 2 Process delay underestimation

Delay is a process whose output lags behind its input [13]. It is usually part of a higher level process which uses its output as new input. Delays are an important source of dynamics in many systems: in organizations, for example, feedback delays are critical, since it may causes relevant aspects of the system to be mispercepted and thus bias erroneously a decision-making process.

The literature provides many examples of processes in which delays are relevant: implementation of a preventive maintenance program [14], implementation of quality programs [6], relevant intervention in traffic network, environmental degradation by human activity [2] or implementation of new policies in society. All such processes involve human interaction and they represent a meaningful change in people's daily activities. It is expected that there will be a decrease in performance, while employees/individuals adapt to these changes, but then the delayed feedback effects finally close the loop and the benefits will show up, as performance enhance beyond earlier levels.

It takes time to measure and report information, as it takes time to make decisions and for these decisions to affect the state of the system. While many thinking models for organizational learning share the assumption that there are no delays between taking an action and perceiving its full consequences [11], learning is significantly hampered if the perceived delays do not match the actual delays. However, correctly perceived delays have only a minor impact on performance [14]. The same study shows that under misperceived delays some organizations may conclude that it has found the optimal payoff and stop exploring other regions of the payoff landscape when they are actually into a local suboptimum.

With the view to better understand delays in the worse-before-better phenomenon we explore different contexts looking for reasons for underestimation of process delays and techniques to improve estimation. Next section presents real data from documented processes that illustrate the phenomenon we described here.

## 3 Examples of process delays

To illustrate the concept of process delays, we present three examples in which the worse-before-better phenomenon is observed from different perspectives. The first example is DuPont's efforts to improve maintenance and equipment availability [1][12]: the first impact of a Preventive Maintenance Program is a decline in equipment availability and an increase in maintenance costs. Only after 15 months will the increase in cumulative cost savings start to show up [5].

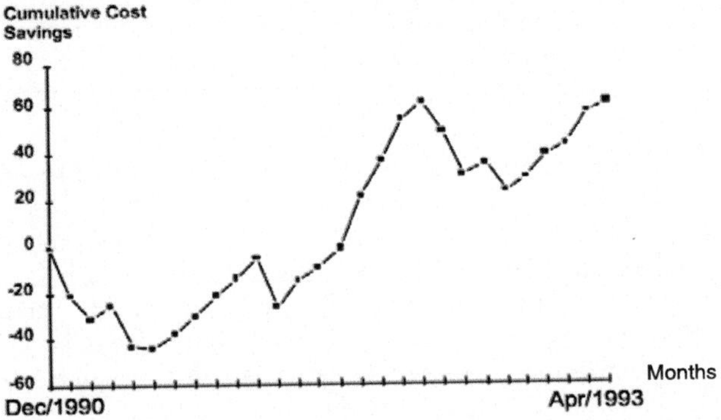


**Figure 1**: Cost saving at a typical DuPont plant after implementing a preventive maintenance program. Vertical Scale disguised [12].

The second example is Tiger Woods' attempt to increase his golf performance by changing his golf swing. At the end of 2002, after the entire season "playing in pain" as he stated, Woods had a knee surgery. Considering his physical situation and the feeling that he "could get better" [4] he decided to change his swing. Figure 2 shows his performance as ranking for prize money. After the change, his performance decreased and after 2 years the improvements became noticeable, as he recognizes at the same article "I'm starting to see some of the fruits now, which is great." [4].

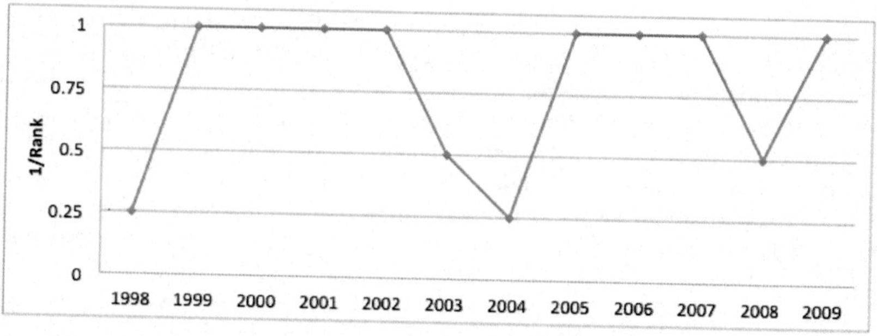


**Figure 2.** Tiger Woods' ranking for Prize Money [3]

The third example applies to software engineering. When Rolls Royce implemented a reuse initiative on a project for embedded software for military jet engines, the cumulative cost of the product line kept higher than the traditional cost for the next 4 projects.

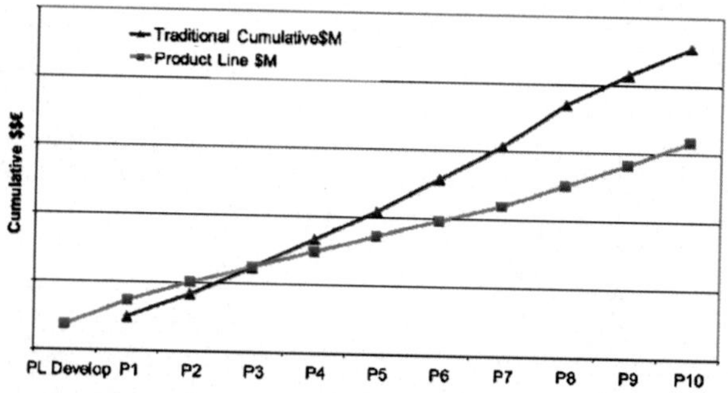


**Figure 3.** Cumulative cost for a typical project for embedded software at Rolls Royce after implementing a reuse initiative [10].

In these three examples, companies (an athlete) succeeded managing the change process and were able to recover the investment with higher performance. In order to make this process more predictable, it is critical to understand the delay and be able to measure it accurately. However, estimating delay and distributions is difficult and requires substantial data and organizational expertise. Also, experimental research suggests people have great difficulty recognizing and accounting for delays, even when information about their existence, length, and content is available and salient [13].

In order to test our hypothesis about people's tendency to underestimate delays in process improvement we conducted a 2-question survey regarding two of the situations presented above, as described in the next section.

## 4 Method

Each question was aimed at testing the same phenomenon of underestimation but in different contexts (manufacturing and software engineering). This helped with external validity, or generalizability, of the results since respondents were asked to think about process improvement delays in vastly different scenarios.

The administration of the survey was done at a professional society conference where participants were in the room at the same time. This allowed us to control for survey reliability because it allowed respondents to request clarification on the questions for their own benefit and the benefit of others.

Although this was not a random selection of subjects, we drew from a homogeneous population of experienced engineers from the automotive, aerospace, IT, and military domains. A total of 24 respondents participated in the survey and, with the exception of one respondent, everyone understood the questions provided.

The following questions were presented to the participants:

- DuPont has implemented a preventive maintenance program at one of its large manufacturing plants. How many months do you think it would take to recover the investment made in the preventive maintenance program?
- Rolls Royce has implemented a reuse initiative (i.e., platform engineering) on a project for embedded software for military jet engines. How many subsequent projects must they reuse the software to get a payback in their investment?

## 5 Results

In order to depict the results from the survey, a histogram was created to show the accuracy of the 23 participants in their estimation of process delays. The answers from each participant were compared to the real case and their relative error has been assessed, Figure 4. The histogram shows the frequency of the relative error for each given answer (bars) and the average error relative to the true answer (triangle).

From Figure 4, we observe that 67.39% (31 out of 46) answers were underestimated and the average relative error of the estimation was -14.71%. The answers are distributed in a very large range; the standard deviation of the relative error is of 49.39%. Also, considering only the underestimated answers, the average relative error is 50%.

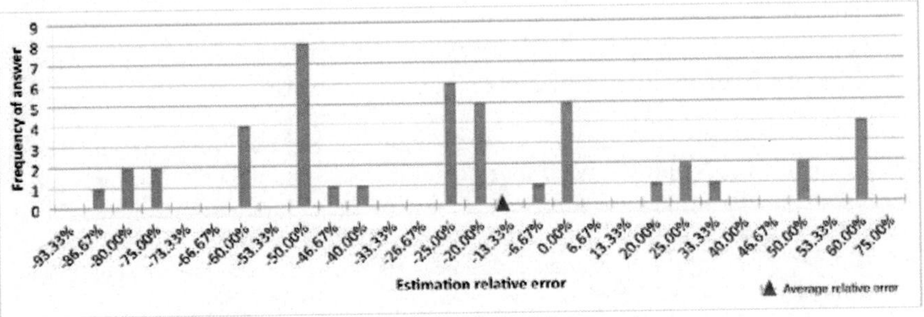


**Figure 4.** Results from Process Delay Estimation Survey

The high standard deviation suggests that there is no consensus about a single time interval, which we interpret as a result of lack of understanding of the intervention process and lack of a uniform approach to make estimations.

The information about Tiger Woods's change in his swing wasn't available at the moment when we conducted the survey, thus we lack this view to enrich our conclusions. Nevertheless, the results from this survey support our general hypothesis that people underestimate the length of delays in consequence of relevant changes in a process. The human element involved in estimating is poorly understood and thus reasons for this behavior or techniques for conducing better estimations still need to be developed.

There is a significant effect towards reducing the delay time if the perceived delays match the actual delays, in the next section we discuss reasons for underestimation and then methodologies for improvement.

## 6 Reasons for underestimation

Usually, managers estimate durations of a project using some statistical method if data are available, or they make predictions based in their own experience or judgment, if data are not available. Judgmental estimates of aggregate delays can be quite unreliable and usually underestimate their duration. It has been shown that the longer the delay, the greater the degree of underestimation [13]. Prediction based in past experiences or judgment will often lead to inaccurate estimates as a result of bias in human judgment or if recent changes in the process are not considered.

Valerdi and Blackburn [15] have studied the influence of optimism bias in human judgment and concluded that systems engineers are inherently optimistic in estimation practice. This same study shows there is a significant increase in accuracy if techniques were used to "calibrate" engineers.

Yet, enterprises are complex systems and any changes in these systems will alter its level of complexity, thus the original baseline of the system's properties do not extend their linearly, making it more difficult to predict their future states. The next section suggests techniques to help engineers make better estimates.

## 7 Mechanisms for improvement

The average length of a delay can be estimated by two principal methods: statistical techniques and firsthand investigation [13]; statistical techniques are not considered in this article because of the lack of data. Firsthand investigation is useful when data are not available because it involves investigation of processes in the field and also considers human judgments. Some strategies are presented here but it is up to individuals to decide which are most appropriate for their circumstances. A combination of these strategies is likely to work the best.

The most useful strategy is decomposition: instead of estimating the total length of the delay, decompose the process into its various stages, then estimate the time required for each stage [13].

A practical technique is betting money or pretending to bet money. A study showed that people do not necessarily have to bet money, as long as you pretend to bet money their accuracy automatically improves [15]. Pretending that there is money involved tends to help individuals be more realistic judges of events because they associate their accuracy with a financial reward.

Another technique is to separate the observation from the task itself so that judgments can be made independent of outcome. One study found that separating "doing" from "observing" a task helps reduce the level of optimism [7].

It is clear that personality plays a significant role, but professions – and associated rewards structures – also determine how well a person is calibrated. With respect to the engineering community, one might implement best practices similar to those of more calibrated professions such as meteorologists and bookies: (I) constantly providing mechanisms of feedback about prior estimates, (II) creating an incentive structure that values accuracy in estimation, and (III) ensuring there is no overreaction to atypical events [8][9].

## 8 Conclusions

In this paper we discuss why humans tend to underestimate delays in process improvement across a variety of circumstances. To illustrate this we conducted a survey about three well-documented scenarios of process improvement: the implementation of a preventative maintenance program at DuPont and the implementation of a platform engineering initiative for the embedded software product line at Rolls-Royce.

Since these are relevant interventions in relatively stable processes, it is important to understand them and estimate accurately the impact of the performance dip (improvement delay) on change management strategy. This research contribution to concurrent engineering is twofold: first, it helps calibrate future planning by leveling expectations for process improvements; second, it facilitates discussion of dynamic behavior of systems and enables visualization of complex behaviors so that decision makers can make more informed decisions.

Results from this survey support our general hypothesis that people underestimate the length of delays as a consequence of relevant changes in a process. The human element involved in estimating is poorly understood and thus reasons for this behavior still need to be developed.

When estimating delays, usually data are not available. We comment why inaccurate estimates result in biases in human judgment and if complexity increased due to recent changes in the process are not considered. Finally, we present techniques to improve estimation accuracy that are known to have a significant impact on quality of estimate.

# 9 References


[1] Carroll J., Sterman J., and Marcus A. (1998). Playing the maintenance game: How mental models drive organizational decisions. In Halpern, J. and Stern, R. (Eds.), Debating rationality: Nonrational elements of organizational decision making (pp. 99-121). Ithaca, NY: Cornell University Press.

[2] Fiddaman Thomas S. Exploring policy options with a behavioral climate–economy model. System Dynamics Review. 18, 243–267, (2002).

[3] PGA Tour Money Leaders. Available at: <http://espn.go.com/golf/moneylist>. Accessed on: Feb. 14th 2011.

[4] GOLF; Change Is Good as Woods Shoots Best Score of the Day (The New York Times). Available at: <http://query.nytimes.com/gst/fullpage.html?res=9B03E0D6153CF935A35752C1A9629C8B63&scp=9&sq=tiger%20woods%20change%20swing&st=cse&emc=eta1>. Accessed on: Feb. 14th 2011.

[5] Jones Drew, Keating Elizabeth K, Oliva Rogelio, Repenning Nelson P, Rockart Scott, Sterman John D. The Improvement Paradox: Designing Sustainable Quality Improvement Programs. Systems Dynamics Conference, 1996.

[6] Keating Elizabeth K, Oliva Rogelio, Repenning Nelson P, Rockart Scott, Sterman John D. Overcoming the Improvement Paradox. European Management Journal, Vol. 17, No. 2, pp. 120-134, 1999.

[7] Koehler D. J., Harvey N. (1997). Confidence judgments by actors and observers. Journal of Behavioral Decision Making, 10, 221-242.

[8] Lee M. and Smith G. (2002). Regression to the mean and football wagers. Journal of Behavioral Decision Making, 15(4), 329-342.

[9] Murphy A. H. and Winkler R. L. (1984). Probability forecasting in meteorology, Journal of the American Statistical Association, 79, 489-500.

[10] Nolan A. Building a Comprehensive Software Product Line Cost Model. (Internal report, Rolls-Royce).

[11] Rahmandad Hazhir, Repenning Nelson and Sterman John. Effects of feedback delay on learning. System Dynamics Review. 25, 309-338, 2009.

[12] Sterman J., Banaghan E., and Gorman E. (1992) Learning to stitch in time: Building a proactive maintenance culture at E.I. Du Pont de Nemours and Co. Case study available from author. MIT, Sloan School of Management, Cambridge, MA 02142.

[13] Sterman John D. Business Dynamics: Systems Thinking and Modeling for a Complex World. McGraw-Hill Higher Education 2000.

[14] Thun Jörn-Hendrick. Maintaining preventive maintenance and maintenance prevention: analysing the dynamic implications of Total Productive Maintenance. System Dynamics Review. 22, 163-179, 2006.

[15] Valerdi Ricardo and Blackburn Craig. The Human Element of Decision Making in Systems Engineers: A Focus on Optimism. 2009 INCOSE International Symposium, Singapore, July 19-23, 2009.

# Automation Tools Supporting the Development of Information Standards for Complex Systems

Eric Simmon[a], Sam Dana[b], Arthur Griesser[b]

[a]National Institute of Standard and Technology, USA
[b]Prometheus Computing LLC

**Abstract.** As information systems continue to become more complex, the data standards to support them grow in complexity as well. To meet the needs of today's systems, not only must the standards change, but the standards development process must change as well.

Most standards development organizations still manually develop their standards. The rising complexity of information standards may soon render this approach infeasible. In the future manually developed standards will likely not meet the needs of the users. Additional reasons for automation are reducing opportunities for human error, and speeding up the consensus building process.

The higher level artifacts that structure information standards depend on the type of standard. Standards that specify basic information structures are well suited to generation of documents from UML class diagrams: we describe a freely available tool for doing so. Standards that describe particular uses for those information structures are better off starting from use cases. We describe a freely available web-based tool for creating modular use cases supporting the re-use of actors and scenarios.

By focusing on the modeling of data exchange standards and automating the process of creating the standards document directly from the model errors can be reduced, the standards development time can be decreased and the revision process is simplified. In this paper we describe how standards could be built directly from analysis artifacts.

**Keywords.** Use cases, system design, complex systems, standards development

## 1 Introduction

Systems, such as those used for manufacturing and infrastructure, continue to become more complex. At the same time they are becoming more distributed and the number of stakeholders participating in the development and use of the systems are increasing. Causes of these trends include both external stakeholders added by outsourcing and the transition from vertical to horizontal integration. A more important reason is the extra efficiency obtainable through new levels of cooperation, when information exchanges are well specified: the SmartGrid is an example.

Without understandable, well-constructed data standards, interoperability between the heterogeneous sub-systems that make up today's meta-systems is very difficult to achieve. Ideally, standards would be constructed so well that thorough compliance testing would ensure interoperability. In practice this is rarely the case, and vendors (sometimes through an Standards Development Organization (SDO) or trade association) must decide how to handle interoperability issues. The more complicated systems get, the harder it is to develop useable standards. This comes partly from the complexity of the system itself and the difficulty to understand this

D. D. Frey et al. (eds.), *Improving Complex Systems Today*,
Advanced Concurrent Engineering, DOI: 10.1007/ 978-0-85729-799-0_2,
© Springer-Verlag London Limited 2011

complexity, and partly from the expanded diversity of stakeholders and the difficulty in resolving business goals.

The first step of developing a standard (after settling on goals [1]) should be the construction of use cases. Properly constructed and managed use cases narrow down the goal-determined scope of a project, they ensure that the standard satisfies stakeholder needs articulated in the use cases, and can serve double duty as compliance and interoperability tests (with the addition of test vectors and mocks [2]). ).Use cases can isolate standards efforts from scope creep and conflicting scope perceptions, but badly developed use cases can create more problems than they solve. It's possible to refine use cases forever, pack them with irrelevant details that tie the hands of implementers, and make so many use cases that implementers lose hope.

There are several kinds of data related standards, and development bodies do not always distinguish between them. Semantic standards define nomenclature and relationships between data. Semantic standards should be defined using languages that are designed for expressing semantics, such as UML [3], Express [4], or OWL [5]. Syntactic standards specialize a semantic standard to express data in a particular format, such as Extensible Markup Language (XML): they could re-express the entire semantic standard in a language appropriate for that purpose (such as XML Schema), or they could specify the use of a standardized set of rules for transforming the semantic description into a syntactic description. Usage semantic standards specify how to use a base semantic standard for a particular purpose or use. These standards may extend, refine, and further constrain the base standard: a language like OCL [6] would be appropriate. Usage syntactic standards such as the sectionals used in the IPC 2580 series of standards [7] specialize an usage semantic standard for a given syntax: in the XML example given earlier, Schematron [8] would be an appropriate language for an usage syntactic standard. These different kinds of standards are often not perceived: people often focus on the syntactic level, some times under the impression that the semantic level is an academic abstraction. Actually the semantic level clears the air for a simpler, more focused, more careful consideration of fundamentals. By not clearly differentiating between base and usage standards, usage information may be mixed into the base standard, diluting the usefulness of the base standard.

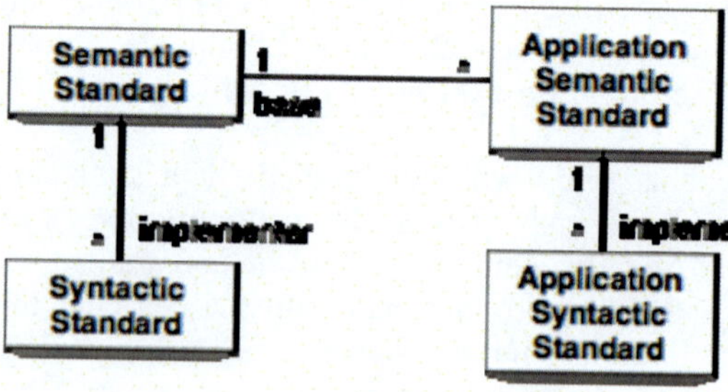


**Figure 1 - Kinds of Standards**

The usage standards amount to expressions of use cases and should be defined before (or in parallel with) the base standard. One of the biggest problems with deriving the usage standards from the use cases is that use cases are traditionally based on word processing documents, which are difficult to parse and extract information from. Another big problem is lack of consistency between different use cases – the larger the effort is, the more likely this is going to happen. For example, different authors may unwittingly use a different name to refer to the same Actor or describe the same step or scenario using slightly different language.

Resolving these problems after the fact is a lot harder than resolving them during construction of the use cases. Purely automated normalization of things appearing in use cases does not work very well, because it requires a machine to understand the meaning of natural language. A more workable approach is to gather the data under machine guidance in the first place, giving (for example) the use case author a list of all the actors already in the system, or giving the author the ability to explicitly specify that a given string is a reference to anothera standard, or a description of an issue that needs to be resolved.

## 2 Use Case Editor

To eliminate the parsing and natural language interpretation problems above, NIST has developed an open source tool for defining and editing use cases. The tool, called Use Case Editor or UCE, provides a browser-based interface that allows users to create and store use cases in a relational database located on a central server. Running the Use Case Editor will launch a server on the local machine which may then be accessed by pointing a browser to the correct address.

A UCE client initially displays a listing of the high-level components of a use case. A user can then choose to create a new component or edit an existing one. When editing a component, the user is presented with an editable form showing the attributes of the component. For any relationships the component has, the user may choose to create or remove associations to other components in the system. Once a component has been modified to the user's satisfaction, the form can be submitted, saving the changes to the server. Once saved, these changes will be available to all users of the Use Case Editor. This client/server model allows multiple users to collaborate simultaneously on the creation and editing of use cases. Each collaborator need only to have access to an internet-connected web browser in order to make changes.

Figure 2 shows an overview of the Use Case Editor's architecture. The client uses a web browser to access the UCE which in turn uses the relational database, Ruby code, and a Graphical User Interface (GUI) specification to produce an appropriate response. The database schema, GUI specification, and Ruby code are each generated from a Unified Modeling Language (UML) class diagram.

The tool has the option of using one of several models for a use case. In the most complex model, we took a broad view and allowed for white box use cases, even though most books on requirements capture agree that black box use cases are more suitable for this purpose. However, in the simplest use case model, entries are more restricted. These use case models are generated from UML diagrams, giving you the ability to easily change the way use cases are structured. Figure 3 shows the UML diagram for the simplest use case model. Because the use case editor is model driven, you can easily change the way use cases are structured by making a change to the UML diagram and regenerating the use case model.

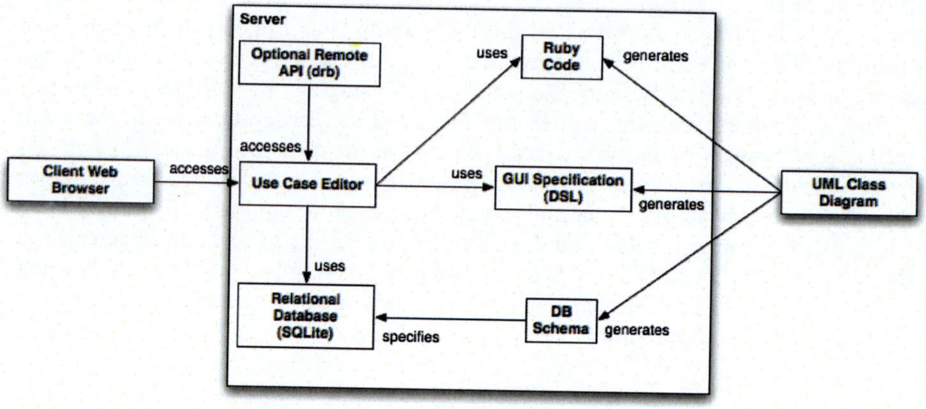


**Figure 2 – Overview of the UCE architecture**

As seen in Figure 3, an Actor can be associated with a number of UseCases which will have associated a number of ordered Steps. Each Step can be associated with an Action and can have pre- and post-conditions. So for example, if using this model with the UCE to create an Actor, you would have the option to define a name string and then define which UseCases you would like the Actor to associate with (with separate headings for 'Usecases Executed' and 'Usecases Describing'). When associating with UseCases, you would be able to select from a list of pre-existing UseCases or choose to create a new one.

One interesting feature of the UCE is the ability (for one of the use case models) to merge scenarios together into a bigger picture. Use case scenarios, as they are usually defined, are linear sequences of steps. It can be hard to envision how these scenarios fit together. They actually define a kind of state machine. Many people think of state machines only as specifying internal states of objects; state machines are indeed often used to represent internal details. The UML standard, however, does acknowledge and explicitly model state machines that do not represent internal details but instead represent how a system interacts with its environment: these are called protocol state machines.

## 2.1 Construction

The UCE is a browser based Asynchronous JavaScript and XML (AJAX) tool that stores use case data in a relational database. A distributed object API also lets you manipulate use case data remotely. The use case models used by the UCE are generated directly from UML class diagrams (such as in Figure 3). A model generated for the Use Case Editor is comprised of a collection of Ruby objects (specified by Ruby code), a database schema with which to store those objects, and a GUI specification. The information for these constructs is pulled directly from the UML model they were generated from, but each piece can be modified directly if desired. This is most useful in the case of the GUI specification file which is a Ruby “internal domain specific language” [9] that is interpreted to provide a layout for objects on the server. By tweaking the specification it is possible to customize the appearance of the UCE.

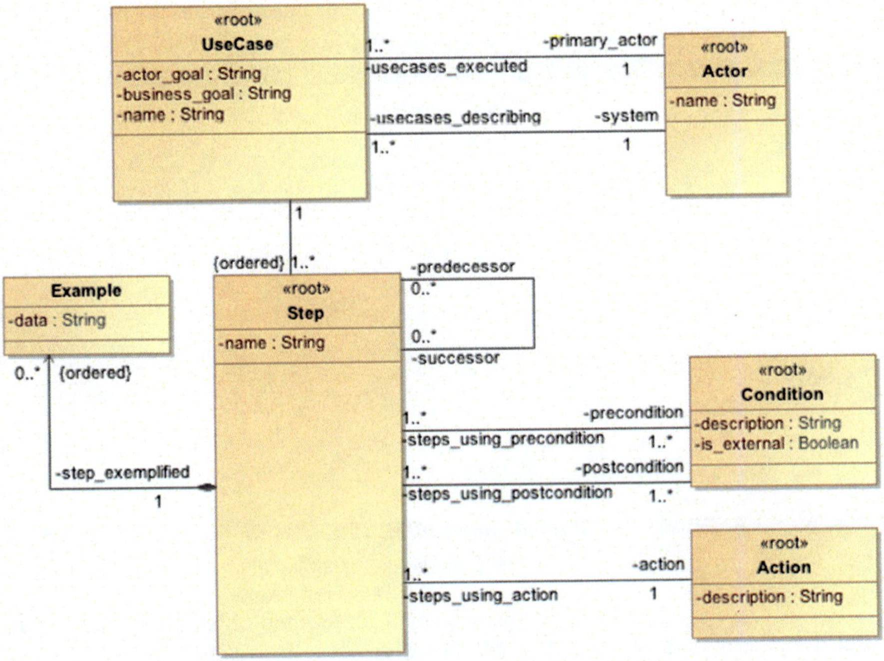


**Figure 3 - UML class diagram of the simple use case model**

The UCE is influenced by the "naked objects" [10] pattern. Like naked objects, the use case editor has a relatively generic user interface that directly represents objects. Some customizations are possible, including the ability to compose more complicated views that edit data from several objects. This approach allows us to dynamically view and edit data for any given use case model. As an example of the resulting interface, a page representing a use case creation (from the complex use case model) is shown in Figure 4.

There are numerous advantages of using normalized data to store use cases. Two features that take advantage of this are the ability to track changes to any object, and report generation. Within the UCE, a user may choose to see the history of an object (such as an actor or use case) in the system. The user is then presented with the changes that have been made to the object along with the authors of those changes. For textual fields, a visual comparison of the differences is displayed. Report generation automates the creation of several report types based on the current state of the system.

The UCE supports several complex data types for use in the UML model. By importing a specific UML profile that is provided with the Use Case Editor, a user may use RichText, Code, or File attributes types in their model. A RichText attribute behaves similarly to a String field, but provides the user with standard markup options (such as headers, font styles, and list types) for their entry. A Code attribute allows code of a specified language to be entered directly. Fourteen languages are currently supported, with syntax highlighting available for Groovy, Javascript, Python, and Ruby. The File attribute type allows users to upload (and subsequently download) a file from their local machine.

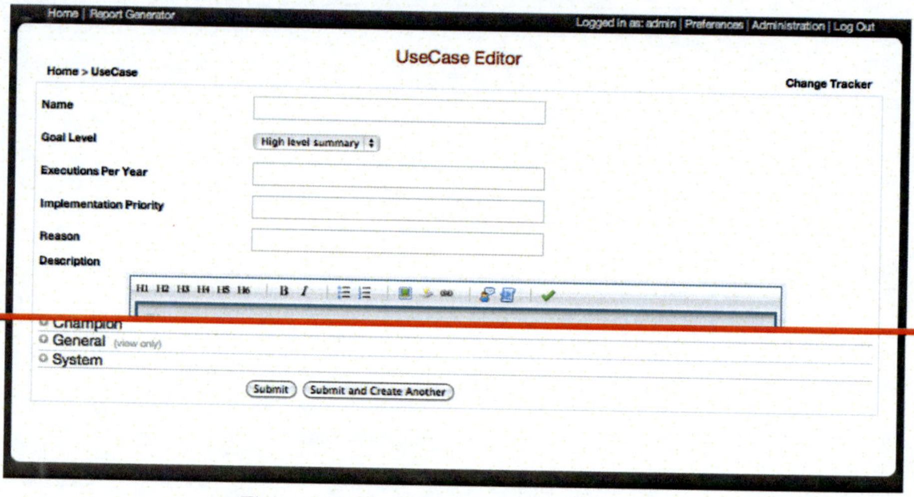


**Figure 4- Screenshot of a use case creation page**

The Use Case Editor also provides a remote distributed object API that allows create, retrieve, update, and delete (CRUD) operations. Programmatic remote data manipulation is a powerful feature but also creates a security hole, so that functionality is not enabled by default.

## 3 Document Generator

Once the use cases are finished, the development of the standard models and documentation can begin. Currently most standard documents are developed in the form of word processor documents. To describe the design model, screen shots of fragments of diagrams representing the syntax are pasted into the document. With this process, two separate documents (the models and the documentation) need to be synchronized. During standards meetings, the discussion is typically focused on one of these documents, with the understanding that the other document will be brought into synch. It's quite easy for some of those changes to slip though the cracks, and never make it into the other document. Even when the ideas in the two documents are flawlessly synchronized, it's difficult and time consuming to update the images and generate and maintain the text descriptions in the word processor document. To simplify this process, we created a tool that automates the generation of images of small focused diagrams and generates the text from a larger UML class diagram. Each image focuses on one class, to show the structure of that class, and neighboring navigable classes.

There is still a problem, however, synchronizing the textual description of the data with the ideas of the model. An easy solution is to generate the text from the model. Many UML tools support the inclusion of comments and some tools support HTML comments, including features such as tables and links. We used this capability to automate the construction of standards, in the form of an HTML document that includes the automatically generated images, as well as extensive cross linking. Similar functionality could be developed for RTF and other formats.

Building the complete standard with this tool starts by building a complete UML class model of all the domain objects used by the standard, and embedding the text of the standard in the UML model element that the text describes. Our generator currently supports HTML text describing packages, interfaces, classes, attributes, primitives, enumerations, and enumeration values. The generator uses UML packages to organize the standard. The generator presumes that a package has a high level discussion of the contents: this description is then followed by descriptions of every primitive, enumeration, class, and interface.

### 3.1 Construction

Once the text has been associated with the model elements, diagrams that will be embedded in the standard are automatically generated and saved as images. The Builder pattern [11] then extracts the description of each UML element, and assembles the final HTML document. An example snippet from the IPC-2581 "Offspring" standard (which describes printed circuit boards) is shown in **Error! Reference source not found.**

**FontDefEmbeddedType**

Implements FontDef

A description of a font using individual characters that are defined in accordance with the Glyph element and are contained in the DictionaryFont. Fonts must be predefined and may not be instantiated as part of the data in the file.

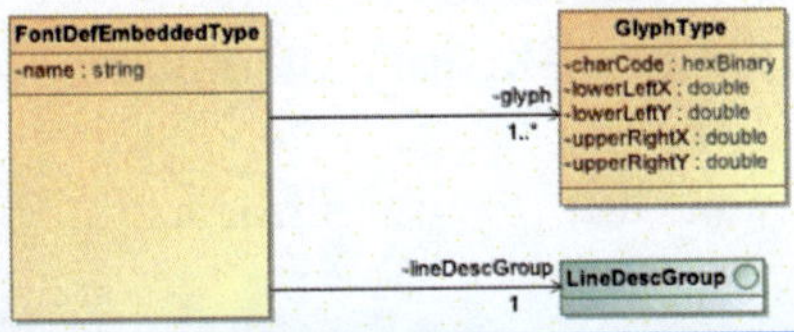


| Name | Multiplicly | Type | Description |
|---|---|---|---|
| glyph | 1..Infinity | GlyphType | The element that contains the description of all the character definitions defined in the user developed font |
| lineDescGroup | 1..1 | LineDescGroup | A substitution group that specifies the LineWidth and LineEnd characteristics of the Font by reference to a predefined LineDesc or specified when the font is instantiated. |
| name | 1..1 | string | A unique name related to the charCode of the Glyph character |

**Figure 5- An example showing how UML classes are represented**

In the Builder pattern a director object traverses data of interest (in our case UML model elements), and issues instructions to a builder object. The builder object is responsible for generating the output. In this case, the builder generates the HTML for the standard. The same director could just as easily guide a different builder the construction of an XML Schema. A third builder could generate an SQL schema, and a fourth could generate Java source code, each one working with the same director.

## 4 Conclusion

Today's complex systems are creating new challenges both in the design and development of systems and the development of supporting information standards. Not only is the complexity of the system an issue, but the amount and diversity of stakeholders are an issue.

To assist in the development of these systems NIST has created freely available software tools. The Use Case Editor helps experts develop improved normalized use cases in a collaborative environment while reducing development time. The Document Generator generates a standards document based on the design model, reducing misalignment and errors between the model, the syntactic standards and the standards documents and again reducing development time.

## 5 Reference List

[1] E. D. Simmon, A. Griesser, *Use Case Based Testing to Improve Smart Grid Development*, Proceedings of the 17th ISPE International Conference on Concurrent Engineering, Krakow, Poland

[2] A Brief History of Mock Objects, Available at http://www.mockobjects.com/2009/09/brief-history-of-mock-objects.html Accessed March 28th, 2011

[3] UML Resource Page, Available at: http://www.uml.org/ Accessed February 27th, 2011

[4] Industrial automation systems and integration - Product data representation and exchange Part 11: Description methods: The EXPRESS language reference manual, International Organization for Standardization, 1994

[5] OWL Overview, Available at: http://www.w3.org/TR/owl2-overview/ Accessed February 27th, 2011

[6] Catalog Of OMG Modeling And Metadata Specifications, Available at: http://www.omg.org/technology/documents/modeling_spec_catalog.htm#OCL Accessed February 27th, 2011

[7] IPC-2581, Available at: http://webstds.ipc.org/2581/2581.htm Accessed February 27th, 2011

[8] Schematron, Available at: http://www.schematron.com/
Accessed February 27th, 2011
[9] Domain Specific Language Available at:
http://www.martinfowler.com/bliki/DomainSpecificLanguage.html
Accessed February 27th, 2011
[10] Pawson, R., *Naked Objects*, Ph.D Thesis, 2004, Trinity College, Dublin, Ireland
[11] Erich Gamma, *Design patterns : elements of reusable object-oriented software* Reading Mass., Addison-Wesley, 1995

# Use of Structured Total Approach in Developing a System for Emergency Power Generation for Aircraft with Fly by Wire

Dinah Eluze Sales Leite[a], José Lourenço de Oliveira[b,1] and Geilson Loureiro[b]

[a]Embraer SA.
[b,1]Embraer SA.
[b]INPE – Instituto Nacional de Pesquisas Espaciais.

**Abstract.** This paper presents a Systems Engineering (SE) approach for development of the Emergency Electrical System (EES) for aircraft with Fly by Wire. After creating the model of the EES and its components relationship, these components were analyzed using the method Total View Framework, which has proved to be an important tool in the design and development of complex products. The proposed method, which includes aspects of the product, the processes and the organization, is justified in light of the failure of performance of traditional project management. In traditional approaches, the inherent complexity in product development is not taken into account. The authors believe this approach will promote the identification of items that meet the requirements related to quality, safety and reliability of multiple factors at the stage of project design.

**Keywords.** System Engineering, Emergency Electrical System, Aircraft, Fly by Wire

## 1 Introduction

The contemporary world is characterized by the development of extremely complex technologies and products and, therefore, an increasing number of variables and attributes to meet the requirements. Among the main features that can be mentioned: reliability, security, maintainability, robustness, precision and durability. These technologies and products of high complexity have the outset of the development marked by a need, desire or expectation of the stakeholders, which are defined by requirements.

An important and competitive market of the aviation industry is occupied by vendors of aircraft manufacturers. In many cases a given system can contribute up to 30% of the final cost of the product and subsystems may have an even higher

---

[1] Embraer SA, Desenvolvimento Tecnológico, Av Brigadeiro Faria Lima, 2170 - Putim, CEP: 12227-901, São José dos Campos, SP, Brazil Tel: +55 (12) 8116-6443; Email: dinah.leite@embraer.com.br

D. D. Frey et al. (eds.), *Improving Complex Systems Today*,
Advanced Concurrent Engineering, DOI: 10.1007/ 978-0-85729-799-0_3,
© Springer-Verlag London Limited 2011

technological level of complexity than the product that receives it. This scenario requires strong competition among system suppliers, demanding that you understand fully, the aircraft manufacturer's business, as well as the company operating the aircraft. In this context, this paper discusses the development of an Emergency Electric System (EES) for an aircraft also under development. The model anticipates the needs of the system with the breadth of product lifecycle host, considering the simultaneous development of the aircraft and emergency system.

## 2 Objective

One of the inspiring aspects of System Engineering (SE) is its applicability across a broad range of industries. This paper aims to present the main concepts of SE, including the Total View Framework, a method for managing the complexity of the relationship among products, processes and organization elements, their interactions, as illustrated in Figure 1, and its application during the development of an Emergency Electrical System (EES).

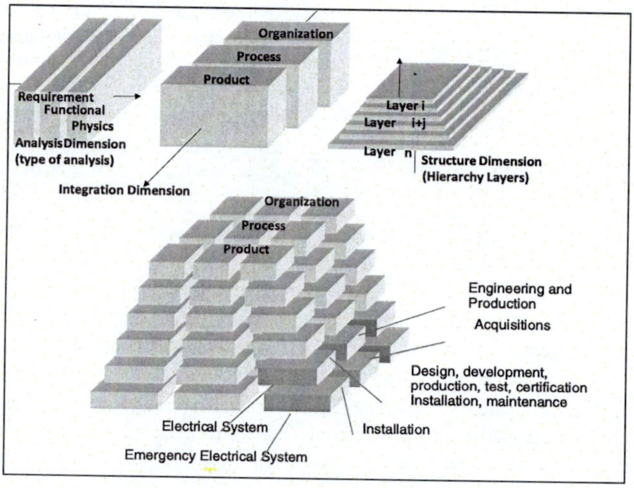


**Figure 1.** Total View Framework. Source: [5]

## 3 The Traditional Model

The reference model for comparison, which in this article will be called "The Traditional Model", is supported in the application cited by [1], in which the focus on the product is clear. An enhancement of the original model was described by [2] who suggests the introduction of new elements in the product design team with the modification of the previous approach procedure.

The reduction of development time with the adoption of teams composed by simultaneity was treated by [3], among others. The performance teams in the environment resulting from the overlap of the development model with the culture of the organization was handled by [4]. The Traditional Model of reference has achieved significant cost and time reduction; however its effectiveness is impaired for major development projects. This model cannot deal with the increased complexity of a new product development, and provide only a partial view of the system elements and their interactions.

The proposed scenario for the model is that a small team colocalized and with great technical ability and authority, can resolve any problems found. In a scenario of large projects, for which a larger number of specialists should interact to find solutions, the communication quality is impaired as well as the effectiveness of actions. The model of total structured approach instrumentalizes the development process with tools to mitigate the problem.

## 4 Total Structured Approach Model

The sequence of work with The Total Structured Approach Model to the EES is shown in the Figure 2. For each job step it will select a case or scenario that may explain the use of the model.

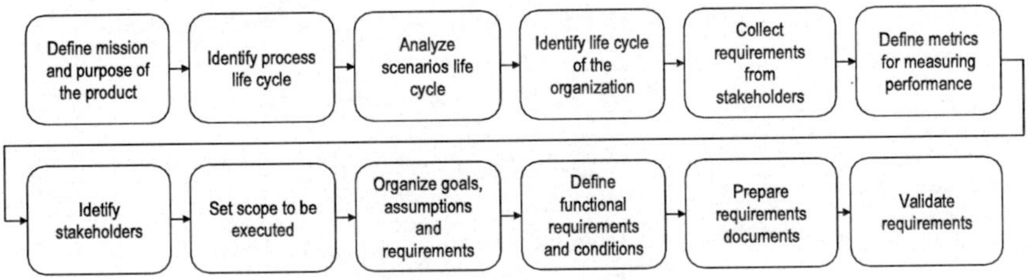


**Figure 2.** Sequence of Work

The work was based on the methodology of [5] and aims to gather all necessary information in advance for product development through a systemic view of organizational, process and product. The methodology involves analysis of context, stakeholder survey, requirements definition and product architecture. This method allows the analysis of interference and interactions of the Emergency Electrical System (EES) with other aircraft systems.

## 5 Emergency Electrical System (EES)

Emergency Electrical System (EES) for aircraft must be designed to ensure electrical power enough to a minimum set of equipment enabling the pilot to make the landing maneuver safely. In the Fly by Wire aircraft, the flight control is made of electronic and hydraulic systems. The best technological solution developed is a wind generator, which combines weight with an adequate availability of power proportional to the size of the turbine and aircraft speed. The turbo generator is housed in the body of the aircraft and it is ejected for receiving the air flow when in an emergency situation. The system has a unit called Transformer Rectifier Unit (TRU), which produces electric power in direct and alternating current as required by the manufacturer of the aircraft.

The product mission should establish clearly the function of the product that should guide the actions during the project. For the product analyzed, the mission was well established: "Emergency Electrical System (EES) is capable of generating electrical power in a general emergency scenario of the main system aircraft to keep the minimum functionality of the operation in flight until landing."

### 5.1 Product Life Cycle Identification

It is very important to have the time to market reduced due to market competitiveness, while maintaining the same quality standard and reducing the product development cost. Facing this scenario, it is worth emphasizing the importance of inter and multidisciplinary activities, with the need for involvement of various functional business areas at the stage of product development.

To perform the product development analysis is important at this stage to deploy a set of processes, with their flow of inputs and outputs. From the complete landscape of the life cycle of the product, it is possible to understand the interactions and the requirements to deploy widely. The Product Life Cycle and its phases, as well as the scope considered in this study are represented on Table 1.

**Table 1.** Product Life Cycle

| Marketing | Development | Production | Support | Closure |
|---|---|---|---|---|
| Technical-commercial proposal | Architecture definition | Production Planning | Correction and improvement | Products and services issue |
| Negotiation | Supplier definition | Components manufacturer | Maintenance | Products and services acceptance |
| Engineering alternative | Preliminary design | Assembly | Technical support | Contractual obligations verifications |
| | Drawing preparation | Electrical Test | | Financial obligations verifications |
| | Qualification documents | Hydraulic Test | | Documentation and contractual obligation |
| | Production guide | Integrated Test | | |
| | **Organization** | **Production** | | |

## 6 EES System Structure Diagrams

This section proposes an approach to analyse the EES life cycle representation through the Structure Diagram, as shown in Figure 3. Functional analysis of the product life cycle processes was made from the IDEF0 diagram that describes the system functionality, control mechanisms and data flows across all functions of the system, highlighting the integration of the whole.

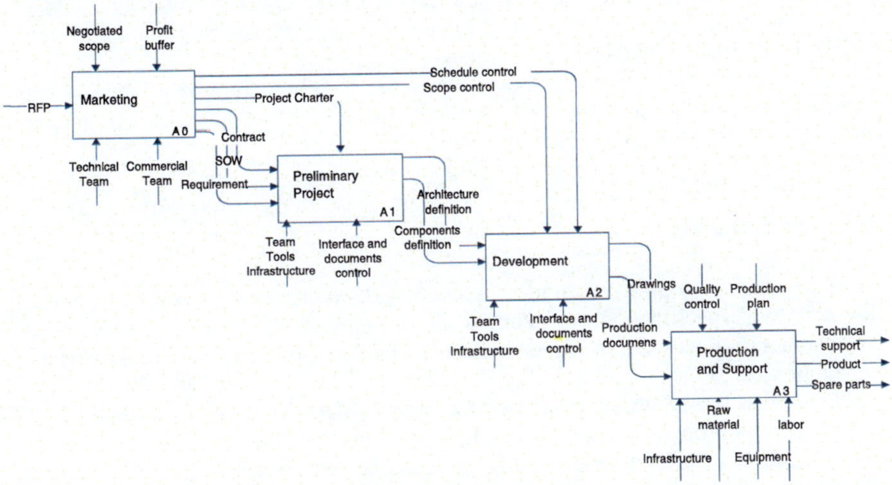


**Figure 3.** Process - Lifecycle Structure Diagram (IDEF0)

The marketing step involves Project, Proposal Preparation and Negotiation. This step has as input the Request for Proposal (RFP) and as output, loss of the business, a Contract Signed or a Statement of Work (SOW). The preliminary stage includes the Conceptual Definitions, Initial Settings and Joint Definitions phases. The entrance is marked by the Contract and the outputs are: Product Definition, Product Requirements and Work Plans.

After the preliminary design stage there is the Project Development. The development has as input the Detail Design, Construction, Prototype Field Test and Validation Tests, and has as output the Validated Product.

The last stage of the product life cycle is the production and support stage. This stage, which takes as input the drawings, production documents and production plans at the Production Preparation can follow three alternative paths. The Logistics Product phase is the product produced as output. The phase Spare Parts Logistics has as its output, parts sold. And during Engineering Support, the output is characterized by technical information.

## 7 Stakeholders Analysis

According to [6], stakeholder is any group or individual who affect or is affected by the business of an organization. Therefore, stakeholder analysis, done by a systemic view of the product, process and organization, involves the analysis of all life cycle process performing organization environment, i.e. how product, process and organization can influence the environment in which organization operates. The perfect understanding of stakeholder needs is essential in applying the Systems Engineering concepts.

Organization Traditional models suggest the main function of the firms to maximize the profits and the business return on investment. But, stakeholder theory asserts that organizations need to consider the interests of groups that may affect or be affected by these organizations [7].

In this work the authors considered only the stakeholders whose interests are directly linked to the phases of the Development, Production and Support, among the steps identified in the product lifecycle.

## 8 Requirements

According to [8], the requirement is a condition to drive a system or component to satisfy a specification, a standard or a contract. Therefore, meeting the stakeholders' needs is a fundamental factor in the product requirements development.

In this paper, all the demands of the stakeholders have been translated into technical requirements to ensure meeting their needs. Firstly, the authors structured the Measures of Effectiveness analysis, as indicated in Table 2. Effectiveness Measures aimed at assessing the level of stakeholder satisfaction, and from these measurements, the product and organization can be evaluated.

**Table 2.** Measures of Effectiveness - Stakeholders Product Operation

| Stakeholders | Interest | Measures of Effectiveness |
|---|---|---|
| Passengers | Reliable and safe product | Product reliability (failure rate) |
| Crew | Reliable and safe product<br>Ease operation<br>High dispatchability | Product reliability<br>1 min between reading the instruction and operation<br>SR - Schedule Reliability |
| Insurer | Premium reducing | Risk of loss |
| Customer: Aircraft carrier | EES to provide safe landing | Success landing rate in EES operation |

From the Measures of Effectiveness analysis, the requirements were defined.

All stakeholders' requirements must be grouped, organized and classified. It is important that the requirements established from stakeholders needs are temporary, as there are many factors that contribute to change them throughout the product life cycle. Considering the requirements process analysis as a static is just a big mistake [9]. Therefore, the requirements should be reviewed, corrected, changed, and revalidated continuously.

## 9 System Functional Analysis

In this section the EES system will be analysed according to their functions and interactions with other plane system interfaces. To model the product functional analysis, a Data Flow Diagram (DFD) was used. The DFDs of Product in operation is represented by Figure 4.

Figure 4 presents a context diagram that shows the elements of the environment for the product or organization. The arrows represent the flow of data, materials or energy. The direction of the arrows represents these elements into or out of the product or organization.

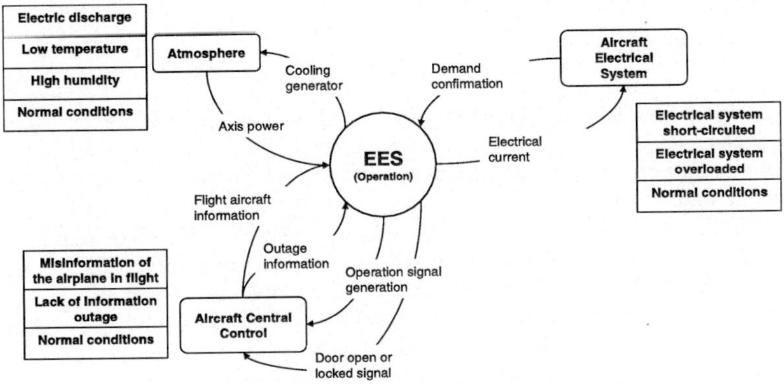


**Figure 4.** Product Operation Functional Diagram

For the product in operation, the system operation modes, the operating conditions, the states and the environmental factors identification are shown in Table 3.

**Table 3.** Operation Modes, Operating Conditions and Identifying of the Environment Elements of Product in Operating

| | | |
|---|---|---|
| | **Modes** | |
| | In flight - electrical malfunction | In maintenance |
| | **Condition** | |
| | Operational | Non Operational |
| **Environment** | **State** | |
| Aircraft electrical system | Short circuit or overload | Short circuit or overload |
| Atmosphere | Special conditions (electrical discharge, low temperature, high humidity) | Normal conditions |
| Aircraft Central Control | Wrong information<br>Lack of information | Normal conditions |

According to [5], from the technical requirements, you can do the functional analysis and develop functional product architectures and organization that are fundamental to the physical analysis.

In the physical analysis, the product and the organization physical architecture are identified and developed. The architectures provide functional and physical elements, their decompositions, and internal and external interfaces, which represent the process of requirements defining alternatives.

The need of a manual override was raised in the setting of electrical failure in flight with open failure information.

The system also features a standalone unit that provides locking and ejection of the turbo generator that can be triggered manually or with systemic identification of severe power problem. The system must meet the requirements of the approved body, in which the host aircraft will operate.

## 10 Conclusions

This paper has as main objective to show the importance of the systemic approach, stakeholder analysis, their interests, needs and expectations, and the anticipation of life cycle process requirements necessary to meet their needs in the development of a project.

Traditional methods applied to product development also address relevant issues, but are too much focused on product operation and development, overlooking the other crucial life cycle processes necessary to meet the expectations of stakeholders. The method presented in this paper provides a structured and integrated approach to complex product, its life cycle processes and their performing organization integrated development.

The conclusions are that the method, used to suit the environment of product development, provides a way to overcome the deficiencies of traditional planning, it is feasible, produces good results, especially when covered with high complexity, performs the unfolding of the requirements to the level of components from suppliers and ensures the integration of the solution and treatment of internal and external requirements in the organization, processes and product.

## 11 Citations and References


[1] Andersen, M. M.; Hein, L. Integrated Product Development. New York: Springer-Verlag, 1987.

[2] Wall, S. D., Use of Concurrent Engineering in Space Mission Design. EuSec2000 Proceedings, September 2000.

[3] Clark, K. B., and Fujimoto, T., Overlapping problem solving in product development, Harvard Business School Working Paper 87-049, 1987.

[4] West, M. A., Effective Teamwork, Leicester: British Psychological Society, 1994.

[5] Loureiro, G. A Systems Engineering and Concurrent Engineering Framework for the Integrated Development of Complex Products. 1999. – PhD Thesis. Loughborough University.

[6] Freeman, C. Innovation and the strategy of the firm. In: FREEMAN, C. The economics of industrial innovation. Harmondsworth: Penguin Books Ltda, 1988.

[7] Andriof, J.; Waddock, S.; Waddock, H.; Rahman, B. Unfolding Stakeholder Thinking: Theory, Responsibility and Engagement, Greenleaf Publishing, UK. 2002.

[8] IEEE Std 1233 Guide for Developing System Requirements Specifications, 1998.

[9] Robertson, S.; Robertson, J. Mastering the Requirements Process. Addison Wesley, 1999.

# System Concurrent Engineering of Space Payload Design Aquarius Instrument

**Luiz Alexandre da Silva[1], Paulo Vinicius Jeronimo[2], Geilson Loureiro[3]**

**Abstract.** This paper presents a System Concurrent Engineering approach applied to the development of the space payload design Aquarius instrument with upfront life cycle process considerations. This approach anticipates the life cycle process issues, identifying and solving problems in advance. The approach was developed by Prof. Geilson Loureiro and has been used, since 1999, in more than 200 academic and industrial examples. The paper starts by presenting the approach, the Aquarius instrument overview as part of the Argentine SAC-D satellite and the application of the approach to this space payload example.

**Keywords.** System Concurrent Engineering, Aquarius Instrument, SAC-D, Satellite.

## 1.1 Introduction

Satellite payloads are complex systems that have the function of fulfilling the space mission core purpose. It is composed of instrument and/or equipment depending on the objectives. They are considered multidisciplinary products. They must cope with extreme environmental conditions over their life cycle. They must undergo very strict assembly, integration and testing (AIT) procedures. There are many opportunities to improve productivity over satellite payload life cycle if a concurrent engineering approach takes place from the beginning of the satellite architecting stage. Traditional systems engineering approaches do not provide an overall view of the system during its various life cycle processes. They focus on an operational product development starting from product concept of operations. They also focus on the development organization that must be put in place in order to assure that the product meets its operational requirements. A product has life cycle processes other than operations and it must be recognized from the outset in order to promote gains in productivity in the product development organization, by the avoidance of late changes, and in other product life cycle process organizations, as the product will be developed taking into consideration their requirements. Life cycle process organizations themselves can be developed simultaneously to product development, when they are part of the scope of the whole system development effort. [9]

[1] Pos Graduate Students at Brazilian Institute for Space Research/INPE
Av. dos Astronautas 1758, São José dos Campos, Brasil; 12227-010
e-mail: luizalex@lit.inpe.br

[2] Pos Graduate Students at Brazilian Institute for Space Research/INPE
Av. dos Astronautas 1758, São José dos Campos, Brasil; 12227-010
e-mail: ojeronimo@lit.inpe.br

[3] Technologist and Professor at the Brazilian Institute for Space Research/INPE and at the Technological Institute of Aeronautics/ITA
Av. dos Astronautas 1758, São José dos Campos, Brasil; 12227-010
e-mail: geilson@lit.inpe.br

D. D. Frey et al. (eds.), *Improving Complex Systems Today*,
Advanced Concurrent Engineering, DOI: 10.1007/ 978-0-85729-799-0_4,
© Springer-Verlag London Limited 2011

This paper aims to present a systems concurrent engineering approach applied to the development of a satellite payload. The approach is different from traditional systems engineering approach because it anticipates to the early stages of system architecting the product life cycle process requirements. It proposes to simultaneously develop, from the outset, the product and its life cycle processes performing organizations. Aquarius instrument of SAC-D satellite was chosen as an example of application. It was already at the stage D (Production /Qualification and Testing) of its life cycle at the time of the preparation of this paper. The main purpose is to present the System Concurrent Engineering approach along with the life cycle process of the Aquarius instrument showing the main concepts of the approach.

The paper is organized as following: Section 2 presents the Aquarius/SAC-D and a brief introduction about Aquarius instrument. Section 3 presents the system concurrent engineering approach. Section 4 discusses the advantages and opportunities for improving the proposed approach. Section 5 concludes this paper.

## *1.2 Aquarius/SAC-D*

Aquarius/SAC-D mission is a partnership between the USA space agency (NASA) and Argentine space agency (CONAE) with launch scheduled for early 2011. CONAE is providing the service platform, SAC-D, the fourth spacecraft in the Scientific Application Satellite (SAC) program, and a complement of instruments. In addition to Aquarius, NASA will provide the launch, from NASA's Western Test Range at Vandenberg AFB, and the launch vehicle, Delta-II. Aquarius data will be sent from the MOC (Mission Operation Center) to the Goddard Space Flight Center for processing salinity related information. The salinity maps will be distributed to the public and eventually archived at the Physical Oceanography DAAC at the Jet Propulsion Laboratory. [5]

### 1.2.1 Aquarius

Aquarius is a combination radiometer developed by NASA's Goddard Space Flight Center, consists of three highly sensitive, temperature-stable radiometer receivers and is the primary instrument for measuring the microwave emissivity of the ocean surface and scatterometer (radar) developed by the Jet Propulsion Laboratory (JPL), will make co-aligned, polarimetric radar backscatter measurements of the ~100-150 km sea surface footprint to correct for the effects of surface roughness in the radiometer's brightness measurement, operating at L-band (1.413 GHz for the radiometer and 1.26 GHz for the scatterometer). The prominent feature of this instrument is the antenna, a 2.5-m offset parabolic reflector with three feed horns. Under NASA's Earth System Science Pathfinder (ESSP) program, the Aquarius instrument designed to map the surface salinity field of the oceans from space, together with the SAC-D bus and several instruments provided by CONAE and its partners. [5, 6 e 11]

The principal objective of the Aquarius instrument is to provide monthly measurements of Sea Surface Salinity (SSS), providing the research community data to better understanding of interaction between ocean circulation and the global water cycle. The knowledge in the salinity field is important wherefore with the changes and variations in SSS (Sea Surface Salinity) may have an impact on climate. The main data-related goals of Aquarius are to provide the first global observations of SSS, covering surface of Earth once every 7 days and delivering monthly 150-kilometer resolution SSS maps over a 3-year mission lifetime. [1]

## *1.3 Step by step of development*

The systems concurrent engineering approach has the following steps: Model the system life cycle, using for example behavior and activity diagrams (IDEF0);

Identify stakeholders and their interest in each life cycle process scenario identified;

Capture and engineer stakeholder and system requirements;

Identify and analyze of system functional and architecture context at every life cycle process scenario;

Deploy down functionally and physically the functional and architecture contexts identified composing the system architecture.

The approach starts by stating the mission or main purpose of the system together with the model of its life cycle processes. The purpose in defining the system life cycle is to establish a framework for meeting the stakeholders' needs in an orderly and efficient manner, along the life cycle. This is usually done by defining life cycle stages, and using decision gates to determine readiness to move from one stage to the next. Skipping phases and eliminating "time consuming" decision gates can greatly increase the risks (cost and schedule), and may adversely affect the technical development as well by reducing the level of systems engineering effort. [4]

### 1.3.1 Behavior Diagram

The Behavior Diagram is a graphical representation with control sequencing from top to bottom. While it is not shown on the graphical construct, the Behavior Diagram model allows data inputs to a function to be characterized as either triggering (a control capability) or data update (not a control implementation). The

Behavior Diagram specification of a system is sufficient to form an executable model allowing dynamic validation via discrete event simulation methods. [7]

A behavior diagram was prepared to demonstrate the main steps of the life cycle of the Aquarius instruments, from initial development to disposal of the product and then down to a second level of detailing. Each stage of the lifecycle (development, production, AIT, launch, operation and disposal) was broken down into another level of detail. Taking as an example the step of AIT, it can break down into the following lower level steps: Aquarius pieces reception, Aquarius integration, Aquarius functional test, Aquarius integration on SAC-D, SAC-D/Aquarius functional and environmental test.

### 1.3.2 IDFE0 Diagram

For each stage of the life cycle of the product developed the IDEF 0 diagram with the inputs, outputs, controls and mechanisms, shows the key features and information exchange between the phases of the life cycle.

The primary content of the IDEF0 Diagram is the specification of data flow between system functions, allow the specification of control as an input to a function but does not have the capability to characterize that control in terms of constructs, as the Behavior Diagrams do. The specification of control with the IDEF0 notation is incomplete and, therefore, not executable. The IDEF0 Diagram also represents the mechanism (usually the component to which the function is allocated) which performs the function. [7]

Using the stages generated in the behavioral diagram, was elaborated the IDEF0 diagram, so this diagram have the same steps, or function meaning in this diagram, with flow inputs, flow outputs, controls and mechanism. In the case of AIT function, it has as inputs Aquarius pieces, SAC-D container, SAG (Solar Array Generator) container, others pieces SAC-D; as outputs the PSR (Preliminary Shipment Review), the SAC-D / Aquarius Integrated and Tested, the SAG; as control the AIT plan, the requirements for the SAC-D transport, the transportation routes; and finally as mechanism the facilities and human resources.

### 1.3.3 Life Cycle Process Identification

Diagrams presented in Figure 1 and 2 define the structure and behavior of life cycle processes. These diagrams decompose the whole life cycle process into life cycle processes scenarios (composition or alternative scenarios). Stakeholders, functional context and architecture context will be identified at each life cycle process scenario.

For approach demonstration purposes, only four life cycle processes will be considered throughout this work. They are: Integration and Data Processing processes for deriving product stakeholders, functional context and architecture context;

Functional testing and design processes for deriving organization stakeholders, functional context and architecture context.

### 1.3.4 Stakeholders Identification

The identification of stakeholders is performed by identifying the people or organizations who are affected by the attributes of the end product, its life cycle processes and their performing organizations within the scope of the development effort. A way of identifying the stakeholders is to separate the system into product and organization elements and investigate who are the people or organizations directly interacting with each of them.

For product, a question that can be made, in order to identify stakeholders, is: 'who are the people who directly interact with the product during its potential life cycle scenarios?' Observe that the question covers the entire life cycle and not only the end-product use.

For organization, i.e., each life cycle process performing organization within the scope of the development effort, stakeholders are the people outside that organization who can play a role in relation to the business

Using the IDEF0 notation in section 1.3.2, stakeholders can be obtained by answering the questions 'who are the sources of inputs', 'who are the mechanisms or sources of mechanisms', 'who are the controls or sources of control', 'who are the destination of outputs'.

Product and organizations may have stakeholders in common. The aim, at this stage, is to obtain a list of system stakeholders as complete as possible no matter how each stakeholder interacts with the system. [8]

Fig.1.1 and Fig. 1.2 present the stakeholders identified for the chosen processes.

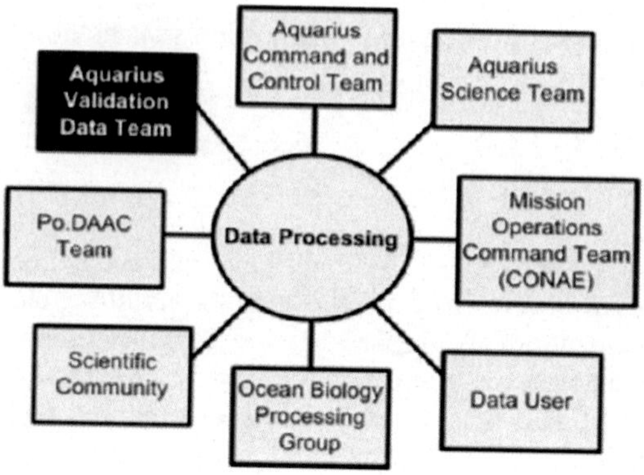


Fig. 1.1 Stakeholders of the product during the data processing process

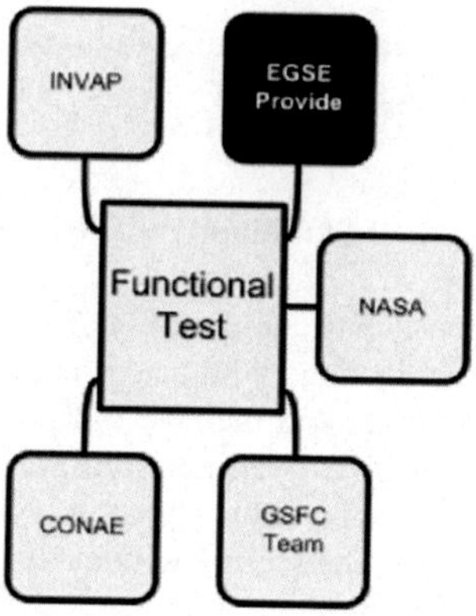


Fig. 1.2 Stakeholders of the functional testing organization

### 1.3.5 Interests, Metric and Measures of Effectiveness (MOE)

For each stakeholder identified, requirements are captured and this can be made of many ways. For example, adapting the Yourdon's approach, cited by Loureiro [8], it can be derived a systematic way of capturing stakeholder requirements. It uses the concept of 'events'. 'Events' can describe how stakeholders interact with the product, process and organization elements of the system. 'Events' have: 'stimulus' and 'response'. 'Stimulus' will derive the 'condition' information related to a requirement. 'Response' will derive the 'function' information of the requirement.

The term 'metrics' is defined as: 'A measure of the extent or degree to which a product possesses and exhibits a quality, a property or an attribute'. [13]

These metrics are used to assess the stakeholders' needs satisfaction in order to assist in defining system requirements. Cited by Loureiro [8], the IEEE-1220 standard defines measures of effectiveness as the metrics by which the customer will measure satisfaction with products produced by the technical effort. Measures of effectiveness reflect overall customer expectations and satisfaction. Key measures of effectiveness may include performance, safety, operability, reliability, and maintainability, or other factors.

A way of identifying measures of effectiveness is to identify the stakeholder concerns towards product, processes and organization. Concerns may also be used as a criteria for grouping requirements. The table 1.1 show interests, metrics and measures of effectiveness identified from stakeholder needs.

**Table 1.1** Interests of GSFS Team

| Goal | Metris | MOE |
|---|---|---|
| Visibility at NASA | 1. Project delivered on schedule<br>2. Project delivered within cost<br>3. Project delivered in compliance with all requirements | 1. Ready before the activities manufacturing<br>2. Cost exceeded from budget<br>3. Number of waves |
| Get the requirements of radiometer | 1. Maximum Weight<br>2. Working range Radar<br>3. Lifetime | 1. kg value<br>2. GHz value<br>3. years amount |

### 1.3.6 Stakeholders and System Requirements

The stakeholder requirements govern the system's development; and they are an essential factor in further defining or clarifying the scope of the development project. If an enterprise is acquiring the system, this process provides the basis for the technical description of the deliverables in an agreement – typically in the form of a system level specification and defined interfaces at the system boundaries. [4]

Table 1.2 shows the requirements that stakeholders must be able to perform and what the system must respond. Stakeholder and system requirements are captured with the life cycle perspective as mentioned earlier in this paper.

**Table 1.2** Stakeholder and system requirements

| # | Stk should be able to | The System should | Concern | Type | Compliance | Status | PPO | Constraint | T/I/D | Procedure |
|---|---|---|---|---|---|---|---|---|---|---|
| R03 | Define the number of screws for system integration in the SAC-D | Have no more than 40 screws to be integrated into the SAC-D | Integ. | C | M | OK | Prod | | I | Pr.3 |
| R04 | Measure the total time of mechanical integration to SAC-D | Promote the integration of mechanical agility to SAC-D in a time not exceeding 48 hours | Integ. | C | M | OK | Prod | | I | Pr.4 |

### 1.3.7 Context Diagram

For each life cycle process, a context diagram is derived. The context diagram shows the system during a given life cycle process scenario in the central bubble, the elements in the environment outside the system and the flows of material, energy and information between system and environment. Fig. 1.3 presents an example of such context diagram adding states of the elements of the environment that can drive modes of performing the process related to the central bubble.

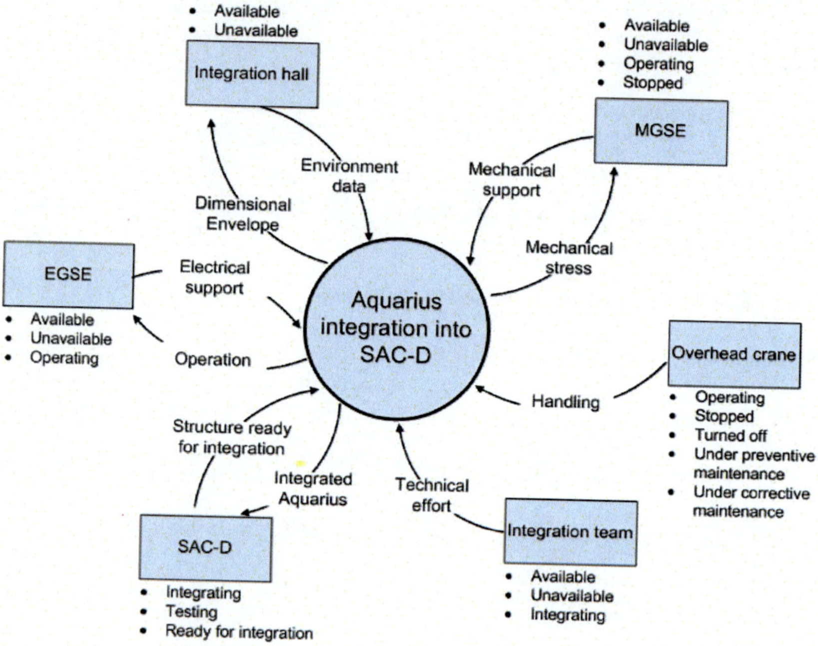


**Fig. 1.3** System Context Diagram

Table 1.3 relates the states of the elements in the environment with modes of integrating the Aquarius payload into the SAC-D platform.

**Table 1.3** Modes of 'Aquarius Instrument Integration on SAC-D'

| **Normal Mode** [1] | | **Setup Mode** [2] | |
|---|---|---|---|
| Environment | Status | Environment | Status |
| MGSE | Operating | MGSE | Stopped |
| Overhead crane | Operating | Overhead crane | Under preventive maintenance |
| Integration team | Available | Integration team | Unavailable |
| SAC-D | Ready for integration | SAC-D | Ready for integration |
| EGSE | Operation | EGSE | Unavailable |
| Integration hall | Available | Integration hall | Unavailable |

[1] The environment is ready to perform the integration activity of the SAC-D Aquarius

[2] The environment is being prepared for integration activity

The functional context diagram is the starting point for identifying the functions that the system must perform within each mode of performing a given life cycle process. From Fig. 1.3 and Table 1.3, functional and performance requirements can be derived for the product so that it can cope with the needs of the integration organization represented by the elements in the environment outside the system and interacting with it.

### 1.3.8 Architecture Context Diagram

The architecture context diagram depicts the system during a given life cycle process scenario, the elements in the environment outside the system interacting with it and the physical connections between the system and the environment. Instead of the flows in the context diagram (see Fig. 1.3), the architecture context diagram depicts the physical interconnections that will support those flows. The architecture context diagram allows that physical external interfaces to be identified early in the development process and with a life cycle perspective. Fig. 1.4 shows an example of such an architecture context diagram for the Aquarius instrument during its integration process.

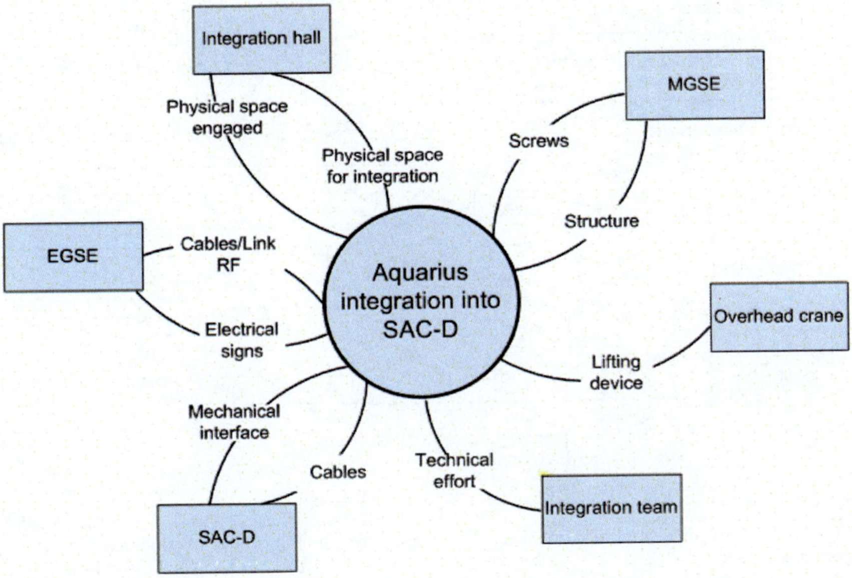


**Fig. 1.4** Architecture context diagram

### 1.3.9 Hierarchical Decomposition of the Context Diagrams

High level system requirements are progressively decomposed into subsystem and component requirements. This is done in any systems engineering exercise. However, using the systems concurrent engineering approach, as the context diagrams were produced for every system scenario, requirements on the system will be systematically captured for every life cycle process. System functional and physical architectures must represent the structure and behaviour of the system.

Fig. 1.5 and Fig 1.6 represent the functional structure and behaviour of the system during its integration. Fig. 1.7 represent the physical structure of the system during integration.

Of course there are elements in the approach, that is also part of the traditional systems engineering approach, that were not shown in this paper. For example, hazard and risk analysis are also performed from each context at every scenario. Table. 1.4 show an allocation matrix that is also an element of the traditional systems engineering

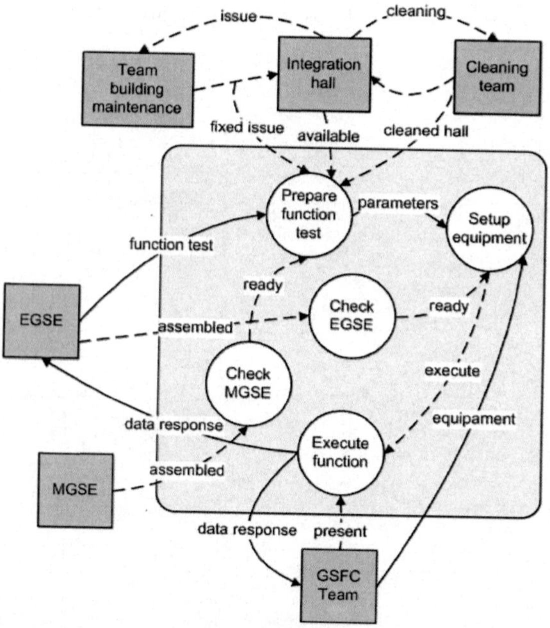


**Fig. 1.5** Functional structure of the Aquarius during integration

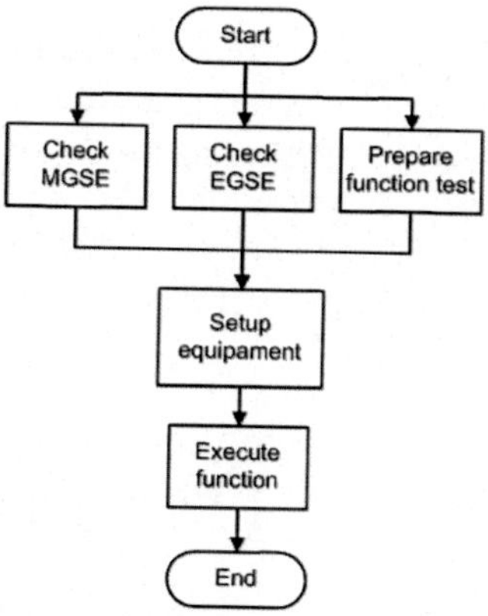


**Fig. 1.6** Functional behavior of the Aquarius during integration

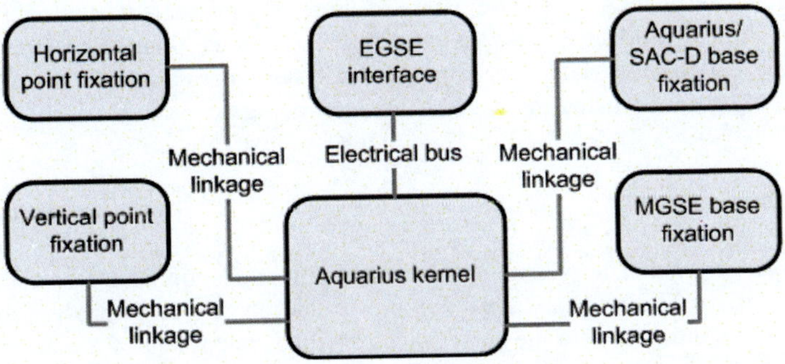


**Fig. 1.7** Physical structure of the Aquarius during integration

**Table 1.4** Allocation Table (As an example the physical part "horizontal clamp point implements the functions of Hoisting Aquarius and Moving Aquarius" as show below).

| | | Physical Parts | | | | |
|---|---|---|---|---|---|---|
| | | MGSE base fixation | Aquarius / SAC-D base fixation | EGSE interface | Horizontal point fixation | Vertical point fixation |
| Function | Allocate Aquarius | ■ | | | | |
| | Lift Aquarius | | | | ■ | ■ |
| | Move Aquarius | | | | ■ | ■ |
| | Connect cables | | ■ | | | |
| | Torque screws | | ■ | | | |
| | Execute electrical tests | | | ■ | | |

## 1.4 Discussion

In this section we discuss the difference between traditional and system concurrent engineering approach for a better understanding. We should not consider only customer and user as stakeholder of interest as in the traditional approach, but stakeholders related to all process of product life cycle must be taken into consideration.

In the traditional systems engineering approach the functional context analysis are performed only for operational scenarios of the product and for product development organization processes. However a system solution is composed by product and organization elements and any other elements that must be considered for the mission success.

In the systems concurrent engineering approach requirements of entire product life cycle can be anticipated to the initial phase of system architecting process, where the stakeholder and system requirements are captured for all product life cycle process scenarios. The anticipation of those requirements will allow for less changes during product development and life, reduced time to delivery and, therefore, reduced cost. The more complex the system, the greater the potential for productivity gains by using the systems concurrent engineering approach.

## 1.5 Conclusion

This paper presented a system concurrent engineering approach for a satellite payload design. The proposed approach presented how life cycle process requirements can be anticipated to the early stages of the development of a system. Concurrent engineering has been traditionally successful for the development of parts and this approach shows a way of using it also for the development of complex systems. The paper focused on the product elements of the Aquarius payload but the same approach is also used to develop the organization elements of the system. Details on how to develop simultaneously the product and organization elements can be found in [9].

## 1.6 References


1. Aquarius Mission Overview. Available: <http://aquarius.gsfc.nasa.gov/overview-mission.html> .
2. Aquarius Validation Data Segment (AVDS) to Aquarius Data Processing Segment (ADPS): Interface Control Document. Available: <http://oceancolor.gsfc.nasa.gov/AQUARIUS/DOCS/AQ-332-0331_AVDS_ADPS_ICD.pdf>.
3. Argo: part of the integrated global observation strategy. Available: <http://www.argo.ucsd.edu/>.
4. Edited by Haskins, Cecilia. Systems Engineering Handbook: A Guide for system life cycle processes and activities. INCOSE, v.3, June 2006.
5. Le Vine, David M., At All - Aquarius Mission Technical Overview, 2006.
6. Le Vine, David M., At All. The Aquarius/SAC-D Mission and Status of the Aquarius Instrument. IEEE, 2008.
7. Long, Jim. Relationships between Common Graphical Representations in System Engineering. Vitech Corporation, 2002.
8. Loureiro, G. A systems engineering and concurrent engineering framework for the integrated development of complex products. Loughborough University, Loughborough, 1999.
9. Loureiro, G. Lessons learned in 12 years of space systems concurrent engineering. 61st International Astronautical Congress, Prague, CZ. 2010.
10. Loureiro, G. System Engineering. Lecture notes in the course of Engineering and Space Technology - Institute for Space Research - INPE, 1° period 2010.
11. M. Fischman, At All. Development and Integration of the Aquarius Scatterometer Processor/Control Electronics for Achieving High Measurement Accuracy. Jet Propulsion Laboratory, California Institute of Technology, Pasadena, CA USA. IEEE, 2009.
12. NASA Systems Engineering Handbook, 1995.

13. Sprole, Noel - Coming to Grips with Measures of Effectiveness. Australian Centre for Test and Evaluation University of South Australia, 1998.
14. TAO: Tropical Atmosphere Ocean project. Available: <http://www.pmel.noaa.gov/tao/>.
15. The Aquarius SAC-D Mission: Designed to Meet the Salinity Remote-Sensing Challeng. Oceanography, v.21, n.1, March, 2008.

# Systems Concurrent Engineering for the Conception of an Attitude and Orbit Control System

Ariana Cristina Caetano de Souza[b] , Lauro Paulo da Silva Neto, Leonardo Leite Oliva[b] and Geilson Loureiro[a,1]

[a]Technologist and Professor at the Integration and Testing Laboratory, Brazilian Institute for Space Research, INPE (São José dos Campos), Brazil.
[b]Post graduate student at Brazilian Institute for Space Research - INPE.

**Abstract.** This paper presents a systems concurrent engineering approach for the conception of an Attitude and Orbit Control System for a Satellite (AOCS). Systems Engineering uses a multidisciplinary approach in order to better understand complex systems and its processes, its products and its organization. The main goal is to satisfy all stakeholders involved in the product and its processes. In this paper several concepts of Systems Engineering, as well as its methodology, were applied in the main processes of the life cycle of an AOCS of an artifitial satellite, in order to give a better understanding of the development of this kind of product, from requirements analysis to system operation.

**Keywords.** Systems concurrent engineering, systems engineering, stakeholders, attitude and orbit control system

## 1 Introduction

Concurrent Engineering is a systematic approach to the concurrent and integrated development of the product and its related processes, including support and manufacture. This approach is essential to bring the requirements from all people involved in the product life cycle process implementation to the early stages of product development. All measurable elements in the product life cycle – from the conception to the disposal – such as costs, deadlines, quality and requirements, are essential in this study. Concurrent Engineering requires an environment of development, with interrelationship between several types of professionals that are important for the product life cycle processes. To accomplish this goal it is essential to use diagrams and a methodology that allows all people involved in the

[1] Technologist and Professor at the Integration and Testing Laboratory, Brazilian Institute for Space Research, Av. dos Astronautas 1758, São José dos Campos, Brasil; 12227-010; Tel: +55 (12) 39456317; Fax: +55 (12) 39411884; Email: geilson@lit.inpe.br

D. D. Frey et al. (eds.), *Improving Complex Systems Today*,
Advanced Concurrent Engineering, DOI: 10.1007/ 978-0-85729-799-0_5,
© Springer-Verlag London Limited 2011

project to understand each other. The philosophy is to build a map of the product life cycle so that everyone has a general view and understands his role.
The development of complex systems, such as those found in space industry, usually requires effort of several technical areas such as computer, electrical, mechanic, control, thermal and materials engineering, physics, chemistry, etc. During a satellite development phase, specific technical teams are called in order to build a successful product. Those teams usually don't communicate with each other, ignoring important information produced by their work neighbors, which leads to a rise in the cost of the project during the development phase, the assembly phase, the integration and testing phase and the operational phase. This can lead to an early and unexpected failure. Therefore, space systems need to use methods that are capable of giving a holistic view of the scope of development. This paper was made in order to give such a vision applying concepts of Systems Engineering, where the product and the life cycle process organizations are analyzed as a part of the same whole.

## 2 The AOCS

The mission of the AOCS is to point and maintain satellite pointing with previously established precision, regarding its mission within all its life cycle. To achieve this objective, it is essential to have a broad view of the system life cycle, with all its relationships with the ground segment, the space segment, the launcher and the user. Therefore, several relations between development environments were created. Those relations considered human elements or entities (technicians, engineers, users, administrators, government, companies, etc), physical effects (radiation, particle collision, electric current, etc) and information (data, knowledge, projects, etc).

For the AOCS there were identified the following processes during the system life cycle: development, production, integration and tests, launch, operation and toss (figure 1).

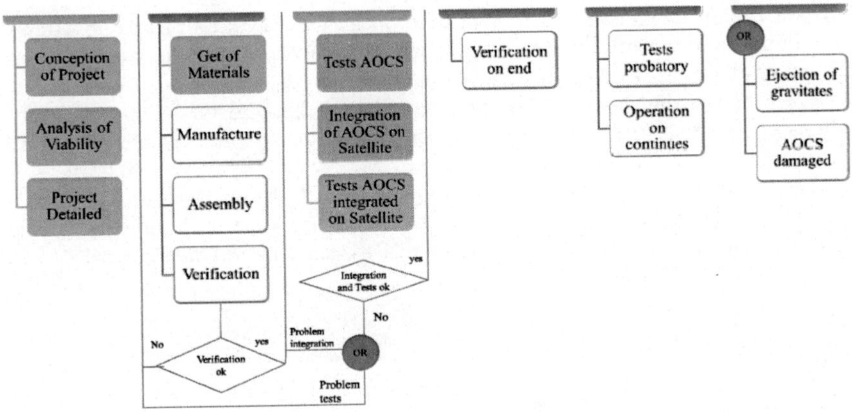


**Figure 1.** Processes in the life cycle of the AOCS.

The Development Process involves the project conception, where the first ideas are given and gathered together in order to meet the basic needs (requirements) established by the stakeholders. These ideas will pass through several reviews, whose objective is to improve the system design before any material is acquired. There is also the Viability Analysis, which is responsible for studying the viability of the whole process taking into consideration the chronogram and the financial availability for the project.

In the Production Process the material acquisition takes place. These materials must be qualified, attending all norms in order to guarantee a safe and satisfactory operation for the product during launch and in space environment. In the fabrication all materials acquired become modules of electronic circuits, mechanical devices, structures, thermal ducts and so on. The assembly is the integration of those systems, linking the different subsystems so that they can operate as a whole (for example, the electronic circuits are placed inside a structure that has a cooling system to maintain certain temperature so that the circuits operate without loss of energy). After that the verification begins. It consists in several tests that permit identify if and how the system works, using simulation to measure performance. When the system meets all norms in this phase, the next step takes place.

The Integration and Tests phase is responsible for tests of static and dynamic nature of the subsystem AOCS individually, as well as those same tests with the AOCS integrated to the entire system (satellite). Those tests are four: thermal, vibration, acoustic and electromagnetic compatibility (EMC). After the system proves that it attends all requirements of those tests it is ready for launch. In this phase, another integration and test take place. This will guarantee that the AOCS and the satellite are integrated to the launcher, and therefore are ready for launch and operation.

Before the full operation of the system there are some probatory tests that must take place – with the satellite already in orbit. Those tests are conduced by the Track and Control Center whose goal is to check the operation of all subsystems that compose the satellite. After this phase, the system enters in a continuous operation phase, in which each subsystem begins to perform its specific task. The AOCS uses the data collected by its sensors and those given by the on board computer to adjust the vehicle attitude and orbit so that the vehicle is able to perform its main task.

The final phase (orbit ejection) occurs either if the subsystem (AOCS) is damaged or if another important subsystem suffers the same. In a more optimistic case, this phase will take place at the end of the mission, when the satellite has already fulfilled all expectations.

## 3 Method

The analysis method was separated in seven parts as follows:

1. Process Structure;
2. Stakeholder Analysis;
3. Context and Architecture Diagrams;
4. Functional structure (Data Flow Diagrams (DFD)) and behaviour (State Transition Diagrams);
5. Hazard Analysis;
6. Architecture Flow and Interconexion Diagrams ;
7. Allocation matrix.

Each part is responsible for a specific view of the product and its life cycle processes.

The activity diagram (1), also called IDEF0, is a function modelling methodology for describing manufacturing functions, and shows all the system life cycle in study with rectangles (processes) that are connected by arrows (flows of material, energy and information).

The stakeholder analysis (2) was divided in four parts: product stakeholders (operation process / production process); and organization stakeholders (development process / integration and tests process). Due to the different emphasis of each part, the views of the same stakeholder may be different (or not).

The context and architecture diagrams (3) show the exchange of material, energy and information between the elements of the environment and the objects in the product and life cycle process organizations. A this stage some circumstances for the environments elements are identified.

The DFD and transition diagrams (4) are tools that enable a better understanding of the data flow in the system. The DFD shows the messages sent and received by each physical component, creating an information net that links all subsystems, while the transition diagrams shows all the possible states of the system (rectangles) and the events responsible by its changes (arrows).

There is a hazard analysis (5), that is a process used to assess risk. The result of such analysis is the identification of unacceptable risks and the selection of means for mitigating them.

Other diagrams include those of interconnection and flow (6), which represent, respectively, the flows of material, energy and information and the physical connections between the inner components of a product or life cycle process organization.

Finally, there is the allocation matrix (7), the final stage of application of systems concurrent engineering. This matrix relates the functions and all parts responsible by them.

## 4 Results

The IDEF0 diagram was constructed considering all life cycle process of the AOCS, since the initial sketches. Figure 2 shows the flow of the product AOCS.

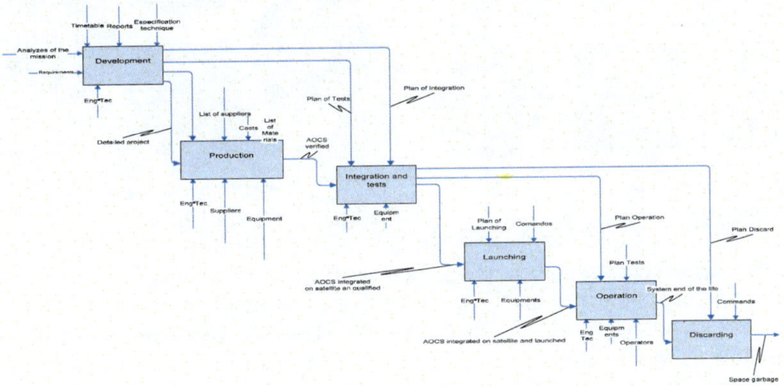


**Figure 2.** IDEF0 diagram os the AOCS

As it can be seen, the flows that enter in the upper side of the rectangles are controls (information) that allow coordinate the process. They work as a guide element. In the development phase, they are reports, chronograms and technical specifications.

The left side entries are inputs and the right ones – that go out – are outputs. They are used to show the materialization of the product, specifying what will be the basic material – the product in a stage of development – to start the process, and how the product or information (output) comes out after the process has been executed.

All flows that come from below are necessary in most processes. They represent the entry of energy, basic material and human resources in the process.

Stakeholder analysis took into consideration four points of view (see Method section). Each one of them has a particular look at the process/product. Figures 3 to 6 show the stakeholders in each of these cases.

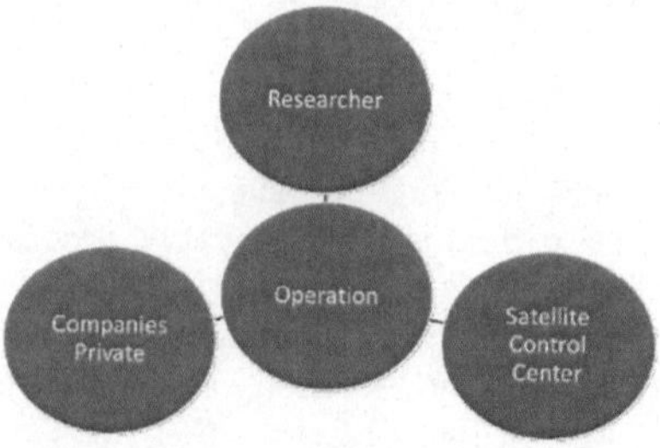


**Figure 3.** Stakeholders of Product: Operation Process

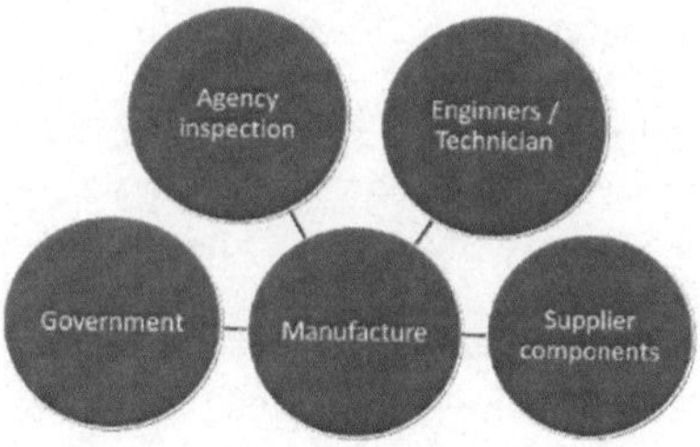


**Figure 4.** Stakeholders of product: Production Process.

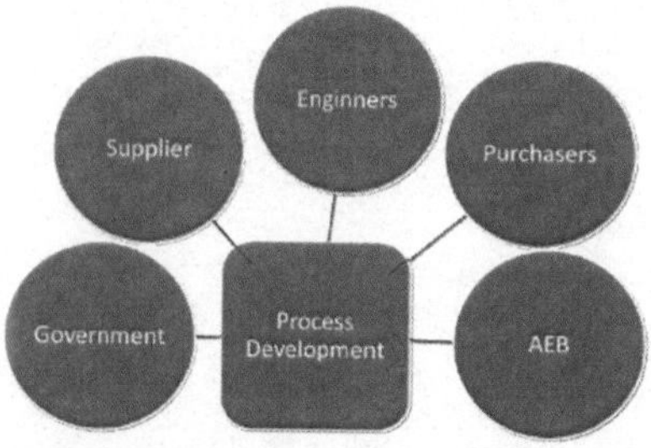


**Figure 5:** Stakeholders of Organization: Development Process.

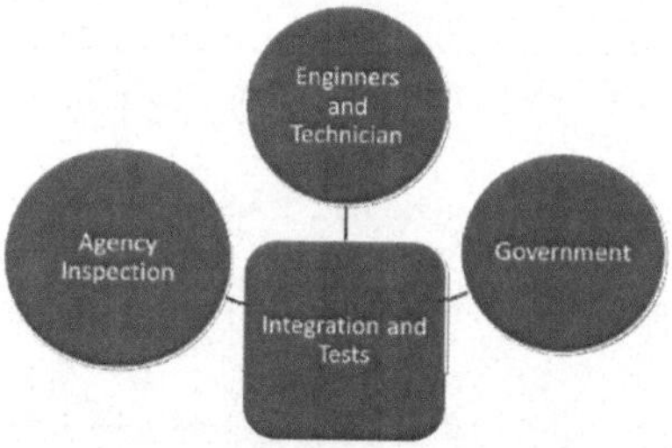


**Figure 6:** Stakeholders of Organization: Integration and Tests Process.

After identifying the stakeholders, their interests were defined in each process. And for one stakeholder in each analysis metrics – something that affects directly the interest – and measures – for measuring the metrics – were pointed out. Tables 1 to 4 show those relations.

**Table 1.** Metrics and measures for the process of operation (Control Center)

| Stakeholders | Interests | Metrics | Measures of effectiveness |
|---|---|---|---|
| Control Center | Periodic flow of information (attitude and orbit); Control of attitude and orbit. | Positioning | Altitude (km); Orientation (rad) angular velocity (rad / s); Velocity around three axes (pitch, roll, yaw). |

**Table 2.** Metrics and measures for the process of production (Regulamentary Institutions).

| Stakeholders | Interests | Metrics | Measures of effectiveness |
|---|---|---|---|
| Regulamentary Instituitions | Standard conformity | Mechanical architecture | Height (m); Length (m); Width (m); Weight (kg); Decomposition time (years); Natural frequency (Hz); Relationship strength / weight. |
| | | Index of conformity | Percentage of compliance (%) |

**Table 3.** Metrics and measures for the process of integration and tests (Engineers and technicians).

| Stakeholders | Interests | Metrics | Measures of effectiveness |
|---|---|---|---|
| Engineers / technicians | Assessment of the product | Efficiency; Readiness; Accreditation. | Time (years); Reliability; Cost. |

**Table 4.** Metrics and measures for the process of development (Government).

| Stakeholders | Interests | Metrics | Measures of effectiveness |
|---|---|---|---|
| Government | Technological innovation | Innovative products | Number of products |
| | Appropriate application of resources | Accomplishment budget and execution plan | Total spending on planned spending; Completed on planned. |

The context and architecture diagrams were built for the product in operation. Both diagrams complement each other, and show the physical and informational relations of the AOCS with the environment and other subsystems.

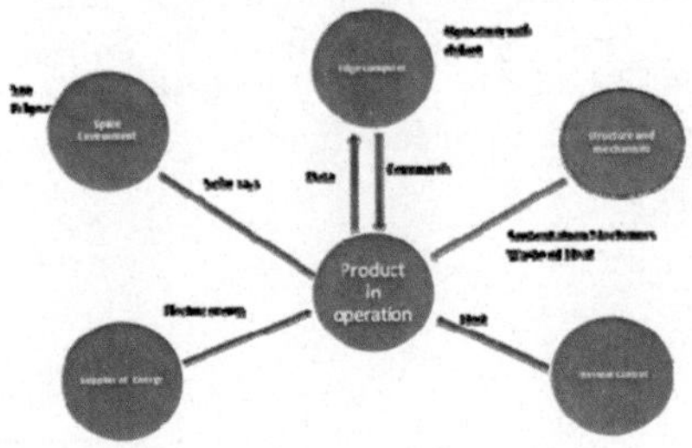


**Figure 7.** Context diagram for AOCS in operation.

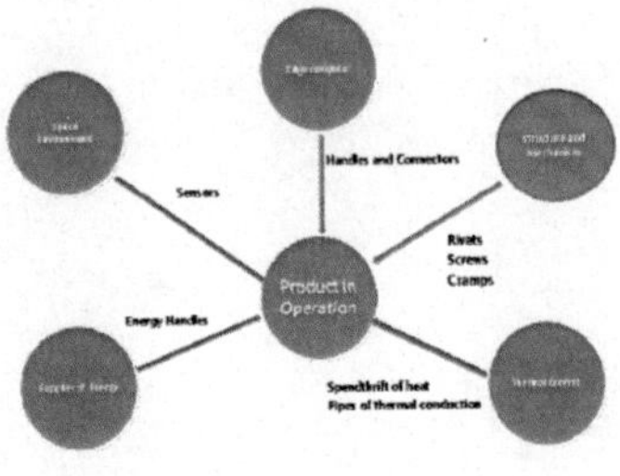


**Figure 8.** Architecture diagram for AOCS in operation.

Following the same line of thought of the context and architecture diagrams, the state transition and data flow diagrams elaborated considered the product (AOCS) in operation (figures 9,10).

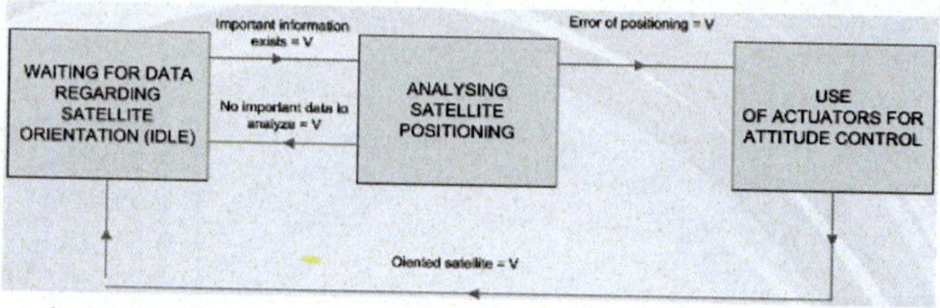


**Figure 9.** State transition diagrams for product in operation.

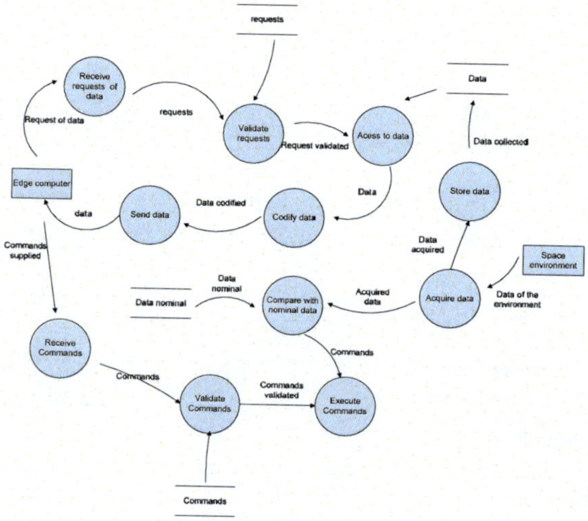


**Figure 10.** Data flow diagrams for the AOCS in operation.

The hazard analysis takes in consideration the physical connections in the context of architecture diagrams, the circumstances defined in the context diagram, and the data flow diagram. The main goal of such study is to point flaws and dangers that may occur in the life cycle of the product – AOCS. And also define the gravity, probability, risk and detection function for each case (tables 5, 6).

**Table 5.** Failure Analysis: AOCS in operation.

| | Failure | Hazard | Gravity | Probability | Risk = G*P | Function | Function description |
|---|---|---|---|---|---|---|---|
| Power supply | Wear solar panel;<br>Circuit failure;<br>Defects in solar cells. | Lack of energy to power the subsystems. | 4 | 1 | 4 | Prevention | Control of energy distribution;<br>Orientation of the solar panel. |
| Tracking errors | Data not identified. | Data loss;<br>Command loss;<br>Execution of unsolicited tasks. | 1 | 3 | 3 | Prevention | Redundancy in data storage. |
| Supports | Broken;<br>Poor fixation. | Collision between the equipments;<br>Subsystems loss. | 2 | 1 | 2 | Prevention | Use double protection system. |
| Thermal control | Breakdown in the thermal circuit;<br>Damage to the sensor sensitivity. | Equipment failure due to fluctuation of temperature. | 3 | 2 | 6 | Detection | Notify changes in temperature fluctuation. |
| Electrical connection | Electromagnetic interference. | High tension;<br>Curt circuit. | 3 | 1 | 3 | Protection | Disable equipment or subsystem where the crash occurred. |
| Heat sinks | Poor contact with heat sinks;<br>Obstruction in the tubing of the fluid thermal. | Overheating;<br>Leak. | 2 | 1 | 2 | Detection | Fluid level control. |

The flow diagram emphasizes the material, energy and information flow between physical components of the subsystem AOCS (figure 11).

The interconnection diagram focus only in the physical relations between the components of the AOCS subsystem (cables, connectors, etc), giving a physical view of the product in operation (figure 12).

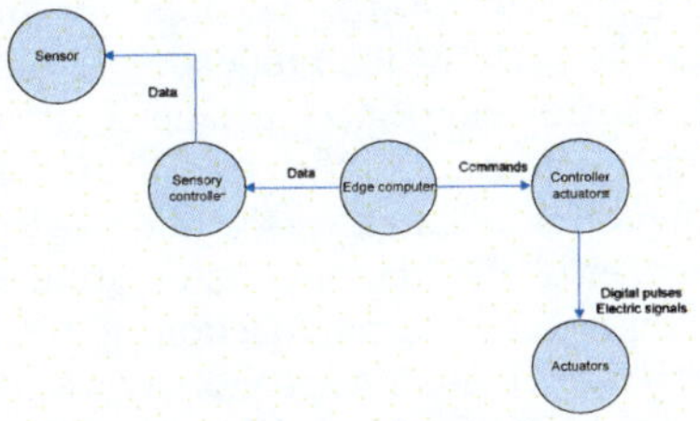


**Figure 11.** Flow diagram: product in operation.

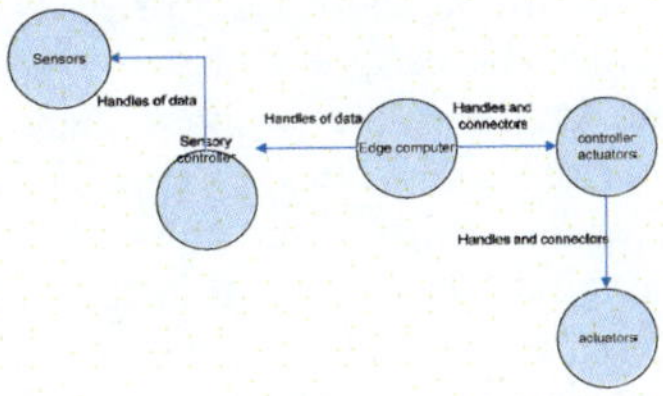


**Figura 12.** Interconnection diagram: product in operation.

The final step of the method consists in elaborating an allocation matrix. This matrix relates all functions and parts responsible for them. Table 6 shows the allocation matrix for the product in operation.

**Table 6.** Allocation matrix for product in operation

| | Components | | | | |
|---|---|---|---|---|---|
| Function | Sensor | Sensor controller | UCP | Actuators | Actuators controller |
| Receive data request | | | X | | |
| Validate request | | | X | | |
| Access data | | | X | | |
| Encode data | | | X | | |
| Send data | | | X | | |
| Receive commands | | | X | | |
| Validate commands | | | X | | |
| Execute commands | | | | X | X |
| Acquiring data | X | X | | | |
| Store data | | | X | | |
| Compare with nominal data | | | X | | |

As can be seen in table 6, each component of the AOCS has one or more functions. This table helps to elaborate relations between functions and components, making easier to study failures that may occur in the system.

The method shall be applied for the product in each of its life cycle process scenarios. In each scenario the organization that implements that scenario is also analyzed and its relationship with the product is captured. In this paper, for the sake of demonstration only, only the processes of operation, production, integration & testing and development were considered, but the method must be applied to all other life cycle processes and their scenarios.

## 5 Discussion

The parallelization of tasks permits a better view of the development process, since the early sketches to the end of the life cycle of the product. And different diagrams allow a more clear view of a particular aspect of the system. For example, a software engineer might be interested only in the flow of information for the design of the embedded software, ignoring other aspects, while an electrical engineer is interested in the relations between the system and its environment of operation, due to the relation between temperature and malfunction of electronic equipments. But, at the same time, it is important that all personnel involved in the development of the product realize that their job affects (and is affected) by the other professionals.

Stakeholder analysis is another point to be considered in concurrent engineering. The stakeholders provide important information that helps the developers to make clear requirements. Also, it is important that they are identified as soon as possible, as well as its metrics and measures, because this information will serve as a guide to all developers (engineers, technicians, administrators, suppliers, etc).

## 6 Conclusion

Through the use of Systems Concurrent Engineering methodology in the description of the AOCS the authors realized that it is essential to work in a systematic manner in order to obtain correct tables and diagrams that will allow engineers, techicians and administrators to take important decisions in the product and/or process development. Most important, the methodology makes possible the dialog between professionals with different backgrounds, thanks to relations built through several kinds of diagrams and tables.

One can detect easily flaws or critical points if the model of the process is built considering all interests involved. And finally, the use of the methodology and concepts of systems engineering makes it easier to eliminate unecessary costs and accomplish deadlines.

## 7 Acknowledgement

The authors would like to thank INPE (the Brazilian Institute for Space Research, www.inpe.br) for the post graduate course opportunity.

## 8 References

[1] Elfving, A., The Attitude and Orbit Control of XMM, ESA bulletin 100- December 1999, Directorate for Scientific Programmes ESTEC, Netherlands;

[2] IEEE Guide for Developing System Requirements Specifications, IEEE Std. 1233-1996.

[3] IEEE Recommended Practice for Software Requirements Specifications, IEEE Std. 830-1998

[4] Larson, Wiley J. and Wertz, James R., Space Mission Analysis and Design 1999, Version 3, ed. Kluwer Academic Publishers;

[5] Loureiro G. A systems engineering and concurrent engineering framework for the integrated development of complex products. Ph.D. thesis, Loughborough University, 1999.

# Systems Concurrent Engineering of a Hypersonic Accelerator Vehicle

Roberto da Cunha Follador[a,1], Andrea de Oliveira Netto Follador[b], Paloma Ribeiro dos Santos[c], Geilson Loureiro[d]

[a] Post Graduate student at Technological Institute of Aeronautics – ITA and Technologist at the Aerothermodynamics and Hypersonic Division (EAH) of Advanced Studies Institute, IEAv, (São José dos Campos), Brazil.
[b,c] Post Graduate students at Technological Institute of Aeronautics - ITA.
[d] Technologist and Professor at the Technological Institute of Aeronautics, ITA and at the Integration and Testing Laboratory, Brazilian Institute for Space Research, INPE (São José dos Campos), Brazil

**Abstract.** This paper presents a systems concurrent engineering approach for the conception of a Hypersonic Accelerator Vehicle (Veículo Acelerador Hipersônico -VAH) to be used in the flight test campaign of the first Brazilian Aerospace Hypersonic Vehicle named 14-X. The 14-X project objective is to develop a higher efficient satellite launch alternative, using a Supersonic Combustion Ramjet (SCRAMJET) engine for its propulsion. As it is a new technology under development and using systems concurrent engineering approach it is possible to perform stakeholder analysis, requirements analysis, functional analysis and implementation architecture analysis, for product and organization simultaneously. From the analysis, requirements and attributes are captured for the product and its organizations and the relationship among them are identified. Requirements to the early stages were based on anticipation of the needs identified for different life cycle process and then late changes are expected to be avoided, reducing development costs, avoiding delays and risks and increasing satisfaction of stakeholders over product life cycle.

**Keywords**. systems concurrent engineering, systems engineering, complex product, integrated product development, hypersonic.

## 1 Introduction

The development of the VAH is inter-dependent on the development of the 14-X, because those two complex systems will operate as a single system once the flight test occurs. In this way the approach for its development must be different from traditional systems engineering and it must take into account both product life cycle process requirements and use them since the early stages of development.

Post Graduate student at Technological Institute of Aeronautics – ITA and Technologist at the Aerothermodynamics and Hypersonics Division (EAH) of Advanced Studies Institute, IEAv, Trevo Cel Av José Alberto Albano do Amarante, n° 1 – Putim; 12.228-001;São José dos Campos – SP – Brasil; Tel: +55 (12) 39475434; Fax: +55 (12) 3944-1177; email:follador@ieav.cta.br; http://www.ieav.cta.br

D. D. Frey et al. (eds.), *Improving Complex Systems Today*,
Advanced Concurrent Engineering, DOI: 10.1007/ 978-0-85729-799-0_6,
© Springer-Verlag London Limited 2011

This paper aims to present a systems concurrent engineering approach for the conception of the VAH. The approach is different from traditional systems engineering approach because it anticipates to the early stages of system architecting the product life cycle process requirements. It proposes to simultaneously develop, from the outset, the product and its life cycle processes performing organizations [1].

The paper is organized as following: Section 2 presents the Hypersonic Accelerator. Section 3 presents the systems concurrent engineering approach framework and method. Section 4 presents the models derived for the VAH using the approach. Section 5 discusses the advantages and opportunities for improving the proposed approach. Section 6 concludes this paper.

## 2 The Hypersonic Accelerator Vehicle

A Hypersonic Accelerator Vehicle (VAH) is basically a modified sounding rocket used to provide the conditions needed to perform a test flight and to collect accurate data from the hypersonic aerospace vehicle 14-X, that is under development by the Institute of Advanced Studies (IEAv) to the Department of Science and Aerospace Technology (DCTA) of the Brazilian Air Force (FAB).

The IEAv's Hypersonic Aerospace Vehicle, named 14-X (after the 14-Bis developed by aviation pioneer Alberto Santos Dumont), initiated in 2005, is the first Brazilian project with the objective of designing, developing, constructing and demonstrating a Mach 10 wave rider in free flight with its required scramjet technology[2]. It is a product that needs tools to provide a safe development process, with compromise with quality and schedule, and at a minimum cost.

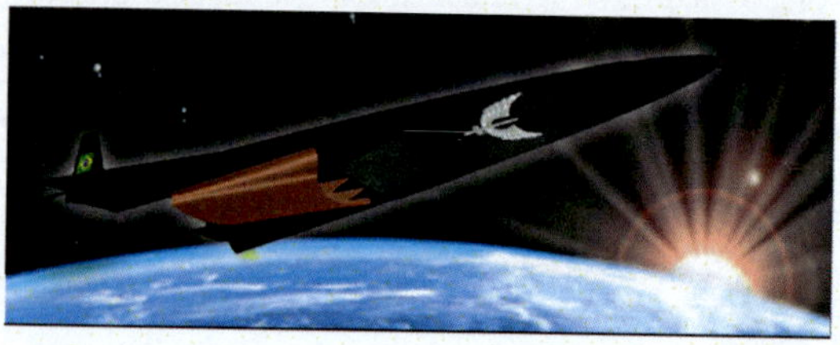

**Figure 1.** Artistic conception of Brazil's Hypersonic Aerospace Vehicle 14-X (Source:[2])

Aerospace and hypersonic vehicles are complex products. During its development process, one of the greatest concerns is safety all over its life cycle. A failure in the design of a safety requirement can lead to problems that may involve the loss of a huge amount of financial resources or even human lives. In the case of a flight test for the development of a new technology, it must be guaranteed the return of flight data, which will provide the information necessary to the continuous development process of a future product. The development organizations need a clear view of the whole life cycle process in order to understand the requirements for a successful and safe test flight that will take place after, at least, six years of development effort. There are many opportunities to improve safety, economy and chances of success over VAH life cycle if a concurrent engineering approach takes place from the beginning of its design stage.

## 3 The systems concurrent engineering approach

The development of complex products has in systems concurrent engineering a powerful tool. Hitchins [3] states that complexity can be understood by what he calls complexity factors. These factors are variety, connectedness and disorder.

Loureiro [4] presents a framework to address complexity in product development – the Total View Framework presented in figure 2. It has three dimensions. Each dimension addresses one of the complexity factors mentioned above. The analysis dimension addresses the variety factor. Along the analysis dimension, it is deployed what must be analyzed in order to develop a complex product. A systems engineering process consists of stakeholder analysis, requirements analysis, functional analysis and implementation or physical analysis. The integration dimension addresses the connectedness factor. It defines what must be integrated along an integrated product development process: product elements and organization elements. Organization here refers to the organizations that perform product life cycle processes. Product elements and organization elements are the system elements. The structure dimension addresses the disorder factor.

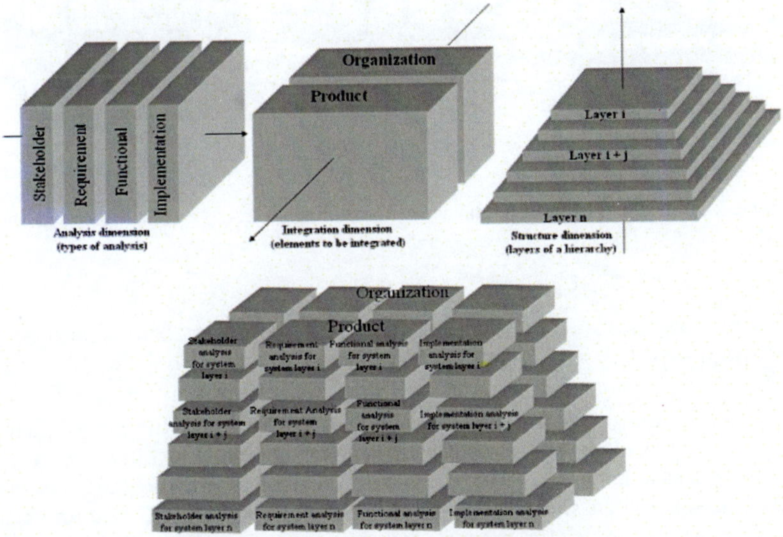


**Figure 2.** A framework to address complexity in complex product development – the total view framework (Source:[4])

The method within the total view framework is called Concurrent Structured Analysis Method evolved from Loureiro [4]. Stakeholder analysis, requirements analysis, functional analysis and implementation (or physical) analysis are performed, for the product under development and its performing organizations simultaneously.

Figure 3 details the concurrent structured analysis method showing the steps to incorporate the concurrent engineering concept in the systems engineering process. The analysis processes are performed at each layer of the system breakdown structure. For example, if a car is the product under development, the analysis processes are performed at the car layer, at the powertrain layer, at the engine layer and so on [1].

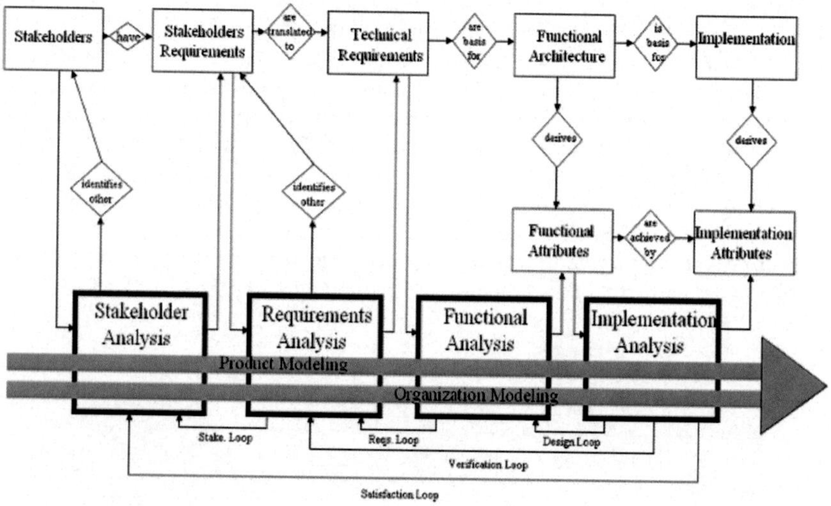


**Figure 3.** A method within the total view framework – the concurrent structured analysis Method (Source:[1])

## 4 The Hypersonic Accelerator Vehicle system concurrent engineering

This section illustrates the steps showed in Section 3 highlighting where the proposed approach is different from traditional approaches. First, the proposed approach is stakeholder driven whereas traditional approaches are customer or user driven. In the various steps listed in Section 3, analyses are performed for each life cycle process scenario, for product and organization simultaneously. Traditional approaches focus on product operation and development organization [1]

The mission statement is a document established by the customer, which reflects the users needs, and is used as input to Phase 0 of a space system project [4].The mission established for the VAH is: *"To provide flight conditions within the speed, flight altitude, flight attitude and dynamic pressure specified in 14-X project and to return valid flight test data".*

Successfully understanding and defining the mission objectives and operational concepts are keys to capturing the stakeholder expectations, which will translate into quality requirements over the life cycle of the project [5].

The life cycles processes and scenarios for the VAH are shown in Table 1.

**Table 1.** VAH life cycle processes and scenarios

| Processes | Development | Manufacturing and Assembly | Operation |
|---|---|---|---|
| Scenarios | Conception | Components Manufacturing | Launching |
| | Detail Project | Assembly | Flight test |
| | Components Project | Integration | Data recording and telemetry |
| | | Qualification test | |
| | Simulation | Acceptation tests | recovery |

The highlighted cells 'conception', 'Detail Project', 'Assembly', 'Integration', 'Qualification test' and 'Acceptation tests' are considered the scope of the development effort. Stakeholder analysis, requirements analysis, functional analysis and implementation architecture analysis will be exemplified for the processes of the life cycle. In practice the methodology explained in Section 3 must be run for all life cycle process scenarios. Figures 4 to 7 just exemplify some steps for selected processes.

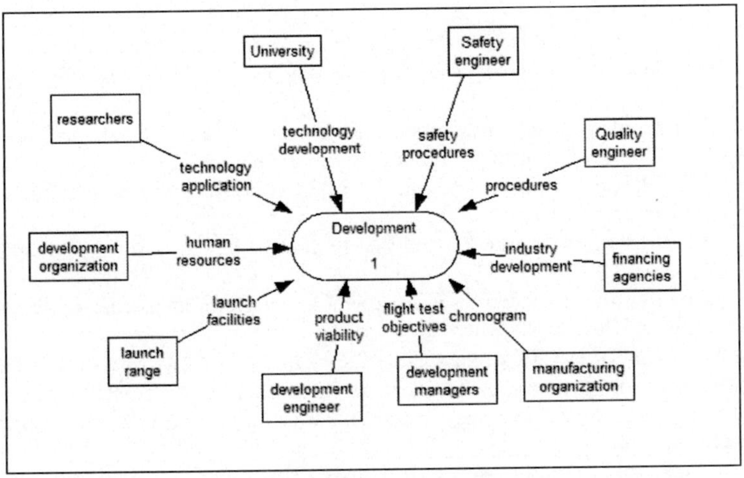


**Figure 4.** Stakeholder analysis - 'Development' life cycle process.

In Figures 4 is exemplified the organization stakeholders concerns for life cycle process of 'Development'. The stakeholder concerns are represented by the connection labels between the stakeholders and the center bubble, indicating the process of the life cycle. This cycle process is a scenario of the 'scope of development effort'.

This pictorial view allows the systems engineering team to identify and rank stakeholders and their needs over that particular process. This done, in a concurrent manner, to all life cycles process and scenarios, allows the accurate capture of the needs as part of the product and organization requirements specification.

Figure 5 presents the product stakeholders identified and their needs for the 'Operation' life cycle process.

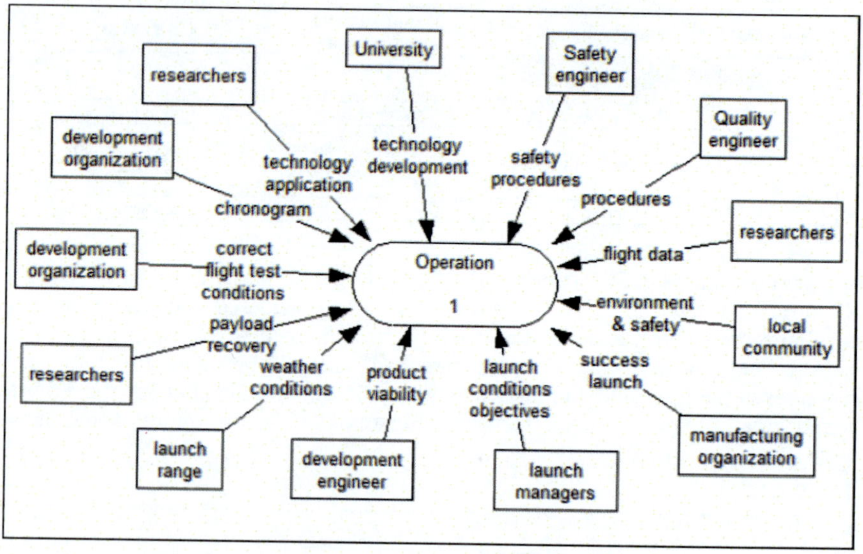


**Figure 5.** Stakeholder analysis - 'Operation' life cycle process.

From stakeholder concerns, requirements are identified and measures of effectiveness (MoEs) are derived. Examples of 'development organization' MoEs, in 'Operation' life cycle process, about the flight conditions during the flight test, can be stated as:

1) The maximum variation for the VAH angle of attack during the flight test was below 3°?

2)The maximum rate of variation for the VAH angle of attack during the flight test was below 6°/s?

Based on identified MOEs the stakeholder requirements will be stated. This is of fundamental importance because it is necessary to understand what the stakeholders want, or believe they want, and translate it in clear and irrefutable characteristics that will compose the final product.

From stakeholder requirements, functions, performance and conditions are identified. Requirement analysis transforms stakeholder requirements into system requirements. System requirements will be met not only by product elements but also by organization elements, changing the traditional focus on systems engineering the product. This approach recognizes that the system solution is not only made of product elements but also of organization elements [1]. Another important point of analysis and source of requirements is the environment where the system life cycle occurs. Each environment element interacts with the system in three ways: exchanging energy, information or material. The clear observation of these factors may lead to relevant requirements. Figures 6 represent an example of context analyze for product in operation.

The context diagrams give a pictorial view of this relationship between environment and system in its life cycle processes. The links between the center and the elements of the context diagram show the kind of information, material or energy exchanged between the environment and the system.

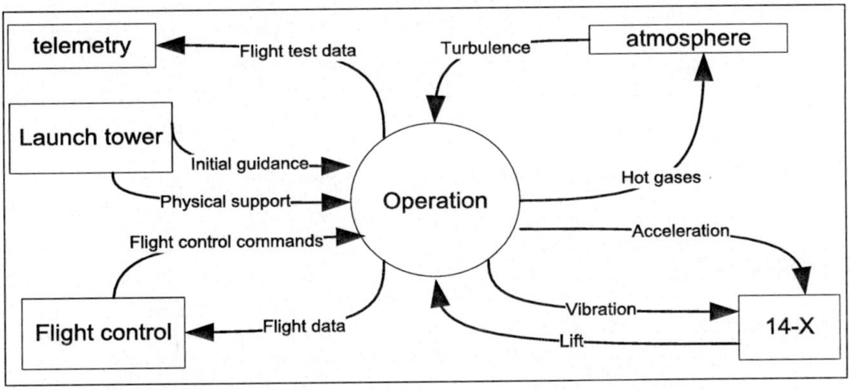


**Figure 6.** Context analysis – product in 'Operation' life cycle process.

Figure 7 presents VAH physical architecture and describes the structural elements and the physical connections, where information, material or energy flows between them.

Concurrent engineering presented here was restricted to dealing with stakeholders, measures of effectiveness, context analysis and physical architecture. But the comprehensive approach covers the analysis of circumstances from which states the system allows the identification and analysis of hazards and risks from the circumstances, thru a FMEA (Failure Mode and Effect Analysis) observing failures and non functions in flows between the elements: product, process and organization with environment, in addition to presenting the behavior of the system and allocation matrix that allows better visualization of the relationship function versus element / subsystem.

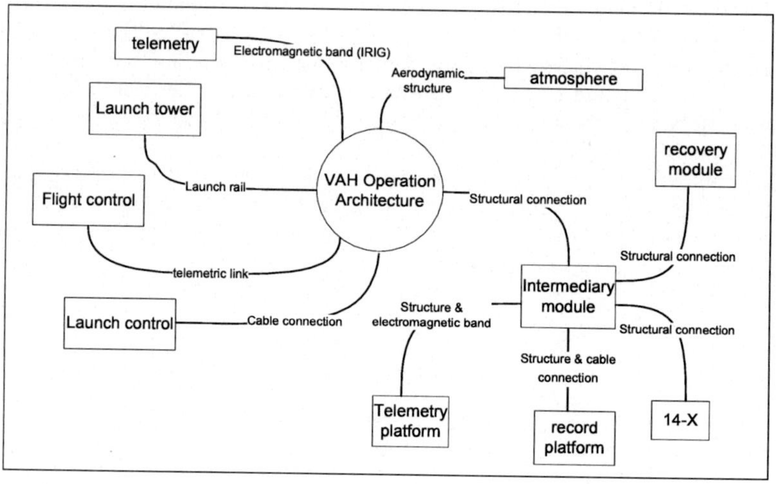


**Figure 7.** VAH Operation architecture: structural elements and physical connections

## 5 Discussion

Concurrent Engineering applied in this work has the advantage of generating broad understanding by looking in parallel at each stage of system development, focusing not only in the product but also in the processes and organizations. The method enables the interaction between multidisciplinary teams reducing failures of non-conformity between one and another stage of product development. Another advantage is the methodological approach and the ease of recognition and consideration among others, the stakeholders involved, as well as their needs, increasing the chances of developing the system required in a efficient and effective way.

Although the method is extremely laborious in the beginning of a project, it shows that, as the study progressed and some new relevant items appeared, most of the time is spent before the actual development of the system, providing a confident progress through the subsequent process, where changes must be avoided and safety must be increased, providing a concurrent safety to the system. It is extremely important to apply this method since the cost advantages are considerable. Since a product developed, without proper planning of its development stages, is likely to present failures not envisaged at some stage, among other situations liable to happen any time during product development may cause rework, schedule delays, generating unnecessary costs, and may even derail completion of the development.

## 6 Conclusions

This study aimed to apply systems engineering to concurrent system design of the VAH, observing the life cycle from the point of view of product, process and organization, the stakeholders that influence this development and the context where the process take place. Concurrent engineering was able to detail the system's development from conception until the operation. Through a vision of parallel processes, the methodology allowed to plan all stages of the life cycle of the system in an integrated and thorough manner.

The paper also described the approach as a way to provide a additional maturity on safety, once the complex product in case has hazardous potential, not allowing failures at any process of its life cycles. This concurrent safety point of view provides a robustness that may guarantee the final objective of a flight test system: to provide the valid flight test data and the information necessary to the continuous development process of a future product.

## 7 Acknowledgements

The authors would like to thank ITA (the Technological Institute of Aeronautics, www.ita.br) for the post graduate course opportunity. The authors would like to thank 3SL (www.threesl.co.uk) for providing a trial version of software Cradle used in this work.

## 8 References

[1] G. Loureiro, J. B. Fulindi, L. A. O. Fideles, D. Fernandes, R. Semabukuro, C. O. Lino. Systems Concurrent Engineering for the Conception of a Hybrid Vehicle, New World Situation: New Directions in Concurrent Engineering Advanced Concurrent Engineering, 2011, Part 7, 555-567.

[2] D. Musielak, High Speed Air Breathing Propulsion 2010, Moving faster towards the Future. AIAA HSABP TC Communications, 2010.

[3] Hitchins DK. Getting to Grips With Complexity. In: proceedings of the 2nd annual conference of the incose – UK chapter, 1996.

[4] Loureiro G. A Systems Engineering and Concurrent Engineering Framework For the Integrated Development of Complex Products. Ph.D. thesis, Loughborough University, 1999.

[5] European cooperation for space standardization. ECSS-E-ST-10C: Space Engineering: System Engineering General Requirements. Noordwijk: ESA, 2009.

[6] National aeronautics and space administration. Nasa Systems Engineering Handbook. Washington, 2007.

[7] Loureiro, G. Lessons learned in 12 years of space systems concurrent engineering. IN: Proceedings of the 61st IAC, Prague, Czech Republic, 27/09-01/10/2010. IAF(International Astronautical Federation), Paris, 2010. Paper IAC-2010-D1.5.2.

# ATTRIBUTES BALANCE ON THE ADOPTION OF PLATFORM BASED SOLUTIONS FOR SATELLITES

Otávio Luiz Bogossian[1,2], Geilson Loureiro[1,3]

[1]Instituto Nacional de Pesquisas Espaciais – INPE, Brazil

**Abstract.** This paper aims to propose the principles of a systems architecting method to assess and guide the conceptual design of platform based solutions for satellite space applications. The method helps the developers to assess the platform *changeability* to achieve future missions with a minimum impact in terms of modifications and to assess the platform *comprehensiveness* in terms of possible missions. *Changeability* attributes applicable to the platform aims to implement parameters like robustness, flexibility, agility, etc. showing the capacity of the platform to comply, fast and with minimum modifications, the future missions. *Comprehensiveness* aims to balance platform comprehensiveness attributes in terms of missions and the platform efficiency based on the additional mass necessary to cover space environment requirements like altitude (drag, radiation and torque), satellite pointing, etc. The more over dimensioned, the less efficient is the satellite. Conclusions are that the method promotes a great enhancement on the productivity of platform based solutions conception while increasing the quality of the conceptual phase results.

**Keywords.** Platform, satellite, multi-mission, changeability, product family.

## 1 Introduction

The family of products concept became relevant with the transformation of the concept of mass production into mass customization aiming to comply with individual client needs [1]. The segmentation market grid based on platform was introduced [2] as the way to leverage the family of product across different market niches. *Meyer and Utterback* [3] attempt to map the evolution of the product family based on platform by means of extensions and upgrades. The family is a set of similar products obtained from a common platform, given to each product the functionalities required by specific clients [2].

The space context has specific characteristics such as the complexity of the products and the very low production volume. It was remarked [4] that space products are designed to comply with a particular mission, being the product, designed for this objective contrasting with other general application in which they are designed for market niche. The space product referenced in this paper corresponds to the satellite, usually established as a system. The satellite is composed of the payload (components to implement the specific satellite mission) and the bus (*i.e.* house-keeping or services functions). The bus is usually divided into sub-systems, each one for a specific discipline (structure, power, communication, on board data handling, etc.) [5-8, 10-11]. Each sub-system is

[1] Av. dos Astronautas, 1758
CEP 12227-010
São José dos Campos, SP
Brazil
HP www.inpe.br

[2] INPE/LIT
Senior Technologist
Phone +551232086153
Fax =551239411884
otavio.bogossian@lit.inpe.br

[3] INPE/LIT
Senior Technologist
Phone +551232086317
Fax +551239411884
geilson@lit.inpe.br

D. D. Frey et al. (eds.), *Improving Complex Systems Today*,
Advanced Concurrent Engineering, DOI: 10.1007/ 978-0-85729-799-0_7,
© Springer-Verlag London Limited 2011

composed of several equipments (*e.g.* power sub-system: one or two batteries, solar panels, regulators, DC-DC converters, etc.).

The space products designed according to *Meyer and Lehnerd* [2] premises are those composed of a common platform that includes usually the same components to all space products (satellites) and a set of specific components that characterize each particular product and mission. The platform for satellite is usually composed by all components necessary to guarantee the satellite operation (structure, thermal, power, on board data handling, attitude control, communication for control purpose, propulsion, etc.). The mission specific components include scientific experiments, cameras (Earth or Sun observation), communication for the specific application, sensors, etc. [5-9, 11].

This paper aims to present and justify the elements of a method for assessing space platform development while it is being developed, to guide the various decision making points during space platform design. This paper has the following specific objectives:

a) To analyze up-to-date efforts on general application and space platform development;
b) To explain the changeability basic elements of the method applicable to the space platforms;
c) To explain the comprehensiveness basic elements of the method;
d) To provide an initial idea on how these elements will be used in the method.

In order to achieve these objectives, this paper is organized as following. Section 2 presents the general application platform effort and the current status of space platform development. Section 3 presents the *Changeability* elements. Section 4 presents the *Comprehensiveness* elements. Section 5 shows how the method can be built from these elements. Section 6 draws some conclusions and sets up some further work.

## 2 Development Process for Satellite Family Based on Platforms

The product family in the general context applications could be developed based on a set of modular components [4, 12-13], referred as *configurational* product family design by some authors. Each product of the family is developed adding, replacing or subtracting one or more functional modules.

Another approach is applying the scalability [14] or parametric configuration in which the capacity or performance of each module could be increased or reduced according to the customer needs.

It was proposed a methodology for product portfolio definition maximizing the use of common components [15]. An approach from a marketing and sales perspective was proposed with the product family definition based on the functional features associated to customers groups [16]. Additional methods based on the technical characteristics [17-18], with the family members defined based on technology, manufacturing process, design parameters, assembly process, etc were presented.

As it was shown, some papers define the product family based on design methods (modularity, platform based, *configurational* or scalable), others define the product family based on variety generation methods to produce product variety to target markets and, finally, others based on technical aspects improving the product process, stock reduction, component reutilization promotion, etc.

In the space context, the most common approach is the independent development [4], which means a product developed for the specific mission requirements. The number of recurring products is very limited, usually of one or two units. There are some exceptions like the GPS (24 satellites) and Galileu (27 satellites) constellations [19], where the satellites are placed in different orbits and phases.

The satellite platform concept was adopted by some space programs to explore common aspects of the products, from one mission to another. They do not have a complete view of the satellite family to be generated. However, they aim to increase the reutilization of the common part (platform) as much as possible when future missions are defined.

*Boas and Crawley* [20] presented a very particular example of simultaneous development of a family of fighter planes from the requirements of the products (case of the Joint Strike Fighter program) and the definition of a common platform. They presented a second example based on the Boeing 777 family of products in which the initial product (the first plane of the family) is developed at the same time as the platform that will be the core of future planes. This approach is clearly the sequential development process. According to the authors, this approach has the inconvenience of making the first plane a strong reference for the platform design that could cause problems for the future planes. This inconvenience is mainly due to long development process and difficulties to define the different members of the family.

The sequential approach is often applied to the development of multi-mission satellite platforms. It is demonstrated by missions like Jason 1 using CNES PROTEUS platform [5, 9, 21], Demeter mission using CNES Myriade Product Line platform (called previously Ligne de Produits Micro-satellite [7]) [8, 22]. and 1998, SkyMed/COSMO mission using ASI/Alenia PRIMA platform [10].

The development approach for a satellite family based on a platform imposes constraints in some mission parameters (*e.g.* orbits, pointing accuracy, available launchers, lifetime, mass and power limit for the payload, etc.) [9-11, 21].

## 3 Changeability

It was proposed a *Design for Changeability* – DfC to challenge the technological and market dynamism [23] with a constant inclusion of new clients, as well as, the changing of the products environment (cell phones services, GPS, Wi-Fi, etc.). During the design phase, the possibility to change the design should be kept open until as late as possible. They considered also modify the product during the operation/utilization phases.

The defined aspects (DfC term) [23] that were considered applicable to the satellite platform design taking into account the sequential development process are the following:

a) **Robustness** – System ability to be insensitive to the environment changes. This aspect is applicable considering that the platform will be used in different mission with different launching, space environment and mission requirements.
b) **Flexibility** – System ability to be easily changed. This aspect is applicable taking into account that it is expected the platform, that is the core of all satellite products, to have the ability to be easily modified for different missions of the same category (e.g. excluding Solar System missions, GEO, etc.). This concept was applied in PRIMA platform [10] considering yes/no and scalability options for the platform equipments.
c) **Agility** – System ability to be quickly changed. This aspect is applicable considering that it is expected that products (satellites) based on a platform, recover the development time spent during the first mission (platform and first product) at each developed product and it is also expected that the time spent on each satellite to be lower than an independent development to make viable the use of a platform.

The aspects considered in the *Design for Changeability* were proposed to be implemented by Basic and Extended Principles. Some of these principles correspond to axioms previously established [24]. The principles considered applicable to the satellite platform design taking into account the satellite family development approach, are the following:

a) **Ideality/Simplicity** – This principle aims to reduce the system complexity. This principle is applicable considering that all satellite projects are complex systems with a lot of interfaces, functions and components integrated in a limited room, with reduced mass and power consumption.
b) **Independence** – This principle aims to minimize the impact of change in design parameters. This principle is applicable considering that for each mission the satellite is composed of mission specific components and platform common components. It is expected that specific component and specific environment do not affect the platform components and the interfaces. This principle is adopted in the PRIMA platform design [10] with the objective of thermally decoupling the specific components (payload module) from the platform components.
c) **Modularity/Encapsulation** – This principle aims to build a system architecture that clusters the system functions into various modules (components) while minimizing the coupling among them (loose coupling) and maximizing the cohesion within the module. This principle is applicable considering the specific and platform components need to be decoupled as much as possible.
d) **Scalability** – Ability to change the components increasing or reducing their capacity or performance (*e.g.* amount of data storage, capacity of angular momentum in the reaction wheels, etc.). This principle is applicable to

increase the efficiency of the standard platform for the different mission requirements. This concept was applied in the PRIMA platform [10].

e) **Integrability** – Characterized by the compatibility and interoperability among the interfaces (proprietary or open systems) by adopting standards that enable to change easily the interconnected components. This principle is applicable mainly for the power bus and on-board data interfaces that are necessary for almost all the components. It will facilitate the scalability implementation. This principle is implemented on the on-board data handling subsystem that interfaces with almost all components using a standard bus [9-10, 21].
f) **Decentralization** – This principle is characterized by the distribution of control, information, resources, architecture attributes or properties among the components. This principle is applicable mainly to the power bus and on-board data handling.

## 4 Mission Comprehensiveness and Platform Efficiency

In order to assess the mission *Comprehensiveness* it is necessary to reduce the scope of possible orbits. The considered orbits were based on the Myriade (CNES), Proteus (CNES), PRIMA (ASI/ALENIA) and PMM (INPE) multi-mission platforms [7, 10, 11, 25]. The following criteria were considered:

a) Only circular orbits (low eccentricity) with altitudes between 400 and 1500 km;
b) Sun-synchronous orbits (SSO) with equator crossing time of 6:00 am and 10:00 am (also almost SSO);
c) Equatorial and low inclination orbits up to 25°;
d) Three orbit inclinations and three SSO with three different altitudes will be considered.

Four possible satellite configurations will be considered all with a parallelepiped shape. Two configurations with fixed solar panels with one or two wings and two with rotating solar panel with one or two wings. The hardware alternatives are the following: 1) up to three torque rods sizes for each axis, 2) up to three propellant reservoir sizes and 3) up to three reaction wheels sizes for each axis.

With respect to the satellite pointing, nadir will be considered for SSO orbits considered and nadir and Sun for other orbits.

The following space environmental effects are considered:

a) **Cumulated radiation** at each possible orbit with impact on the component hardness in terms of krads. Mission lifetime will be considered with respect to the maximum lifetime of the platform. An indirect impact on mass will be considered based on the additional price of the components and the cost to transport a unit of mass to the space;

b) **Aerodynamic torque due to the residual atmosphere** – This will affect the reaction wheels dimensioning in terms of angular momentum and the increment of required mass;

c) **Drag due to the residual atmosphere** – This will affect the propellant and reservoir to keep the orbit error lower than a specified value;

d) **Use of the magnetic field to unload the reaction wheels** – Dimensioning of the torque rods and its increment in terms of mass;

e) **Static and dynamic launcher environment** – The increment in the structure mass necessary to consider several launchers;

f) **Power required** – For each orbit, pointing and satellite solar panels configuration, it will be determined the amount of solar panel necessary to provide the minimum amount of power established for the platform as well as the battery necessary to overcome the eclipse periods. The increment of mass necessary to cover all the orbits will be considered for the solar panels and batteries.

The thermal dimensioning and the structure are not considered as part of the platform [8] due to the necessity to design for each specific mission (payload module for the structure), therefore it does not induce any inefficiency.

The launching window with respect to the solar flux cycle was not considered as inefficiency due to the necessity of the platform to be ready to launch at any time in the solar cycle.

## 5 Elements of the method to be developed

Considering the basic development approach, the method shall implement *Changeability* in such a way the platform will be designed with the capacity to be quickly adapted for future and unknown missions. *Changeability* will provide to the platform designer, at the initial design phase, enough information about the robustness, flexibility and agility of the platform. Based on the *Changeability* aspects and principles, the method will implement objective parameters based on the satellite configuration.

The method shall implement also the *Comprehensiveness* (in terms of missions) and the corresponding platform efficiency to provide to the platform designer, at the initial design phase, enough information to decide how much to pay in terms of efficiency, to implement the *Comprehensiveness*.

For *Comprehensiveness* implementation, the method shall implement a limited number of cases in terms of orbits, pointing and satellite configurations, considering those more applicable. It shall also consider some premises (orbit maximum error, reaction wheels and torque rods dimensioning, etc.) and the implementation of the scalability to give to the designer, an easy way to increase the platform efficiency. The method will consider as objective parameter to implement the *Comprehensiveness* the worst/least case of equipment mass, directly or indirectly,

The method also shall receive inputs, from the platform designer, such as lifetime, mass, inertia, surface of the solar panels, for Comprehensiveness and satellite architecture definitions and the qualification process for *Changeability*.

The method to be developed shall be easily implemented (spreadsheet or simple program) without the necessity to perform simulations to assess the platform. The simulations will be performed during the method design phase for all considered cases.

## 6 Conclusions and further work

Section 2 presented the effort to develop the platform concept for general application and also concluded the development approach applied on the development of space platforms. Section 3 presented the *Changeability* basic elements applied to the space platforms derived from the general application platform concept. Section 4 presented the basic *Comprehensiveness* elements developed, based on the application scope of some existing or in development satellite platforms. Section 5 presented the initial idea of the method implementation. The Sections 2 to 5 have presented and justified the elements of the method to assess space platform during the development phase. As conclusion, the main and specific objectives presented in the Section 1 were achieved by this paper.

To finalize the method development,further work is necessary. With respect to *Changeability*, it is necessary to define objective parameters to assess the platform. These parameters will cover the platform architecture (*e.g* for the independence principle a bonus will be given to the architecture that implements a standard bus for data handling) and the qualification plan (*e.g.* for robustness a bonus will be given to the qualification plan that covers a significant number of different environment scenarios). With respect to Comprehensiveness, it is necessary to perform several simulations and verify the mass impact on the platform components.

Finally it is necessary to produce two measures, one for *Changeability* and another for the *Comprehensiveness*.

## 7 References

[5] **Aerospatiale and Sextant** – Aerospatiale/Sextant Avionique brochure. *Filière Proteus CNES. Octobre 1995.*

[16] **Agard, B. and Kusiak, A** – Data-mining-based methodology for the design of product families. *International Journal of Production Research (2004), 42(15), 2955-2969.*

[6] **Alary, D. and Lambert, H.** - The Myriade product line, a real success story. *ACTA Astronautica (2007) 61, 223-227.*

[20] **Boas, R. C. and Crawley, E. F.** - Extending Platforming to the Sequential Development of System Families. *INCOSE 2006 – 16th Annual International Symposium Proceedings.*

[8] **Bouzat, C.** – CNES Microsatellite Product Line, an approach for innovation - *Small Satellites Systems and Services, 5th International Symposium, 19-23 June 2000, La Boule France.*

[7] **Buisson, F., Cussac, T., Lassalle-Balier, G., Laurens, A., Ledu, M., Llorens, J-C and Chadoutaud, P.** – La ligne de produits Micro-satellite du CNES. *Small Satellites Systems and Services, 4th International Symposium, 14-18 September 1998, Antibes – San Juan Les Pains France.*

[22] **Cussac, T.;Buisson, F. and Parrot, M.** – The Demeter Program: Mission and Satellite Description – Early in Flight Results. *55th International Astronautical Congress 2004 IAC-04-IAA.4.11.2.04. Vancouver, Canada.*

[15] **Dahmus, J. B., Gonzales-Zugasti, J. P. and Otto, K. N.** – Modular Product Architecture. *Design Studies 22 (2001) 409-424.*

[9] **Dechezelles, J-J and Huttin, G.** – PROTEUS: A Multimission Platform fo Low Earth Orbits. *Air & Space Europe (2000), Vol. 2 N° 1, 77-81.*

[17] **De Lit, P. G. and Delchambre, A.** – Integrated design of a product family and its assembly system (2003). *Massachusetts: Kluwer Academic Publishers.*

[12] **Du, X., Jiao, J. and Tseng, M. M.** – Architecture of product family: Fundamentals and Methodology. *Concurrent Engineering: Research and Application (2001), 9(4), 309-325.*

[19] **Forrest, W. M.** – Interoperability of the GPS and Galileo Timescales for Positioning and Metrology. *European Frequency and Time Forum (2004), 18th ( 468-475).*

[23] **Fricke, E. and Schulz, A. P.** – Design for Changeability (DfC): Principles To Enable Changes in Systems Throughout Their Entire Lifecycle. *System Engineering, Vol. 8, N° 4, 2005.*

[10] **Galeazzi, C.** - Prima: A new, competitive small satellite platform. *Acta Astronautica 2000 Vol. 46, Nos. 2-6, 379-388.*

[4] **Gonzalez-Zugasti, J. P.; Otto K. N. and Baker J. D.** - A Method for Architecting Product Platforms. *Research in Engineering Design (2000) 12:61-72.*

[21] **Grivel, C., Doullet, F., Huiban, T., Sainct, H., Bailion, Y., Terrenoire, P. Schrive, J. and Lazard, B.** – Proteus: European Standard for small satellites, *Small Satellites Systems and Services, 5th International Symposium, 19-23 June 2000, La Boule France.*

[11] **INPE** - Multimission Platform: Data Package for System Requirement Review, August, 10th 2001, INPE's internal document.

[3] **Meyer, M. and Utterback, J.** - The product family and the dynamics of core capability. *Sloan Management Review, Spring 1993, 29-47.*

[2] **Meyer, M. and Lehnerd, A. P.** - The power of product platform – building value and cost leadship (1997). *New York: Free Press.*

[1] **Pine, B. J.** - Mass customization: *The new frontier in business competition (1993). Boston: Harvard Business School Press.*

[25] **Rougeron, M.** – CNES Minisatellite Missions / Les Missions Proteus - *Small Satellites Systems and Services, 5th International Symposium, 19-23 June 2000, La Boule France.*

[18] **Simpson, T. W.** - Product platform design and customization: Status and promise. *Artificial Intelligence for Engineering Design, Analysis and Manufacturing (2004), 18(1), 3-20.*

[14] **Simpson, T.W., Maier, J.R.A. and Mistree, F.** – Product platform design: Method and Application. *Research in Engineering Design, 2001 13(1), 2-22.*

[24] **Suh, N. P** – Principles of Design (1990). *Oxford University Press New York.*

[13] **Ulrich, K.** – The role of product architecture in the manufacturing firm. *Research Policy, 1995 24(3), 419-440.*

# Petri Nets for Systems Concurrent Engineering

Giuliani Paulineli Garbi[a,1], Geilson Loureiro[b]

[a]Engineering Department, College Anhanguera of São José (São José dos Campos), BRA.
[b]Brazilian Institute for Space Research (INPE) and Technological Institute of Aeronautics (ITA).

**Abstract.** The paper proposes a generic model to represent the workflow based on Petri Nets theory of the activities used in the Systems Concurrent Engineering process and to use Petri Nets to support decision making when architecting a complex system. Systems Concurrent Engineering is a method that integrates systems engineering and concurrent engineering in the same integrated product development framework. The framework is applicable to the development of complex products. The properties of the Petri Nets allow the design of the generic model through formal language, semantic formalism and techniques to supporting analysis of process and architecture performance through graphical and algebric tools.

Currently to maintain the competitiveness the main productive segments must seek to apply methods to innovate and develop their complex products and services with lower costs, improved productivity and quality, and in less time. Addressing the needs of productive segments, the paper presents a generic model to support and encourage the development of complex products and services. The main benefit of the generic model is that despite the actual various ways of implementing a system and of performing the systems concurrent engineering process, Petri Nets would allow to assess the process and architecture alternatives at very early stages of a complex product development process, based only on the process and product Petri Net models.

**Keywords.** Systems Concurrent Engineering, Petri Nets, Systems Engineering Process, Concurrent Engineering, System Architecture

---

[1] Engineering Department, College Anhanguera of São José, avenue Dr. João Batista de Sousa Soares, 4121, Jardim Morumbi, São José dos Campos, São Paulo, Brazil, CEP: 12223-660; +55 (12) 3512-1300; Fax: +55 (12) 3512-1316; Email: giuliani.garbi@unianhanguera.edu.br; http://www.unianhanguera.edu.br

D. D. Frey et al. (eds.), *Improving Complex Systems Today*,
Advanced Concurrent Engineering, DOI: 10.1007/ 978-0-85729-799-0_8,
© Springer-Verlag London Limited 2011

## 1 Introduction

This paper concerns a generic model to support decision making in architectural reasoning for complex socio-technical systems. The architecture term denotes the stable properties of the system of interest. The architectural reasoning is defined as a transformative process that utilizes knowledge about stable properties in a system to achieve certain global objectives. The complex socio-technical systems refers to systems involving multiple stakeholders and requiring multiple knowledge domains [6].

In the process of architecting complex socio-technical systems that involves multiple stakeholders and knowledge domains, to assess the architecture alternatives at very early stages of a development process often becomes a considerable challenge. This challenge presents two interrelated opportunities. First, a domain-independent architectural reasoning techniques that can be implemented computationally over multiple disciplines and second, identifying a single formal language and the techniques analysis tools to support Systems Concurrent Engineering process.

Therefore this paper proposes a generic architecture that represents a workflow based on Petri Nets theory to Systems Concurrent Engineering process. The main purpose of workflow is to support the definition, execution, registration and control processes, and the development with Petri Nets allows the construction of a single formal language and the techniques analysis tools to support analysis of process performance because it is a combination of specification of oriented events and states with excellent graphics tools [3, 5].

The paper presents in Section 2 the Systems Concurrent Engineering approach that integrated systems engineering and concurrent engineering process for integrated complex product development. Section 3 presents the main concepts of Petri Nets. Section 4 presents the generic architecture that represent a workflow based on Petri Nets theory to the Systems Concurrent Engineering process and Section 5 draws some conclusions.

## 2 Systems Concurrent Engineering

The Systems Concurrent Engineering is a modeling framework that integrates the product and their performing organizations [1, 2]. Stakeholder analysis, requirements analysis, functional analysis and implementation or physical analysis processes are carried out through the simultaneous modeling of product and organization, at all levels of the product hierarchy, deriving attributes as emergent properties of a whole integrated system [7, 8, 9].

Figure 1 presents the total view framework, it has three dimensions. Figure 2 provides an overview of the stakeholder analysis, requirements analysis, functional analysis and implementation (or physical) analysis is performed, simultaneously, for the product under development and its life cycle process performing organizations. The analysis processes are performed at each layer of the system breakdown structure. Figure 3 details the concurrent structured analysis method showing how to incorporate the concurrent engineering concept in the systems engineering process.

Step 1: identify the product mission, the product life cycle processes and their scenarios and, the scope of the development effort. The scope of the development effort consists of the life cycle processes or their scenarios that the development organization is also responsible for accomplishing.

Step 2: identify product stakeholders and their concerns for each product life cycle process scenario. Identify organization stakeholders and their concerns for each process within the scope of the development effort. From stakeholder concerns, stakeholder requirements are identified and measures of effectiveness (MoEs) are derived. MoEs must measure how the system meets the stakeholder requirements. Requirement analysis transforms stakeholder requirements into system requirements.

Step 3: identify functional context for product at each life cycle process scenario and for organization at each life cycle process scenario within the scope of the development effort. For each function, performance requirements are identified. Circumstances, flows between the system and the environment and function failures are sources of hazards. Risk analysis is performed on each identified potential hazard and exception handling functions are also identified at this stage.

Step 4: identify implementation architecture context for product at each life cycle process scenario and for organization at each life cycle process scenario within the scope of the development effort. Physical connections between the system and the environment elements define the physical external interface requirements. Physical parts are identified.

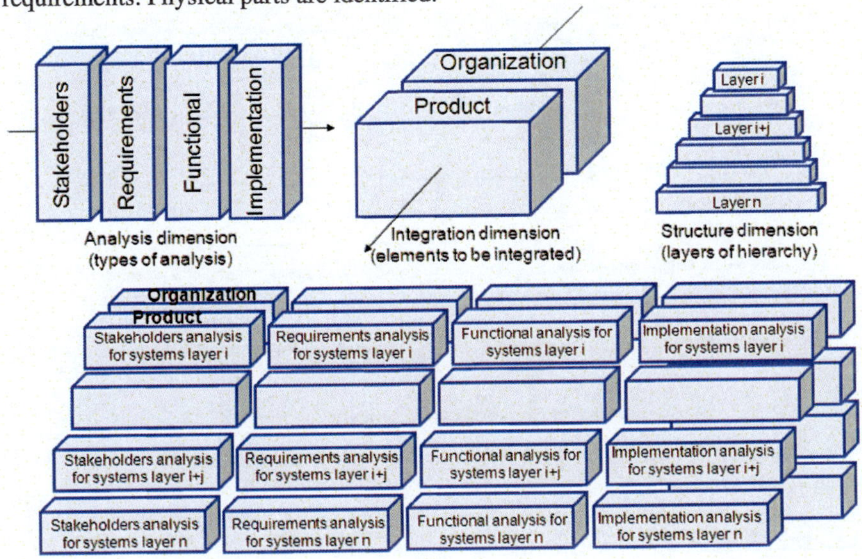


**Figure 1.** A framework to address complexity in complex product development – the total view framework. Source: [7], [8]

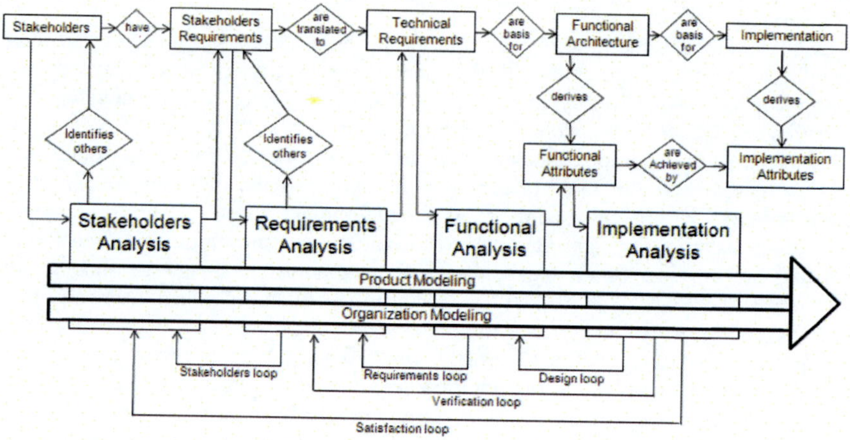


**Figure 2.** A method within the total view framework – the concurrent structured analysis method. Source: [7], [8]

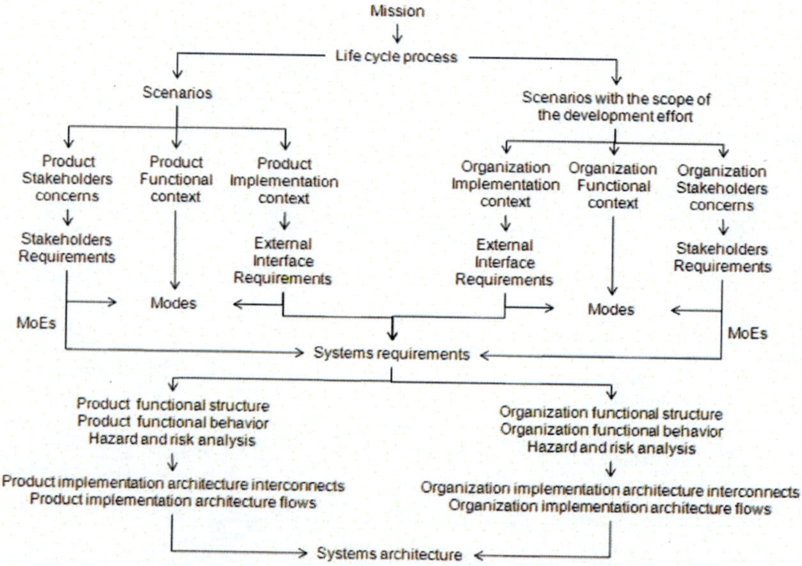


**Figure 3.** The system concurrent engineering method in detail. Source: [7], [8]

## 3 Petri Nets

The concept of Petri Nets was introduced by Carl Adam Petri in his doctoral thesis in 1962. It is a modeling technique that allows the representation of systems through its graphical and algebric formalism. The technique has properties that allow to model parallel systems, concurrent, asynchronous and non-deterministic, and has mechanisms that treat the hierarchy design and high level of abstraction that are fundamental to the development of complex systems. During the past 20 years, Petri Nets have been applied in many applications in different areas, currently there are many commercial and academic tools for design, simulation, and analysis system based on Petri Nets [4, 5].

Petri Net is a model of the state-event type, where each event possesses daily pre-conditions to allow its occurrence and pos-conditions of this event, illustrated in Figure 4. It is also seen as a particular type of guided graph that allows modeling the static properties of a system to the discrete events: transitions (events that characterize the changes of state in the system), and the places (conditions against which the events must be certified in order to happen) linked by directed weighed arcs. The transition is triggered only if there is at least one marking or fiche (token) in place proceeding of transition.

Petri Net is, therefore, a formalism that allows the modeling of discrete dynamic systems with great power of expressiveness, allowing to represent with easiness all the relations of causalities between the processes in situations of sequence, conflict, parallelism and synchronization. Figure 4 provides an overview of Petri Nets graphical tools.

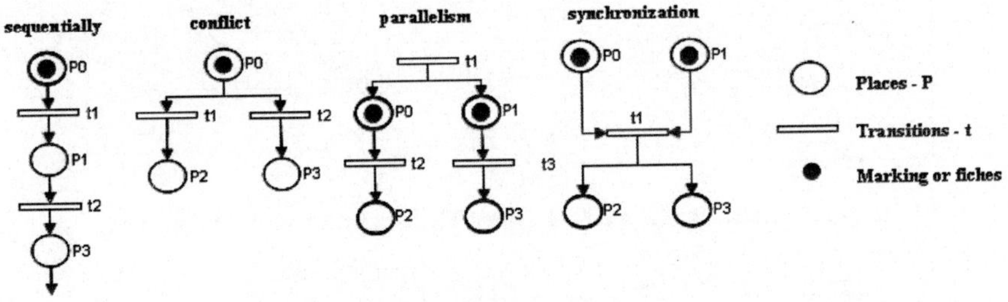


**Figure 4.** Petri Nets graphical tools. Source: [3]

A Petri Net (simple or autonomous) is composed of five parts: a set of places *P*, a set of transitions *t*, an application of input *I*, an application of exit *O* and a set of markings *M* that represent the markings of places *P*, illustrated in the Equation 1.

$$R = (P, T, I, O, M) \tag{1}$$

## 4 Petri Nets for Systems Concurrent Engineering

Figure 5 presents the Petri Net graph for the Systems Concurrent Engineering process. The Figure 5 represents the generic model of the concurrent structured analysis method workflow using Petri Nets notation. The stages of the system architecting process illustrated in Figure 3 are defined by the workflow of the places and the transitions.

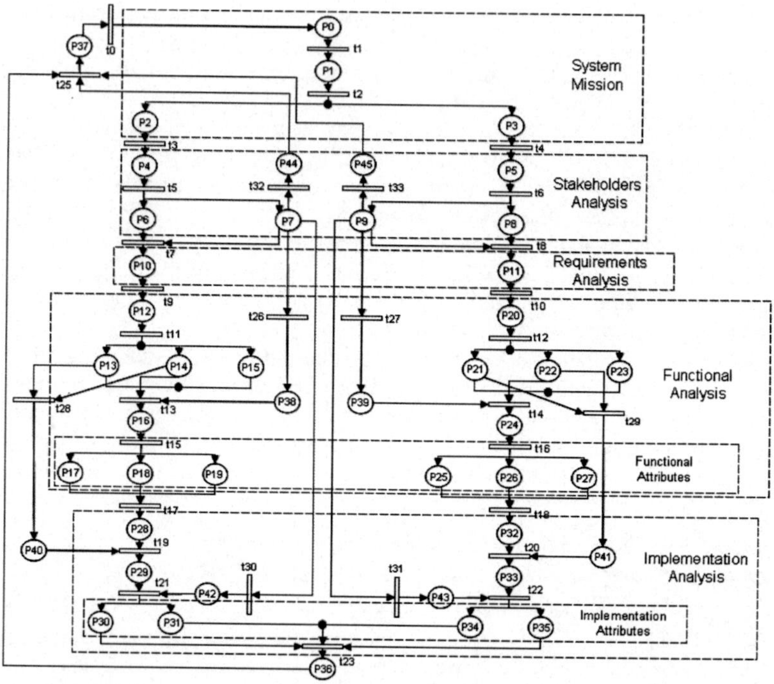


**Figure 5.** Petri Nets graph for Systems Concurrent Engineering process

**System Mission Places**
P0 = Mission
P1 = Life cycle process
P2 = Scenario
P3 = Scenarios with the scope of the development effort

**Stakeholders Analysis Places**
P4 = Product stakeholders concerns
P5 = Organization stakeholders concerns
P6 = Product stakeholders requirements
P7 = Product measures of effectiveness
P8 = Organizations stakeholders requirements
P9 = Organization measures of effectiveness

**Requirements Analysis Places**
P10 = Product technical requirements
P11 = Organizations technical requirements

**Functional Analysis Places**
P12 = Product functional context
P13 = Product system element
P14 = Product environment element
P15 = Product behavior modeling
P16 = Product mode
P17 = Product functional structure
P18 = Product functional behavior
P19 = Product hazard and risk analysis
P20 = Organization functional context
P21 = Organization system element
P22 = Organization environment element
P23 = Organization behavior modeling
P24 = Organization mode
P25 = Organization functional structure
P26 = Organization functional behavior
P27 = Organization hazard and risk analysis

**Implementation Analysis Places**
P28 = Product implementation context
P29 = Product physical external interface requirements
P30 = Product implementation architecture interconnects
P31 = Product implementation architecture flows
P32 = Organization implementation context
P33 = Organization physical external interface requirements
P34 = Organization implementation architecture interconnects
P35 = Organization implementation architecture flows

**Loops Places**
P36 = System architecture
P37 = Satisfaction loop
P38 =Product requirements loop
P39 =Organization requirements loop
P40 =Product design loop
P41 =Organization design loop
P42 =Product verification loop
P43 =Organization verification loop
P44 =Product satisfaction loop
P45 =Organization satisfaction loop

**Figure 6.** Description of the places in the Petri Nets graph for the Systems Concurrent Engineering process

**System Mission Transitions**
t0 = Identify the product mission
t1 = Identify the product life cycle process
t2 = Identify their scenarios and scenarios with the scope of the development effort

**Stakeholders Analysis Transitions**
t3 = Identify product stakeholders and their concerns
t4 = identify organization stakeholders concerns and their concerns
t5 = Identify product stakeholders requirements and define product measures of effectiveness
t6 = Identify organizations stakeholders requirements and define organization measures of effectiveness

**Requirements Analysis Transitions**
t7 = Transform product stakeholders requirements in technical requirements
t8 = Transform organizations stakeholders requirements in technical requirements

**Functional Analysis Transitions**
t9 = Identify product functional context
t10 = Identify organization functional context
t11 = Define product system element, identify product environment element and product behavior modeling
t12 = Define organization system element, identify organization environment element and organization behavior modeling
t13 = Identify product modes
t14 = Identify organization modes
t15 = Define product functional attributes
t16 = Define organization functional attributes

**Implementation Analysis Transitions**
t17 = Identify product implementation context
t18 = Identify organization implementation context
t19 = Define product physical external interface requirements
t20 = Define organization physical external interface requirements
t21 = Define product implementation attributes
t22 = Define organization implementation attributes

**Loops Transitions**
t24 = Define system architecture
t25 = Identify satisfaction loop
t26 = Identify product requirements loop
t27 = Identify organization requirements loop
t28 = Identify product design loop
t29 = Identify organization design loop
t30 = Identify product verification loop
t31 = Identify organization verification loop
t32 = Identify product satisfaction loop
t33 = Identify organization satisfaction loop

**Figure 7.** Description of the transitions in the Petri Nets graph for the Systems Concurrent Engineering process

Figure 6 and 7 present the semantic formalism that describes the function of places and transitions in the Petri Nets Systems Concurrent Engineering process generic model. From the Petri Nets graph, it can be applied the Petri Net analysis tools. For example, Figure 8 presents the reachability tree. The reachability tree is basic to study the dynamic properties of any system modeled with Petri Nets. The triggered transition modifies the distribution of marks or tokens on the graph of Petri Nets. In the definition of Petri Nets, it is called reachability of a mark *Mn* the set of all the markings generated from *M0*.

| | Places | | | | | | | | | | | | | | | | | | | | | | | | | | | | | | | | | | | | | | | | | | | | | |
|---|---|---|---|---|---|---|---|---|---|---|---|---|---|---|---|---|---|---|---|---|---|---|---|---|---|---|---|---|---|---|---|---|---|---|---|---|---|---|---|---|---|---|---|---|---|---|
| | 0 | 1 | 2 | 3 | 4 | 5 | 6 | 7 | 8 | 9 | 10 | 11 | 12 | 13 | 14 | 15 | 16 | 17 | 18 | 19 | 20 | 21 | 22 | 23 | 24 | 25 | 26 | 27 | 28 | 29 | 30 | 31 | 32 | 33 | 34 | 35 | 36 | 37 | 38 | 39 | 40 | 41 | 42 | 43 | 44 | 45 | |
| M0 | 1 | 0 | 0 | 0 | 0 | 0 | 0 | 0 | 0 | 0 | 0 | 0 | 0 | 0 | 0 | 0 | 0 | 0 | 0 | 0 | 0 | 0 | 0 | 0 | 0 | 0 | 0 | 0 | 0 | 0 | 0 | 0 | 0 | 0 | 0 | 0 | 0 | 0 | 0 | 0 | 0 | 0 | 0 | 0 | 0 | 0 | t1 |
| M1 | 0 | 1 | 0 | 0 | 0 | 0 | 0 | 0 | 0 | 0 | 0 | 0 | 0 | 0 | 0 | 0 | 0 | 0 | 0 | 0 | 0 | 0 | 0 | 0 | 0 | 0 | 0 | 0 | 0 | 0 | 0 | 0 | 0 | 0 | 0 | 0 | 0 | 0 | 0 | 0 | 0 | 0 | 0 | 0 | 0 | 0 | t2 |
| M2 | 0 | 0 | 1 | 1 | 0 | 0 | 0 | 0 | 0 | 0 | 0 | 0 | 0 | 0 | 0 | 0 | 0 | 0 | 0 | 0 | 0 | 0 | 0 | 0 | 0 | 0 | 0 | 0 | 0 | 0 | 0 | 0 | 0 | 0 | 0 | 0 | 0 | 0 | 0 | 0 | 0 | 0 | 0 | 0 | 0 | 0 | t3 and t4 |
| M3 | 0 | 0 | 0 | 0 | 1 | 1 | 0 | 0 | 0 | 0 | 0 | 0 | 0 | 0 | 0 | 0 | 0 | 0 | 0 | 0 | 0 | 0 | 0 | 0 | 0 | 0 | 0 | 0 | 0 | 0 | 0 | 0 | 0 | 0 | 0 | 0 | 0 | 0 | 0 | 0 | 0 | 0 | 0 | 0 | 0 | 0 | t5 and t6 |
| M4 | 0 | 0 | 0 | 0 | 0 | 0 | 1 | 1 | 1 | 1 | 0 | 0 | 0 | 0 | 0 | 0 | 0 | 0 | 0 | 0 | 0 | 0 | 0 | 0 | 0 | 0 | 0 | 0 | 0 | 0 | 0 | 0 | 0 | 0 | 0 | 0 | 0 | 0 | 0 | 0 | 0 | 0 | 0 | 0 | 0 | 0 | t7, t8, t26, t27, t30, t31, t32 and t33 |
| M5 | 0 | 0 | 0 | 0 | 0 | 0 | 0 | 0 | 0 | 0 | 1 | 1 | 0 | 0 | 0 | 0 | 0 | 0 | 0 | 0 | 0 | 0 | 0 | 0 | 0 | 0 | 0 | 0 | 0 | 0 | 0 | 0 | 0 | 0 | 0 | 0 | 0 | 0 | 1 | 1 | 0 | 0 | 1 | 1 | 1 | 1 | t9, t10, |
| M6 | 0 | 0 | 0 | 0 | 0 | 0 | 0 | 0 | 0 | 0 | 0 | 0 | 1 | 0 | 0 | 0 | 0 | 0 | 0 | 0 | 1 | 0 | 0 | 0 | 0 | 0 | 0 | 0 | 0 | 0 | 0 | 0 | 0 | 0 | 0 | 0 | 0 | 0 | 1 | 1 | 0 | 0 | 1 | 1 | 1 | 1 | t11 and t12 |
| M7 | 0 | 0 | 0 | 0 | 0 | 0 | 0 | 0 | 0 | 0 | 0 | 0 | 0 | 1 | 1 | 1 | 0 | 0 | 0 | 0 | 0 | 1 | 1 | 1 | 0 | 0 | 0 | 0 | 0 | 0 | 0 | 0 | 0 | 0 | 0 | 0 | 0 | 0 | 1 | 1 | 0 | 0 | 1 | 1 | 1 | 1 | t13, t14, t28 and t29 |
| M8 | 0 | 0 | 0 | 0 | 0 | 0 | 0 | 0 | 0 | 0 | 0 | 0 | 0 | 0 | 0 | 0 | 1 | 0 | 0 | 0 | 0 | 0 | 0 | 0 | 1 | 0 | 0 | 0 | 0 | 0 | 0 | 0 | 0 | 0 | 0 | 0 | 0 | 0 | 0 | 0 | 1 | 1 | 1 | 1 | 1 | 1 | t15 and t16 |
| M9 | 0 | 0 | 0 | 0 | 0 | 0 | 0 | 0 | 0 | 0 | 0 | 0 | 0 | 0 | 0 | 0 | 0 | 1 | 1 | 1 | 0 | 0 | 0 | 0 | 0 | 1 | 1 | 1 | 0 | 0 | 0 | 0 | 0 | 0 | 0 | 0 | 0 | 0 | 0 | 0 | 1 | 1 | 1 | 1 | 1 | 1 | t17 and t18 |
| M10 | 0 | 0 | 0 | 0 | 0 | 0 | 0 | 0 | 0 | 0 | 0 | 0 | 0 | 0 | 0 | 0 | 0 | 0 | 0 | 0 | 0 | 0 | 0 | 0 | 0 | 0 | 0 | 0 | 1 | 0 | 0 | 0 | 1 | 0 | 0 | 0 | 0 | 0 | 0 | 0 | 1 | 1 | 1 | 1 | 1 | 1 | t19 and t20 |
| M11 | 0 | 0 | 0 | 0 | 0 | 0 | 0 | 0 | 0 | 0 | 0 | 0 | 0 | 0 | 0 | 0 | 0 | 0 | 0 | 0 | 0 | 0 | 0 | 0 | 0 | 0 | 0 | 0 | 0 | 1 | 0 | 0 | 0 | 1 | 0 | 0 | 0 | 0 | 0 | 0 | 0 | 0 | 1 | 1 | 1 | 1 | t21, t22 |
| M12 | 0 | 0 | 0 | 0 | 0 | 0 | 0 | 0 | 0 | 0 | 0 | 0 | 0 | 0 | 0 | 0 | 0 | 0 | 0 | 0 | 0 | 0 | 0 | 0 | 0 | 0 | 0 | 0 | 0 | 0 | 1 | 1 | 0 | 0 | 1 | 1 | 0 | 0 | 0 | 0 | 0 | 0 | 0 | 0 | 1 | 1 | t23 |
| M13 | 0 | 0 | 0 | 0 | 0 | 0 | 0 | 0 | 0 | 0 | 0 | 0 | 0 | 0 | 0 | 0 | 0 | 0 | 0 | 0 | 0 | 0 | 0 | 0 | 0 | 0 | 0 | 0 | 0 | 0 | 0 | 0 | 0 | 0 | 0 | 0 | 1 | 0 | 0 | 0 | 0 | 0 | 0 | 0 | 1 | 1 | t25 |
| M14 | 0 | 0 | 0 | 0 | 0 | 0 | 0 | 0 | 0 | 0 | 0 | 0 | 0 | 0 | 0 | 0 | 0 | 0 | 0 | 0 | 0 | 0 | 0 | 0 | 0 | 0 | 0 | 0 | 0 | 0 | 0 | 0 | 0 | 0 | 0 | 0 | 0 | 1 | 0 | 0 | 0 | 0 | 0 | 0 | 0 | 0 | t0 |

**Figure 8.** Reachability of the Petri Nets graph for Systems Concurrent Engineering process

## 5 Conclusion

The generic model is represented by graphical tools and semantics formalism of Petri Nets that allows a visualization of high abstraction of the concurrent activities for integrated development of complex products by Systems Concurrent Engineering process. Dynamic analysis, simulation, verification, implementation and design analysis of iteration throughout the integrated development process can be analyzed by graph of Petri Net, for example using a reachability tree. From the generic model proposed, it is possible to develop models specific to a domain of application including, for example, the various decision making points during the complex product architecting process.

For a space satellite development, for example, decisions to be made are: which stakeholders to satisfy, which requirements to meet, which concept to choose along the life cycle processes, which functions the product and organizations shall perform, which alternative reference architecture models to choose, which solutions to choose in order to implement the chosen architecture. Further steps of this work are to demonstrate how to move form the generic model to a given application domain and in that domain develop a tool that anticipates to the early complex product development stages, the choices and decisions, and therefore their consequences. Also, the tool will provide support along the system life cycle process and will incorporate the lessons learned.

This will allow a gain in productivity in the system architecting process, will allow a common language to be shared among different stakeholders along the system life cycle process and will allow focus of product development in alternative solutions of greater potential.

## 6 References

[1] Bandecchi M. The ESA Concurrent Design Facility (CDF): concurrent engineering applied to space mission assessments. ESA - 2nd Nordic Systems Engineering Boat Seminar. ESA/ESTEC Noordwijk, NED, 2001.

[2] Bandecchi M, Melton B, Ongaro F. Concurrent engineering applied to space mission assessment and design. ESA bulletin, NED, 1999.

[3] Garbi GP. Application of Petri Nets in System Identification and transport of three different types of products. Master in Engineering Thesis, Taubaté, BRA, 2004.

[4] Garbi GP, Grandinetti FJ. Intelligent Transporter System Modeling by Petri Nets. 24° International Congress of the Aeronautical Sciences - ICAS, JPN, 2004.

[5] Girault C, Valk R. Petri Nets for Systems Engineering. A Guide to modeling, verification, and applications. Berlin Heidelberg New York: Springer, 2002; 473–539.

[6] Koo HYB. A Meta-language for Systems Architecting. PhD Thesis. Massachusetts Institute of Technology, Massachusetts, USA, 2005.

[7] Loureiro G. A systems engineering and concurrent engineering framework for the integrated development of complex products. PhD Thesis. Loughborough University, Loughborough, UK, 1999.

[8] Loureiro G. Lessons learned in 12 years of space systems concurrent engineering. 61st International Astronautical Congress. Prague, CZ, 2010.

[9] Systems Engineering Handbook. International Council on Systems Engineering - INCOSE, USA, 2010.

[10] Wall SD. Use of concurrent engineering in space mission design. Jet Propulsion Laboratory, California Institute of Technology, Pasadena, USA, 1999.

[11] Zhou J, Carmi S; Lau A, Koulas S. Concurrent engineering concepts applied to concurrent education and research. Third ISPE International Conference on Concurrent Engineering. Ontario, CAN, 1996.

# Systems Support Engineering: Looking Beyond The Physical

Luke Webb[1] and Cees Bil

*School of Aerospace Mechanical and Manufacturing Engineering, RMIT University, Melbourne, Australia*

**Abstract.** The aerospace and defence industries are currently undergoing a significant, although sometimes subtle, transformation in the way it operates and the way it is structured. What was once an industry that was predominately product focused now is morphing into a supplier of products and services designed to give an operator a capability to perform their mission. This paper is drawn from research into how aircraft manufacturers are slowly changing themselves from pure design-and-build agencies, to whole-of-life, service-driven enterprises that are centred on solving customer problems. It examines some of the challenges aircraft manufacturers are facing in moving away from a product-centred approach, to a service-driven approach. It will examine examples of a 'servitized' approach in aerospace, and detail some of challenges this changing environment poses for manufacturers. In the conclusion it will present a unique postgraduate program for system support engineering that is being developed by the Australian Department of Defence in partnership with academia and industry.

## 1 Introduction

The growth in Original Equipment Manufacturer (OEM) support services (such as logistics, maintenance and in-service engineering) in recent years has been remarkable, especially with the growing number of defence customers outsourcing to OEMs and third parties the technical management of many of their technologically advanced fleets of aircraft and advanced weapon systems. Traditionally, aircraft programs have had a strong emphasis on the design and build of an aircraft, reflected through the strength of technical expertise residing within aircraft manufacturers. However, as whole-of-life approach's to developing, implementing and delivering ongoing, cost-effective capability solutions for operators becomes an increasingly popular business strategy, high levels of design and manufacturing expertise are no longer sufficient of their own in delivering on an 'airborne solution' program. These capabilities now have to be harnessed as part of a deeply rooted service ethos and architecture that enables manufacturers to become capability partners.

This paper focuses on how aircraft manufacturers are changing from pure design-and-build organisations, to whole-of-life, service-driven enterprises that are centred around solving customer needs. It examines examples of a 'servitized'

[1] Postgraduate research candidate, School of Aerospace, Mechanical and Manufacturing Engineering, RMIT University, PO Box 71, Bundoora, VIC 3083, Australia.

D. D. Frey et al. (eds.), *Improving Complex Systems Today*,
Advanced Concurrent Engineering, DOI: 10.1007/ 978-0-85729-799-0_9,
© Springer-Verlag London Limited 2011

approach in aerospace, and details some of challenges this changing environment poses for manufacturers.

## 2 Changing World

Traditional manufacturing business models have had a very strong product-centric dominant logic [3]. Such a mindset saw value as been transacted in an exchange of money for ownership of a physical asset. The core competency of a manufacturer was to design and build technically superior products and systems. Organisations that offered "services" along with their products often did so reluctantly, often perceiving it as an activity that distracted from the 'main-game' of product manufacturing [4].

The production-centric nature of aircraft manufacturers is well reflected in the development of the Boeing 777 [5]. The Boeing 777 was a program that marked a significant departure from previous production-centric approach to aircraft development and one that was dominated by a "working together" ethos. Under the 777 program potential customer airlines were embedded in the design teams, providing feedback to designers about aspects of the aircraft's design.

There is a significant amount of research that suggests that we not longer live in a product-centric world. According to the Organisation for Economic Co-operation and Development (OECD), we now live in a services-economy [6]. By 2002, the share of the service sector amounted to about 70% of total value added in most OECD economies, and has experienced significant growth ever since the 1970s. It's important to note that is it more accurate to suggest that we live in a services-dominated economy, as clearly there are still product-commodities available to purchase. However, the term helps crystallise an understanding that the exchange of products for monetary value no longer dominates the global economy. OECD's concept of services is described as "a diverse group of economic activities not directly associated with the manufacture of goods, mining or agriculture. They typically involve the provision of human value added in the form of labour, advice, managerial skill, entertainment, training, intermediation and the like".

The "service-economy" is not just restricted to mean personal-services. Consider industries such as health, telecommunications, transport, finance, education, professional services, hospitality, and public administration. Much of the economic activity of these industries is based around services, or at least a combination of services and products, and not just products alone. A telecommunications firm sells conversations over a distance, not just handsets. They are a means to an end, but not the end.

Synonymous with the growth of the services sector is also the concept of the "knowledge-based economy" which OCED defines as "economies which are directly based on the production, distribution and use of knowledge and information" [7]. There are a number of similarities between the knowledge-based economy and the services economy. The OECD observes that just like the services economy requires "human value added" (ie, human innovation), so too does the knowledge-based economy.

## 3 Towards a Services-Centric World

Henry Ford once made the assertion that "a business absolutely devoted to service will have only one worry about profits. They will be embarrassingly large." [8]. It is important to realise that a services-centric worldview is not incompatible with a manufacturing operation. The meaning of "services" in this context is a transcendent one (there are two meanings, one of which has a 'higher' meaning – the verb "service" describes the process or activity of delivering a service, whereas the noun "service" refers to the fundamental shift in worldview about how businesses relate to their customers.

A key work [3] in defining this new service-centric thinking paradigm describes "services" (noun) as the "application, of specialised competence (knowledge and skills) though deeds, processes, and performances for the benefit of another entity or the entity itself". They identify a dichotomy of "product marketing" and "service marketing" and make the argument that, effectively, marketing is always about selling benefits, rather than goods/products. In defining the new paradigm, they identify six comparisons between a traditional goods/product-centred view, and the emerging service-centred view, highlighted in table 1.

**Table 1.** Characteristics of Product-versus-Service Focused Thinking (Adapted from [3])

| | Traditional Goods-Centred Dominant Logic | Emerging Service-Centred Logic |
|---|---|---|
| *Primary Unit of Exchange* | Goods are exchanged for monetary value | An acquisition of the benefits of specialised competencies or services |
| *Role of Goods* | Goods are the end-product | Goods are transmitters of 'embedded knowledge' or functionality when used |
| *Role of Customer* | The customer is the recipient of goods, and marketing researches and promotes to them | The customer is a co-producer of service, and marketing involves interaction with the customer |
| *Determination and Meaning of Value* | Value is determined by the producer | Value is perceived and determined by the consumer on the basis of "value in use". |
| *Firm-Customer Interaction* | Customers transact with a firm | Customers are active participants in relational exchanges |
| *Source of Economic Growth* | Wealth consists of owning, controlling and producing resources | Wealth is obtained through the application and exchange of specialised knowledge and skills |

Under the service-centric worldview, the concept of services (noun) is dominant. Customers are not purchasing products, but rather the benefit that products and/or services (verb) can provide for a customer. In effect, products deliver benefits (services), and become a subset of services [9]. In other words, the value of products is in their use, and not in their ownership.

Another key compelling foundation of the emerging services-centric view is that of customer focus. They make the argument that "over the past 50 years, marketing has been transitioning from a product and production focus to a consumer focus and, more recently, from a transaction focus to a relationship focus." Another related work describes how this approach is very similar to the pre-Industrial Revolution concept of providers been close to their customers, and how such relationships saw the offering of customised services [10].

## 4 The Realm of Industrial Services

Whilst industrial services have often been ascribed to those which are offered by the systems manufacturer [11], this is not always the case. Independent system integrators, or organisations who integrate products from a broad range of suppliers and then support the whole system, can provide just as valid services for complex systems as can system manufacturers [12]. Many of the points of this argument are important to consider and have implications for manufacturers.

Aircraft manufacturers are offering ever broader and deeper levels of services to operator-customers [1]. Aircraft manufacturers are increasingly selling outcomes via services, not just products or equipment. However, this approach is not just limited to the aerospace and defence sectors. The whole concept of manufacturers morphing into service focused organisations that offer integrated solutions is the basis of a field of research called "Servitisation". The term was originally in 1988 by Vandermerwe & Rada [13], and itself has emerged from the field of competitive manufacturing strategy research. Ref. 14 defines servitisation as "the innovation of an organisation's capabilities and processes to better create mutual value through a shift from selling product, to selling Product-Service Systems".

Other definitions include "a trend in which manufacturing firms adopt more and more service components in their offerings" [15]; and, "the emergence of product-based services which blur the distinction between manufacturing and traditional service-sector activities" [16].

One of the general observations of servitization research is the spectrum of ways in which a manufacturer can become more servitized. This spectrum ranges from the offering of a pure tangible good, right through to a pure-service, with combinations thereof (such as a tangible good with accompanying services, or a major service with accompanying minor goods). An alternate spectrum is shown in figure 1.

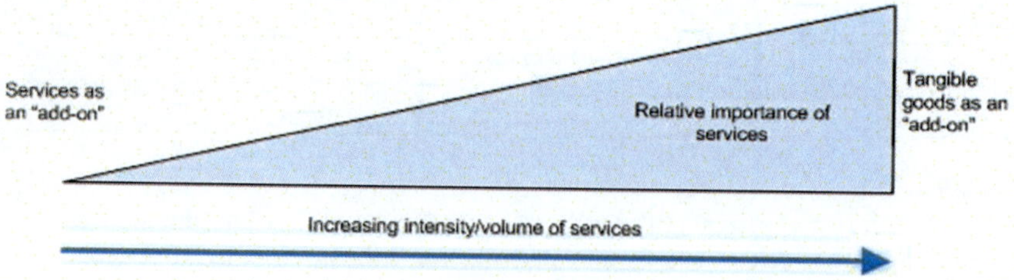


**Figure 1.** Product-Service Continuum (adapted from ref. 17)

The concept of Product-Service Systems (PSS) is argued to be a special case of servitization [18,19], and is a specific manifestation of more integrated approach consistent with a hybrid of products and services. Whereas the concept of servitization tends to deal more with the manufacturer undergoing transformation to offer more services, the PSS field identifies specific ways in which services are bundled (and more importantly, integrated with) products to provide enhanced benefits to a given customer. In other words, it is more focused on the customer, and the way in which the manufacturer delivers an integrated service to them, rather than focusing solely on the manufacturer alone.

However, research from both streams is in many ways identical, and thus this discussion will from now on discuss both concepts at once. Much of the research has come from the USA and Western Europe, with studies of examples of traditionally manufacturing-orientated firms adopting a more servitized approach, including the rail transport, power generation and distribution, communications, pilot-training, documentation management, aerospace and defence industries. Across all these sectors, a number of similar themes have been identified in the journey from products-centric business, to a service-centric approach. Some of the more significant observations include issues within customer organisations that ultimately boil down to the relationship with the service provider, and the way in which the manufacturer organises itself to provide a more integrated solution. From the customer's perspective, issues include control, ownership, and trust. The issue of "ownerless consumption" (or transversely, ownership without control) has been identified by a number of researchers; so too has the issues of past attitudes by manufacturers towards customers, acting opportunistically, generating levels of mistrust within the customer organisation [Brax].

It has been identified by some authors that there exists less research on servitisation from the customer perspective than there is from the manufacturers. However, that does generate more applicable insights for this paper. Observations particularly relate to the need for a 'business not as usual' attitude, design of product service systems, and strategies for organising and transforming the organisation in a more service-centric manner. Ref. 19 asks the question about how 'traditional' manufacturing firms make the journey to servitized organisations. Ref. 1 highlights a number of more specific issues. They include a disconnect between a product-service strategy, and a dominant product-centric mindset and culture within the organisation; the disconnect between the parts of the business that interacted daily with the customer, and the parts of the business that supported this more front-line operation, and; the implications that a product-centric organisational strategy had on processes and operations of the business (and how this actually impaired the company from tactically focusing on customer needs. Ref. 4 also offers that "becoming a provider of industrial services is not just a matter of the offering; the whole organisation needs to re-focus its attention."

## 5 Application to Real Through Life Support Programs

Two examples of Through-Life Support as a service can be found in the "Performance-Based Logistics" contracts on the F-22 and C-17. A significant policy direction laid out in the US Department of Defense 2001 Quadrennial Defense Review specified that performance-based logistics was to become the preferred method of system sustainment for the future [20]. Under the contracts, the USAF pays for the availability of the fleet, rather than the acquisition of spares or repair-services. For the F-22 (Figure 3), Lockheed Martin coordinate all the maintenance, engineering support and spares logistics activities, integrating them into a solution whereby the USAF is guaranteed of a minimum level of availability. A similar arrangement exists for the C-17 (Figure 4), with Boeing playing an identical role in a program which it has branded as the "Globemaster Sustainment Partnership", which was initiated in 2001.

**Figure 3.** F-22 Raptor

**Figure 4.** C-17 Globemaster

These particular examples have been chosen because of recent events that have transpired over the future of the PBL programs for these aircraft types. In April 2010, Flight International [21] reported that the USAF will be taking back control of the maintenance programs for both aircraft types, commencing in 2012 for the C-17, marking a significant departure from what has been US Department of Defense policy for about the past decade. A number of reasons were put forward for the annulling of the contracts [22]. A political issue, particularly relating to Federal laws stating a minimum level of maintenance work that must be carried out at Government depots, was flagged. However, in a follow-up article, it became more apparent that the USAF was having concerns regarding the value-for-money of the programs, especially in the case of the C-17. A business case analysis was performed on the aircraft type, and it was concluded that it was more cost effective for the USAF to have the aircraft maintained by a Government depot facility rather than with the manufacturer. A similar situation was evident for the F-22.

There are some lessons that can be learnt. Firstly, when a customer outsources an activity, there is normally a good reason behind it. Often it is because the customer wants something done better, or more efficiently, than what current arrangements can deliver. Therefore, when an operator-customer outsources, there is a belief that the contractor (in this case, the aircraft manufacturer) can offer a better solution. However, it cannot be automatically assumed that merely outsourcing the same work, to be performed in the same way, to a different organisation is going to deliver better outcomes. There needs to be a demonstration of how a manufacturer can "add value" to a sustainment operation. As discussed in a previous section, a definition of services (as a noun) is "application, of specialised competence (knowledge and skills) though deeds, processes, and performances for the benefit of another entity or the entity itself". What specialised skills and knowledge does an aircraft manufacturer have that 'adds value' (ie, provides more valuable benefit than what can anyone else do)? This is a key question that needs to be asked by both operator-customers, and manufacturers.

## 6 Master of System Support Engineering

The Australian Department of Defence follows the rest of world in performance – based contracting for its defence materiel acquisition programs. For example, in 1997, the Department of Defence awarded a contract for a new fighter jet for lead-in pilot training and air support role for the Royal Australian Air Force (RAAF) The contract was awarded to BAE SYSTEMS with its Hawk 127. In total 33 aircraft, plus support equipment, were purchased with the last delivered in 2001. The In-Service Support Contract requires BAE SYSTEMS to provide deeper maintenance support throughout the 25-year life of type, in five-year renewable terms. The service provider guarantees access to a minimum number of serviceable aircraft at any time.

This new environment of providing a service or capability, rather then hardware only, has prompted the Defence Materiel Organisation (DMO) to support the development of a new university program to train and educate industry personnel in a more service-focused ethos. The Masters Degree in Systems Support Engineering (MSSE) is a new postgraduate program managed by RMIT University in partnership with the University of South Australia, BAE SYSTEMS, SAAB and ASC. Systems Support Engineering (SSE) is an emerging field which requires extensive knowledge, skills and competency in the following areas:

- Systems and service thinking
- Performance based logistics support
- Supply chain
- Asset management and capability enhancement

This unique program provides the students with the ability to architect and deliver service solutions for complex systems as a business solution. The complete Masters program is three years (part-time) and is delivered essentially by distance. The industry partners provide valuable practical input through case studies and lecturing material. The program commences in 2011.

## 7 Conclusion

It has been shown that twenty-first-century society no longer lives in a product-dominated economy. Rather, it is one that is dominated by services. This holds true also in the aerospace industry as well. Whilst products have not been done away with, it is clear that it is their value-in-use that is critical, and less so their ownership. Aerospace manufacturers who wish to pursue further business in the integrated-solutions space need to identify their core competencies, and how these increase the benefit to an operator-customer.

Current training and education programs do not prepare graduates well for this new environment by focusing mostly on technical excellence. With regard to the Australian environment, the Department of Defense has initiated the development of a new postgraduate program to expose industry personnel to a more service-thinking ethos.

## 8 References


[1] Johnstone, S., A. Dainty, and A. Wilkinson, Integrating products and services through life: an aerospace experience. International Journal of Operations & Production Management, 2009. 29(5): p. 520-538.

[2] Ward, Y. and A. Graves, Through-life management: the provision of total customer solutions in the aerospace industry. International Journal of Services Technology and Management, 2007. 8(6): p. 455-477.

[3] Vargo, S. and R. Lusch, Evolving to a new dominant logic for marketing. Journal of marketing, 2004. 68(1): p. 1-17.

[4] Brax, S., A manufacturer becoming service provider–challenges and a paradox. Managing service quality, 2005. 15(2): p. 142-155.

[5] Sabbagh, K., 21st century jet: the making and marketing of the Boeing 777. 1996: Scribner, New York.

[6] Organisation for Economic Co-operation and Development (OECD), The Service Economy, Business And Industry Policy Forum Series, Paris, 2000

[7] Organisation for Economic Co-operation and Development (OECD), The Knowledge Economy, Paris, 1996

[8] Glueck, J., Joudal, P., Vaessen, W., The Service Revolution: Manufacturing's Missing Crown Jewel, Deloitte Review, 2007

[9] Nonaka, I. and H. Takeuchi, The knowledge-creating company: How Japanese companies create the dynamics of innovation. 1995: Oxford University Press, USA.

[10] Hauser, J. and D. Clausing, The house of quality. Harvard business review, 1988. 66(3): p. 63-73.

[11] Homburg, C. and B. Garbe, Towards an improved understanding of industrial services: quality dimensions and their impact on buyer-seller relationships. Journal of Business-to-Business Marketing, 1999. 6(2): p. 39-71.

[12] Kapletia, D. and D. Probert, Migrating from products to solutions: An exploration of system support in the UK defence industry. Industrial Marketing Management, 2009.

[13] Vandermerwe, S. and J. Rada, Servitization of business. European Management Journal, 1988. 6(4): p. 314-324.

[14] Baines, T., et al., The servitization of manufacturing. Management, 2009. 20(5): p. 547 - 567.

[15] Desmet, S., R. Van Dierdonck, and B. Van Looy, Servitization: or why services management is relevant for manufacturing environments. Services Management: An Integrated Approach, Harlow: Pearson Education, 2003: p. 40-51.

[16] White, A., M. Stoughton, and L. Feng, Servicizing: The quiet transition to extended product responsibility. Tellus Institute, Boston, 1999: p. 97.

[17] Oliva, R. and R. Kallenberg, Managing the transition from products to services. International Journal of Service Industry Management, 2003. 14(2): p. 160-172.

[18] Baines, T., et al., Towards an operations strategy for product-centric servitization. International Journal of Operations and Production Management, 2009. 29(5): p. 494 - 519

[19] Baines, T., et al., State-of-the-art in product-service systems. Proceedings of the Institution of Mechanical Engineers, Part B: Journal of Engineering Manufacture, 2007. 221(10): p. 1543-1552.

[20] Rumsfeld, D., Report of the Quadrennial Defense Review. Washington DC, Department of Defense, 2001.

[21] Trimble, S., Upkeep u-turn, Flight International [Magazine], 20-26 April 2010

[22] Trimble, S., Service Conflict, Flight International [Magazine], 18-24 May 2010

# Domain-Centred Functionalities Discovering for Testing Systems Functionalities

German Urrego-Giraldo[a] and Gloria Lucía Giraldo[b]
[a] Universidad de Antioquia, Dpto. Ingeniería de Sistemas., Medellín, Colombia.
[b] Universidad Nacional de Colombia, Escuela de Sistemas, Medellín, Colombia.

**Abstract.** The System Definition Phase, modelled in terms of Context Objective, gathers the basic knowledge for the system development. This knowledge is incorporated in the Structure of Domain Services and Object (SDSO) and in the Solution Use Context Model (SUCM). A pattern of Agent Intervention, extended and instantiated with highest categories of the SDSO, is applied for deriving domain-centred system functionalities, which are confronted with known system functionalities previously obtained from system requirements. Thus, the consistency of domain-centred system functionalities and requirements based system functionalities is tested. A system for selecting co-created contributions aiming to innovative products is used for illustrating the proposed approach.

**Keywords.** System Definition Phase, Agent Intervention pattern, Domain Model, Testing

## 1 Introduction

Although context and domain concepts are in general considered in system development methodologies, the characterization of both concepts is rather informal and concepts' scopes and limits are not always comparable among methodologies, which combine and interpret these concepts in a different way. Domain concept is widely treated, while context is intuitively considered as external concept in a similar way as the environment concept. A known formal domain meta-model which considers object, activity, and goal as fundamental concepts is presented by Ramadour and Cauvet, in [1]. In the same article, context is considered as the domain knowledge driving the choice of goal components, activity organizations or object descriptions.

One of the more accepted domain concepts is known as Domain Analysis. This one, introduced by Neighborn [2] and complemented -among others- by McCain [3], Prieto-Diaz [4], and Maiden [5] involves the identification, capture and organization of essential knowledge for systems developing.

A short overview of the system development literature identifies two high trends for modelling the domain: Functionalities-Based Models and Information-Based Models. In the latter, two categories are recognized: Hierarchical Information Structures and Information Units Graph. Hierarchical Information Structures include Domain Ontologies and Structures of Domain Services and

D. D. Frey et al. (eds.), *Improving Complex Systems Today*,
Advanced Concurrent Engineering, DOI: 10.1007/ 978-0-85729-799-0_10,
© Springer-Verlag London Limited 2011

Objects (SDSO). This structure, introduced in [6], offers domain concepts for instantiating a pattern of Agent Intervention containing nine categories of functionalities. The functionalities of the instantiated pattern are compared against system requirements-based functionalities in the consistency testing model proposed in this paper.

Domain Ontological Models are based in germinal works of Bunge [7], Gruber [8], and Borst [9] -among others.

Information Structures Graphs contain Entity-Relationship Diagrams developed by Chen [10], objects classes graphs treated -among others- by Shlaer and Mellor [11], Booch [12], and Frames created by Minsky [13].

Functionalities-Based Models correspond to classical approaches, proposed by Constantine [14] and Ross [15] -among others- which use the Data Flow Diagram (DFD) as the System Conceptual Model. First abstraction level of the DFD is named context model, where the external agents interact with the system. Requirements elicitation is based on low abstraction levels of the DFD combining domain and context concepts.

For supporting a line of model consistency test, *domain* is defined as a real or imaginary activity field integrated by products, services, objects, and relationships among them, containing the knowledge on which agents from different contexts intervene aiming to satisfy specific objectives.

In spite of the progress in methodologies and technical aids for supporting the different phases of systems lifecycle, software quality problems and evolution difficulties still remain. Thus, the need of deeper research and additional efforts of development and implantation, aiming to develop and improve testing models for verifying the correspondence of the knowledge of the system definition along other system development phases, is recognized. In order to contribute to overcome the previously mentioned lacks, the SDSO is used, in this research, for discovering system functionalities and for testing the requirement set of functionalities established from agent requirements previously identified.

In this paper, a domain-centred approach for functionalities instantiates an Agent Intervention extended pattern with conceptual categories of the SDSO. The functionalities of this instantiated extended pattern are tested against system functionalities obtained from agent's requirements. The use of the SDSO and the testing process is the objective of this research.

In addition to the introduction, this article includes the following sections: Section 2: a brief description of the Agent Intervention extended pattern. The Structure of Domain Services and Objects (SDSO), is explained in Section 3. Section 4 describes the functionalities consistency test, which is the objective in this paper. Conclusion and future work are stated in Section 5. Bibliography corresponds to Section 6.

## 2 Agent Intervention Extended Pattern

Agent Intervention Pattern, included in [6], decomposes a system goal in nine goals of following lower abstraction level. Abstraction level decreases from service abstraction level to action/state abstraction level, through process abstraction level and activity abstraction level, sequentially. In the goal operationalization process, the pattern is applied for each reduction in the abstraction level. This pattern agrees with the model developed by Alexander in [16], constitutes the core of a set of operationalization patterns and it is composed of six simple patterns: initiation, selection, authorization, execution, transference, and termination. In this Section, only the part procedure of the pattern, Figure 1, is transcribed.

| Procedure of the agent Intervention pattern | | |
|---|---|---|
| | Actions of participant agents | |
| | Actions of the Principal agent | Actions of the interacting Agent |
| 1 | Get active and put in relationship services and objects involved in an agents interaction. | |
| 2 | Present alternative services or objects on which agents interact. | |
| 3 | | Activate one particular service or object. |
| 4 | Demand information, elements or confidential data confirmations, for effecting one chosen service or object | |
| 5 | | Confirm or present demanded information, elements or confidential data confirmations. |
| 6 | Validate demanded information, elements or confidential data confirmations and take the corresponding decision. | |
| 7 | Execute the selected services or object. | |
| 8 | Transfer, update, and communicate results. | |
| 9 | End the interaction. | |

**Figure 1. Procedure of the Agent Intervention Pattern**

Nine previously depicted goals are considered, in an agent intervention extended pattern, as nine (9) goal categories and each one disaggregated in more concrete goals, following the structure of the process concept. Hence, the goals categories are classified in three groups of *goals: Preparation or input, Transformation or Execution, and Transference or Result, or Output*. In Figure 1, the first six goals are *Preparation* goals; the seventh is a *Transformation or Execution* goal and the eighth and ninth are *Transference or Output* goals.

The agent intervention extended pattern is instantiated with the concepts of the SDSO in order to obtain the Domain-Centred System Functionalities, which are then tested for consistency against the System Operationalizable Goals. These results are the objective of this article.

# 3 Structures of Domain Services and Objects

Considered as an activity field, the domain may be characterized in terms of products or services and their involved objects. In this research the domain is modelled as a Structure of Domain Services and Objects (SDSO). In this structure the concepts expressing products or services and objects are, in general, connected among them by four types of relationships: generalization/specialization, composition, characterization (attributes), and those specific of a domain.

Top concepts of the SDSO are essential elements treated by the envisioned solution. These elements are extracted from system context objectives and complemented with knowledge and experiences about activities and objects of the domain. Subordinated concepts and relationships also result from system context objectives, knowledge and experiences related to the domain, from analysis and

reasoning about the completeness, pertinence, integrity, and coherence of domain elements, as well as, by analogy with structures and elements of other domains. A generic domain for a system in charge of supporting to selecting contributions (ideas) from a set of co-created contributions (ideas) is represented in the SDSO, in Figure 2. For explaining the construction of the SDSO and the development of the application case, the following system context objective is used:
*The system for selecting co-created ideas curries out the selection of agents' contributions (ideas) for developing innovative products from agents' contributions (ideas) recovering to delivering of selected and classified agents' contributions (ideas) using a hardware-software platform for information processing applying methods and resource for software programming and for information and communications management, oriented to innovative persons and organizations and for scientists and knowledge managers.*

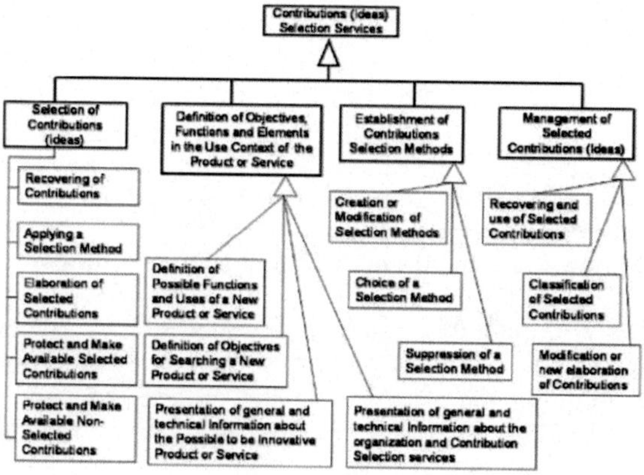


**Figure 2. Structure of Domain Services and Objects (SDSO) for a Contributions Selection System**

Four top concepts are identified in the domain of selecting co-created contributions (idea): *Selection of Contributions (ideas), Definition of Objectives, Functions and Elements in the Use Context of the Product or Service, Establishment of Contributions Selection Methods, and Management of Selected Contributions (Ideas).* First top concept has five subordinated concepts connected by composition relationships, arranged in the first branch of the structures. In the other three branches the top concept is specialized in four, three, and three subordinated concepts, respectively. All concepts are used for instantiating the agent intervention extended pattern for finding domain-centred system functionalities, in the next section.

## 4 Functionalities Consistency Testing

Exploiting context and domain concepts, defined in introduction section, some approaches for testing the consistency among models of lifecycle phases, developed in the research project "Knowledge and Information Integration", are considered in [17]. References of correction and consistency testing may be also found in [17].

### 4.1 General Description of the Testing Model

As it appears in the general model depicted in Figure 3, system definition phase gathers existing and emerging knowledge and may be modelled as context objectives. Knowledge of definition phase could be found in social, organizational and system context and expressed in objectives of each context. Anyway, using an adjustment pattern, objectives of one context may be transformed in objectives of the following lower context, and finally all knowledge of the definition phase is embedded in objectives of the system context.

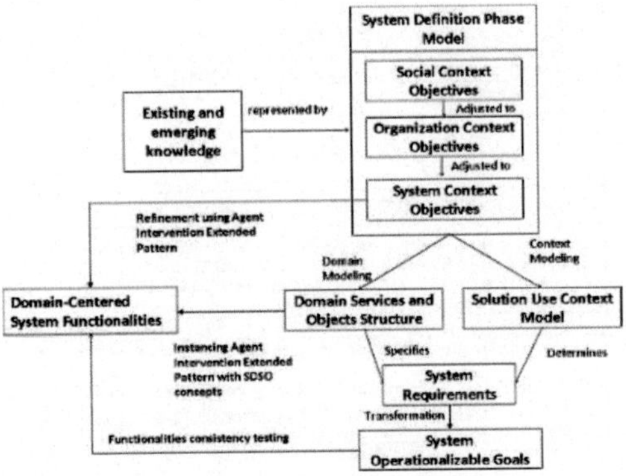


**Figure 3. Scope of Consistency Testing of System Functionalities**

System context objectives are refined in system functionalities using the Agent Intervention Extended Pattern, introduced in the previous section, instantiated with the concepts of the SDSO.

The Structure of Domain Services and Objects, and the Solution Use Context Model are built with knowledge of the definition phase, as represented in Figure 3. Both models support the elicitation of agent's requirements, which are then transformed in System operationalizable goal. Requirements and system operationalizable goals were introduced in [17].

### 4.2 Application Case for the Testing Model

Guided by the elements of the model of the Figure 3, the agent intervention extended pattern applied for refining the system context objective is instantiated with the concepts of the SDSO depicted in Figure 2. Domain-centred functionalities are detailed in the third column of the Table 1, grouped by highest categories of the SDSO. In other precedent works, the consistency testing based on the domain concepts is supported on the concepts of the SDSO.

These approaches verify the inclusion of SDSO concepts in the System Context Objectives and in the System Operationalizable Goals obtained from system requirements.

**Table 1.Domain-Centred Functionalities of the Contribution Selection System**

| Code | | Domain-Centred System Functionalities: Contributions (ideas) Selection System |
|---|---|---|
| 1 | | Selection of Contributions (ideas) Context of the Product or Service |
| | 1.1 | Recovering of Contributions |
| | | 1.1.1 Finding generated and elaborated contributions<br>1.1.2 Accessing generated and elaborated contributions<br>1.1.3 Presenting generated and elaborated contributions |
| | 1.2 | Applying a Selection Method |
| | | 1.2.1 Searching a selection method<br>1.2.2 Executing the selection method<br>1.2.3 Giving the selected generated and elaborated contributions. |
| | 1.3 | Elaboration of Selected Contributions |
| | | 1.3.1 Taking selected generated and elaborated contributions.<br>1.3.2 Intervening on selected generated and elaborated contributions.<br>1.3.3 Accepting elaborated contributions |
| | 1.4 | Protect and Make Available Selected Contributions |
| | | 1.4.1 Colleting selected contributions<br>1.4.2 Storing selected contributions<br>1.4.3 Guiding the access to selected contributions |
| | 1.5 | Protect and Make Available Non-Selected Contributions |
| | | 1.5.1 Colleting non-selected contributions<br>1.5.2 Storing non-selected contributions<br>1.5.3 Guiding the access to non-selected contributions |
| | | Definition of Objectives, Functions and Elements in the Use Context of the Product or Service |
| | 2.1 | Definition of Possible Functions and Uses of a New Product or Service |
| | | 2.1.1 Offering boxes for introducing possible functions and uses of a new product or service.<br>2.1.2 Appling mechanisms introducing possible functions and uses of a new product or service.<br>2.1.3 Generating functions and uses of a new product or service.<br>2.1.3.1 Introducing functions and uses of a new product or service.<br>2.1.3.2 Sending functions and uses of a new product or service. |
| | 2.2 | Definition of Objectives for Searching a New Product or Service |
| | | 2.2.1 Offering boxes for Objectives for Searching a New Product or Service.<br>2.2.2 Appling mechanisms introducing Objectives for Searching a New Product or Service.<br>2.2.3 Generating Objectives for Searching a New Product or Service.<br>2.2.3.1 Introducing Objectives for Searching a New Product or Service.<br>2.2.3.2 Sending Objectives for Searching a New Product or Service. |
| | 2.3 | Presentation of general and technical Information about the Possible to be Innovative Product or Service |
| | | 2.3.1 Accessing the files of general and technical information about the possible to be Innovative Product or Service<br>2,3,2 Finding of general and technical information about the possible to be Innovative Product or Service<br>2.3.3 Activating of general and technical information about the possible to be Innovative Product or Service |
| | 2.4 | Presentation of general and technical Information about the organization and Contribution Selection services |
| | | 2.4.1 Accessing the files of general and technical information about the organization and contribution selection services.<br>2,4,2 Finding of general and technical information about the organization and contribution selection services.<br>2.4.3 Activating of general and technical information about the organization and contribution selection services. |
| 3 | | Establishment of Contributions Selection Methods |
| | 3.1 | Creation or Modification of Selection Methods |
| | | 3.1.1 Offering boxes for introducing contributions selection methods.<br>3.1.2 Appling mechanisms introducing contributions selection methods.<br>3.1.3 Generating contributions selection methods.<br>3.1.3.1 Introducing contributions selection methods.<br>3.1.3.2 Sending contributions selection methods. |
| | 3.2 | Choice of a Selection Method |
| | | 3.2.1 Accessing selection methods<br>3.2.2 Showing selection methods<br>3.2.3 Choosing a selection method.<br>3.2.3.1 Marking a selection method<br>3.2.3.2.Activating a selection method |
| | 3.3 | Suppression of a Selection Method |
| | | 3.3.1 Accessing selection methods<br>3.3.2 Showing selection methods<br>3.3.3 Suppressing a selection method<br>3.3.3.1 Marking a selection method.<br>3.3.3.2Eliminating a selection method |
| 4 | | Management of Selected Contributions (Ideas) |
| | 4.1 | Recovering and use of Selected Contributions |
| | | 4.1.1 Finding of general and selected generated and elaborated contributions<br>4.1.2 Accessing of general and selected generated and elaborated contributions<br>4.1.3 Delivering general and selected generated and elaborated contributions. |
| | 4.2 | Classification of Selected Contributions |
| | | 4.2.1 Taking general and selected generated and elaborated contributions.<br>4.2.2 Classifying general and selected generated and elaborated contributions.<br>4.2.3 Presenting general and selected generated and elaborated classified contributions. |
| | 4.3 | Modification or new elaboration of Contributions |
| | | 4.3.1 Taking general and selected generated and elaborated contributions.<br>3.3.2 Intervening on general and selected generated and elaborated contributions.<br>4.3.3 Presenting general and selected generated and elaborated modified contributions. |

The agent intervention extended pattern is a functionalities pattern considering nine functionalities categories arranged in three sub-categories of functionalities: *Preparation, Transformation Execution, and Consequences or Results.*

For sake of clarification, in this article only functionalities of the category *Transformation or Execution* are treated. The fifty five system functionalities detailed in Table 1 are expressed in the second column of the Table 2 with the number of group indicating the corresponding concepts category of the SDSO (Figure 2). These numbers appears in the first and second column of Table 1.

The inclusion of functionalities identified in each row of the second column of Tablc 2 in the corresponding general functionality in the same row in the column 1 of this table, is tested. The eleven general functionalities of the column 1 constitute only the system operationalizable goals of the *Transformation* or *Execution* category, obtained from agent's requirements for the contribution (Ideas) selection system, considered in this application case. The satisfactory test of correspondence between domain-based functionalities and requirement-based functionalities allowed us formulating an agile method for prototyping the contribution (Ideas) selection system.

**Table 2. Consistency Testing between System Functionalities**

| **Operationalizable Transformation goals of the System for Selection of Contributions (ideas) -Derived from Agent's Requirements-** | **Domain-Centered System Functionalities** |
|---|---|
| 1- Carry out ideas selection services: Selection of contributions (ideas), Definition of objectives, functions and elements in the context of use of the product or service; Selection of methods, and management of selected contributions (ideas). | 1, 2, 3, 4 |
| 2- Manage the integration of services, processes, and systems of the Products Innovation Area with the Organization Management, the Strategic Management Area, and the Improving and Development Area of Innovative Services and Products, as well as with the processes of involved third parts. | 1.2, 1.3, 1.4, 1.5, 2.1, 2.2, 2.3, 2.4, 4.1, 4.2, 4.3 |
| 3- Effect procedures for assuring the scope, the pertinence, and the quality of contribution (ideas) selection services of the Products Innovation Area, Organization Management, Strategic Management Area, and Improving and Development Area of Innovative Services and Products. | 1.1, 1.3, 1.4, 1.5, 4.1, 4.2, 4.3 |
| 4- Effect procedures for the demonstration and simulation of Contributions (ideas) selection services aiding the employers training and the development of the service culture in the areas related to Products innovation. | 1.2, 1.3, 3.1, 2.1, 2.2, 2.3, 2.4, 3.2, 4.1, 4.2, 4.3 |
| 5- Effect procedures for researching and making operative and interactive Contributions (ideas) selection services: Selection of contributions (ideas), Definition of objectives, functions and elements in the context of use of the product or service, Selection of methods, and management of selected contributions (ideas). | 1.1, 1.2. 1.3, 1.4, 1.5, 2.1, 2.2, 2.3, 2.4, 3.1, 3.2, 3.3, 4.1, 4.2, 4.3 |
| 6- Manage modifications, renovations, and cancellation of innovative contributions (ideas). | 4.1, 4.2, 4.3 |
| 7- Manage interrogation and handling of registered operation in D.B. | 1.1, 1.4, 1.5, 4.1, 4.2, 4.3 |
| 8- Effect procedures for applying combinations of innovative contributions (ideas) under particular technical and commercial conditions. | 1.3, 2.1, 2.2, 4.1, 4.2, 4.3 |
| 9- Manage services achievement involving third parts such as: developers of systems for supporting the innovation processes, Control Public Organizations, Scientific Organizations, etc. | 2.1, 2.2, 2.3, 2.4. 3.1 |
| 10- Effect procedures for using information related to organizational, technical, promotional, legal, and quality aspects. | 2.1, 2.2,2.3, 2.4 |
| 11- Effect procedures for using information coming from the products innovation area, the Organization Management, the Strategic Management Area, and the Improving and Development Area of Innovative Services and Products, as well as from involved third parts. | 2.1, 2.2, 2.3, 2.4, 4.1, 4.2, 4.3 |

## 5 Conclusion and Future Work

The separately treatment of domain and context concepts leads to generic solutions, resulting from contextual models, using generic instantiable domain models. Domain-centred functionalities discovering, proposed in this paper, is a way for developing knowledge and information systems when there is not an expedite way for previously obtaining systems requirements, as it occurs in forecasting, planning, innovating, and researching. This domain-centred direction is also the way for constructing system development agile methods rapidly leading to prototypes construction, and for supporting this research line focused on model consistency testing.

Domain model represents the knowledge that software systems works on. Thus, domain model becomes an important contribution in direction of system pertinence and completeness.

The use of the domain concept for discovering system functionalities offers another way for consistency testing between models in the early phases of system development. Consistency testing could also be carried out following both a domain-based path and a context based path. Hence, more robust and complete testing models can be built taken advantage of strengths of two ways. This double point of view -on domain and on context- leads the use of concurrent engineering principles, supporting the simultaneous treatment of products and processes.

The generic set of functionalities arranged in the agent intervention extended pattern may use generic SDSO, leading to generic applications and testing models. A generic domain model for a contributions (ideas) selection system is used in this paper for finding domain-centred system functionalities whose correspondence with system functionalities, obtained from agents' requirements, is successfully verified. Ongoing research is looking for elaborating model testing patterns along the system lifecycle.

## 6. References

[1] P. Ramadour and C. Couvet. Modélisation de domaine pour l'ingénierie des SI para réutilisation. 12éme journées francophones d'ingénierie des connaissances. Grenoble, 2001.

[2] J. Neighbors. "Software Construction Using Components". Ph.D. Thesis, Department of Information and Computer Science, University of California, Irvine, 1981.

[3] R. McCain. "Reusable Software Component Construction: A Product-Oriented Paradigm". In Proceedings of the 5th AIAA/ACM/NASA/IEEE Computers in Aerospace Conference, Long Beach, CA, pp.125-135, October 21-23, 1985.

[4] R. Prieto-Diaz. "Domain Analysis: an introduction". ACM SIGSoft Software Engineering Notes". Vol. 15, Nro 2, pp. 47-54.

[5] N. A. M. Maiden. "Analogy as a paradigm for specification reuse", in Software Engineering Journal, Vol. 6, Nro 1. 1991.

[6] G. Urregp-Giraldo. "ABC-Besoins: Une approche d'ingénierie de besoins fonctionnels et non-fonctionnels centrée sur les Agents, les Buts, et les Contextes". Ph.D. Thesis, Universidad Paris 1, Pantéon Sorbona, 2005.

[7] M. Bunge. "La causalidad: el principio de causalidad en la ciencia moderna". Buenos Aires, Editorial Universitaria de Buenos Aires (4.ª edición). Reeditado por Editorial Sudamericana (Buenos Aires), en 1997.

[8] T. Gruber."A translation approach to portable ontology specifications" *Knowledge Acquisition.* Vol. 5, No. 2, 1993, pp. 199-220.

[9] W.N Borst. "Construction of Engineering Ontologies for Knowledge Sharing and Reuse". PhD Thesis, University of Twente, Enschede, 1997.

[10] P. Chen. "The Entity-Relationship Model--Toward a Unified View of Data ACM Transactions on Database Systems, Vol. 1, No. 1, March, 1976, pp. 9–36.

[11] S. Shlaer and S.J. Mellor, "An object-oriented approach to domain analysis". ACM SIGSOFT Software Engineering Notes, v.14 n.5, p.66-77, July 1989.

[12] G. Booch. "Object oriented design with applications". Benjamin-Cummings Publishing Co., Inc., Redwood City, CA, 1990.

[13] M. Minsky. "A Framework for Representing Knowledge". The Psychology of Computer Vision, P. H. Winston (ed.), McGraw-Hill 1975.

[14] L. Constantine. "Structured Design", IBM Systems Journal, Vol. 13, Nro. 2, 1974.

[15] D. T. Ross and K. E. Shoman, Jr., "Structured analysis for requirements definition," IEEE Trans. Software Eng., vol. SE-3, pp. 69-84, Jan. 1977.

[16] Alexander C., Ishikawa, S., Silverstein M. 1977. A patron language. Oxford University Press, New York.

[17] G. Urrego-Giraldo and G.L. Giraldo. "Context model for testing analysis phase of information systems. In 17th ISPE International Conference on Concurrent Engineering, pp 565, 573, Poland, Sept. 2010.

# Comparison of Domain-Centered and Context-Centered Systems Functionalities

Gloria L. Giraldo[a] and German Urrego-Giraldo[b]

[a]*Universidad Nacional de Colombia, Escuela de Sistemas, Medellín, Colombia. glgiraldog@unal.edu.co*
[b]*Universidad de Antioquia, Dpto. Ingeniería de Sistemas, Medellín, Colombia. gaurrego@udea.edu.co*

**Abstract**: The Structure of Domain Services and Objects (SDSO) and the Solution Use Context Model (SUCM) summarize the knowledge collected in the System Definition Phase. This knowledge is expressed in system context objectives adjusted from organizational and social contexts. An Agent Intervention extended Pattern is, in turn, instantiated with the highest categories of the SDSO and then, with Agent Operationalizable
Objectives obtained from the SUCM in order to produce Domain-Centred System Functionalities and Context-Centred System Functionalities, respectively. Both functionalities sets are compared for analyzing the particular possibilities of domain and context models looking for system functionalities, and system analysis agile methods. The two types of functionalities are tested against system functionalities obtained from system requirements in the case of selecting co-related contributions for innovative products.

**Keywords**: system definition phase, agent intervention pattern, system functionalities, domain model, context model, testing

## 1 Introduction

Context and domain concepts models represent the necessary knowledge for constructing a logical model of an information system.

Considerations about the context and the need of research on this topic are treated in [1], [2], [3]. In [4], based on linguistics, *context* is defined as *the space where the knowledge acquires meaning and worth*. Based on this concept, *context of agents intervention* is defined as the space where the knowledge associated to actions and interactions of agents, aiming to specific objectives, acquires meaning and worth.

The *domain*, in relation to knowledge and information systems, is the real or imaginary field on which the system and their related agents intervene. This is the field in which the knowledge associated to related product, services, and objects is managed by knowledge and information systems in order to support activities and conditions on this domain.



D. D. Frey et al. (eds.), *Improving Complex Systems Today*,
Advanced Concurrent Engineering, DOI: 10.1007/ 978-0-85729-799-0_11,
© Springer-Verlag London Limited 2011

In current software development methodologies, the domain concept is widely used in the analysis phase, while context concept is treated as an environment containing external agents and entities. A rich variety of domain models have been proposed, among others, by Ramadour and Cauvet in [5], Neighborn in [6], McCain in [7], Prieto-Diaz in [8], Maiden in [9], Constantine in [10] and Ross in [11].

Solution Use Context Model (SUCM) and Structure of Domain Services and Objects (SDSO) are here adopted as context and domain models, respectively.

SUCM is a *context of agent interventions* defined as the space where a solution may be implemented and the knowledge associated to actions and interactions of agents, aiming to specific objectives and supported by this solution, acquires meaning and worth. SDSO is a hierarchy of concepts representing products, services, objects, and relationships among them, belonging to a work domain of information and knowledge systems, in order to support interventions of agents in a context where they are looking for achieving their own objectives, with which system's objectives must be aligned. Many types of relationships may be involved in SDSO, but, four categories are generally considered: generalization /specialization, composition, characterization, and domain specific relationships.

Going further in system development methods and, in particular, in requirements engineering field, important advances in research and practice are recognized, but it is clear the need of research for improving the requirements elicitation process, where context and domain are central concepts aiming to obtain appropriated agent requirements and systems functionalities which entirely satisfy these requirements. Difficulties to pass without losing the whole knowledge obtained in the definition phase to the analysis phase are recognized in the current system development methodologies. For getting over these difficulties, context and domain models aim to represent this knowledge and to use it for establishing system functionalities. A comparison of functionalities obtained separately by using, in turn, context model and domain model is presented in this research. Thus, an individual weigh of context and domain models for determining system functionalities is highlighted.

In addition to the introduction, in this article the following sections are included: Section 2 presents a short description of the agent intervention extended pattern, used for generating domain-centred and context-centred system functionalities. The Structure of Domain Services and Objects is presented in Section 3. In Section 4 the Solution Use Context Model is explained. Domain-centred and context-centred functionalities are compared in section 5. Section 6 presents the conclusion and future work. References are listed in section 7.

## 2 Agents Intervention Extended Pattern

In the research focused in system analysis patterns and specifically in the line of requirements operationalization, four abstraction levels, top-down, are considered: service, process, activity, and action/state abstraction level. In the goal operationalization process, the agent intervention pattern is applied for each reduction in the abstraction level. The basic agent intervention pattern introduced in [2] decomposes a system goal in nine following lower abstraction level goals.

The nine goals considered originally in the Agent Intervention Pattern depicted in Figure 1, are extended as nine goal categories keeping the same type as the initial goals.

This extended pattern is composed of six simple patterns: initiation, selection, authorization, execution, transference, and termination. In this Section, only the procedural component of the agent intervention pattern is depicted in Figure 1.

| Procedure of the agent Intervention pattern | | |
|---|---|---|
| | Actions of participant agents | |
| | Actions of the Principal agent | Actions of the interacting Agent |
| 1 | Get active and put in relationship services and objects involved in an agents interaction. | |
| 2 | Present alternative services or objects on which agents interact. | |
| 3 | | Activate one particular service or object. |
| 4 | Demand information, elements or confidential data confirmations, for effecting one chosen service or object | |
| 5 | | Confirm or present demanded information, elements or confidential data confirmations. |
| 6 | Validate demanded information, elements or confidential data confirmations and take the corresponding decision. | |
| 7 | Execute the selected services or object. | |
| 8 | Transfer, update, and communicate results. | |
| 9 | End the interaction. | |

**Figure 1. Procedure of the Agent Intervention Pattern**

In according with the process structure (input, activities, output) these categories are organized in three groups: *preparation or input, transformation or execution, and transference, result,* or *output.* Each group constitutes a pattern, which may contains one or more patterns desegregated in turn in the three groups associated to the process model. All patterns are in agreement with the model pattern proposed by Alexander in [12]. This agent intervention extended pattern is instantiated in turn with domain-centred system objectives and context-centred system objectives, and then, the two set of systems functionalities are compared, in section 5.

## 3 Structures of Domain Services and Objects

Domain is modelled in this research as a concepts graph named Structure of Domain Services and Objects (SDSO). The head of each branch of the graph is a representative product or service of this domain, extracted from context objectives and from knowledge and experiences about activities and objects of the domain. In each branch, subordinated concepts are also discovered by considering system context objectives, knowledge and experiences related to the domain and knowledge obtained from analysis and reasoning about the completeness, pertinence, integrity, and coherence of domain elements, as well as, by analogy with structures and elements of other domains.

In Figure 3 the SDSO for a system occupied of selecting contributions (ideas) from a set of registered co-created contributions (ideas) is represented.

Four branches of the domain of a system for selecting generated and elaborated co-created contributions (idea) are headed by: *Selection of Contributions (ideas), Definition of Objectives, Functions and Elements in the Use Context of the Product or Service, Establishment of Contributions Selection Methods, and Management of Selected Contributions (Ideas).* First branch uses composition relationships in order to connect five subordinated concepts. Specialization relationships exist in the other three branches. The execution of all concepts of the SDSO determines the same number of domain-centred system objectives aligned with system context objectives.

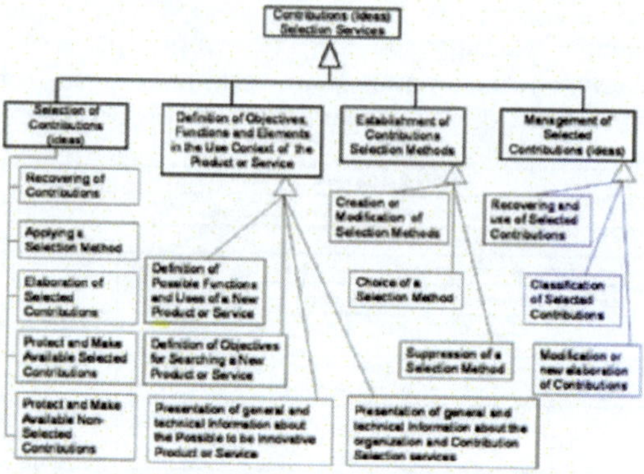


**Figure 2. Structure of Domain Services and Objects (SDSO) for a Contributions Selection System**

## 4 Solution Use Context Model

Knowledge and information systems support the achieving of interventions (actions, and interactions) of agents aiming to satisfy their objectives. The intervention context of these agents, represented in the Solution Use Context Model (SUCM), constitute the space where agent requirements arise and the corresponding functionalities of the envisioned system are explicit. Figure 3 shows a reduced SUCM where the aid of the envisioned co-created contribution selection system, the application case in this work, is placed.

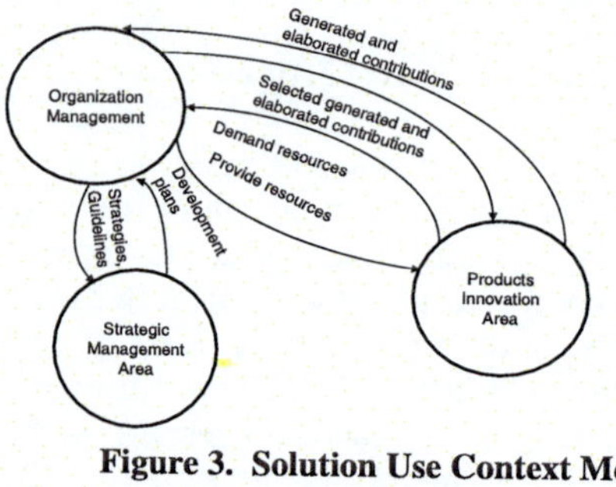


**Figure 3. Solution Use Context Model**

## 5 Comparison of System Functionalities

Context and domain models may be built before the requirements elicitation and applied for discovering system functionalities leading to a system prototype rapid construction. Furthermore, in this research line, the requirements elicitation, testing models, and system analysis agile methods are developed.

In the field of consistency testing, approaches based on context and domain concepts are included in [4] and [13], respectively. Supported on these works, this paper presents the comparison of context-based and domain-based system functionalities established by instantiation of the Agent Intervention Extended Pattern.

## 5.1 Functionalities focused on Domain and Context Concepts

As depicted in Figure 4, in the left branch, system context objectives are decomposed in domain-centred operationalizable objectives, using SDSO concepts. In the right branch, system context objectives are decomposed in context-centred operationalizable objectives, using agent interactions of the SUCM. One objective is defined for each SDDO concept and each SUCM agent interaction.

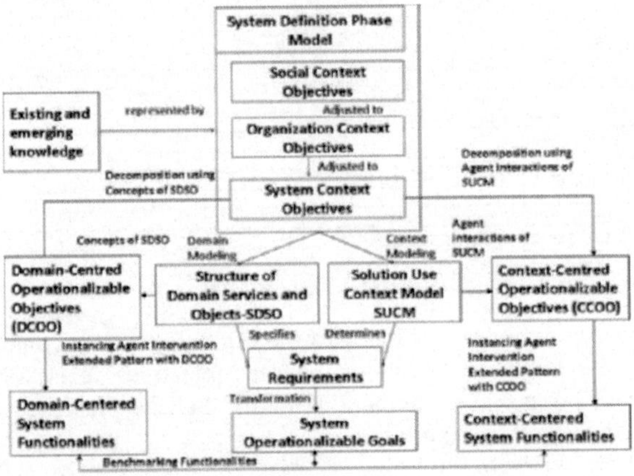


**Figure 4. Alternative Ways for Discovering System Functionalities**

Aiming to domain-centred system functionalities, the agent intervention extended pattern is instantiated with domain-centred operationalizable objectives, while to context-centred functionalities this pattern is instantiated with context-centred operationalizable objectives. Benchmarking for both groups of functionalities and for system operationalizable goals is made. These goals are obtained from system requirements elicited with the support of the Structure of Domain Services and Objects (SDSO) and of the Solution Use Context Model (SUCM). These models enclose the knowledge of the system definition phase, expressed by system context objectives, which summarize social and organizational context objectives.

## 5.2 Benchmarking of Functionalities Centred in Context and Domain

The highest levels of the SDSO are services to be offered by the envisioned system. System context objectives, in Figure 4, are decomposed in domain-centred operationalizable objectives, each one treating one of the nineteen concepts of the SDSO, listed in Figure 5.

The agent intervention extended pattern introduces more goal types in the nine goal categories of the basic pattern contained in Figure 1. In the part *Transformation* o *Execution* of the pattern, there is only the functionalities category: "*To execute the chosen option*" in which objectives described in Figure 5, as instances of options offered in the part *Preparation* of the pattern, are defined as domain centred system functionalities. The agent intervention extended pattern allows decompose objectives of the second highest level in three functionalities following the same process structure: *preparation or input, execution or transformation, transference or output*. For sake of space, the outcomes of this refinement are not included in this paper.

| Code | | Contributions (ideas) Selection System: Domain-Centred System Objectives |
|---|---|---|
| 1 | | Selection of Contributions (ideas) Context of the Product or Service |
| | 1.1 | Recovering of Contributions |
| | 1.2 | Applying a Selection Method |
| | 1.3 | Elaboration of Selected Contributions |
| | 1.4 | Protect and Make Available Selected Contributions |
| | 1.5 | Protect and Make Available Non-Selected Contributions |
| 2 | | Definition of Objectives, Functions and Elements in the Use Context of the Product or Service |
| | 2.1 | Definition of Possible Functions and Uses of a New Product or Service |
| | 2.2 | Definition of Objectives for Searching a New Product or Service |
| | 2.3 | Presentation of general and technical Information about the Possible to be Innovative Product or Service |
| | 2.4 | Presentation of general and technical Information about the organization and Contribution Selection services |
| 3 | | Establishment of Contributions Selection Methods |
| | 3.1 | Creation or Modification of Selection Methods |
| | 3.2 | Choice of a Selection Method |
| | 3.3 | Suppression of a Selection Method |
| 4 | | Management of Selected Contributions (Ideas) |
| | 4.1 | Recovering and use of Selected Contributions |
| | 4.2 | Classification of Selected Contributions |
| | 4.3 | Modification or new elaboration of Contributions |

**Figure 5. Domain-Centred Operationalizable Objectives**

Equally, as it appears in the right branch of the Figure 4, system context objectives are decomposed in context-centred operationalizable objectives, each one expressing a system aid to one of three agent interactions of the SUCM presented in Figure 3 and identified in the first column of Figure 6.

| Code | | Contributions (ideas) Selection System: Context-Centred Operationalizable Objectives | Domain-Centered System Functionalities | Global Domain-Centered Functionalities |
|---|---|---|---|---|
| 1 | | Establish resources for the selection of agents' contributions (ideas) for developing innovative products: selection methods, TICs, contextual knowledge associated to the innovative products, hardware-software platform for information processing and methods and resource for software programming and for information and communications management | | 2.1,2.2,2.3,2.4. 3.1,3.2,3.3 |
| | 1.1 | Propose resources for the selection of agents' contributions (ideas) (for developing innovative products): selection methods, TICs, contextual knowledge associated to the innovative products, hardware-software platform for information processing and methods and resource for software programming and for information and communications management | 2.3,2.4.3.1 | |
| | 1.2 | Adopt resources for the selection of agents' contributions (ideas) for developing innovative products | 3.2 | |
| | 1.3 | Manage resources for the selection of agents' contributions (ideas) for developing innovative products | 3.3 | |
| 1a | | Establishes selected generated and elaborated agents' contributions (ideas) for developing innovative products | | 4.1,4.2,4.3. |
| | 1a.1 | Propose selected generated and elaborated agents' contributions (ideas) for developing innovative products | | |
| | 1a.2 | Adopt selected generated and elaborated agents' contributions (ideas) for developing innovative products | 1.4,1.5 | |
| | 1a.3 | Manage selected generated and elaborated agents' contributions (ideas) for developing innovative products | 4.1,4.2,4.3 | |
| 2 | | Establish 0rganizational strategies, objectives, goals, projects and resources driving to select of agents' contributions (ideas) for developing innovative products | | 1.1,1.2,1.3,1.4,1.5 |
| | 2.1 | Propose organizational strategies, objectives, goals, projects and resources driving to select of agents' contributions (ideas) for developing innovative products. | | |
| | 2.2 | Adopt organizational strategies, objectives, goals, projects and resources driving to select of agents' contributions (ideas) for developing innovative products. | 2.1,2.2 | |
| | 2.3 | Manage organizational strategies, objectives, goals, projects and resources driving to select of agents' contributions (ideas) for developing innovative products. | 1.1,1.2,1.3 | |

**Figure 6. Domain-Centred Functionalities Matching Context-Centred Functionalities**

In order to make comparable both set of operationalizable objectives, the three context-centred objectives are decomposed, in second and third columns of Figure 6, in three context-centred system functionalities which agree with the process structure, as considered in the extended pattern. Fourth column contains the codes of domain-centred functionalities matching the context-centred system functionalities.

This functionalities correspondence indicates the coincidence of envisioned system processes. The context-centred functionalities which do not match are anyway included in the semantic of domain-centred functionalities or in an obvious missed activity for defining a whole process. In the last column, a global coincidence, at process level, between domain-centred functionalities and context-centred functionalities is confirmed.

In Figure 7, the correspondence of system operationalizable goals, obtained from agent requirements with domain-centred and context-centred functionalities, is established. The eleven functionalities, in first column, are *transformation* functionalities extracted from the set of system functionalities obtained from agent requirements. Functionalities discovered by domain-centred and context-centred approaches, detailed in second and third columns, respectively, each one compose the requirement based system functionalities of first column. Thus, the efficacy of domain and context centred approaches for support system analysis agile methods and prototyping is appreciated. The correspondence of two approaches, showed in second and third columns, constitute a mutual confirmation of the capacity of each one for agile system analysis, prototyping, and testing.

| **Extract Transformation goals of a System for Selecting Generated and Elaborated Co-created Contributions (ideas)** **-Derived from Agent's Requirements-** | **Domain-Centered System Functionalities** | **Context-Centered System Functionalities** |
|---|---|---|
| 1- Carry out ideas selection services: Selection of contributions (ideas), Definition of objectives, functions and elements in the context of use of the product or service; Selection of methods, and management of selected contributions (ideas). | 1.1, 1.2, 1.3, 1.4, 1.5, 2.1, 2.2, 2.3, 2.4, 2.5, 3.1, 3.2, 3.3, 4.1, 4.2, 4.3 | 1.1, 1.2, 1.3, 1a.1, 1a.2, 1a.3, 2.1, 2.2, 2.3 |
| 2- Manage the integration of services, processes, and systems of the Products Innovation Area with the Organization Management, and the Strategic Management Area. | 1.1, 1.2, 1.3, 1.4, 1.5, 2.1, 2.2, 2.3, 2.4, 2.5, 3.1, 3.2, 3.3, 4.1, 4.2, 4.3 | 1.1, 1.2, 1.3, 1a.1, 1a.2, 1a.3, 2.1, 2.2, 2.3 |
| 3- Effect procedures for assuring the scope, the pertinence, and the quality of contribution (ideas) selection services of the Products Innovation Area, the Organization Management, and the Strategic Management Area. | 1.1, 1.2, 1.3, 1.4, 1.5, 2.1, 2.2, 2.3, 2.4, 2.5, 3.1, 3.2, 3.3, 4.1, 4.2, 4.3 | 1.1, 1.2, 1.3, 1a.1, 1a.2, 1a.3, 2.1, 2.2, 2.3 |
| 4- Effect procedures for the demonstration and simulation of Contributions (ideas) selection services aiding the employers training and the development of the service culture in the areas related to Products innovation. | 1.1, 1.2, 1.3, 1.4, 1.5, 2.1, 2.2, 2.3, 2.4 | 1.1, 1.2, 1.3, 2.1, 2.2, 2.3 |
| 5- Effect procedures for researching and making operative and interactive (navigable) Contributions (ideas) selection services: Selection of contributions (ideas), Definition of objectives, functions and elements in the context of use of the product or service, Selection of methods, and management of selected contributions (ideas). | 1.1, 1.2, 1.3, 1.4, 1.5, 2.1, 2.2, 2.3, 2.4, 2.5, 3.1, 3.2, 3.3, 4.1, 4.2, 4.3 | 1.1, 1.2, 1.3, 1a.1, 1a.2, 1a.3, 2.1, 2.2, 2.3 |
| 6- Manage modifications, renovations, and cancellation of innovative contributions (ideas). | 4.1, 4.2, 4.3 | 1a.1, 1a.2, 1a.3, 1.1, 1.2, 1.3, |
| 7- Manage interrogation and handling of registered operation in D.B. | 4.1, 4.2, 4.3 | 1a.1, 1a.2, 1a.3, 1.1, 1.2, 1.3, |
| 8- Effect procedures for applying combinations of innovative contributions (ideas) under particular technical and commercial conditions. | 4.1, 4.2, 4.3 | 1a.1, 1a.2, 1a.3, 1.1, 1.2, 1.3, |
| 9- Manage services achievement involving third parts such as: developers of systems for supporting the innovation processes, Control Public Organizations, Scientific Organizations, etc. | 2.1, 2.2, 2.3, 2.4. 3.1, 3.2, 3.3 | 1.1, 1.2, 1.3, 3.1, 3.2, 3.3 |
| 10- Effect procedures for using information related to organizational, technical, promotional, legal, and quality aspects. | 2.1, 2.2, 2.3, 2.4, 3.1, 3.2, 3.3 | 1.1, 1.2, 1.3, 2.1, 2.2, 2.3 |
| 11- Effect procedures for using information coming from the products innovation area, the Organization Management, and the Strategic Management Area, as well as from involved third parts. | 2.1, 2.2, 2.3, 2.4, 4.1, 4.2, 4.3 | 1.1, 1.2, 1.3, 1a.1, 1a.2, 1a.3 2.1, 2.2, 2.3 |

**Figure 7. Matching of Functionalities Based on Context, Domain, and Requirements**

## 6 Conclusion and Future Work

The consistency between domain-centred and context-centred functionalities determines a double way for agile analysis of systems or products when context and domain knowledge is not enough. Domain-centred approaches support preliminary analysis of big or complex systems when the requirements are not established and interested agents and the system scope are not completely defined. Context-centred approaches are a good option when systems or products are not entirely defined or are unknown. That is the case in research, innovation, and forecasting projects.

Ongoing work uses domain-centred and context-centred approaches in constructing consistency testing models. In the line of system analysis agile methods and patterns, new projects are envisioned.

## 7. References

[1] Zave P, Jackson M. Four dark corners of requirements engineering. ACM Transactions on Software Engineering and Methodology, 1997, 6(1): 1-30.

[2] Urrego-Giraldo G. ABC-Besoins : Une approche d'ingénierie de besoins fonctionnels et non-fonctionnels centrée sur les Agents, les Buts, et les Contextes. Ph.D. Thesis, Universidad Paris 1, Pantéon Sorbona, 2005.

[3] Urrego-Giraldo G, Giraldo G. L Estructuras de servicios y de objetos del dominio: una aproximación al concepto de ontología. En: Colombia TecnoLógicas ISSN: 0123-7799 ed: Instituto Tecnológico Metropolitano v.15, 45-67, 2006.

[4] Urrego-Giraldo G, Giraldo G.L.. "Context model for testing analysis phase of information systems. In 17th ISPE International Conference on Concurrent Engineering, pp 565, 573, Poland, Sept. 2010.

[5] Ramadour P. and Couvet C. Modélisation de domaine pour l'ingénierie des SI para réutilisation. 12éme journées francophones d'ingénierie des connaissances. Grenoble, 2001.

[6] Neighbors J.. "Software Construction Using Components". Ph.D. Thesis, Department of Information and Computer Science, University of California, Irvine, 1981.

[7] McCain R.. "Reusable Software Component Construction: A Product-Oriented Paradigm". In Proceedings of the 5th AIAA/ACM/NASA/IEEE Computers in Aerospace Conference, Long Beach, CA, pp.125-135, October 21-23, 1985.

[8] Prieto-Diaz R.. "Domain Analysis: an introduction". ACM SIGSoft Software Engineering Notes". Vol. 15, Nro 2, pp. 47-54.

[9] Maiden N. A. M.. "Analogy as a paradigm for specification reuse", in Software Engineering Journal, Vol. 6, Nro 1. 1991.

[10] Constantine L.. "Structured Design", IBM Systems Journal, Vol. 13, Nro. 2, 1974.

[11] Ross D. T. and. Shoman K. E, Jr., "Structured analysis for requirements definition," IEEE Trans. Software Eng., vol. SE-3, pp. 69-84, Jan. 1977.

[12] Alexander C., Ishikawa, S., Silverstein M. 1977. A patron language. Oxford University Press, New York

[13] Urrego-Giraldo G , Giraldo G. "Contribución del modelado del dominio a la definición de los servicios de una plataforma de apoyo a la co-innovación de productos y servicios. Tendencias en Ingeniería de Software e Inteligencia Artificial. V. 4. Zapata, Carlos Mario, Giraldo, Gloria Lucía y Velásquez, Juan David (Ed).43-50, 2011.

# Concurrent Engineering Applications and Lessons Learnt in Oil Refinery

Amar Abbas [a] , Irfan Anjum Manarvi[b,1]

[a]Department of Engineering Management, CASE, Islamabad Pakistan.
[b]Department of Mechanical Engineering, HITEC University Taxila, Pakwastan.

**Abstract.** Due to volatile environment and dynamic market of oil business, refineries are always looking for new ways to improve efficiency and productivity. In this research, Concurrent engineering (CE) was considered as one approach which could possibly be used to increase the efficiency of processes. Therefore it was investigated wether CE principles of multidisciplinary team, computer supported information sharing, development of team work environment and group decision making could be applied in various process/projects of a refinery and what influence could be expected in production as well as quality of output in terms of savings or added revenues.

**Keywords.** Concurrent Engineering (CE), Multi-disciplinary Team, Computer supported Information sharing, Continuous Improvement.

## 1 Introduction

Due to volatile environment and dynamic market of oil business, refineries are always looking for new ways to improve productivity of system. It had been realized that strong barriers exist among departments of a company which impede essential communication and cooperation. A supportive environment may therefore be considered vital to implement CE. It had also been established through past research [1-3] that the role of top managers was very important to establish a teamwork-based environment. Concurrent engineering principles of multidisciplinary team, computer support, and software tools are required to improve the efficiency, productivity and communication system of the company [4] [5] [6]. Continuous improvement was a major tenet of CE which was necessary to remain competitive in a rapidly changing and volatile climate. This requires management systems which promote continuous improvement [7-8]. Benchmarking also maintains the stimulus for continuous improvement and aims at competitiveness advantage [9].

[1]Professor, Department of Mechanical Engineering, HITEC University, Taxila Education City, Pakistan Tel: +92 336 5025 073; Email: irfanmanarvi@yahoo.com; http://www.hitecuni.edu.com

D. D. Frey et al. (eds.), *Improving Complex Systems Today*,
Advanced Concurrent Engineering, DOI: 10.1007/ 978-0-85729-799-0_12,
© Springer-Verlag London Limited 2011

Refining division of an oil company was selected as a case study to apply principles of concurrent engineering

## 3 Overview of Company and environment of CE application

Refinery Company (PARCO) was actively involved in various facets of oil storage, transportation, refining and marketing of energy products. Company was the key player in the country's strategic oil supply and its logistics. Its refining division was selected as a case study for application of concurrent engineering principles. The Refinery capacity was 4.5 million tons per annum equivalent to a processing throughput of 100,000 barrels per day. It produces a wide range of petroleum products, including Liquefied Petroleum Gas, Gasoline, HOBC, Kerosene, Jet A-1, JP-8, High Speed Diesel, and Furnace Oil. The effectiveness of an organization is also influenced by its working environment. It was recognized that CE cannot be applied without a conducive environment. Role of top managers was considered important to foster that environment and their commitment was vital to achieve this objective. Values of teamwork are considered very high in this environment. A major step was taken to plan rotation of department managers besides other measures.

*Rotation of Department Managers*— A plan was made to rotate managers who were involved in decision making of Technical services, Engineering services, Process, and Utilities & Oil Movement departments. It included stay of at least nine months to maximum of one year at each department. It was ensured that they evaluate employees' performance of that department. Communication flow started in various directions i.e., vertical, horizontal and diagonal. The organization had a system of reporting wrong doings or mistakes through anonymous letters to the CEO every month. After implementation of CE principles, the number of complaints for various departments reduced by over 70%. Thus rotation helped to improve communication among departments.

*Team based Training and Review of Performance Evaluation Method*— During the period of managers' rotation, team based trainings were conducted for various levels of management. Performance evaluation of an employee input weight age of immediate supervisor was reduced from 100% to 60% and remaining weight age came from team leaders if he remained a member of team.

## 4 Applications of CE in various Processes and Projects

### 4.1 Process of Management of Change (MOC)

*Management of Change*— Sometimes engineering work on plant goes beyond maintenance and constitutes modifications. These are proposed to improve company's profit, operational safety, reduce production cost, ease of operation etc. Such modification involves a change in the plant and/or process and can introduce a hazard. It was essential to have a system of identifying and controlling modifications. A management of change (MOC) process was used to define the specific review, authorization requirements and the procedures through which changes are authorized and documented. Questions which may be asked concerning it were: Was it necessary? Was it economical? Was there a better alternative? Was it safe and environment friendly?

*Issues with MOC Methodology*— With above methodology of MOC, proposal took a long lead time to implement. Data indicated that there were around 40% of change proposals which had been recycled in review stage for further study. Lead time to complete proposal also increased due to rework. Concern was shown for high percentage of restudy and long time spent in this process. It was also an opportunity loss due to delay in implementation of proposal. It was observed that generally those aspects which are not related to the expertise domain of their individual departments were missed during study.

*Introduction of Multi-disciplinary Team*— A multi-disciplinary team was introduced to cater this issue. Involvement of team members begins from the first phase of technical study and continues till implementation and closeout report. Consideration had been given to following points for formation of multi-disciplinary teams.

- Leader of team was to be chosen from that department which was responsible for study in earlier MOC scheme.
- The person who initiated the proposal was to be made member of multi-functional team and considered as an internal customer.
- Team members will vary for each proposal representing all departments.
- Conflicts are to be used in positive way and methods to be explored to resolve conflicts.
- Multidisciplinary teams are to be empowered to reject proposal.

*Composition of Multi-disciplinary Team*— Generally, members of middle management group is selected to represent team. However, no restriction was imposed to choose members from lower management group. Other benefits of this group are as follows:

- When members of this group move up, it will reinforce team based culture.
- Helps to improve communication in all directions.

*Introduction of Computer Supported Information Sharing*— After the introduction of multidisciplinary team concept, it was realized that this lead time can be further reduced if engineering data base was easily accessible and retrievable any time by all team members. An Engineering data base was established and placed at information interchange medium of local area network (LAN) and wide area network (WAN) systems. Softcopies of all operating, maintenance, safety manuals were placed on this system with access to everyone so that right information was available to the right people at the right time.

*Finding*— After the introduction of multi-disciplinary team, rework on proposals reduced from 40% to 9%. Lead time to implement the proposal also reduced by 25%. Also by use of computer supported data sharing; lead time to implement the proposal reduced by 37% as shown in figure 1.0. However, rework rate remained same at 9%.

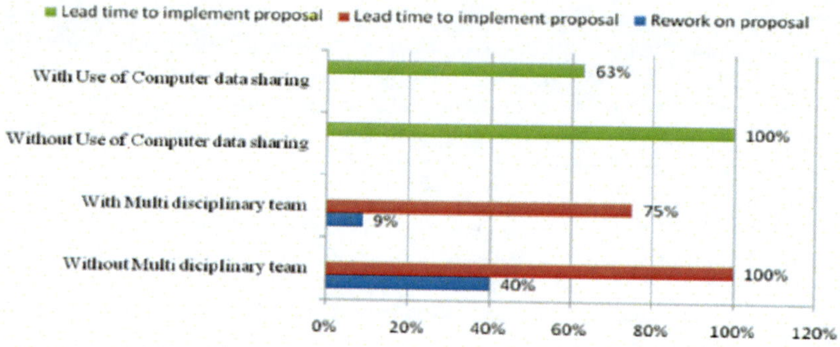


Figure 1 Lead time for Small to Medium size Projects

*Strengthening of computer supported system*—The approach to concurrent engineering had to be through the automatic sharing of knowledge by computerized systems and application of software tools. Various tools were made part of the system. Tools include computerized maintenance management system Maximo, process history data base PHD, Computer Aided Design CAD, project planner Primavera, Financial Accounting System FAS, LP model for crude scheduling etc. These tools aided in the improvement of productivity.

## 4.2 Planning of Turnaround

With the success in MOC process by applying CE principles, it was decided to expand these principles to other projects as well as apply more principles of CE. The Turn-around was a kind of fast track project with unique tasks. It was an orchestrated shut down of the plant, conducted after thorough planning, in a relatively short time frame. Key elements of Turnaround management are usually categorized into six phases:

| | |
|---|---|
| Phase I | Conceptual preparation |
| Phase II | Preliminary preparation |
| Phase III | Detailed Preparation |
| Phase IV | Pre-T/A execution & Final preparation |
| Phase V | Execution |

Phase VI Post Turnaround

First four phases are related to planning of turnaround. Planning section of maintenance department was responsible for planning of turnaround. Its execution phase had a direct link with planning phase. Better planning makes a smooth process of execution. Abbreviations TA-1 and TA-2 will be used for turnarounds.

*Working Methodology for TA-1*— Planning section was responsible for planning of turnaround. All departments sent their inputs to planning section and section managed activities by employing planning and scheduling software. All these planning and scheduling work was again sent back to individual departments for their feedback. Progress was reviewed on monthly basis by senior management. Contractor force joined the project near commencement of turnaround.

*Working Methodology for TA-2*— A core team, consisting of middle management group, had been established for planning of turnaround. Maintenance manager was chosen to lead team who plays a major role in turnaround. Members were chosen from process, utilities and oil movement, engineering services, technical services, maintenance, material, and administration departments. Most of the participants continued their day-to-day plant responsibilities. However, their TA roles had been developed specifically around their T/A process responsibilities. Senior management started progress review meetings on bimonthly basis instead of monthly basis after establish of core team. Contractor's supervisory staff was mobilized earlier and made part of sub core team on each major job. Dry run demonstration was done with available resources at that time to identify any shortcomings prior to actual execution of jobs in TA.

*Finding*— In TA-1, actual duration of turnaround exceeded planned duration by one day. One day delay means a loss of approx. $ 1.4 million. After the introduction of core team concept, TA-2 was completed earlier than plan as shown in figure 2.0 below:

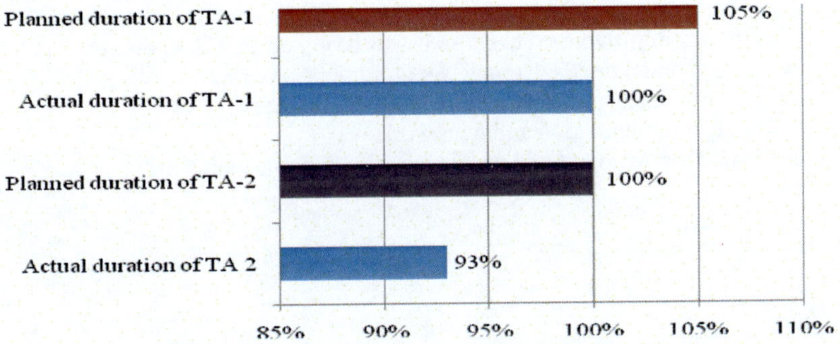


Figure 2 Variation in Turnaround due to CE principles

### 4.3 Implementation of Integrated Management System

Company planned to implement integrated management system which comprises WASO 9001:2000, WASO 14001:2004 and OHSAS 18001:1999. Prior to implementation, an independent group carried out gap analysis and gave a period of 9 months based on gaps in various departments and scope of work. A steering committee was constituted to complete this project and members were selected from all departments. Computer supported system was utilized effectively. A folder on integrated management system was placed on LAN which was accessible by every employee. All supportive material was placed on it. Each completed task had been placed on this folder as soon as it was finished. Minutes of meeting were also placed on it. Progress report was updated on weekly basis with access by everyone. Multidisciplinary team comprising middle management group and computer supported information sharing system facilitated to achieve this project in record time.

*Finding*— Due to introduction of steering committee and computer supported information sharing system; certification was achieved in 7.5 months as compared to planned schedule of 9 months.

### 4.4 Continuous Improvement by Implementation of Integrated Management

CE culture focuses on continuous improvement. For CE changes to be successful, managers and team leaders need to drive continuous improvement of quality, safety, health, environment and productivity. Establishment of quality management system WASO9001, environment management system WASO14001, and occupational health & safety management system was a step in that direction. These management systems integrate all components of a business into one coherent system so as to enable the achievement of its purpose and mission. Implementation of this system helped to achieve following objectives.

- Clear and concise policy on its direction.
- Continuous improvement because it was based on PDCA approach
- Make sure that agreed customer's requirements are met.
- Provide guidance and clarification to all employees
- Enhance communication and teamwork.
- Sustained involvement and participation of top management.

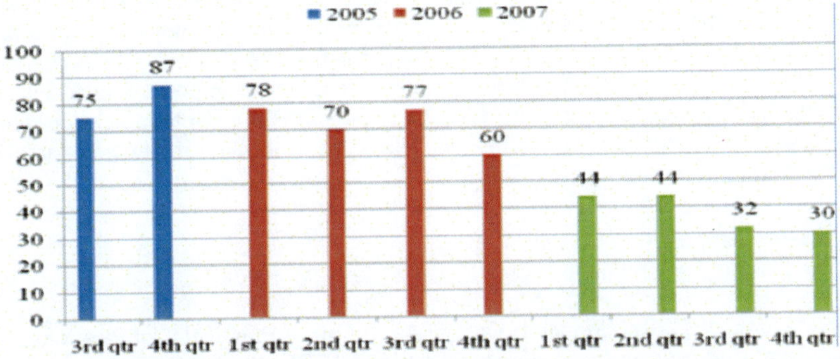


Figure 3 No. of quality related incidents in observed period

### 4.5 **Continuous improvement through employee's participation**

Employee involvement helps to promote culture of continuous improvement. It was decided to establish a system so that voice of employees even at the lowest level of hierarchy was listened and acted upon. A system was developed named Quality related Incidents (QRI). This system was provided in addition to safety related incident reporting system. Any employee can raise a QRI request related to quality related incidents or having potential to incidents. The System was used to improve process system by analyzing the incidents, giving short and long term recommendation to avoid recurrence. This was basically a quality improvement program. Oracle based application was used to develop it as paperless system.

In order to check effectiveness of the system, quality related incidents had been categorized such as operational, mechanical, electrical, instrument, IT, administrative. It had also been categorized by equipment wise. Analysis of three years data aided to identify many problems and their troubleshooting. Analysis indicated which type of incidents frequency are high, major causes of failure etc. It proved to be a diagnostic tool. Strong follow-up of recommendations was done with feedback to top management. Implementation of recommendation helped to improve process system. Number of quality related incidents have reduced by approximately 60%. This system had provided a mechanism for continuous improvement.

*Finding*— Number of quality related incidents have reduced by approximately 60% in two and half years as shown in figure 3 above. This improvement was achieved through involvement of employees.

## 6 Conclusions

Present research was focused on application of CE principles of using multi disciplinary teams, computer supported data sharing, use of a steering committee in projects and involvement of employee's in decision making. It was observed that the rework on proposal development as well as its implementation reduced considerably. The computer sharing of information provided a better insight to all stakeholders regarding various phases of the projects and they were able to plan the resources for future requirements efficiently. The most important and critical operation of Turnaround helped achieve the required tasks completion before time. CE application resulted in saving in terms of millions of dollars for the organization understudy.

# 7. References

[1] Donna H. Rhodes and Carey A. Smith, "Practical Applications of Concurrent Engineering for Systems Integration", IEEE, 1992.

[2] Haihang Wang, Qidi Wu, Caihong Zhou, and Junwei Yan, "Research on Teamwork-based Organization Environment and Enterprise Cultures for Concurrent Engineering", IEEE International Conference on Industrial Technology, 1996.

[3] K.T. Yeo, and K.K. Teo, "Perception and Relevance of Concurrent Engineering in Singapore's Manufacturing Industry", 1994.

[4] Fulvio E. Oliveto "Concurrent Engineering: Evolution and Application", IEEE, 2000.

[5] Stephen Evans, "Concurrent Engineering – Pitfalls and Implementation Successes", Engineering Management Journal, June 1992.

[6] James P. Pennell, Robert I. Winner, Harold E. Bertrand, and Marko M. G. Slusarczuk, "Concurrent Engineering: An Overview for Autotestcon", IEEE, 1989.

[7] P.J. Veness, G. Chidolue and S.S. Medhat, "Concurrent Engineering Infrastructure Tools, Technologies and Methods in British Industry", Engineering Management Journal, June 1992.

[8] Raymond T. Yeh, "Notes on Concurrent Engineering", IEEE Transactions on Knowledge and Data Engineering, Vol. 4, Oct. 1992.

[9] JAG Knight, H.S. Abdalla, and M.S. Ahmed, "A Methodology for Product Life-Cycle Development for Global Manufacturing", 1994.

# Analyzing influence of absence of CE on Pakistan's Economy from Perspective of Developed Countries using it

Syed Zahid Raza Rizvi[a], Irfan Anjum Manarvi[b,1]

[a]Department of Engineering Management, CASE, Islamabad, Pakistan.
[b]Department of Mechanical Engineering, HITEC University Taxila, Pakistan.

**Abstract.** Continuous analysis and improvement are required in concurrent engineering environment. PDCA cycle described by quality guru can be used to improve quality of the processes and systems. Evaluation is sort of Check cycle to critically analyze that whether the planned and functional processes and systems are producing the desired quality results. Decision making is primarily based on data analysis. This research is focused on the analysis of Pakistan's economy with respect to developed countries. 35 different variables for 41 countries have been analyzed (total 1435 records). It provides an opportunity to review the strong and weak areas and take necessary steps for domestic legislation and adapt to ensure that it is in line with developed countries. Thus, the process itself may be seen as a positive one for the progress of Pakistan. GDP per capita may be increased by education, producing skilled manpower, adopting latest technology in agriculture sector, and increasing spending on education and ICT.

**Keywords.** Concurrent Engineering (CE), Macroeconomic, Education, Science and Technology, Economy, Pakistan.

## 1 Introduction

Most impressive platform for developed countries is OECD which was established in 1961 with head office in France. There are total 32 permanent members (Australia, Austria, Belgium, Canada, Chile, Czech Republic, Denmark, Finland, France, Germany, Greece, Hungary, Iceland, Ireland, Italy, Japan, Korea, Luxembourg, Mexico, Netherlands, New Zealand, Norway, Poland, Portugal, Slovak Republic, Slovenia, Spain, Sweden, Switzerland, Turkey, United Kingdom, and United States) who are leading the world in economy and social sector. OECD has also undergoing membership discussion with eight countries (Brazil, China,

[1]Professor, Department of Mechanical Engineering, HITEC University, Taxila Education City, Pakistan Tel: +92 336 5025 073; Email: irfanmanarvi@yahoo.com; http://www.hitecuni.edu.com

D. D. Frey et al. (eds.), *Improving Complex Systems Today*,
Advanced Concurrent Engineering, DOI: 10.1007/ 978-0-85729-799-0_13,
© Springer-Verlag London Limited 2011

Estonia, India, Indonesia, Israel, Russia, and South Africa). The country which is to become member of OECD has to demonstrate its attachment to the basic values including an open market economy, democratic pluralism and respect for human rights. Macroeconomic variables are most crucial as if productivity is increased it improves the employment rate. High productivity also require high skilled work force. Similarly trade balance can be brought in positive or negative zone. It is proven fact that if workers are educated they can push the productivity to higher level by adapting enhanced technology for better yield [1]. Most promising approach to impact measurement of health and economic benefits is the work by society in health research to prevent and treat the illness [2]. Technical aspects effect capacity development of a particular country system. Beyond these system-specific aspects, there lie many challenges in the development of any country. The most important of which are forceful domestic leadership and skilled management [3]. Japan's assistance has contributed to the improvement of the conditions in elementary education and to the promotion and encouragement of education for girls in the North West Frontier Province (Khyber Pakhtoon Kha) of Pakistan. Indus Highway Construction Project and the Kohat Tunnel Construction Project financed by Japan's loan aid have also contributed to the transport development. Therefore the foreign aid can be utilized to uplift the literacy and communication, if monitored and utilized in efficient manners [4]. Japan is also assisting Pakistan in social sector, which includes education/health/water supply; improvement of economic infrastructure, which includes power/transportation; agriculture, which includes irrigation/agricultural research; and environmental conservation). It is very helpful and continuing with the development needs of Pakistan [4]. Pakistan, population growth rates is in excess of 2% a year which is putting increasing strain on the environment. Although similar high growth rate can be seen in Israel but that is only due to immigration and urbanization; over 90% of Israel's population now lives in urban areas. The population of Pakistan is still predominantly rural just as in the case of China, Guatemala, Honduras, Kenya, Thailand and Vietnam. But the high speed of rural-urban migration in Pakistan and these countries is making urban-rural population ratio will quickly result in population concentrated towns and cities [5]. A major conclusion of the World Summit on Sustainable Development in Johannesburg was that science and technology should play a greater role in contributing to sustainable development. Countries need to develop their capacities through inputting education, research and employment to utilize S&T for sustainable development. Countries need to invest in S&T to create needed new technologies [6]. The consumption expenditures, a proxy of well-being, are significantly higher of non-farm households located at head and middle reaches of the water distributary as compared to their counterparts living at the tail-end of water distributary. The households located at middle levels of distributary are well off as compare to others. Its main reason could be flood in head and shortage in tail distributaries, whereas middle levels receive proper water supply [7]. Pakistan introduced reforms in custom department, then it struggled to yield the expected results due to complicating factor in the way of successful implementation by the reluctance of the most important stakeholders, Customs staff and the trading community; interestingly both have different reasons to oppose [8]. This is the general trend, whenever reforms are introduced in a particular department to boost its capabilities and functionality, stake holder's fight for their hidden agendas. Due to international events and Government reforms policies, the macroeconomic situation improved in Pakistan improved since 2000. Both growth and reserves increased, but the situation remains fragile. On the negative side, a very little progress in agriculture and natural resource management can be seen [9]. Pakistan needs to implement reform policies on a good pace to keep macroeconomic trends moving in positive direction. A strong management is also recommended. If look closely at the causes of recent price increases, it will be explored that it is linked with yield shortfalls and climate change. More investments in R&D, technology transfer and extension services, is required is especially in less developed economies, to increase productivity and output. More research is required to be carried for the resilience of crops against stress such as drought. More incentives should be provided to farmers and they may also be educated to produce high yields [10]. Results of different policies implementation suggests that if more aid is injected in primary education sector of poor countries, it benefits in promoting economic growth and also help the countries to achieve the Millennium Development Goal of universal primary education defined by United Nations subsidiaries [11]. Present research is focused on identification of weeknesses and stretghs Pakistan's economy' variables. The analysis results depend on population as well. The approach used in this research may also provide guidelines to other researchers and analysts in their respective country comparison for similar enhancements and analysis.

## 2 Research Methodology

The core aim of the research is to find out the weaknesses and strengths of Pakistan with reference to developed countries. Data included 1435 records, collected with all important and effective variables. Methodology used is to analyze the data by selecting all the major sectors. Results in graphical and tabular formats are shown after quantitative analysis for ease of decision making.

## 3 Analysis and Discussion

Macroeconomic trends and economic globalization are the most important categories on which progress of a country depends. It draws a picture of budgetary condition, GDP related issues, and open market economy. Moreover financial ratios help to evaluate economic condition of any entity. They give better understanding of the financial position of particular country. Reference Table 1, analysis shows that Pakistan's revenue to expenditure ratio is 0.73, which is below than the average ratio of 0.86 and quite lower than 1.22 the maximum ratio, but better than lowest in selected countries which is 0.6. GDP analysis shows that it is very much near the median value of the developed countries, but more realistic comparison of GDP and GNI per capita depicts that Pakistan stands lowest in this very important economic parameter. Analysis of Figure 1 shows that no country other than Pakistan has the highest share of agriculture in GDP, it also indicates that depending on weather & climatic condition, Pakistan's GDP may fluctuate. On the other hand GDP percentage part of industry and services are better than the minimum of developed countries although these values are still quite lower than the average values of developed countries.

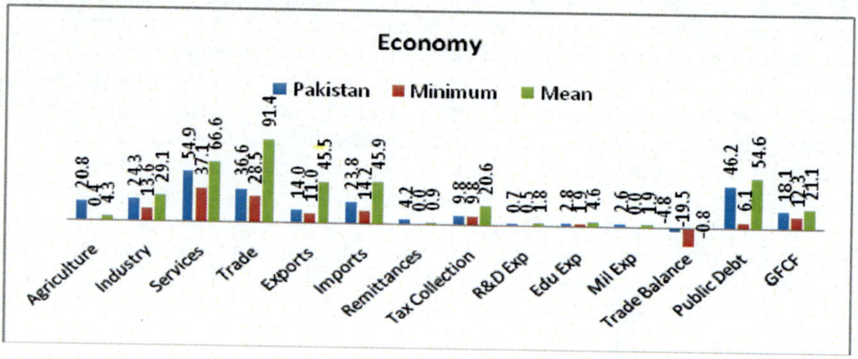


**Fig 1 – Economy variables as percentage of GDP**

Analysis also shows that Inflation rate in Pakistan is too high, more than 6 times of the average value. It can result in loss in stability in the real value of money and other monetary items; uncertainty which discourages investment and saving, and may lead to shortages of goods if consumers begin hoarding out of concern that prices will increase in the future. Figure 1 and table 1 both depicts that GFCF is good at 18.10 % of GDP and it is near mean and median value of developed countries which means Pakistan is holding assets of good value with reference to its GDP. Table 1 also shows real GDP and Industrial growth. Many developed countries are having negative growth rates, but this is mostly due to the fact that their economy is already at best or saturation levels. Therefore, it may not impact worst on their economy. Pakistan's GDP growth rate is quite above the mean, whereas its industrial production growth rate is better than 75% of the developed countries respective growth rate. Pakistan's economy needs its GDP & Industrial growth rates remain positive to strengthen the country's economy conditions. Tax collection is bone for economy to run smoothly and independently and rely on foreign and local debt is minimized. It also depicts the government's policies to boost the economy as well as the condition of local industry. Figure 1 presents the tax collection as percentage of GDP. It shows that Pakistan's standing is below than lowest of the developed countries. There is need to boost tax to GDP ratio. Figure 1 also shows that Pakistan spends a part of GDP which is more than the 75% of developed countries expenditure on their respective defence. Due to Pakistan's internal & border situation in addition to terrorism, Pakistan's military expense is a bit at high end. It also explains the condition of debts as percentage of GDP. Although the debts seem very high but when it is seen in comparison of other developed countries, it is discovered that public and external debts are very much less than the mean values of other 40 countries. External debt per capita USD 318 is very low as compare to average value for developed countries which is

USD 174262. Current account balance at Billion USD (-) 2.9 is significantly above the minimum value at Billion USD (-) 419.9. Table 1 below summarizes the macroeconomic trends.

**Table 1 – Macroeconomic Trends**

| S.# | Variable | Pak | Other Countries | | | |
|---|---|---|---|---|---|---|
| | | | Min | Max | Mean | Median |
| 1 | Revenue to Expenditure *Ratio* | 0.73 | 0.60 | 1.22 | 0.86 | 0.96 |
| 2 | GDP *(Billion USD)* | 439.6 | 12.7 | 14256 | 1441.9 | 441.1 |
| 3 | GDP per Capita (PPP) *USD* | 2661 | 2941 | 78395 | 28531 | 29541 |
| 4 | GNI per Capita *USD* | 2710 | 3230 | 57640 | 28195 | 29885 |
| 5 | GDP: Agriculture *%age* | 20.8 | 0.4 | 17 | 3.8 | 2.7 |
| 6 | GDP: Industry *%age* | 24.3 | 13.6 | 47.6 | 29.16 | 27.2 |
| 7 | GDP: Services *%age* | 54.9 | 37.1 | 86 | 66.95 | 69.5 |
| 8 | Inflation *%age* | 13.60 | -4.5 | 12 | 2.11 | 1.2 |
| 9 | GFCF: *%age of GDP* | 18.1 | 12.3 | 45.2 | 21.1 | 20.3 |
| 10 | Real GDP Growth *%age* | 2.7 | -14.1 | 8.7 | -3.04 | -3.55 |
| 11 | Industrial Growth Rate | -3.6 | -25.2 | 9.5 | -7.16 | -7.85 |
| 12 | Tax as %age of GDP | 9.81 | 10.17 | 48.20 | 21.02 | 20.32 |
| 13 | Military Expenditure *%age of GDP* | 2.6 | 0 | 7 | 1.85 | 1.4 |
| 14 | Public Debt *%age of GDP* | 46.2 | 6.1 | 189.3 | 54.76 | 49.7 |
| 15 | Debt Ext *%age of GDP* | 31 | 7 | 3854 | 236.45 | 97.5 |
| 16 | Debt Ext per capita | 318 | 187 | 402828 | 174262 | 33258 |
| 17 | Education expenditure *%age of GDP* | 2.8 | 1.9 | 8.3 | 4.7 | 4.73 |
| 18 | Expenditure on R & D *%age of GDP* | 0.67 | 0.46 | 4.7 | 1.8 | 1.63 |

Economic Globalization deals with open market economy i.e. all the international transactions and their sub heads are discussed. Pakistan is facing a trade deficit of 20.91 Billion USD which is 4.75% of GDP, it indicates that Pakistan does not produce much to export or even for its own need therefore it has to spend more on import. The maximum trade deficit is faced by Korea which 19.49% of its GDP. Pakistan's share in trade as a percentage of its GDP is better when compared with other countries is listed in Table 2. Its exports are just below the lowest 5% developed countries exports. It is important to note that GDP per capita of other countries is much better than Pakistan's. Remittances analysis shows that the received remittances are at highest level as compare to other countries. It is also an indication of another fact that migration from Pakistan is taking place at higher rate and people send their remittances to their home which brings foreign exchange and improve the reserves level. Moreover, it also improves the quality of life of related population in Pakistan. Table 2 shows FDI flows and its analysis gives mixed results. The main reason of Pakistan's lowest outward FDI stocks could be that Pakistan is not self sufficient in economy. Negative FDI flows outward is still better than many developed countries. However Pakistan's FDI stocks inward and FDI flows inward are better than several developed countries. This exhibits investment opportunities and investor attractive policies in Pakistan. Table 2 also describes the interest rates issue. Discount rate is as maximum whereas long term interest rate is highest than any other country. These could be main hurdle in investment and business activities. Pakistan needs to make policies for open market so that more investment could be brought in, which will improve economic condition and quality of life.

**Table 2 – Economic Globalization**

| S.# | Variable | Pak | Other Countries | | |
|---|---|---|---|---|---|
| | | | Min | Max | Mean |
| 1 | Trade in Goods & Services *%age of GDP* | 36.6 | 28.5 | 330.0 | 93.6 |
| 2 | Exports of Goods & Services *%age of GDP* | 14.0 | 11 | 173 | 46.65 |
| 3 | Imports of Goods & Services *%age of GDP* | 23.8 | 14.2 | 150.7 | 46.9 |
| 4 | Trade Deficit % of GDP | -4.8 | -19.5 | 32.4 | -0.7 |
| 5 | Remittances as %age of GDP | 4.2 | 0.0 | 4.2 | 0.854 |
| 6 | FDI Stocks Outwards | 2201 | 2744 | 4302851 | 425397 |
| 7 | FDI Flows Outward | -14 | -15064 | 248074 | 23649.8 |
| 8 | FDI Stocks Inward | 17789 | 8283 | 3120583 | 356727 |
| 9 | FDI Flows Inward | 2387 | -5575 | 129883 | 19653.3 |
| 10 | Discount Rate | 15.0 | 0.1 | 15.0 | 4.2 |
| 11 | Long Term Interest Rate | 14.0 | 0.5 | 11.1 | 4.9 |

Although expenditure on education as percentage of GDP (figure 1) is better than minimum of compared countries, but due to poor education condition, it has to be improved to achieve greater economic goals and progress of country. Figure 1 also shows that Pakistan expenditure on R&D is poor but better than the minimum. Here it is important to note that Pakistan's GDP itself is not at high level. Pakistan is fighting with poor economic conditions; therefore it is ascertain that it is not investing in this sector. Following table 3 summarizes the comparative index regarding Pakistan.

**Table 3 – 3E Comparative Index**

| S.# | Category | Variables | +ve | -ve | Nil | %age Matching |
|---|---|---|---|---|---|---|
| 1 | Economy: Macroeconomic Trends | 18 | 13 | 4 | 1 | 76.47 |
| 2 | Economy: Economic Globalization | 11 | 8 | 3 | 0 | 72.72 |
| | **Total** | **29** | **21** | **7** | **1** | **75.00** |

## 4 Findings

Revenue to expenditure ratio 0.73 is better than minimum of developed countries but most of them are generating more revenue than their expenditures; therefore they can add more quality of life, investment, make more money and contribute to aid programs for under developed countries. GDP per capita and GNI per capita are lowest. GDP depends on agriculture a lot. Earning percentage from industry and service sector is comparable with other countries. It is having highest inflation rate. Gross Fixed Capital Formation (as percentage of GDP) is better; meaning by Pakistan is holding good value assets. Real GDP growth and industrial production growth rates are better than most of the countries. Military expenditure is relatively in high range. Public debt is not too high and falls below median value of compared countries. Current account balance is about -2.9 Billion USD. External debt is about 31% of its GDP which is high but more than 3 times lower than median value of developed countries. External debt per capita is less than 500 times of average respective value of other countries. Tax as percentage of GDP 9.81 is lowest than any other country, whereas it's maximum value is 48.20. Trade in goods and services is 36.6% of GDP; 14% for export and 23.8% for import; all figures are above the minimum respective value for compared countries. Trade deficit in terms of percentage of GDP is 4.8 which is better in prevailing conditions in developed countries whereas 12 countries are suffering more deficits in terms of their respective GDP. Expatriate Pakistanis from all over the world contribute in remittances as 4.2 percent of GDP which is very high in comparison. FDI stocks outward is lower than the minimum value. However FDI stocks inward, FDI inflows and outflows show relatively better opportunities. Discount rate is as maximum as of the developed countries but long term interest rate is 2.85 times higher than maximum of OECD members' interest rate. Expenditure on education as percentage of GDP is just above the minimum of OECD value. Education index is lowest. R&D expenditure as percentage of GDP is 0.67% which is just above the minimum of OECD value.

## 5 Conclusions

Lowest GDP per capita and GNI per capita are also indicative of economy conditions and earnings. Low income may contribute to add up lower literacy rates, shorter life and poor quality of life. Need to improve its GDP share from agriculture. Latest technology and methodology may be utilized for growing high value crops and increasing yield per unit area. GDP & Industrial growth rates are quite above average with respective rates of OECD countries and it puts positive impact on economy. But economies of most of OECD countries are already at saturation; therefore growth may not affect their development. Pakistan needs to keep the pace of this growth rate to catch up the developed countries. There is lot of room for this growth in every area; agriculture, industry and service sectors. Defence spending although in normal ranges of OECD countries spending, but it is due to internal situation and border conflicts. Due to low income i.e. 73% of total expenditures, current account deficit and public debts increases, but still they are in normal range. External debt is quite high. Debtors dictate their policies, which in one sense could serve well as Pakistan may follow better financial standards. Pakistan matches 76.47% with developed countries level in Macroeconomic Trends category. Pakistan needs to improve its GDP, GDP per capita, GNI per capita to higher levels by introducing financial policies, human resource management, energy management and industrialization. Negative trade balance puts more pressure on economy. One of main source for foreign reserves is expatriate remittances. Expatriates also get skill, exposure and better quality of life in the countries of their work in addition to the remittances. These remittances to their home mates also improve the quality of life in Pakistan. Although FDI stock outward is low but FDI stock inward and inward flows are indications of better investment prospects. Pakistan may form the policies and improve the law and

order situation to bring more foreign investment and increase the FDI inflows level in all sectors. This will create jobs, better quality of life and better economy conditions. Highest interest rates may be one of the reasons for slow business and industrial progress. Needs to devise incentive policies for foreign investors in addition to self industry build up, which could boost economy, GDP and all related factors. Matches 72.72% with developed countries level in Economic Globalization category. Lowest literacy rate may be due to the lowest educational expenditure. Needs to raise S&T spending to produce more engineers & scientists for faster development. More business from foreign countries may be brought in Pakistan and this export level may be increased.
Continuous analysis and improvements are required to build effective and efficient concurrent engineering environment. Further research may be carried in other variables including population, energy, environment, excellence of life and employment.

## 6 References

[1] Bruno Amable, Lilas Demmou and Donatella Gatti, "Institutions, unemployment and inactivity in the OECD countries" *Paris-Jourdan Sciences Economique,* 2006

[2] Alan Bernstein, Vern Hicks, Peggy Borbey, Terry Campbell, "A framework to measure the impact of investments in health research" *OECD Blue Sky,* September 25, 2006

[3] Nils Borsen, "Capacity Development Learning, Capacity Development of Country Systems" *OECD DAC* Issue Brief 1, 2009

[4] Koichiro Tanaka, Yuka Kimura, Ichiro Okumura, "Country Assistance Evaluation of Pakistan, summary report" *International Development Center of Japan* March 2004

[5] Maxime Kennet, Ronald Steenblik, "Environmental Goods and Services, A Synthesis of Country Studies" *OECD Trade and Environment Working Papers* 2005/3.

[6] Secretary General OECD, "Making Poverty Reduction Work: OECD's Role in Development Partnership" *OECD* 2005.

[7] Sajjad Akhtar, M. Naeem Sarwar, "Water Availability and Poverty Among Non-Farm Rural Households: Evidence From Sample Survey of Few Districts in Punjab" *Centre for Research on Poverty Reduction and Income Distribution (CRPRID)* August 2006.

[8] Evdokia Morise, "*Trade Facilitation Reforms in the Service of Development"* OECD Trade Policy Working Papers No. 12 2005

[9] World Bank, "Pakistan: Country Assistance Evaluation" *Independent Evaluation Group, World Bank,* Feb 17, 2006

[10] OECD, "Rising Food Prices: Causes and Consequences" *OECD* 2008

[11] Elizabeth Asiedu and Boaz Nandwa, "The Impact of Foreign Aid in Education Growth" *United Nations University-World Institute for Development Economic Research,* 2007/06

# Part II
# Sustainable Environment

# Green Transportation Strategies and Impact Evaluation Using System Dynamic Modeling

Amy J.C. Trappey[a,b], Gilbert Y.P. Lin[a,*], Jerry J.R. Ou[c], Chih-Tung Hsiao[d], Kevin W.P. Chen[b]

[a] Department of Industrial Engineering and Engineering Management, National Tsing Hwa University, Taiwan
[b] Department of Industrial Engineering and Management, National Taipei University of Technology, Taiwan
[c] Bureau of Energy, Ministry of Economic Affairs, Taiwan
[d] Department of Economics, Tunghai University, Taiwan

**Abstract.**

Owing to fierce global warming, carbon reduction and energy saving have become the common responsibility of the international community. According to the statistical report in 2006 by International Energy Agency (IEA), many developed and developing countries, with manufacturing based economy, generate much higher carbon dioxide ($CO_2$) emissions comparing to the other regions. The IEA reported statistical information reveals that there is much to be improved in reducing $CO_2$ emissions in the industrial countries. Thereafter, Taiwan government also passed the Sustainable Energy Policy on World Environment Day in 2008 and committed to improve energy efficiency, develop clean energy, and ensure stable energy supply. Specifically, Taiwan has planned 10 benchmarking programs with 35 sub-projects in 2009 and announced that 2010 is the year of energy saving and carbon reduction. The objective of this paper is to develop a cost-benefit evaluation methodology to guide low carbon policy development. First, we use the input-output analysis and location quotient methods from top to bottom level to effectively assess the organization's carbon footprint in a given region. In our in-depth case study, we apply the methods to Taiwan's Penghu County's (Taiwan's largest island) low carbon island development project. Based on the assessment of carbon emissions in the given region, this research applies system dynamics (SD) approach to construct the mathematical model showing the causal feedback relationships for the island's green transportation development. The causal feedback loop diagram is defined and used to analyze the effectiveness of required investment cost and the corresponding benefits for carbon reduction via green transportation strategy. Finally, the SD model is constructed and the policy scenarios are simulated to evaluate the time-varying impacts of proposed green transportation strategies.

**Keywords.** Input-output Analysis, System Dynamics, Carbon Footprint, Carbon Emission.

---

* Please send all correspondence to Dr. Gilbert Lin, Department of Industrial Engineering and Engineering Management, National Tsing Hua University, Hsinchu (30013), Taiwan, E-mail: d923834@oz.nthu.edu.tw. Tel: +886-3-5742651; Fax: +886-3-5722204

D. D. Frey et al. (eds.), *Improving Complex Systems Today*,
Advanced Concurrent Engineering, DOI: 10.1007/ 978-0-85729-799-0_14,
© Springer-Verlag London Limited 2011

## 1 Introduction

The effects of global warming and climate change have increased the attentions of the environmental protection consciousness. Release of greenhouse gases (GHG), especially carbon dioxide ($CO_2$), is the primary anthropogenic driver of climate change. According to the report of Intergovernmental Panel on Climate Change (IPCC) [9], if the human does not take deal with this problem, the earth's average temperature will rise 1.4-5.8 degrees Celsius (2.5-10.4 degrees Fahrenheit) during this century. Sea levels will also rise by 0.09-0.88 meters by 2100 which escalates inundation of coastal land. Therefore, it is critical to control and stabilize $CO_2$ emissions. According to the statistical report in 2006 by International Energy Agency (IEA), many developed and developing countries, with manufacturing based economy, generate much higher $CO_2$ emissions comparing to the other regions. For example, the population of Taiwan only accounted for 0.35% of the world population, but the proportion of $CO_2$ emissions is as high as 0.96% of the world emissions. Therefore, Taiwan government passed the Sustainable Energy Policy on World Environment Day in 2008 and committed to improve energy efficiency, develop clean energy, and ensure stable energy supply. Taiwan has planned 10 benchmarking programs with 35 sub-projects in 2009 and announced that 2010 is the year of promoting energy saving and carbon reduction.

One of the 35 sub-projects is called Penghu low carbon island development project. This research develops a benefit evaluation approach of green transportation strategy to support low carbon policy development and uses Penghu Island as the case study to demonstrate the proposed methodology at work. First, we use the input-output analysis and location quotient method to measure the organization's carbon footprint of the transportation sector in Penghu Island. Afterwards, this research constructs the system dynamics model with specific causal feedback loops to analyze the effectiveness of required investment cost and the corresponding benefits for the island's green transportation development. Finally, we run the the SD model simulating the scenarios and evaluating the time-varying impacts of proposed green transportation strategies.

## 2 Literature Review

Carbon footprint is the measurement of the GHG emissions volume caused directly and indirectly by a person, organization, event, or product [3]. Recently, many researches discuss the estimation of the regional carbon footprint, such as low carbon city and low carbon island [11, 15]. The low carbon island is also called renewable energy-island or sustainable energy island. The Samso island in Denmark is a successful and well-known low carbon island of the world which transforms natural resources into useful energy [16] and solves the problem about energy self-sufficient [5]. Scholar pointed out that life cycle assessment (LCA) method can be used to assess the carbon footprint [12]. Currently, two types of LCA, process-based LCA and economic input-output LCA (IO-LCA) approaches are conventional. Process-based LCA models are more precise but time-consuming and difficult to obtain detailed inventory data. For this reason, IO-LCA models are more efficient and address the drawbacks of process-based LCA model to expand the system scope to large scale region [4, 8]. Recently, IO-LCA method has been broadly applied in the calculating of CO2 and GHG emissions [2, 13]. Acquaye and Duffy [1] used the IO-LCA method to estimate energy and GHG emissions intensity of the Irish construction sector and estimated its contribution to Irish national emissions. Ju and Chen [12] also used the IO-LCA method to evaluate the upstream carbon footprint and $CO_2$ emissions of 28 economic sectors of Chongqing area in China and identified the significant sectors that contribute the most to climate change.

The system dynamics (SD) is a modelling approach to describe the activities of a complex system over time. SD employs the various control factors of the system and observes how the system reacts and behaves in trends. Therefore, SD can be used to assist decision making (e.g., policy experiments) when systems are complex and dynamic [6]. SD is also often used to analyze and assess the environmental impact. Jin et al. [10] developed a dynamic ecological footprint forecasting platform to support policy making for urban sustainability improvement. Han and Hayashi [7] took the inter-city passenger transport in China as a case and developed a system dynamics model for policy assessment and $CO_2$ mitigation potential analysis.

## 3 Green Transportation Strategy Evaluation Methodology

The research processes of the green transportation strategy evaluation methodology are divided into three parts as shown in Figure 1. First, we use the input-output analysis and location quotient method from top to bottom level to assess the carbon emissions of the transportation industry for a given region. Afterwards, SD approach is applied to construct the mathematical model with causal feedback relationships based on the proposed green transportation strategy in a given region. Finally, different scenarios of the SD model are evaluated using simulation approach.

### 3.1 Regional Input-Output Model

There are four steps to construct the regional input-output model.

**Step1**: Construct the input-output transactions table of the transportation industry sector and its sub-sectors in a particular region. In table 1, Intermediate output (*O*) and Intermediate input (*I*) represent the flows of sales and purchases between sub-sectors. $X_{ij}$ indicates the output of sub-sector *i* to the sub-sector *j*. The total input is the sum of the Intermediate input (*I*) and the added value (*V*). And, the total output is the sum of the Intermediate output (*O*) and the Final demand (*F*) (i.e., GDP).

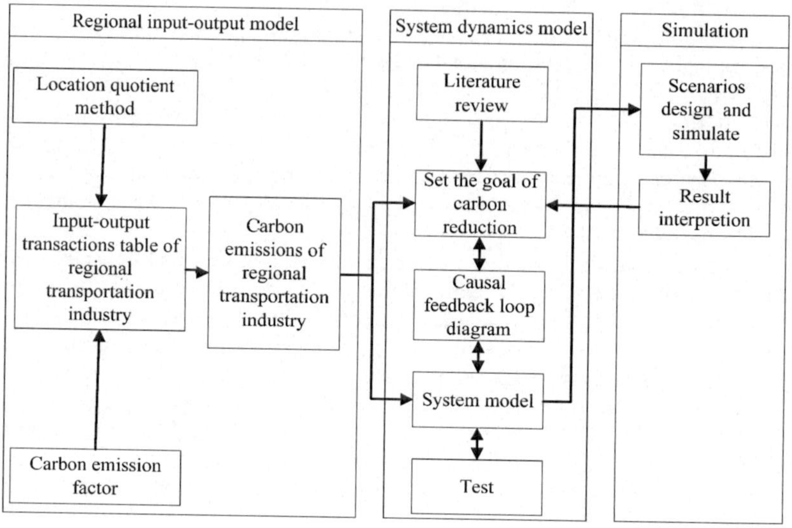


**Figure 1.** Methodology processes of green transportation development

**Table 1.** Input-output transactions table

| | Input to sectors | | | | Intermediate output $O$ | Final demand $F$ | Total output $X$ |
|---|---|---|---|---|---|---|---|
| **Output from sectors** | 1 | 2 | 3 | n | | | |
| 1 | $X_{11}$ | $X_{12}$ | $X_{13}$ | $X_{1n}$ | $O_1$ | $F_1$ | $X_1$ |
| 2 | $X_{21}$ | $X_{22}$ | $X_{23}$ | $X_{2n}$ | $O_2$ | $F_2$ | $X_2$ |
| 3 | $X_{31}$ | $X_{32}$ | $X_{33}$ | $X_{3n}$ | $O_3$ | $F_3$ | $X_3$ |
| n | $X_{n1}$ | $X_{n2}$ | $X_{n3}$ | $X_{nn}$ | $O_n$ | $F_n$ | $X_n$ |
| **Intermediate input $I$** | $I_1$ | $I_2$ | $I_3$ | $I_n$ | | | |
| **Value added $V$** | $V_1$ | $V_2$ | $V_3$ | $V_n$ | | GDP | |
| **Total input $X$** | $X_1$ | $X_2$ | $X_3$ | $X_n$ | | | |

**Step2:** After constructing the input-output transactions table, the technical coefficient matrix $a_{ij}^{\ n}$ is derived from Table 1 using the following formula:

$$a_{ij}^{\ n} = \begin{bmatrix} X_{11}/X_1 = a_{11} & X_{12}/X_2 = a_{12} & \hbar & X_{1n}/X_n = a_{1n} \\ X_{21}/X_1 = a_{21} & X_{22}/X_2 = a_{22} & \hbar & X_{2n}/X_n = a_{2n} \\ \hbar & \hbar & \hbar\hbar & \hbar \\ X_{n1}/X_1 = a_{n1} & X_{n2}/X_2 = a_{n2} & \hbar & X_{nn}/X_n = a_{nn} \end{bmatrix}$$

This matrix $a_{ij}^{\ n}$ is the national technical coefficient matrix and it represents the quantity of input by sub-sector $i$ when sub-sector $j$ ouput a unit product or service.
**Step3:** Use the location-quotient method to effectively assess the regional technical coefficient matrix $a_{ij}^{\ r}$. The location-quotient analysis is used to compare levels of employment between two geographic areas in order to gauge the concentration of a particular good or service. The location-quotient number ($LQ_i$) can be expressed as the following equation:

$$LQ_i = \frac{E_{ir}/E_r}{E_{in}/E_n}$$

$E_{ir}$ and $E_{in}$ are the quantity of employment of $i$ industry in a given region or nation respectively. $E_r$ and $E_n$ are the total quantity of employment in a given region or nation respectively. If the $LQ_i$ is greater than one ($LQ_i$>1), it implies that sector $i$ is more concentrated in the region than nation. Therefore, the regional coefficients are the same as the nation. If the $LQ_i$ is less than one ($LQ_i$<1), it is assumed that the region as being less able to satisfy regional demand for its output. Therefore, the national coefficients are needed to be adjusted by multiplying by $LQ_i$ to acquire the regional coefficients. The formulas are shown as below [14]:

$$a_{ij}^r = a_{ij}^n \quad \textit{if}\ \mathrm{LQ}_i > 1$$

$$a_{ij}^r = a_{ij}^n \times \mathrm{LQ}_i \quad \textit{if}\ \mathrm{LQ}_i < 1$$

$a_{ij}^n$ is the national technical coefficient and $a_{ij}^r$ is the regional technical coefficient matrix. Finally, we can estimate the regional technical coefficient matrix $a_{ij}{}^r$ of the transportation industry sector in a particular region.

**Step4**: Use the Leontief Inverse matrix $(I - a_{ij}{}^r)^{-1}$ and the following formulas to assess the quantity of CO2 emissions by the transportation industry in a particular region.

$$x = (I - a_{ij}{}^r)^{-1} y$$

$$b = Rx$$

Where $x$ is the direct vector of required inputs, $y$ is the vector of the desired output, $I$ is the identity matrix, $b$ is the vector of the environmental burden, and $R$ is a matrix with diagonal elements representing the emissions per output dollar (i.e., the coefficients of carbon emissions) from each sector. The carbon emission coefficients for all industrial sectors ($R$) are obtained from the government's Energy Bureau at the Ministry of Economic Affairs, the IPCC and other relevant secondary sources. Therefore, we use the input-output analysis and location quotient methods from top to bottom level to effectively assess the transport organization's carbon footprint in a given region.

### 3.2 System Dynamics Model and Simulation

In order to understand the $CO_2$ emissions of a administrative region's transportation sectors and the corresponding benefits from the green transportation policy, this research constructs a SD model. The procedure of the SD model is shown as Figure 2. In the preliminary analysis, it is necessary to identify the boundary of the system and define the internal and external variables, especially the feedback casual loops of the variables. Based on the results of preliminary analysis, the system structure is constructed in the specified analysis. And, the coefficients and equations are specified to conduct a simulation process quantitatively. In the comprehensive analysis, the simulation results from different scenarios of the green transportation strategies are estimated and compared, and relevant conclusions and policy suggestions are summarized.

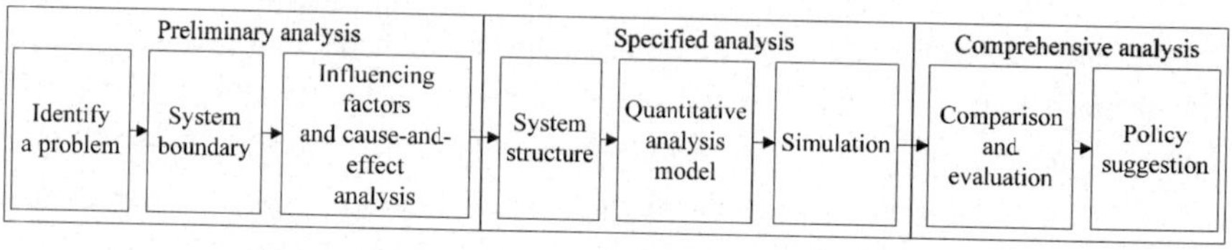


**Figure 2.** The procedure of system dynamics modeling

# 4 Case Study: Penghu Low Carbon Island Implementation

This research use the implementation case in Penghu island (Taiwan's largest offshore island) as the case study to demonstrate the proposed methodology. Penghu is the only government county of Taiwan which, as an island, locates on the Taiwan Strait between Mainland China and Taiwan. In odrer to promote local economic development as an natural resort and an environmental friendly island, Penghu County government carried out a low carbon island development project. This project combines with local features and several kinds of low carbon actions and applications to create a clean energy-saving life. This research focuses on green transportation policy and evaluates its benefit of $CO_2$ emissions reduction.

In Penghu island, the government implements three kinds of green transportation strategies, including (1) changing two-stroke gasoline motorcycles into electric scooters, (2) substituting public diesel buses into hybrid electric vehicles, and (3) promoting biodiesel vehicles with some incentive programs. This research designs the corresponding scenarios according to the above three strategies to constructs the SD model. First, we use the input-output analysis and location quotient method to assess the transportation sectors' carbon footprint in Penghu area before carrying out the green transportation policy. Based on the assessment of current carbon emissions, the SD mathematical model is built. Figure 3 shows a causal feedback loop diagram used to analyze the effectiveness of required investment cost and the corresponding benefits for carbon reduction via green transportation strategy. In Penghu area, the transportation demand is affected by its population and tourist growth rate which influences the quantities of diesel vehicles and motorcycles. Therefore, the total $CO_2$ emissions caused by diesel vehicles and motorcycles are estimated. Afterwards, the SD model is constructed based on the causal feedback loop diagram as shown in Figure 4. Further, the simulation can be implemented by Vensim software to evaluate the time-varying impacts of proposed green transportation strategies. The result of the system dynamics simulation can be used to assist the government for proper resources allocation (e.g., required investment cost) and accurate benefit evaluation (e.g., carbon reduction).

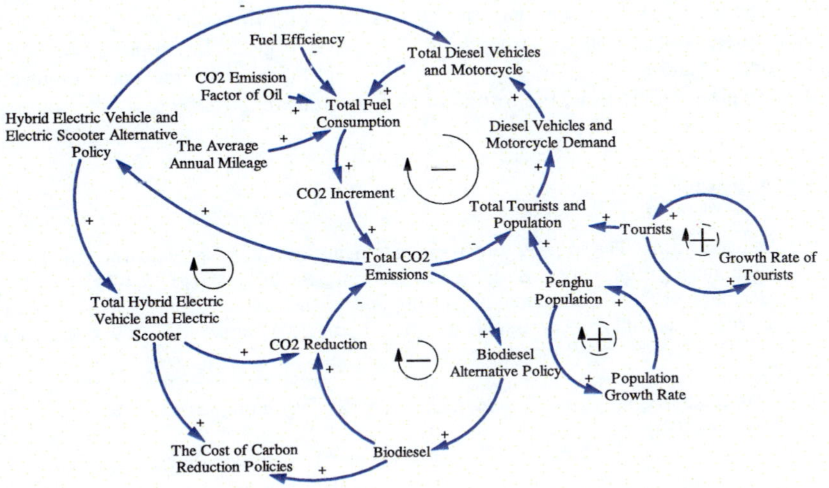


**Figure 3.** Causal feedback loop diagram with respect to a green transportation strategy

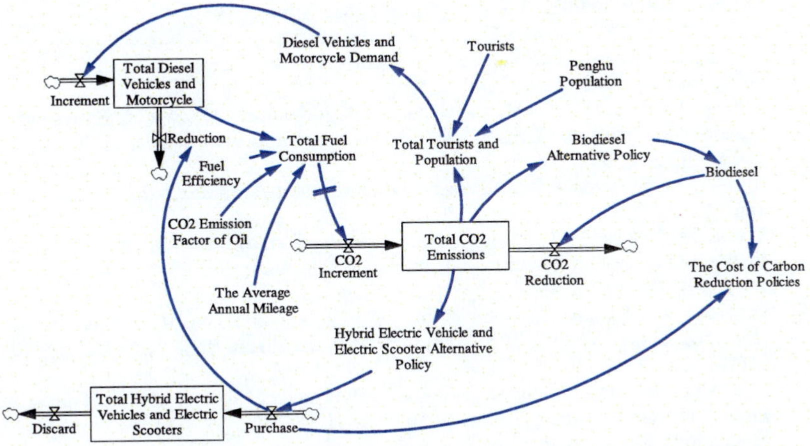


**Figure 4.** A SD model of the green transportation

## 5 Conclusions

This study evaluates the carbon footprint of an administrative region's (e.g., Taiwan's biggest offshore island, Penghu) transportation sector applying a cost-effect macro view. The approach, using the economiv input-output (EIO) analysis and location quotient method, is different from the ISO 14064 GHG microassessment approach. Further, this research develops a specific system dynamics model to assess the carbon emissions influencing and influenced by factors related to the green transportation strategy in the given region. It supports the local government making the right decisions with quantitative analytical data about the cost of implementation and the results of carbon footprint reduction.

## 6 References

[1] Acquaye AA, Duffy AP. Input–output analysis of Irish construction sector greenhouse gas emissions. Building and Environment, 2010; 45(3): 784-791.
[2] Alcántara V, Padilla E. Input–output subsystems and pollution: an application to the service sector and CO2 emissions in Spain. Ecological Economics 2009, 68(3): 905-914.
[3] Carbon Trust. Available at: <http://www.carbontrust.co.uk/cut-carbon-reduce-costs/calculate/carbon-footprinting/pages/carbon-footprinting.aspx>. Accessed on: Nov. 29th 2010.
[4] Chang Y, Ries RJ, Wang Y. The embodied energy and environmental emissions of construction projects in china: An economic input-output LCA model. Journal of Energy Policy, 2010; 38(11): 6597-6603.
[5] Chen F, Duic N, Alves LM, Carvalho MG. Renewislands: Renewable energy solutions for islands. Renewable and Sustainable Energy Reviews, 2007; 11(8): 1888-1902.
[6] Forrester JW. Principles of System. In: MA: MIT Press, 1968.
[7] Han J, Hayashi Y. A system dynamics model of CO2 mitigation in China's inter-city passenger transport. Transportation Research Part D: Transport and Environment, 2008; 13(5): 298-305.
[8] Huang YA, Weber CL, Matthews HS. Categorization of Scope 3 Emissions for Streamlined Enterprise Carbon Footprinting. Environmental science & Technology 2009, 43(22): 8509-8515.
[9] IPCC. Available at: <http://www.ipcc.ch/ipccreports/tar/index.htm.> Accessed on: Sep. 20th 2010.
[10] Jin W, Xu L, Yang Z. Modeling a policy making framework for urban sustainability: Incorporating system dynamics into the ecological footprint. Ecological Economics, 2009; 68(12): 2938-2949.
[11] Liu HY, Wu SD. An assessment on the planning and construction of an island renewable energy system: A case study of Kinmen Island. Renewable Energy, 2010; 35(12):2723-2731.
[12] Ju L, Chen B. An input-output model to analyze sector linkages and CO2 emissions. Procedia Environmental Sciences, 2010; 2: 1841-1845.
[13] Machado D, Schaeffer R, Worrell E. Energy and carbon embodied in the international trade of Brazil: An input-output approach. Ecological Economics, 2001; 39(3): 409-424.
[14] Miller RE, Blair PD. Input–Output Analysis: Foundations and Extensions. In: New Jersey: Prentice-Hall, 1985.
[15] Phdungsilp A. Integrated energy and carbon modelling with adecision support system: Policy scenarios for low-carbon city development in Bangkok. Energy Policy, 2010; 38: 4808-4817.
[16] Samso energy academy: Renewable energy Denmark. Available at: <http://www.onlinepdf.dk/Books/onlinepdf.aspx?onlinepdf=24054>. Accessed on: Nov. 21th 2010.

# System Dynamics Evaluation of Renewable Energy Policies

Charles V. Trappey[a], Amy J.C. Trappey[b,c], Gilbert Y.P. Lin[b,]* and Yu-Sheng Chang[b]

[a]Department of Management Science, National Chiao Tung University, Taiwan
[b]Department of Industrial Engineering and Engineering Management, National Tsing Hua University, Taiwan
[c]Department of Industrial Engineering and Management, National Taipei University of Technology, Taiwan

**Abstract**

The high consumption of fossil fuels (e.g., petroleum and natural gas) followed the industrial revolution and the introduction of new technologies and modern transportations. The high utilization of natural resources increases the environmental impact and climate change become a critical problem all over the world. Taiwan has set an aggressive target and proposed a set of greenhouse gas (GHG) control strategies to reduce carbon dioxide ($CO_2$) emissions. Currently, Taiwan, similar to most developed nations, is highly dependent on thermal power which accounts for 70% of total primary energy supply in 2009. The data show that Taiwan needs to actively seek the renewable energy sources and promote effective policies to reduce carbon emissions, such as policies to promote the development of the solar energy industrial sector and utilization of solar energy products. The objective of this paper is to develop a dynamic cost-benefit evaluation method for administrative regions to review the effectiveness of their renewable energy policies. This paper develops system dynamics (SD) models to construct and simulate causal feedback relationships of solar energy applications (e.g., photovoltaics (PV) systems and solar water heating or lighting systems). The SD models, with causal feedback loops considering the renewable policies, have been developed and verified as workable with environmental and economic development experts. The results describe the relationship between the renewable policies, the impacts of economy, and the effects of carbon emissions. Afterward, the SD models are run with simulated data to analyze the required costs and related effects of carbon reduction policies. This research provides a formal methodology, reference models, and analytical results for evaluating renewable energy policies and strategies in different economic regions with different input and output factors.

**Keywords.** Solar Energy Applications, System Dynamics, Carbon Reduction, Greenhouse Gas (GHG) Emissions.

* Please send all correspondence to Dr. Gilbert Lin, Department of Industrial Engineering and Engineering Management, National Tsing Hua University, Hsinchu (30013), Taiwan, E-mail: d923834@oz.nthu.edu.tw. Tel: +886-3-5742651; Fax: +886-3-5722204



D. D. Frey et al. (eds.), *Improving Complex Systems Today*,
Advanced Concurrent Engineering, DOI: 10.1007/ 978-0-85729-799-0_15,
© Springer-Verlag London Limited 2011

# 1 Introduction

Industrial development has increased the demand for fossil fuels (such as coal, petroleum, and natural gas). Rising consumption of fossil fuel increases greenhouse gas (GHG) emissions and, based on the assessment report by the Intergovernmental Panel on Climate Change (IPCC) in 2007, GHG emissions may be 14 times higher than the level in the year 2000. IPCC predicts global temperature rise is very likely (>90%) linked to GHGs [4]. Therefore, under these conditions, many international conventions were held to discuss these issues including the Kyoto Protocol of United Nations Framework Convention on Climate Change (UNFCCC), which aims to reduce the emission of GHGs and to mitigate climate change [12]. This protocol was initially proposed in December 1997 and came into effect in February 2005. Until 2010, 191 nations have signed and ratified the protocol. In order to participate in the global actions, Taiwan has set an aggressive target and proposed a set of GHG control strategies to achieve a sizable reduction of carbon dioxide ($CO_2$) emissions. The target of the policies is to keep $CO_2$ emissions during the 2016~2020 periods at the 2008 level, to reduce them to the 2000 level in 2025, and to cut that level in half by 2050.

The carbon footprint is a popular issue that has many different definitions and influence ranges. The Taiwan Bureau of Energy in Ministry of Economic Affairs (MOEA) indicates that the carbon footprint measures the total GHG emissions produced by direct and indirect activities in the entire life cycle of a product or service [1]. In general, the carbon footprint considers six categories of GHGs, including carbon dioxide ($CO_2$), methane ($CH_4$), nitrous oxide ($N_2O$), perfluorocarbons (PFCs), hydrofluorocarbons (HFCs) and sulphur hexafluoride ($SF_6$). The carbon footprint is expressed in tonnes of carbon dioxide equivalent ($CO_2e$) and is calculated by summing the quantity of each kind of GHGs emissions which are multiplied by their global warming potential (GWP). This value is then compared on a unified basis. For example, the quantity of $CO_2$ emissions is multiplied by 1, and the quantity of $CH_4$ emissions is multiplied by 25, etc. The carbon footprint (CF) applications are divided into four levels, including personal CF, product CF, enterprise CF, and regional CF (e.g., a given country or a city). The concept of low carbon region means an economy takes 3Es dimensions into consideration (i.e., energy, economic development, and environment) and does not highly depend on the combustion of fossil fuels [6]. Therefore, several countries have focused on developing low carbon islands attempting to utilize renewable energy sources and, reduce $CO_2$ emissions effectively in controlled island regions. Some interesting low carbon island projects are well known, such as Gokceada Islands in Turkey [2], Kinmen Island in Taiwan [5], Dodecanese Islands in Greece [7], and Yakushima Island in Japan [11].

Taiwan, similar to most developed nations, is highly dependent on thermal power which accounts for 70% of total primary energy supply in 2009. The other carbon-free fuels contribute merely 30% of energy supply, such as nuclear power (16%), hydropower (14%), and wind power (< 1%) [1]. These data show that the Taiwan government needs to actively increase renewable energy adoption and promote effective policies to reduce the carbon emissions. For example, the current policies promote the development of solar energy industrial sector and utilization of solar-energy products. In the current international solar market, the prices of solar system modules decline 5% steadily per year which has accelerated the promotion, popularity, and adoption of solar energy applications [9]. The solar energy industry is roughly divided into photovoltaic (PV) industry subsector and solar thermal industry subsector. Related research shows that the top five PV-producing countries are Japan, China, Germany, Taiwan, and the United States [10]. In addition, the manufacturing processes of the solar energy industry are similar to the thin film transistor liquid crystal display (TFT-LCD), light-emitting diode (LED), and semiconductor industries which are mature industries for low-cost mass production in the country such as Taiwan. Therefore, Taiwan is suitable to develop and promote its solar energy applications in areas such as public infrastructure (e.g., green transportations, solar street lightings) and domestic applications (e.g., water heating, solar power generating).

The objective of this paper is to develop a cost-benefit evaluation method based on system dynamics (SD) modelling for given administrative regions to evaluate the feasibility of their renewable energy policies. The SD approach is a method used to describe and analyze dynamical behaviour in the real world [3]. In this research, the SD model with causal feedback loops considering the renewable policies is constructed. The loops describe the relationship between the renewable policies, impacts of economy, and effects of carbon emissions. To prove the concept, this research uses Taiwan's Penghu low carbon island project as the case study to analyze the costs and related effects of the carbon reduction policies, particularly in promoting solar energy applications on the island.

## 2 Methodology

This paper provides a cost-benefit evaluation method for given regions to analyze the effectiveness of their renewable energy policies. First, we review the current organization's carbon footprint due to energy consumption in a given region to develop the as-is model. Subsequently, the to-be model is constructed according to the government's renewable energy policies (e.g., photovoltaics systems and solar water heating systems). This research uses the system dynamics (SD) approach to estimate the results of the proposed to-be model.

The processes of SD approach are divided into two parts, i.e., causal feedback loops and SD model construction. The causal feedback loops is used to identify the relationships between related variables (e.g., the renewable policies, impacts of economy, and effects of carbon emissions) and realize the causality in a system which we concern. Afterward, the SD model is constructed to simulate the dynamic system variation about the renewable energy policies. The SD simulation results support decision makers to evaluate the required costs and related effects of the carbon reduction. Figure 1 shows the methodology and steps for economic region carbon emissions analysis.

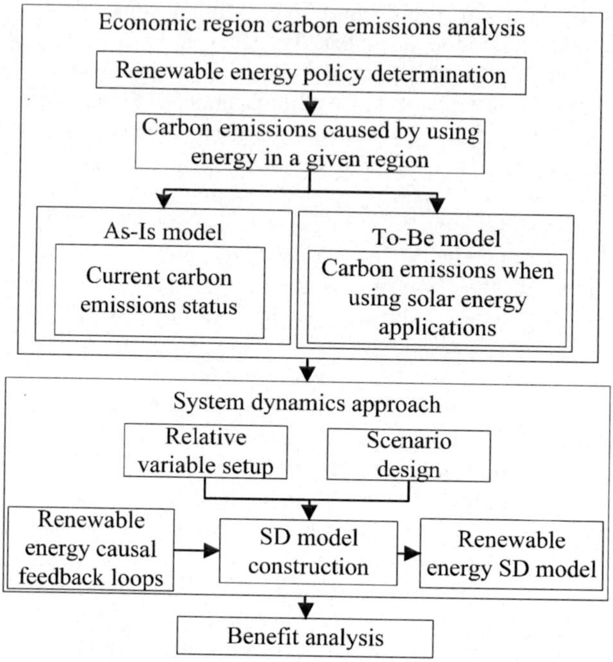


**Figure 1.** Renewable energy policies benefit analysis process

# 3 Case Study

This research uses Taiwan's Penghu low carbon island project as the case study to verify the proposed methodology. There are five major aspects in the Penghu project and one of them is the renewable energy substitute policy. For example, several solar energy applications are installed and implemented on Penghu Island to decrease the carbon emissions, and its dynamic costs and efects will be evaluated using the developed SD model.

## 3.1 Current Carbon Emissions Situation and Renewable Energy Policy of Penghu Island

Before implementing the renewable energy policy, a rough estimate of the regional carbon emissions in Penghu Island is derived from its population, which is positively correlated with current energy uses. The current carbon emissions in Penghu Island can be estimated by the following formula.

Regional carbon emissions per year = regional population × personal carbon emissions per year

According to a statistical report in 2009, Penghu Island has 96,210 residents and the personal carbon emission in Taiwan is about 10.4 tons. Therefore, there are about 1,000,000 tons carbon emissions in Penghu Island per year. However, the industry structure of the Penghu area focuses on fishery, agriculture, and tourism, which is different from the main island of Taiwan. Using Ministry of Economic Affairs (MOEA) statistics, the industry sector contributes 46% of carbon emissions in Taiwan. In order to evaluate accurate carbon emissions, the carbon emissions produced from the industry sector is deduced and the appropriate carbon emission of Penghu Island is about 540,000 tons per year.

For the Penghu low carbon island project, Taiwan plans to promote the utilization of solar-energy products, including PV systems and solar water heating systems. It is encouraged to set up the building-integrated photovoltaic (BIPV) applications allow contribution and achieve the total installed capacity to 1.5MW. In addition, the government increases 50% subsidy for encouraging residents to install solar water heating system on the roof of new or existing houses. The goal of the installation capacity of solar water heating systems is to reach 1,000 household units and the mounting area is about 5,000 square meters.

## 3.2 System Dynamics Construction

### *3.2.1 Problem Definition and System Description*

Before constructing the SD model, the problem and its target system are defined.

- Problem definition: This research focuses on the cost-benefit evaluation of solar renewable policy execution.
- System description: The SD system inputs are of two kinds, and include PV system and solar water heating system. The SD system outputs are the reduction volume of carbon emissions and the cost of policy implementation.

### *3.2.2 Causal Feedback Loops*

Causual feedback loops define the relationship between system variables, direction of variable action, system structure and boundaries. The most important feature is that the causal feedback loops can be used to estimate the causal relationships between related variables [8, 13]. Figure 2 shows the causal feedback loops considering the renewable policies on Penghu Island. The loops describe the relationship between the renewable policies, the impact of economy and effects of carbon emissions. The population using electricity on Penghu Island will be affected by its growth rate. The population using electricity and GDP will affect the demands of electric equipment (e.g., refrigerator and 3C products) which increase generated energy and carbon emissions. In order to control the carbon emissions, Penghu Island's government promotes the renewable energy policies which stimulates the demands of PV and solar water heating systems. The carbon emission volume is reduced due to the installation of the solar energy applications. However, the installation of the solar energy equipments will deplete the policy funds and decrease GDP in the future. In Figure 2, the causal feedback loops are balanced and then converged over a period of time. These results help determine whether the renewable energy policies can help Penghu Island to achieve a stable carbon emissions situation.

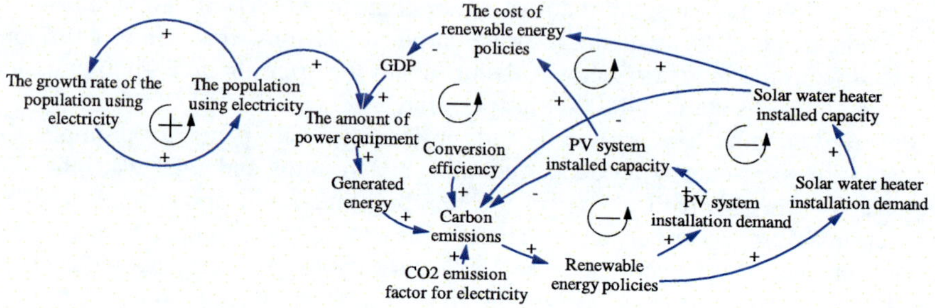


**Figure 2.** The renewable energy policies causal feedback loops

*3.2.3 System Dynamics Model Construction*

The SD model is constructed based on the above-mentioned causal feedback loop diagram to evaluate the renewable energy policies. The SD model includes stocks, flows, connectors and auxiliaries. The renewable energy policies in the SD model has four key stocks, including the poplation using eletricity, the installation capacity of the solar energy applications, policy costs, and $CO_2$ emissions volume. The development status of the renewable energy policies are evaluated by the variation of stocks and their flows in the SD model. For example, the population using electricity is a stock which is influenced by the growth rate of the population. This affects the demands for solar applications which is also the target of the installation capacity in the policy. A cycling effect on the implemented costs and the $CO_2$ emissions in Penghu Island is produced. Figure 3 shows the SD model for renewable policies which can be used to analyze the required implemented costs and related effects of the carbon reduction through simulation.

## 3.3 Proposed Solar Application Policies

Taiwan proposes the Penghu low carbon island project to demonstarte low carbon emmission policies and actions. In respect to the implementation of the solar application policy, this research plans a time schedule to match the phase goals of the national sustainable energy policy as shown in Table 1.

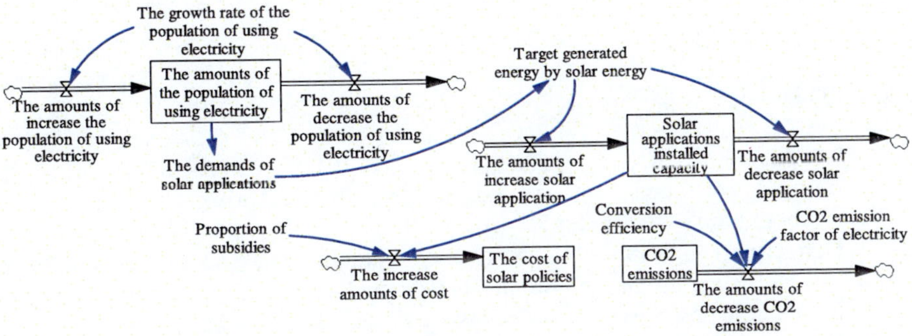


**Figure 3.** The renewable energy policies SD model

**Table 1.** Schedule plan for solar application policies

| Term | Policy target |
|---|---|
| Short-term (2011~2015) | Installing PV system (e.g., BIPV) to achieve the total installed capacity to 1.5MW. In addition, installing the solar water heating systems which can be integrated into the building to achieve the installed capacity to 1,000 household units. The solar water heating system mounting area is about 5,000 square meters and the subsidy increase 50% to encourage residents to install solar water heating system. |
| Medium-term (2016~2020) | Increasing the installed quantities of PV and solar water heating systems, and determining the installation capacity that will keep the 2016~2020 carbon emissions at the 2005 level. |
| Long-term (2021~2025) | Increase the installed quantities of PV and solar water heating systems continually, and determine the capacity that will reduce carbon emissions to the 2000 level by 2025. |

## 4 Conclusions

This research provides a formal methodology, reference models, and solid analytical results in evaluating renewable energy policies and strategies for given economic regions. We construct a SD model with causal feedback loops to analyze the implementation costs and the effects of carbon emissions reduction for the low carbon island project. The scope of the solar applications focuses on PV system and solar water heating system. The results of the proposed methodology can be used to support decision making for the government to implement the renewable energy policies in different regions with changing factors.

## 6 References


[1] Bureau of Energy (BOE). Available at: <http://www.moeaboe.gov.tw/English/english_index.aspx?Group=4>. Accessed on: Feb. 9th 2011.

[2] Demiroren A, Yilmaz U. Analysis of change in electric energy cost with using renewable energy sources in Gökceada, Turkey: An island example. Renewable and Sustainable Energy Reviews, 2010; 14(1): 323-333.

[3] Forrester JW. Industrial Dynamics. MIT Press, 1961.

[4] IPCC, 2007, IPCC Fourth assessment report (AR4), Available at: < http://www.ipcc.ch/publications_and_data/publications_ipcc_fourth_assessment_report_synthesis_report.htm >. Accessed on: Dec. 29th 2010.

[5] Liu HY, Wu SD. An assessment on the planning and construction of an island renewable energy system: A case study of Kinmen Island. Renewable Energy, 2010; 35(12): 2723-2731.

[6] Nakata T, Silva D, Rodionov M. Application of energy system models for designing a low-carbon society. Progress in Energy and Combustion Science, 2010; (in press).

[7] Oikonomou EK, Kilias V, Goumas A, Rigopoulos A, Karakatsani E, Damasiotis M., et al. Renewable energy sources (RES) projects and their barriers on a regional scale: The case study of wind parks in the Dodecanese islands, Greece. Energy Policy, 2009; 37(11): 4874-4883.

[8] Shin KH, Kwon IH, Lee JH, Kim CO. Performance trajectory-based optimised supply chain dynamics. International Journal of Computer Integrated Manufacturing, 2010; 23(1): 87-100.

[9] Srinivasan S. The Indian solar photovoltaic industry: A life cycle analysis. Renewable and Sustainable Energy Reviews, 2007; 11(1): 133-147.

[10] Susman GI. Evolution of the solar energy industry: Strategic groups and industry structure. Portland International Conference on Management of Engineering and Technology, 2008: 2489-2498.

[11] Uemura Y, Kai T, Natori R, Takahashi T, Hatate Y, Yoshida M. et al. Potential of renewable energy sources and its applications in Yakushima Island. Renewable Energy, 2004; 29(4): 581-591.

[12] United Nations. Kyoto protocol to the united nations framework convention on climate change. UNFCCC, 1998.

[13] Yim NH, Kim SH, Kim HW, Kwahk KY. Knowledge based decision making on higher level strategic concerns: System dynamics approach. Expert Systems with Applications, 2004; 27(1): 143-158.

# Initial sizing model of PV/battery system with battery scheduling algorithm

Kazuki Maeda[a], Yuki Imanishi[a,1], Kenji Tanaka[a]

[a] Graduate School of Engineering, The University of Tokyo, Dept. of Systems Innovation

**Abstract**

This study presents an optimal sizing model for PV/battery systems with battery scheduling algorithms. We propose multistep leveling algorithms to determine the half-hourly charge/discharge schedule of a battery. The purpose of these scheduling algorithms is to utilize the available solar energy and to reduce the electricity cost for dwelling houses. The model calculates the optimal size of a PV/battery system in terms of payback time. The following three cases are simulated; more than 25% reduction of $CO_2$ emission, more than 50% reduction of $CO_2$ emission and unconstrained $CO_2$ reduction.

**Keywords.** Smart grid, Initial design, Storage, Battery, Renewable energy

## 1. Introduction

For several years, there has been a growing interest in renewable resources because they are continually available nearly anywhere in the world. Accordingly, there have been a number of studies on the optimization and sizing of renewable resources systems. Borowy et al.[1] proposed graphical construction technique that changes the number of panels and battery systems. Protogeropoulos et al.[2] presented an optimization model for PV-wind-battery systems that varies the size of the battery until autonomy is ensured.. Koutroulis et al.[3] used genetic algorithms for optimizing the size of these systems. Tina et al.[4] presented a probabilistic approach based on the convolution technique. Yang et al.[5][6] proposed an iterative optimization technique based on the probability for loss of power supply.

Many researchers have studied control strategies too because of the necessity to determine when a battery will be charged or discharged and which battery or supply on the grid should have priority to deliver electricity when the demand surpasses the energy generated by renewable resources. Barley et al.[7] proposed three control strategies for PV-diesel-battery systems, namely, the zero-charge strategy, full cycle-charge strategy and predictive control strategy. Barley et al.[8] developed the operational models and proposed four strategies, namely, the frugal dispatch strategy, load following strategy, SOC_setpoint strategy and diesel

1 Corresponding Author E-mail: imanishi@triton.naoe.t.u-tokyo.ac.jp

D. D. Frey et al. (eds.), *Improving Complex Systems Today*,
Advanced Concurrent Engineering, DOI: 10.1007/ 978-0-85729-799-0_16,
© Springer-Verlag London Limited 2011

operation strategy. Seeling-Hochmuth[9] applied the genetic algorithm to realize an optimal operation system. Bo et al.[10] used a Lagrangian relaxation-based optimization algorithm for grid-connected PV/battery system. Jeong et al.[11] used a fuzzy logic algorithm to decide when the fuel cell ought to supply electricity.

This study proposes an initial sizing model for a PV/battery system with a battery scheduling algorithm. Specially, a multistep leveling algorithm is proposed for scheduling. This algorithm seeks to level the electrical supply and decrease the peak demand from the grid.

## 2. Proposed algorithm

### 2.1 Model of grid-connected PV system with battery storage

The model of a power supply system with a battery scheduling algorithm is shown in Fig.1.

The grid supplies electricity to the distributor. At the same time, electrical power generated by PV is supplied to both the distributor and battery, while the battery is charged /discharged by the distributor. Some electricity loss should be taken into account by the schedule of charging/discharging. $e_{PB}$ denotes the electrical efficiency from PV to the battery; $e_{PD}$, the electrical efficiency from PV to the distributor ;and $e_{BD}$, the electrical efficiency between the battery and the distributor.

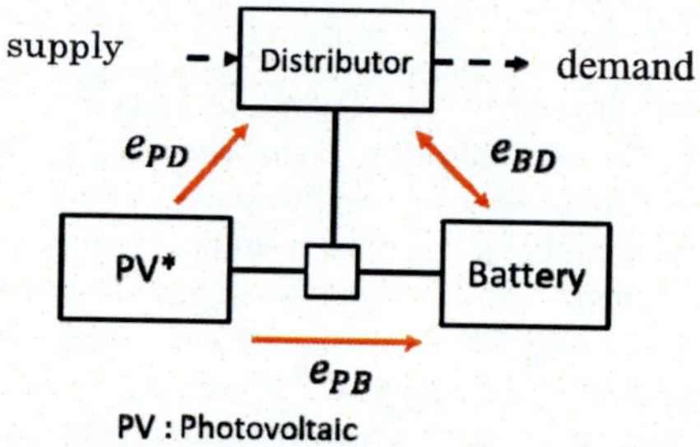


Fig.1 Model of grid-connected PV system with battery storage

### 2.2 Multistep leveling algorithm

The multistep leveling algorithm is a battery scheduling algorithm to level the electricity supply and decrease the peak demand from the grid. Fig.2 shows transitions of the electricity generated by PV, the electricity consumption and the electrical supply from the grid, before introducing the multistep leveling algorithm. Fig.3 shows the same transitions after introducing the algorithm. Fig.2 and fig.3 demonstrate how well the multistep leveling algorithm levels the electrical supply from the grid by charging/discharging the battery.

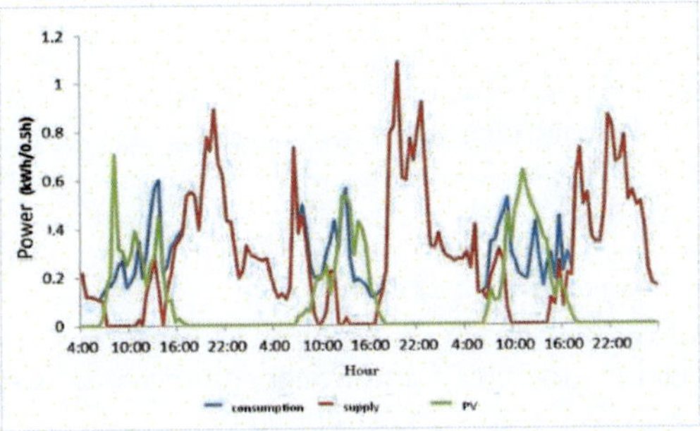


**Fig.2.** Production and grid dependence before introducing the algorithm

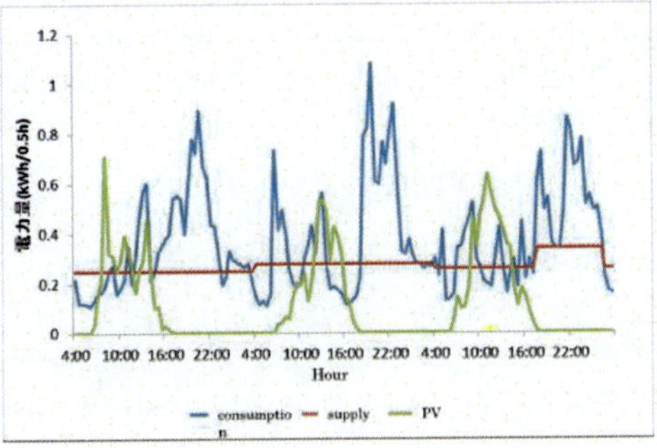


**Fig.3.** Production and grid dependence after introducing the algorithm

The multistep leveling algorithm decides the battery schedule according to the following three steps:

1.Setting up the target leveling line

We define the net demand as the electrical demand subtracted by the electricity generated by PV, taking electrical efficiency into account. This is expressed at time $t$ as:

$$netDemand_t = demand_t - e_{PD} \cdot reSupply_t \quad (1)$$

where $netDemand_t$ is the net demand, $demand_t$ is the electrical demand and $reSuppy_t$ is the electricity generated by PV. We assume the prediction of the electrical demand and PV output are given.

Next, the targeted leveling line is set as the average net demand over a scheduling span. This can be described by:

$$line = \frac{1}{span} \sum_{t=0}^{span-1} netDemand_t \quad (2)$$

where *line* is the targeted leveling line(kWh).

2.Calculating the divided steps

We consider the net demand as wavelike and define the net demand above the targeted leveling line as the crest of the wave and that below the targeted leveling line as the trough. $f_t$ is defined as the net demand subtracted by the targeted line, as follows:

$$f_t = netDemand_t - line \quad (3)$$

Accordingly, the time $t_k$ [$k$ = 0, 1, ..., n; $t_0 < t_1 < \cdots < t_n$] at which the magnitude of net demand changes with respect to its targeted leveling line is given by

$$f_{tk-1} \cdot f_{tk} < 0 \quad (4)$$

3.Leveling each crest and trough of a wave

The net demand is divided into the crests and troughs at time $t_k$, and the battery schedule is planned for the crests and troughs to reach the targeted leveling line. The span for the $k$-th crest or trough is from $t_{k-1}$ to $t_k$.

A) In the case of a crest

In the case of leveling crests, the objective is for the maximum power supply from the grid(high) to decrease as much as possible and ideally be equal to the targeted leveling line. After the maximum power supply has been planned, supply at time t, $supply_t$, is described as

$$supply_t = \begin{cases} netDemand_t \; [netDemand_t \leq high] \\ high \; [netDemand_t > high] \end{cases} \quad (5)$$

$nd_i[i \in N, 1 \leq i \leq t_k - t_{k-1}]$ is the $netDemand_t$ $[t_{k-1} \leq t \leq t_k - 1]$ arranged in ascending order. The values of $nd_i$ are then checked, the least possible maximum power supply within the capability of the battery to discharge is calculated.

B) In the case of a trough

In the case of leveling troughs, the objective is for the minimum power supply from the grid(low) to increase as much as possible and ideally be equal to the targeted leveling line. After the minimum power supply has been planed, the power supply at time t, $supply_t$, is described as

$$supply_t = \begin{cases} netDemand_t \; [netDemand_t \geq low] \\ low \; [netDemand_t < low] \end{cases} \quad (6)$$

$nd_i[i \in N, 1 \leq i \leq t_k - t_{k-1}]$ is the $netDemand_t$ $[t_{k-1} \leq t \leq t_k - 1]$ arranged in ascending order. The values of $nd_i$ are then checked and the highest possible minimum power supply within the capability of the battery to charge is calculated.

## 3. Assessment criteria

Payback time is used to assess the size of a PV/battery system and battery scheduling algorithm. The payback time is defined as the initial cost of the entire system divided by the annual savings, which in turn equal the annual merit subtracted by the annualized cost. The payback time is expressed as

$$PaybackTime = \frac{C_f}{(R_v - C_v)} \quad (7)$$

where $C_f$ is the initial cost: $R_v$. the annual merit: and $C_v$. the annual cost.

### 3.1 Annual merit

The introduction of a PV/battery system and scheduling algorithm has three prominent merits: decrease in usage based rate, decrease in contract rate, and decrease in $CO_2$ emissions

The $CO_2$ emissions and the usage based rate of electricity are reduced when renewable resources are used. Further, when the electricity demand is leveled, its peak demand drops; hence, there is a decrease in the contract rate, which is proportional to the maximum electrical supply from the grid in a period of time decrease because the peak demand drops by leveling the electricity demand. The annual merit is expressed as

$$R_v = R_m + R_c + R_{co2} \tag{8}$$

where $R_m$ is the merit of decreasing the usage based rate, $R_c$ is the merit of decreasing the contract rate and $R_{CO2}$ is the merit of decreasing $CO_2$ emissions.

#### 3.1.1. *The merit of decreasing the usage based rate*

The merit of decreasing the usage based rate is defined as the usage based rates before and after introducing the PV/battery system and scheduling algorithm. The usage based rate is calculated with fee schedule B of Tokyo Electric Power Company in Japan. This fee schedule has three price setting levels, depending on the amount of electricity used. According to fee schedule B, the function *MF(E)* used to calculate the usage based rate is expressed as

$$MF(E) = \begin{cases} E\,P_1\,, \ E \leq E_1 \\ E_1P_1 + (E - E_1)P_2\,, E_1 < E \leq E_2 \\ E_1P_1 + (E_2 - E_1)P_2 + (E - E_2)P_3\,, E > E_2 \end{cases} \tag{9}$$

where $E_1$ and $E_2$ are threshold values, while $P_1$ ,$P_2$ ,and $P_3$ are price settings levels.

The merit of decreasing the usage based rate can thus be calculated as

$$R_m = MF(E_b) - MF(E_a) \tag{10}$$

where $E_a$ and $E_b$ are the values of electric supply before and after introducing the entire systems respectively.

#### 3.1.2 *The merit of decreasing the contract rate*

The merit of decreasing contract rate is defined as the difference between the contract rates before and after introducing the entire systems. The function $CF(E_M)$ used to calculate the contract rate is described by

$$CF(E_M) = a_c E_M \tag{11}$$

where $E_m$ is the maximum electric supply and $a_c$ is constant.

The merit of decreasing the contract rate can be thus calculated as eq(12).

$$R_c = CF(E_{Mb}) - CF(E_{Ma}) \tag{12}$$

$E_{Mb}$ and $E_{Ma}$ are the values of maximum electric supply before and after introducing the entire systems.

#### 3.1.3 *The merit of decreasing $CO_2$ emissions*

The merit of decreasing $CO_2$ emissions is described by

$$R_{CO2} = S_c\,P_{CO2}(E_b - E_a) \tag{13}$$

where $S_c$ is $CO_2$ emissions intensity (kg-$CO_2$/kWh) and $P_{CO2,}$ the emissions trading costs (JPY/kg-$CO_2$).

### 3.2 Cost

The initial cost, which is the sum of the capital and replacement costs of the PV/battery system, can be expressed as

$$C_f = C_e + C_s \tag{14}$$

Where $C_e$ is the capital cost and $C_s$ is the replacement cost.

The annual cost is simply equal to the maintenance cost, i.e.,

$$C_v = C_m \tag{15}$$

where $C_m$ is the maintenance cost.

In this study, both the replacement cost and the maintenance cost are proportional to the capital cost. However, the annual maintenance cost is defined as the capital cost divided by the lifespan of the PV/battery system. Hence, the replacement cost is expressed as

$$C_s = a_s C_e \tag{16}$$

and the annual maintenance cost is expressed as

$$C_m = a_m C_e / Y_d \tag{17}$$

where $a_s$ and $a_m$ are constants and $Y_d$ is the lifespan of the system.

## 4. Case Study

### 4.1. Data input

#### 4.1.1. *Load data*

The Load data for this study are derived from the half-hourly electrical demand for a dwelling house in Japan from June to November. Table 1 shows the average demand and maximum demand as calculated for data recorded during the period.

**Table 1.** The load data for a dwelling house

| Region | Period | | Time | Average load | Maximum load |
|---|---|---|---|---|---|
| | Start | End | days | kWh | kW |
| Tokyo | 6/28 | 11/25 | 151 | 14.3 | 3.5 |

#### 4.1.2 *Environmental data*

We formulate that the electricity generated by PV is directly related to the available solar energy. The relevant solar irradiance data and temperature data were obtained from the Japan Meteorological Agency.

### 4.2. Analysis

#### 4.2.1 *Effect of battery size on the multistep leveling algorithm*

Fig.4 and Fig.5 show the result of the multistep leveling algorithm for a 6-hour-storage bank and a 18-hour-storage bank respectively. Clearly, the figures indicate that increased battery size implies superior load leveling.

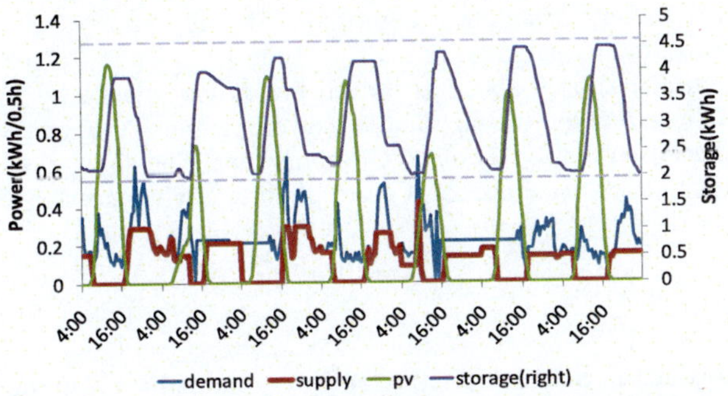


**Fig.4.** Result of the multistep leveling algorithm for a 6-hour-storage battery bank

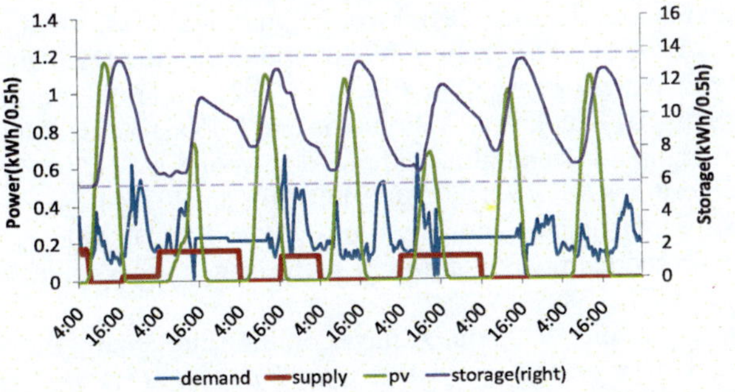


**Fig.5.** Result of the multistep leveling algorithm for a 18-hour-storage battery bank

### 4.2.2 *Optimizing the sizes of PV/battery systems*

Three sizing optimization operations were conducted in terms of the reduction of CO2emissions; an unconstraint reduction, reduction greater than 25% reduction and reduction greater than 50%. Table 2 shows the results of the three cases. The payback time is shortest when the house introduces no battery. Table 2 explains that battery capacity becomes more important when greater $CO_2$ reduction is required.

**Table 2.** Results for three cases of $CO_2$ emissions

| | Unit | No constraint | Greater than 25% | Greater than 50 % |
|---|---|---|---|---|
| PV | kWp | 0.7 | 2.8 | 4.9 |
| Battery | kWh | 0 | 0 | 10.4 |
| Payback time | year | 15.7 | 19.4 | 25.2 |
| The reduction of $CO_2$ | % | 9.3 | 31.3 | 50.5 |

## 5. Conclusion

This study presents an optimal sizing model for a grid-connected PV/battery system with a battery scheduling algorithm. The system configuration can be determined in terms of payback time. The battery scheduling algorithm aims to level the electricity supply and thus reduce the cost of electricity for a dwelling house. This design method was applied to a dwelling house in Japan, with different constraints placed on the reduction of $CO_2$ emissions. The simulation results show that the optimal battery size becomes more important when greater $CO_2$ reduction is required.

## References


[1] B. Borowy and Z. Salameh, Methodology for optimally sizing the combination of a battery bank and PV array in a wind/PV hybrid system, *IEEE Trans Energy Conver* **11** (2) (1995), pp. 367–375.

[2] C. Protogeropoulos, B.J. Brinkworth and R.H. Marshall, Sizing and techno-economical optimization for hybrid solar photovoltaic/wind power systems with battery storage, *Int J Energy Res* **21** (6) (1998), pp. 465–479.

[3] E. Koutroulis, D. Kolokotsa, A. Potirakis and K. Kalaitzakis, Methodology for optimal sizing of stand-alone photovoltaic/wind-generator systems using genetic algorithms,*Solar Energy* **80** (9) (2006), pp. 1072–1088.

[4] G. Tina, S. Gagliano and S. Raiti, Hybrid solar/wind power system probabilistic modeling for long-term performance assessment, *Solar Energy* **80**(2006), pp. 578–588.

[5] H. Yang, L. Lu and W. Zhou, A novel optimization sizing model for hybrid solar-wind power generation system, *Solar Energy* **81** (1) (2007), pp. 76–84.

[6] H. Yang, W. Zhou and C. Lou, Optimal design and techno-economic analysis of a hybrid solar–wind power generation system, *Apply Energy* **86** (2) (2009), pp. 163–169.

[7] C.D. Barley, C.B. Winn, L. Flowers and H.J. Green, Optimal control of remote hybrid power systems. Part I. Simplified model, *Proceedings of WindPower'95* Washington, DC (1995).

[8] C.D. Barley and C.B. Winn, Optimal dispatch strategy in remote hybrid power systems, *Solar Energy* **58** (4–6) (1996), pp. 165–179.

[9] G.C. Seeling-Hochmuth, A combined optimization concept for the design and operation strategy of hybrid-PV energy systems, *Solar Energy* **61** (2) (1997), pp. 77–87.

[10] B. Lu, M. Shahidehpour, Short-term scheduling of battery in a grid-connected pv/battery system, *IEEE Trans Power Sys* **20**(2)(2005), pp. 1053-1061

[11] K. Jeong, W. Lee and C. Kim, Energy management strategies of a fuel cell/battery hybrid system using fuzzy logics, *Power Sources* **145**(2005), pp. 319–326.

# Part III
# New ICT for Concurrent Engineering

# Feature Performance Metrics for Software as a Service Offering

Avi Latner[a,1] and Ricardo Valerdi[b]

[a]MIT, System Design and Management Program, Cambridge, MA, USA.
[b]MIT, Lean Advancement Initiative, Cambridge, MA, USA.

**Abstract.** This paper provides a framework to measure the performance of software as a service (SaaS) product features to prioritize development efforts. Firstly, relative value is measured by the impact of each feature on customer acquisition and retention. Secondly, feature value is compared to feature cost and specifically development investment to determine feature profitability. Thirdly, feature sensitivity is measured. Feature sensitivity is defined as the effect a fixed amount of development investment has on value in a given time. Fourthly, features are segmented according to their location relative to the value to cost trend line into most valuable features, outperforming, underperforming and fledglings. Finally, results are analyzed to determine future action. Maintenance and bug fixes are prioritized according to feature value. Product enhancements are prioritized according to sensitivity with special attention to fledglings. Underperforming features are either put on "life-support," terminated or overhauled. This framework is applicable to SaaS companies for the purpose of prioritizing features, an important consideration for concurrent engineering of software systems.

**Keywords.** Product development, performance metrics, software as a service, customer value

## 1 Introduction

In the past decade, with the increase of internet bandwidth, decrease in hardware costs, and a paradigm shift in business models, software delivery has been increasingly done through the software as a service (SaaS) model [3]. The software application is hosted at the vendor, licensed for a recurring subscription fee and accessed through a web browser. SaaS is not only a change in the business model but also a change in the product development paradigm [4].
Since software typically resides on vendors' servers, it is easier for vendors to release updates at more frequent intervals, and with agile development practices, applications are updated almost continuously without traditional version control. Software hosting also allows vendors to collect valuable information about customers' usage patterns.

[1]Massachusetts Institute of Technology, System Design and Management Program, 77 Massachusetts Avenue building E38-642 Cambridge, MA USA 02139; Tel +1 215 301 6037; Email: latner@sloan.mit.edu



D. D. Frey et al. (eds.), *Improving Complex Systems Today*,
Advanced Concurrent Engineering, DOI: 10.1007/ 978-0-85729-799-0_17,
© Springer-Verlag London Limited 2011

The available information is unprecedented in scope and immediate in availability. With a continuous deployment model and immediate customer response, the feedback loop between development and customers has never been faster.

However, in order to fully leverage the fast feedback loop, companies must use the right performance metrics. Most software suites are a collection of several features or applications. It is important to focus development efforts on the features in which the investment will make the most impact on software usage and company profitability [5, 9]. For startups, often with limited cash flow and high uncertainty, this focus is all the more important.

In a SaaS model, much like other subscription based models, vendors use relationship marketing where relationships with customers are viewed as long term instead of a series of discrete transactions [1]. Hence, the marketers' objective is to acquire and retain customers [6].

Other studies created performance measurements for R&D projects in a concurrent engineering environment [12]. This study focuses on measurements used in selection phase of a product life cycle and is meant to be part of broader concurrent engineering practices.

## 2 Methodology

This research was done using data collected at HubSpot, an online inbound marketing service. The service includes features that help websites get found online by more qualified visitors, show customers how to convert more visitors into leads, give customers tools to close those leads efficiently, and provide analytics to help make smart marketing investments. HubSpot offers several products that differ in level of support. All products offer the same 17 features. HubSpot is an interesting case study as it has a diversified service with many features, a rapidly expanding customer base and open management, and the company implements the latest development methodologies such as lean startup[11].

Usage information was collected from HubSpot's 3,000 customers over a period of four months. For each feature the percentage of users that used the feature was calculated. If a user accessed a feature at least once in the week prior to the measure then the user counted as an active user. In order to eliminate seasonal volatility the usage was based on the average of four measures taken at the end of four consecutive months.

Development cost was collected from evaluating sprint logs (short development activities) from the earliest point available. In total, 27 sprints were taken into account. The research involved the analysis of over a thousand user stories (high level functionality requirements) for which development effort was indicated in story points, an agile development measure for effort [2]. For feature sensitivity the study looked at two measurements six months apart, from sprint 21 and 27.

# 3 Feature Performance Framework

This paper proposes a framework through which to prioritize product development efforts. By following the proposed steps a company can gain awareness of relative importance of product features and examine how well past decision-making is aligned with feature importance. A step beyond that would be for a company to make future prioritization decisions in light of these findings.

Two hypotheses are examined as a base for the framework. The first hypothesis investigates customer feature usages as a predictor of customer retention. This premise is essential for capturing feature value. The second hypothesis explores the correlation between feature investment and feature value.

When this methodology is followed, a company's focus may shift causing a change in performance assessment. Features that were once underinvested may receive more investment and, as a result, may increase in value. Another feature may mature and exert its full potential suggesting a needed shift in investment. Therefore, it is - recommend that a company repeat this framework and re-evaluate the situation every time there is a significant change in the business environment (new competitor entering the market, new uses by customers, new capability available). Feature value, which mainly captures external shifts in the way the product is used, should be calculated as frequently as once a month. Measures that are also dependent on internal shifts in investment such as feature profitability and sensitivity should be calculated less frequently, perhaps once a quarter. Since investment is based on cumulative data a few months have to pass before significant change may be observed.

## 3.1 Usage As a Predictor of Customer Retention

If customers that frequently use a given product are less likely to discontinue service we can use usage data of a given feature as a proxy for the impact that the feature has on customer retention. This connection is tested by the following hypothesis:

> $H_0$: No difference exists between retention of customers who use a feature and those who do not
> $H_1$: Customers who use a feature are more likely to retain service subscription

## 3.2 Correlation Between Development Effort and Value

It is assumed a company, even without having implemented this framework, would have some understanding of feature value and would therefore strive to align development effort to feature value. This assumption is formulized as a hypothesis and tested.

> $H_0$: No correlation between development effort and feature value
> $H_1$: A positive correlation exists between development effort and feature value

### 3.3 Measuring Feature Value

SaaS product revenue is a function of the customer base and the product price. Customer base is a function of the number of new subscribers and the attrition rate. Therefore a measure of a feature value should actually be a measure of the impact that a feature has on these key parameters: customer acquisition and attrition rate.
Studies suggest that retaining existing customers is more important than acquiring new ones [7, 8]. One reason for that is that service discontinuers create a negative word of mouth that is more powerful than the positive word of mouth of continuing customers. Another reason is that new customers come at high adoption cost of sales and marketing [10]. Hence our research gives more value to retention rate impact than to acquisition impact using a weighted average of 70% and 30% respectively. The exact weight should be a managerial decision and is a way to focus a company's priorities. A company concerned with slow adoption should choose a more evenly weighted measure, closer to 50% for each parameter.
In order to measure feature effect on retention we evaluate usage data. This choice of parameter is validated through the hypothesis articulated in section 3.1. Our other measure of feature retention value is expert opinion survey done among business development and support staff within the company. Surveyed employees were asked to rate the top five most valuable features in the product to the customers. With no preference to either measure they were given equal importance.
Feature effect on customer acquisition is computed using two equally weighted measures. The first is the customer usage data in the first 30 days after subscription to the service started, since the usage in this early period reflects the features that led the customer to subscribe to the service. The second is the expert opinion of the sales representatives. The salespeople were asked to rank the five most important features in closing a sale. The equations below summarize the measure calculations.

$$value_i = 0.7 \cdot retention\ score_i + 0.3 \cdot adoption\ score_i$$
$$retention\ score_i = 0.5 \cdot usage_i + 0.5 \cdot support\ poll_i$$
$$adoption\ score_i = 0.5 \cdot early\ usage_i + 0.5 \cdot sale\ poll_i$$

Where $i$ denotes a feature out of $n$ features in a given SaaS offering.

Analyzing usage data in this way is valid in cases where all the features are client-facing, meaning that customers utilize the features by actively accessing them. When a SaaS product contains back office features, such as a product for network maintenance that has an automatic remotely triggered fix feature, a different measure must be used. Another example is user-automated reports that run without user interference. For example, Salesforce.com found a strong connection between user adoption of automated reports and retention. In their case a feature's value formula should also measure the amount of user automation.

### 3.3 Measuring Feature Profitability

Capturing cost is in principle straightforward as many of the costs are feature specific. The lion share of the cost of a feature is development costs spent on building, enhancing or debugging a given feature. Other feature-specific costs include costs of goods sold for things such as servers and storage. Cost is the accumulation of investment on a feature over the product history.
Since this paper aims at giving a practical measure we kept value and cost at relative terms by dividing 100% of value and cost amongst the features. This is the simplest way to compare value and cost without compromising accuracy. In that case profitability is the difference between a feature's value and the breakeven value. Breakeven value is defined as feature cost divided by the product gross margin. The equation set below summarizes the profitability measure.

$$profitablity_i = value_i - breakeven\ value_i = value_i - \frac{cost_i}{Gross\ Margin}$$

Where *i* denotes a feature out of *n* features in a given SaaS offering.

### 3.4 Measuring Feature Sensitivity

Feature sensitivity is defined as the effect a fixed amount of development investment has on value in a given time. It is a measure of how effective recent development investments have been in improving features. Sensitivity is a dynamic measure that captures change between two time periods. One should use two measures of value and cost that are significantly apart, perhaps four or six months. Since feature value and cost described in sections 3.2 and 3.3 are measured in relative terms the average sensitivity would be zero and many of the features will have negative sensitivity. It is our experience that overall zero sensitivity can be counter-intuitive to some business managers. To prevent that, the value in a given time can be multiplied by the growth in customer base. This way the average sensitivity score would be equal to the customer base growth rate. The overall sensitivity score will then reflect the product performance as a whole. The equation set below summarizes the sensitivity measure.

$$sensitivity_i = \frac{(N)_t}{(N)_{t-1}} \cdot \frac{(value_i)_t - (value_i)_{t-1}}{(cost_i)_t - (cost_i)_{t-1}}$$

Where *i* denotes a feature out of *n* features in a given SaaS offering, *N* denotes the number of costumers and *t* and *t*-1 denote consecutive measures.

## 4 Results

The hypothesis testing connection between customer usages and retention was applied to five features separately and on the product as a whole. The total number of customers sampled in each test was 2,843. In all cases but one, the null hypothesis was rejected at a 90% confidence level. For the product as a whole the null hypothesis was rejected at a 99% confidence level. These results mean that users who do not use a feature are much more likely to discontinue service than active users. The "List Manager" feature is an exception and that is due to a very low usage population. It is worth noting that although features were tested separately they are not necessarily independent variables. Table 1 presents the results. To protect sensitive business information, retention is stated in relative term to 'x' the attrition rate for active users of the product as a whole.

**Table1.** Usage Connection to Retention

| **Feature** | **Attrition Rate Active Users** | **Attrition Rate Non-Active Users** | **% Usage** | **Distribution Under $H_0$** | **P Value** |
|---|---|---|---|---|---|
| Blog | 0.47x | 2.89x | 30.53% | $\hat{p}$~B(2.86%, 0.005%) | ~0 |
| Content Management | 1.3x | 2.51x | 39.71% | $\hat{p}$~B(2.72%, 0.004%) | 0.0376 |
| Leads | 2.03x | 3.53x | 44.50% | $\hat{p}$~B(3.19%, 0.004%) | ~0 |
| Landing Page | 0.76x | 2.9x | 23.41% | $\hat{p}$~B(3.04%, 0.006%) | ~0 |
| Lead Nurturing | 0.77x | 2.22x | 10.15% | $\hat{p}$~B(2.76%, 0.01%) | 0.054 |
| List Manager | 0.97x | 2.58x | 2.29% | $\hat{p}$~B(3.72%, 0.05%) | 0.356 |
| Product | x | 2.69x | 82.35% | $\hat{p}$~B(1.76%, 0.004%) | 0.0003 |

The value and development effort scores in table 2 are used to test the second hypothesis presented in 3.2. The relation between value and cost is *value=0.59*development effort + 0.024*; $R^2=0.33$. The null hypothesis is rejected at a 90% confidence level showing that there is a positive relationship between value and development effort. The total profitability is 60% and is equal to the company's gross margin at the time. The sum of all sensitivity scores is 0.24 which is the growth of customer base between the two sensitivity measurements.

**Table 2.** Feature Scores

| Feature | Value | Development Effort | Profitability | Sensitivity | Segment |
|---|---|---|---|---|---|
| Leads | 14.49% | 17.08% | 8% | 0.25 | Most Valuable |
| Sources | 14.44% | 8.72% | 11% | 0.29 | Most Valuable |
| Content Management | 9.73% | 8.63% | 6% | 0.21 | Outperforming |
| Landing Page | 9.47% | 2.22% | 9% | 0.80 | Outperforming |
| Keyword Grader | 9.28% | 7.96% | 6% | 0.03 | Outperforming |
| Blog | 7.56% | 4.21% | 6% | 1.11 | Outperforming |
| Social Media | 5.14% | 8.07% | 2% | (0.05) | Underperforming |
| Competitors | 4.81% | 10.03% | 1% | 0.09 | Underperforming |
| Page Grader | 3.91% | 6.89% | 1% | (0.06) | Underperforming |
| Prospects | 3.10% | 1.86% | 2% | (0.06) | Fledglings |
| Blog Grader | 3.30% | 3.66% | 2% | 0.21 | Fledglings |
| Lead Nurturing | 3.62% | 9.75% | 0% | (0.13) | Underperforming |
| Link Grader | 2.88% | 1.96% | 2% | 0.48 | Fledglings |
| Visits by Page | 2.78% | 1.36% | 2% | 0.28 | Fledglings |
| Reach | 2.23% | 1.60% | 2% | 0.18 | Fledglings |
| Email | 1.98% | 3.74% | 0% | 0.14 | Fledglings |
| List Manager | 1.28% | 2.26% | 0% | 0.24 | Fledglings |
| **Total** | **100%** | **100%** | **60%** | **0.24** | |

## 5 Conclusions

Based on the results we segment the features. This is most easily done by looking at a scatter plot of cost on the horizontal axis and value on the vertical axis. The scatter plot should also have a line representing the gross margin and the linear regression line.

The most valuable features are a segment of features that are high in value and investment. These features are recognized as important by the company. This group should be the highest in priority for bug fixes and regular up-keep. As long as the sensitivity is positive they should also be considered for enhancements.

Outperformers are a segment of features that are doing very well relative to the investment in them. In the scatter plot described above they will appear closest to the top left corner. By contrast, underperforming is a segment containing features that in

retrospect do not justify their investment. Out of this group the features with zero or negative profitability need re-examination. If value covers the cost of goods sold and there is little maintenance development anticipated the feature could be kept on 'life support'; that is kept alive while avoiding investment as much as possible. Otherwise the feature should either be terminated or overhauled.
In the fourth segment, fledglings are features that have had little investment and provide little value. This group holds the most potential as it may include promising features that have yet to mature. For example, link grader and visits by page have an above average sensitivity and therefore they are ideal candidates for future investment.

## 6 Reference


[1] Berger D., Nasr N. Customer Lifetime Value: Marketing Models and Applications. Journal of Interactive Marketing 1998; 12:17-30.

[2] Cohn M. User Stories Applied For Agile Software Development. Boston MA. Addison Wesley, 2004: 87-96

[3] Dubey A., Wagle D. Delivering Software as a Service. The McKinsey Quarterly May, 2007.

[4] Vidyanand C. Software as a Service: Implications for Investment in Software Development. IEEE System Sciences. 2007.

[5] Hubbard D. How to Measure Anything: Finding the Value of Intangibles in Business. Hoboken New-Jersey: John Wiley and Sons, Inc. 1962; 119-136.

[6] Kotler P., Armstrong. G., Principles of Marketing, 7Th ed., Englewood Cliffs New-Jersey: Prenlice-Hall. 1996.

[7] Oliver L.. Satisfaction: A Behavioral Perspective on the Consumer. New York. McGraw-Hill, 1997

[8] MahajanM V., Muller E., Kerin R. Introduction strategies for new products with positive and negative word-of-mouth. Management Science 1984: Vol. 30, December.

[9] McGarry J., Card D., Layman B., Clarck E., Dean J. et al. Practical Software Management: Objective Information for Decision Makers. Indianapolis Indiana. Addison-Wesley, 2001; 13-26.

[10] Madhavan P., Bhattacherjee A. Understanding Post-Adoption Behavior in the Context of Online Services. Information System Research 1998: Vol. 9, No. 4, December.

[11] Ries E., Lessons Learned: What is Lean about the Lean Startup? Available at: < http://www.startuplessonslearned.com/2009/12/what-is-lean-about-lean-startup.html>. Accessed on: Feb 16, 2011.

[12] Sivathanu P., Joshi A., Srinivasa K., Performance measurement of R&D projects in a multi-project, concurrent engineering environment. International Journal of Project Management 2002: Vol. 20, February.

# Automating Services for Spacecraft Conceptual Design via an Enterprise Service Bus

Ariana C. Caetano de Souza[a], Walter A. dos Santos[b,1]

INPE – The Brazilian Institute for Space Research, Brazil
[a] MSc Student in Space Technology and Engineering - INPE
[b]Space System Division - DSE

**Abstract.** As space system designs are growing more complex and market demands bigger, technical issues related to concept development become more important and difficult. Moreover, due to pressure on cost and time, many space projects nowadays demand distribution as they undergo their development divided among several project partners, sometimes temporally and geographically separated hampering the information exchange and causing adding risks to its conception. This key phase maps client needs to product use functions and is where functional architecture (and sometimes the physical architecture) is decided upon. Typically, the design specifications and constraints impose a heavy burden on systems-of-systems engineering. This paper shows how some processes of the space mission conceptual phase can be standardized and automated using a Service-Oriented Architecture (SOA) paradigm. By employing an enterprise bus service (ESB), named SpaceESB, applications become distributed and its reuse promoted.

**Keywords.** Service-Oriented Architecture, Enterprise Service Bus, Systems Engineering

## 1 Introduction

Recent interest in collaborative environments for promoting cloud-enabled systems engineering has risen in the literature [10]. Space systems engineering demands this type of environment as it requires essentially multidisciplinary expertise ranging from onboard computing to launcher vehicle interfacing nevertheless sometimes experts may not be temporally and / or geographically co-located.

Difference factors such as time zone, language barriers, numbering and units of measurement conventions and different IT platforms may all promote an adverse effect on the project timeline and budget.

In parallel, there is a growing demand generated by clients of services provided by space systems which increases pressure for good space system engineering on

[1]INPE-DSE, Lambda Building, Av. dos Astronautas 1758, São José dos Campos, Brazil - 12227-010; Tel: +55 (12) 3208 6622; Fax: +55 (12) 3941 1890; Email: walter@dss.inpe.br

D. D. Frey et al. (eds.), *Improving Complex Systems Today*,
Advanced Concurrent Engineering, DOI: 10.1007/ 978-0-85729-799-0_18,
© Springer-Verlag London Limited 2011

delivery time reduction, higher quality and performance as well as cost reduction [1].

As space system designs are growing more complex, technical issues related to concepts development, also referred as Phase A, become more important and difficult. Conceptual design maps client needs to a functional architecture, and sometimes, even the physical architecture.

Typically, the design specifications and constraints impose a heavy burden on systems-of-systems engineering and particularly on requirements engineering which drives the whole system's life cycle. Henceforth, taking suitable decisions at this project phase ends up paying dividends on schedule, performance, cost and risk. Therefore agility and flexibility in the execution of intrinsic processes are necessary.

This highly-coupled and distributed scenario can be tackled thanks to the availability of open standards to reduce barriers between different platforms as well as infrastructure to support the creation of services. This abstraction is possible via SOA and web services [3] which are being adopted to make business processes more efficient and effective. These technologies contribute to shape business processes, create solutions, design, develop and deliver services.

This work uses the concept of Enterprise Service Bus [11], here customized and named SpaceESB, with a set of budgeting services to support the conceptual design phase of a satellite project. This allows systems engineering partners to consume services regardless of platforms. For illustration, the set of budgeting services here considered comprises of three simple services expressed by their WSDL interfaces [11] to the SpaceESB. Ultimately, this realizes a prototype environment for collaborative and distributed space systems engineering.

This paper is organized into the following. Section 2 presents a brief introduction to SOA, web services and, the concept of SpaceESB. The creation of budgeting services is briefly described in section 3. The implementation of services is shown in section 4. Finally, section 5 closes it with conclusions.

## 2 Background

As organizations grow they acquire an ever-increasing number of applications distributed across a number of departments and sites as well as sharing information between these applications. SOA arose out of this need to allow intercommunication between applications [10] as it entails developing loosely coupled interfaces to applications (services). By combining these services, it is possible to develop adhoc applications (mash-ups) as required.

### 2.1 Service-Oriented Architecture

SOA is a software architecture style whose basic principle advocates that the functionalities implemented by the applications should be delivered as services enabling greater reuse, redundancy reduction and, greater efficiency in maintenance [3]. Frequently these services are organized through a "service bus" [11] that provides interfaces, or contracts, accessible via web services or another form of communication between applications. The SOA is based on principles of distributed computing paradigm and uses the request / reply to establish the communication between client systems and the systems that implement these services [14].

## 2.2 Web Services

The web services are one of the possible ways of realizing the SOA abstraction as they can integrate applications through messages based on XML (eXtensible Markup Language) usually employing HTTP (Hypertext Transfer Protocol).

There are many types of web services, but the most known and well-used are: RPC, WS-* (WSDL / SOAP) and REST. The WS-* architecture [9] is the mostly used nowadays. The main specifications for this architecture are SOAP (Simple Object Access Protocol), WSDL (Web Services Definition Language) e UDDI (Universal Description, Discovery and Integration) [3].

The service description, and how to access it, is defined by a document that uses the WSDL language. The registration of a service and its location is defined by the UDDI. Thence for service publication and consumption, as Figure 1 illustrates, client and service can exchange messages enveloped on the SOAP protocol and transmit data represented in XML.

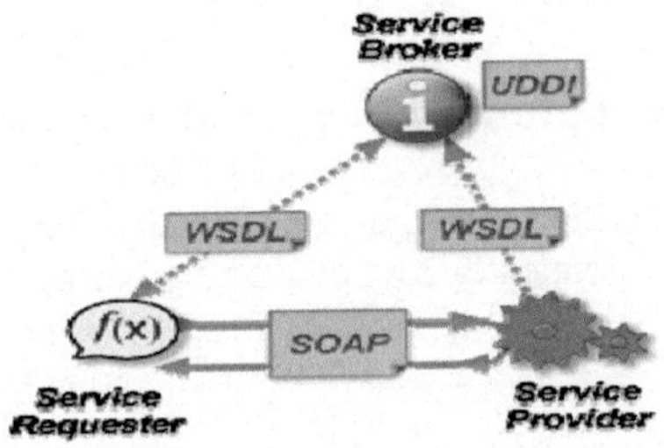


**Figure 1.** Basic scheme of web services and its messaging system

## 2.3 Enterprise Service Bus and the SpaceESB Concept

An ESB is an infrastructure that enables high interoperability between services, allowing exposed services to be consumed by clients. Its layout is sketched in Figure 2. The main responsibilities of an ESB are: (1) Data transformation; (2) (Intelligent) routing; (3) Dealing with reliability; (4) Service management; (5) Monitoring and logging; (6) Providing connectivity and; (7) Dealing with security, among others.

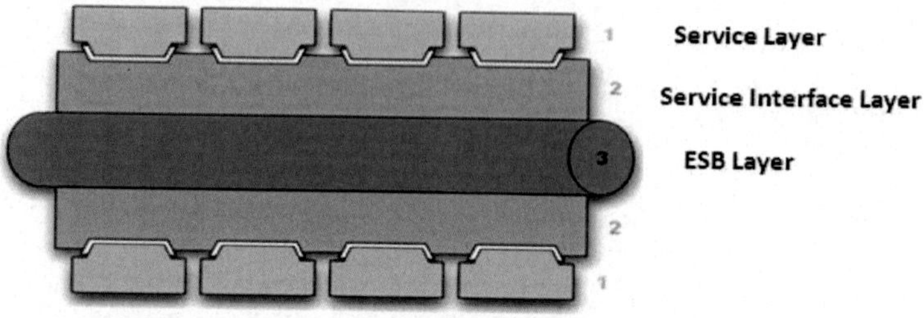


**Figure 2.** Generic structure of an Enterprise Service Bus [8]

As some systems engineering activities for space systems are becoming more complex and distributed, due to issues previously mentioned, one possible solution for this problem is the creation of a dedicated ESB, the SpaceESB, for this part of the project life cycle. Thereby, information related to the conceptual design would be available as services which could be invoked from anywhere and regardless of the underlying platform. This increases team communication as impacts on decisions taken by one team are rapidly evaluated by the other team members involved thus precluding misconceptions.

## 3 Budgeting Service Automation for Satellite Conceptual Design

One of the first steps to a SOA deployment is to identify which activities will be provided as services independently executed and generating well defined results. The business functionalities are mapped into these services and they are composed by parts, named operations, which encapsulates the complexity of existing business rules.

Typically in satellite conceptual design, suitable architectures are sought that successfully matches mission objectives [6] just like any space design exploration. As a simple illustration of activities at this early phase, this paper presents the budgets required to evaluate the amount of thermistors, number of direct commands for critical onboard functionalities and, solar panel area. Briefly discussed, each one of these estimates is implemented as a service based upon systems engineering business rules. A simplified set of business rules from [5] were used to program the web service logic.

### 3.1 Budgeting the Number of Thermistors

The satellite thermal subsystem is generally responsible for keeping all on-board equipment in a suitable operating temperature range at all times.

This subsystem may have two alternatives for control strategies, either a passive or active control, see [6] for details. As shown in Figure 3, the passive strategy employs materials, coatings, surface finishing, thermal properties whereas the active approach employs thermostatic heaters, heat pipes, and pumping systems with radiators and heat exchange.

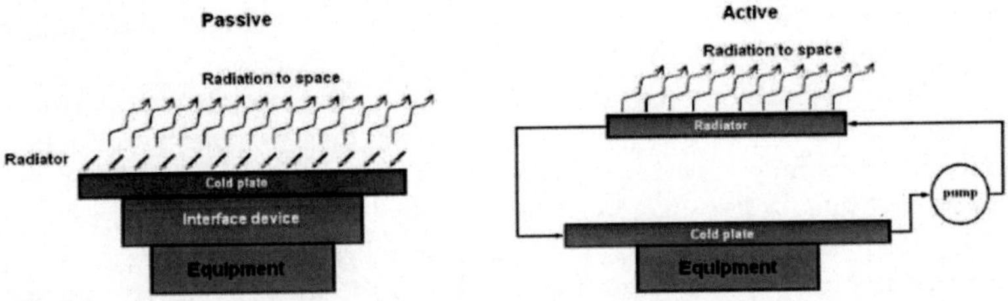


**Figure 3**. Some active and passive thermal control schemes [12]

One key component in the Thermal Control Subsystem is the sensing element, usually a thermistor. During the conceptual phase it is necessary to estimate the required number of thermistors assigned for each satellite component that needs thermal monitoring. This budget affects other coupled subsystems design for example, on-board processing, power and harnessing.

### 3.2 Budgeting the Number of Direct Command

Direct Commands are not processed and executed by the on-board computer, but are directly hard-coded and executed. These are mainly dedicated for mission-critical command execution like the following equipment items: On-board computers, batteries and telecommunications transceivers. This particular budget affects the satellite reliability and other coupled subsystems design like on-board processing, commuunications and harnessing, for example.

### 3.3 Budgeting the Solar Panel Area

The power subsystem, see [6] for details, is sketched in Figure 4 and is mainly responsible for (1) the provision of power mainly acquired from solar panels; (2) energy storage generally via batteries; (3) power conditioning, conversion, regulation and distribution to all subsystems and equipments.

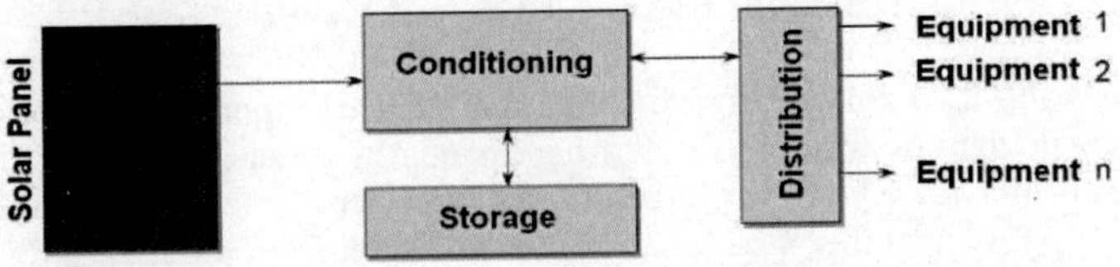


**Figure 4.** A typical block diagram for the power subsystem [12]

In order to meet the power demands, the required solar panel area needs to be properly evaluated and an extra margin for power generation should be taken into account allowing for battery charge and discharge cycles. This particular budget is highly vital and it affects almost all subsystems design.

## 4 Implementation of the Budgeting Web Services

The realization of a SOA abstraction, needs a set of tools and a development environment in order to create its business models, its services and, its overall derived elements.

After the service definition and operations models, one has to perform the transition of the models to computer systems, in this case to a web service. For this work, the programming language chosen is Java and the development environment chosen is Netbeans, a free integrated development environment (IDE) which has SOA plug-ins resources for service creation and orchestration.

The first implementation step is the creation of the web services just mentioned previously. At the end of the web services creation, a WSDL file is also generated. The out coming WSDL file contains details on the service description, its operations and the access conditions.

The second implementation step is realizing the underlying service orchestration. The web service creation only implements the service operations, but it is essential to implement also the operation execution flow.

The orchestration is responsible for defining the services invocation sequence and conditions. BPEL (Business Process Execution Language) is the chosen strategy for service orchestration [2] which application is depicted in Figure 5. BPEL is an XML dialect that defines services interactions.

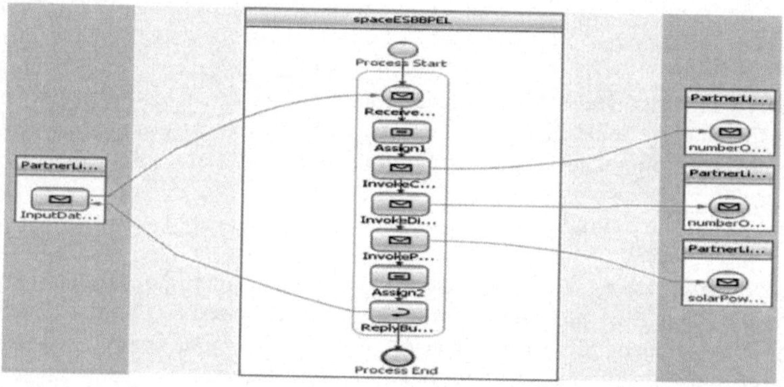


**Figure 5.** BPEL diagram for 3 budgeting web services

At the right-hand side of Figure 5 lies the service responsible for performing budgeting calculations while at the left-hand side, the requesting WSDL file. The center part displays the workflow taken for all 3 service consumptions. In this case, the execution flow is rather simple: after the data input from the requesting WSDL file, all services are called and after their completion, results are sent back to the requestor.

Afterwards, service creation is complete and it can be coupled to the SpaceESB which makes it available for all project partners who may need that functionality. Figure 6 shows the SpaceESB execution scheme where all data exchange is performed via XML files wrapped inside SOAP messages. The creation methodology used here can be extended to any other services. A trial is being currently planned that will populate the SpaceESB adding more functionalities.

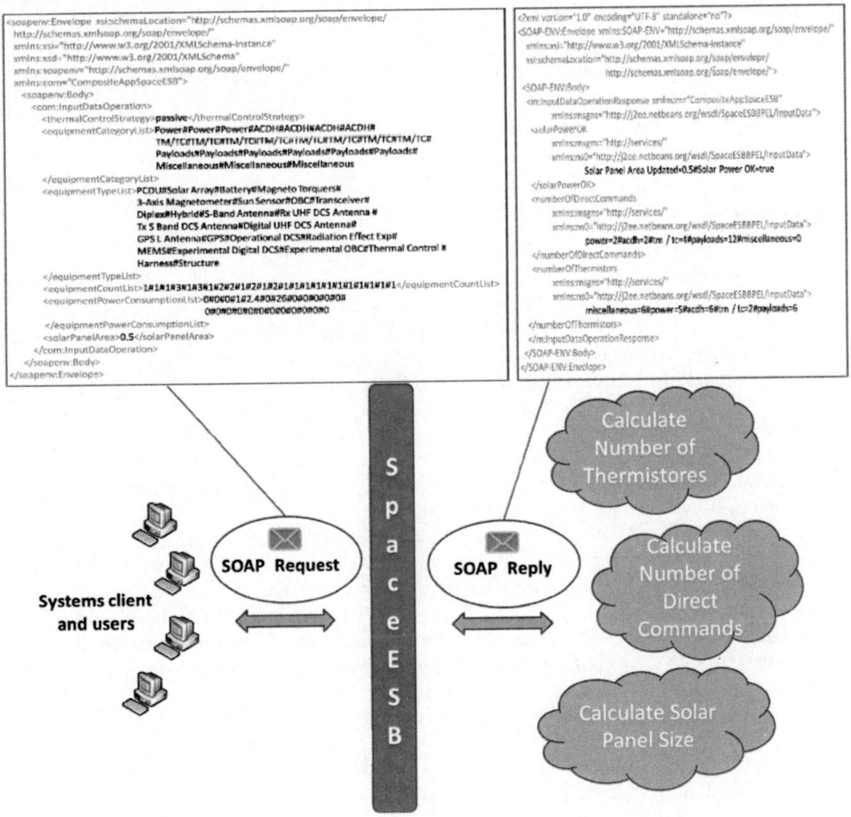


**Figure 6.** Scheme for service execution using the SpaceESB

## 5 Conclusions

Due to its complex and multi disciplinary nature, the flawless design, construction and launching of space systems is quite challenging. Experts from many fields are required to cooperate to ensure project success. Sometimes these tasks are geographically and/or temporally distributed demanding a cloud enabled environment.

In order to support the conceptual design phase of a satellite project, this paper shows how a set of three simple budgeting services can be coupled to a customized Enterprise Service Bus, named SpaceESB. The selected services where all related to budgeting activities concerned with thermal control, commandability and power generation which affects also other subsystems.

The mapping of a project activity to a Java service which implements these functionalities have been briefly presented using BPEL and a WSDL file was generated which describes the interface to the SpaceESB. This approach is general enough for the inclusion of new services populating the SpaceESB. The implementation has been done using only open source tools on all its steps.

Automation at this level is desirable as it allows project partners to actively participate in the decision-making process having greater agility and flexibility thus coping fortunately to pressures on costs and time.

## 7 References


[1] Bandecchi M., Melton B.; Ongaro F. Concurrent Engineering Applied to Space Mission Assessment and Design. In ESA Bulletin, 1999.

[2] BPEL tutorial. SearcSOA.com Available at <http://searchsoa.techtarget.com/generic/0,295582,sid26_gci1330911_mem1,00.html?ShortReg=1&mboxConv=searchSOA_RegActivate_Submit&> Access on: Jun., 14th 2010.

[3] Erl, T. Service Oriented Architecture: A Field Guide to Integrating XML and Web Services, The Prentice Hall, 2004.

[4] Grosskopf, A.; Decker, G.; Weske, M. The Process: Business Process Modeling using BPMN. Meghan Kiffer Press, 2009.

[5] Hardwick, A. A Webservice-based collaborative systems engineering for space applications. School of Computer Science and Information Systems, Birkbeck College, University of London, 2009.

[6] Larson W. J. Wertz J. R. Space Mission Analysis and Design, second edition, 1991.

[7] Leonor, B.B.F A Knowledge-Based and Model-Driven Requirements Engineering Approach to Conceptual Satellite Design (in Portuguese). INPE MSc Report, 2010.

[8] Longo J. S. C. SOA with OpenESB, Sun Microsystems. Available at <http://api.ning.com/files/Sn6i83r-auY8oBXNNYPn61ZsEMqGd3sU-421*dPFgL0a9aX*MnF0rHU0uZzcxDGOxOq8WCIDX5pk6XchShvtTKV*Y-L3lfEc/SOAcomOpenESB.pdf>. Access on: Oct., 10th 2010.

[9] Moro, T. D., Dorneles, C. F., Rebonatto, M. T. Web services WS-* versus Web Services REST. Available at <http://www.upf.br/computacao/images/stories/TCs/arquivos_20092/Tharcis_Dal_Moro.pdf>. Access on: Nov., 11th 2010.

[10] Ross, J. and et al. Enterprise Architecture as Strategy: Creating a Foundation for Business Execution. Cambridge, Harvard Business School Press, 2006.

[11] Shin S. SOA using Open ESB, BPEL, and NetBeans. Sun Microsystems. Available at <http://docs.huihoo.com/openesb/soabpelopenesb.pdf>. Access on: Nov., 13th 2010.

[12] Souza, P.N. Introductory Course on Satellite Technology. Lecture Notes. Slides – INPE, 2008.

[13] W3C. Simple Object Access Protocol (SOAP) 1.1: W3C Note 08 May 2000, available at <http://www.w3.org/TR/2000/NOTE-SOAP-20000508>. Access on: Oct., 12th 2010.

[14] Wu Q.; Zhou C., Jing T. Research on SOA based framework of collaborative design system for complicated product. In International Conference on Computer Science and Software Engineering, 2008.

# An approach to adapt collaborative architectures to cloud computing

Moisés Dutra[a], Minh Tri Nguyen[b], Parisa Ghodous[b,1], Youssef Roummieh[b], Chirine Ghedira[b], and Djamal Benslimane[b]

[a] Catarinense Federal Institute of Science and Technology, Brazil.
[b] University of Lyon 1, France.

**Abstract.** Companies keep trying various strategies in order to reduce the cost of their computer systems. The trend through which "everything is as a service" seems like the most effective solution used. By benefiting from the advantages of Cloud Computing, companies can minimize the cost of their systems and focus on their core businesses, by placing their IT parts into cloud providers. New companies can build their entire systems based on the clouds from the scratch. For legacy systems, however, the adaptation to the Cloud Computing paradigm remais the more effective solution. By proposing a methodology to adapt collaborative architectures to Cloud Computing, we intend to contribute to the first application of this technology in the industrial world. In order to validate the proposed approach, we have adapted our team's collaborative system to the clouds.

**Keywords.** Cloud Computing, Collaborative Systems

## 1 Introduction

The year 2009 saw the emergence of the term "Cloud Computing" in publications. Historically, the term "Cloud Computing" was first used in 2002 by Amazon [13], a leading e-business, who had invested in a vast machinery, sized to handle the heavy load of orders made on their website at Christmas, but unused for the rest of the year. Sub-sizing their fleets would have caused downtime of their website at the time of the peaks, thus jeopardizing their work during the holidays (a big part of their turnover business). Their idea is to open all these unused resources to businesses to hire them on demand. Since then, Amazon is investing heavily in this area and continues to expand its fleet and services.

---

[1] ghodous@liris.cnrs.fr – Head of Collaborative Modelling Theme, SOC Group
Laboratory of Computer Graphics, Images and Information Systems (LIRIS)
University of Lyon 1, Bâtiment Nautibus, 43, bd. du 11 Novembre. 1918,
69622 Villeurbanne cedex, France
Tel: +33 (0) 4 72 44 58 84
Fax: +33 (0) 4 72 43 13 12

D. D. Frey et al. (eds.), *Improving Complex Systems Today*,
Advanced Concurrent Engineering, DOI: 10.1007/ 978-0-85729-799-0_19,
© Springer-Verlag London Limited 2011

One difficulty is that there is not between providers a single definition of cloud computing [4]. In this paper, we present the most popular and widely understood definition, given by NIST [5]: Cloud computing is a model for enabling convenient, on-demand network access to a shared pool of configurable computing resources (e.g., networks, servers, storage, applications, and services) that can be rapidly provisioned and released with minimal management effort or service provider interaction. This cloud model promotes availability and is composed of five essential characteristics, three service models, and four deployment models.

With many benefits of cloud computing, experts foresee a major trend in this century when companies apply this technology in their system information [1]. Currently, the major actors in IT world have built their own cloud like Azure of Microsoft, IBM Blue House or Amazon Web Services. There are two main ways to apply the technology of cloud computing for businesses. First, the owners of the company may decide to build a new cloud computing-based system. This decision is handy with startup businesses that have not yet built their local system. However, this solution is not suitable for most companies that have already built their own information system. In this latter scenario, the migration of legacy systems into the clouds is considered a better solution. So the problem is how can we adapt existing systems – especially business applications – to the clouds?

The remainder of this manuscript is organized as follows: Section 2 proposes a methodology for migrating legacy systems into the clouds; Section 3 presents the validation of the proposed methodology, by using a classic collaborative architecture scenario; and Section 4 presents the conclusions of this work and some perspectives on this matter.

## 2 Methodology

Currently, there are not yet neither developed standards for cloud computing, nor effective strategies for moving business applications into the clouds. Throughput in recent research, users are often lost in the clouds development, which makes moving an application to the clouds quite a challenge [2][3]. We consider it is important to have a methodology for systematizing this process. In this section, we present this methodology for transforming legacy systems – especially commercial applications – into cloud-based systems. In order to achieve so, we propose the following steps.

**i) Analyzing the existing system**

Analyzing the existing system means determining the system's structure by investigating data, business applications and other system components, in order to identify what will be brought into the clouds and what will be kept locally. We should determine an order through which the externalization will be done.

- Data: Have the highest level of priority for migration, because they are usually the first part thought/built in a system. Business applications rely strongly on data. However, depending on the system architecture, this can be considered differently.

- Business Applications: Many people think we should take everything into the clouds, but the reality is quite the opposite. To ensure the success of the externalization, we consider that it is better to start doing it from the simplest to the most complicated strategies.
- Other components: In general, any system component can be migrated to the clouds. Nevertheless, we can also chose to use an already existing cloud service – in the form of SaaS (Software as a Service) – instead of undertaking the migration.

After this step, we will have a mapping of data, business applications and components, representing those which will be taken to the clouds and those which will be kept locally. The next step consists of finding the suitable clouds for migration.

**ii) Choosing the suitable cloud configuration**

There are a number of clouds available for migration, each one of them suitable for a specific need. Typically, there are three main types of cloud [7]:

- Public Cloud: Available to the general public or large industries, and it is owned by an organization that sells cloud services. With this solution, we put all the data and business applications on a public cloud, such as the Amazon or Microsoft cloud.
- Private Cloud: Exclusively operated by just one organization. This operation can be managed directly by the organization or be outsourced. This model is very suitable for companies that have great resources available and want to have full control of their data, as they can enjoy the benefits of the cloud as a whole.
- Hybrid Cloud: The infrastructure of the hybrid cloud consists of the composition of two or more private or public clouds, which remain single entities and are bounded by a standard – or proprietary technology – that allows the portability of data and applications.

**iii) Designing the new architecture**

We can now start conceiving a new architecture for our cloud-based system, which will generally have the following structure.

- Databases: databases can set the clouds in the form of DaaS (Database as a Service). Currently, there are available services [12] for simple data provided by Amazon Simple Database [8], and services for relational data, provided by FathomDB [9] and Amazon [10].
- Business Applications: For business applications, we propose the use of the IaaS (Infrastructure as a Service) or PaaS (Platform as a Service) technology. Currently, there are a number of platforms that support multiple languages, such as .Net, Java, Python, PHP, or Ruby on Rail. These platforms provide APIs to facilitate the application deployment. Moreover, they support several tools to manage those applications on the clouds.
- Other components: this architecture provides us with the possibility of using different clouds for different components, according to our needs.

After determining the new system's architecture, we will perform the migration and some tests, which must consist of a detailed process of checking and validation, in order to ensure the system's workability. These processes will be done according to the following steps.

**iv) Choosing the clouds for system deployment**

During the creation of the new architecture, it is defined what the system's needs are and what technologies (e.g., PaaS, Saas, and DaaS) will be used to build it. The next step consists of choosing the most suitable clouds to deploy the new system. There are some popular commercial clouds that can be used to this purpose, as told before.

**v) Building the new system**

We will work here with the same order of components depicted in the first stage, namely databases, business applications, and other components.

- Databases: for simple data, such as XML/OWL files, we can put them on the same place as the business applications. However, for specific databases, such as relational databases, we must put them on a DaaS cloud. Normally, the clouds will provide us with Web interfaces to facilitate the import of databases, as well as provide us with specific information for accessing data from another service.
- Business Applications: There are two ways of deploying business applications on the clouds. First, we create a new project on a PaaS cloud. Then we keep carrying out the migration of the required files until we have the whole project deployed on the clouds. Second, we can create services from local applications and make these services available on the clouds. By using the information database, which has already been deployed, we can establish connections between applications on the clouds and databases elsewhere.
- Other components: for modules such as communication modules, which already exist on the clouds in the form of SaaS, we should just pay for these services according to our needs and obtain information from them, in order to make the connections between these services and business applications (before the deployment of these business applications on the PaaS cloud is done).

In order to reduce the appearance of problems, we should systematically test if everything works well after each step taken. Once we have guaranteed the workability of the system, we can continue externalizing the remaining components.

**vi) Defining governance strategies**

What should we do if some day the service supplier of the clouds is replaced? How could our system be accessed? We can imagine a situation where we do not know what these changes are, neither how they affect our system. In other words, there would be no governance – or the ability to monitor service changes – and service utilization. In the corporate world, governance means controlling the system. On the clouds, we should control our data and our services to ensure the success of the system. We must know who has access to our data and who can create, delete or change data and services. The idea of governance is to provide command, control and surveillance services, including local services and cloud services. We must clarify the responsibilities of governance with our cloud suppliers. Normally, IaaS clouds provide customers with more level of governance than other clouds do. On the contrary, SaaS clouds provide very little level of governance to their customers. As a consequence, customers must deeply investigate contracts between cloud buyers and suppliers, in order to clearly determine what their policies of governance are.

## 3 Application of the Methodology

### Existing system

Our team has developed a generic and synchronous ontology-based collaborative platform [11]. This platform is based on the collaboration of all actors involved in a design project (designers, customers, engineers, vendors, distributors, etc.), and also comprises a multilayer architecture. First, each participant is called an agent. Agents are grouped according to their knowledge and skill levels. Each group of agents is called an agency. All agents share and publish data through a workspace called "Agency Shared Workspace" (ASW). Similarly, all agencies share and publish their data through a workspace called "Project Shared Workspace" (PSW) [11].

The PSW consists of a blackboard with an ontology module and a knowledge-based reasoning module. The ASW has the same structure that the PSW (Figure 1).

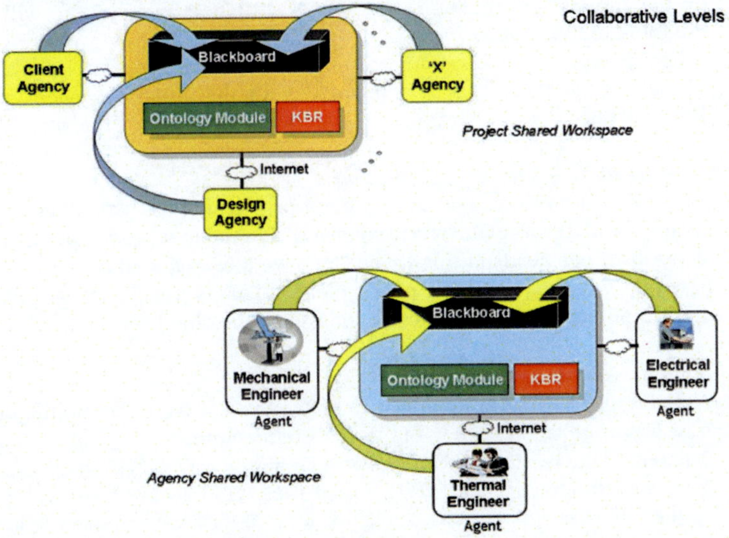


Figure 1. Collaborative platform.

The architecture of this platform is based on a range of agencies collaborating in the process of constructing complex artifacts (e.g., aeronautical products, their processes and services). Each agency is composed of several agents, who work in this process. In the current system, all resources and services are located locally. To take advantage of the cloud computing paradigm, we will put this system on the clouds, by using the methodology presented in Section 2.

**New system**

We will now apply the proposed methodology to transform our collaborative system into a cloud-based system. As the proposed methodology stands for, we will follow the following steps to build the new system.

*1) Analyzing the existing system*

The first step during the building of the new system is to analyze the existing system, in order to identify the components to be taken or not to the clouds. In our architecture, we have identified the following components to be taken to the clouds.

- Data: The existing system uses MySQL relational database and OWL-based files. We have then decided to preferentially transfer all these data to the clouds.
- Business applications: business applications consist of the existing applications used by the project manager and users, and the existing applications designed to deal with ontologies. Due to the complexity of the ontology applications, we have decided to transfer the project manager and user applications first. In a second step, further on, after a successful processing of the first one, we will then transfer the ontology applications to the clouds. In the end, the whole collaborative system will have migrated to the clouds.
- Other components: Existing communication services, as the Instant Messaging Service, can be integrated with the clouds as a Service for Enrichment.

*2) Choosing the suitable cloud configuration*

Currently, we do not have material to build private clouds. Moreover, the cost of construction of clouds remains very expensive. On the other hand, public clouds can perfectly meet our needs at this time. Thus, we have decided to build the new system based on the architecture of public clouds. That is to say, we will put data and business applications onto the clouds and use public cloud services.

*3) Designing the new architecture*

After selecting the general architecture of the system based on public clouds, we will now associate the system with the new architecture.

- Databases: the relational database will be put on a DaaS cloud, as proposed by our methodology. The OWL-based files will be put on the same place as the business applications, since this kind of data is very specific and this solution will simplify its access by the design application.
- Business Applications: According to the methodology, we will put the business applications on a PaaS cloud. In addition, as the existing system was developed in Java, we have chosen a cloud that contains this feature.

- Other components: In this scenario, modules for communications are the other components. We will use existing SaaS services for this purpose.

*4) Choosing the clouds for system deployment*

The appropriate clouds must be chosen according to the needs of each system component:

- Database: currently, there are two large DaaS clouds available: the one provided by Amazon and the FathomDB relational database. For this system, we have decided to use FathomDB, due to its simplicity and lower cost.
- Business Applications: for these components, we have decided to choose the PaaS Amazon Web Services cloud, due to three main reasons: it supports Java, it is a large cloud, and it is free for the first use.
- Other components: It has been decided that the communication component will work with mail services, Google chat, Amazon Simple Queue Service, Amazon Simple Notification, and Zoho's "Instant Messaging as a Service".

*5) Building the new system*

At this stage, we will perform the migration according to the architecture proposed in the previous step.

- Databases: for specific databases such as relational databases, we will put them on a specific DaaS cloud. After seeing details of the information services on the FathomDB website [9], we have decided to choose the "Small Instance" – 150 MB of memory – for the first try. The DaaS will provide us with the information required to access the databases. We now can use the just provided login information to create our database.
- Business Applications: we use Eclipse IDE along with the Eclipse AWS Plugin to implement the business applications and to deploy them onto the Amazon clouds. For the first step, we have chosen to launch an instance of Amazon EC2. We have chosen a small instance for the first try. Through the use of FathomDB database connections, we can establish connections between applications on the cloud and databases outside it.
- Other components: in addition to the existing mail applications, we can integrate Zoho's "Instant Messenger Application as a Service" [6].

After the migration, the new system will work on the clouds as well as they did locally, previously.

*6) Security, Governance and elasticity*

We have defined very clear security and governance policies to be applied on the chosen clouds. For example, the policies concerning data are very rigid; we can only access data by a specific IP or by an Amazon cloud. Amazon EC2 defines very clear security policies for each specific connection type. Considering governance, clouds provides us with the ability to easily manage our system through Web interfaces. In addition, they provide us with automatic backup services to be used by the system's instance on the clouds.

A big advantage of migration is to extend the system automatically, without human interaction. When necessary, we can also easily expand the database or even pay for more instances of Amazon EC2.

## 4 Conclusions and Perspectives

Cloud computing technology is rapidly becoming more and more known/used. Possessing benefits such as elasticity and self-service, along with the cost to transfer an entire system to the clouds, companies have now at their disposal a very effective option to reduce the cost of their systems' maintenance. This becomes even clearer in the world's present context, with all the economic difficulties that companies have been facing. In such a context, cloud computing plays a very important role, as it provides a whole new simpler way of designing and managing architectures of enterprise systems. Moreover, building a new system from the scratch is not always an option well seen by companies. Considering this scenario, this work proposes a methodology for transferring companies' legacy systems into the clouds. This methodology is composed of well defined steps of externalization, in a rising level of complexity. We have chosen to validate this methodology by using it during the migration of our team's collaborative system to the clouds.

Next steps of this work consist of filling the existing gaps in the methodology, especially the security and governance issues. Moreover, a prototype that can help transfer the system automatically to the clouds is a goal that we want to achieve as well.

## 5 References


[1] M. Armbrust, A. Fox, R. Griffith, A. D. Joseph, R. Katz, A. Konwinski, G. Lee, D. Patterson, A. Rabkin, I. Stoica, and M. Zaharia – Above the Clouds: A Berkeley View of Cloud Computing, Armburst, 2009.

[2] A. Chandra and J. Weissman, University of Minnesota – Nebulas: Using Distributed Voluntary Resources to Build Clouds, 2009.

[3] A. T. Velte, T. J. Velte, and R. Elsenpeter – *Cloud Computing - A practical approach*, 2009.

[4] *Twenty-One Experts Define Cloud Computing*: http://cloudcomputing.sys-con.com/node/612375, in February 2011.

[5] Peter Mell and Tim Grance – A NIST Notional Definition of Cloud Computing.

[6] http://chat.zoho.com, in February 2011.

[7] Cloud Computing Use Case Discussion Group – *Cloud Computing Use Cases.*

[8] http://aws.amazon.com/simpledb/, in February 2011.

[9] http://fathomdb.com, in February 2011.

[10] http://aws.amazon.com/rds/, in February 2011.

[11] M. Dutra, P. Ghodous, O. Kuhn, N. M. Tri – *A Generic and Synchronous Ontology-based Architecture for Collaborative Design.* Concurrent Engineering, Research and Applications 18(1):65-74, Sage, ISSN 1063 293X, 2010.

[12] Luis Felipe Cabrera, Christopher Kurt, Don Box (Microsoft Corporation) – *Introduction à l'architecture de services Web et ses spécifications WS.* Octobre 2004.

[13] http://www.cloudconsulting.fr/index.php/fr/le-cloud-computing/historique, in February 2011.

# MimoSecco: A Middleware for Secure Cloud Storage

Dirk Achenbach, Matthias Gabel, Matthias Huber

Institute of Cryptography and Security, Karlsruhe Institute of Technology
{dirk.achenbach, matthias.gabel, matthias.huber}@kit.edu

**Abstract.** The biggest impediment for the adoption of cloud computing practices is the lack of trust in the confidentiality of one's data in the cloud. The prevalent threat in the cloud computing model are so-called insider attacks. Full data encryption can only solve the problem in the trivial case of backups. Any sophisticated service provided on data requires insight into the structure of that data. One purpose of encryption is to prevent such insights. We introduce the MimoSecco[1] project. In MimoSecco, we are investigating reasonable compromises. We employ two techniques, separation of duties and secure hardware. With separation of duties, we fragment a database and separate the fragments geographically. The goal is to make it infeasible to reconstruct the database from one fragment alone. The secure hardware tokens we employ are hard-to-copy devices which offer encryption, decryption and cryptographically signing of data. The keys used are stored in the tamper-proof hardware device and never leave it. We are in the process of developing a prototypical database adapter that behaves like a SQL database, but stores data securely.

## 1 Introduction

Cloud Computing promises huge benefits especially for small and medium enterprises (SMEs). Pay-per-use payment models, dynamic scaling of services and the outsourcing of infrastructure lead to a better resource utilization. Also, the risk of fatal data loss is minimized due to specialization of the individual providers.

There is a growing market for mobile computing devices like cellphones and tablet PCs. They are usually used as auxiliary devices to a personal computer and are limited in capacity and computing power, but are equipped with an always-on internet connection. Consequently, mobile computing devices profit most from accessing cloud services. Most manufacturers of mobile computing hardware already bundle their offer with Software-as-a-Service products like mail, calendar, cloud storage, or social network applications.

[1] This work has been funded by the Federal Ministry of Economics and Technology, Germany (BMWi, Contract No. 01MS10002). The responsibility for the content of this article lies solely with the authors.



D. D. Frey et al. (eds.), *Improving Complex Systems Today*,
Advanced Concurrent Engineering, DOI: 10.1007/ 978-0-85729-799-0_20,
© Springer-Verlag London Limited 2011

A huge impediment for the adoption of cloud computing for small and medium enterprises is the lack of security guarantees it offers. Because they have to rely on third-party infrastructure, SMEs lose control over the data they store in a public cloud. Service Providers offer a high level of security against so-called external attacks and natural disasters. They put sophisticated physical and electronic measures in place to reduce technical failures of their infrastructure to a minimum and offer a high level of protection against external attackers. A new threat in the cloud computing model are insider attacks—an employee of the cloud provider might steal sensitive data and sell it to a competitor, for example.

There are legislative means to ensure a certain degree of privacy and security in cloud services. Laws are restricted to specific countries and the compliance of a service with a Service Level Agreement (SLA) is hard to supervise in general. Therefore, in the MimoSecco project we ensure data privacy in cloud computing applications through technological means. MimoSecco is a cooperation of the Karlsruhe Institute of Technology, WIBU-SYSTEMS AG and CAS Software AG, all located in Karlsruhe, Germany. We aim to implement approaches that are feasible with current hardware. We are also in the process of developing and extending a prototype of a secure cloud data storage middleware. This paper describes our current approach to secure database outsourcing as it is implemented in our prototype. We presented an early version of our prototype on the CeBIT, the world's largest IT expo in Hannover, Germany.

Our methods are the separation of duties and the use of secure hardware. Separation of duties means partitioning a service and distributing its parts to different servers. The scheme we present impedes adversaries from learning valuable information by reading data that is stored on a single server. The secure hardware we employ are specialized tamper-proof tokens which offer encryption, decryption and digital signatures. They perform operations without the need for the cryptographic keys to leave the device.

In the next section, we discuss Related Work. In Section 3, we introduce the methods we employ in our prototype. Section 4 explains the database scheme we use. We describe the implementation of our prototype in Section 5. Section 6 concludes.

## 2 Related Work

Many services rely on databases. Consequently, the problem of secure database outsourcing emerged early, in 2002. Since then, many schemes have been proposed that try to solve this problem. All approaches try to find a tradeoff between security, performance and efficient support for as many query types as possible. We identified two clases of approaches, which we term coarse indices approach and distribution approach, that potentially suppport searching a database in sublinear time.

In [4] Hacigümüs et al propose to encrypt a database tuple-wise and create coarse indices. This enables for efficient exact-match and range queries. This scheme, however, does not support like queries efficiently.

In [2] Aggarwal et al propose distributing a database to different servers in order to meet so-called privacy constraints. If a privacy constraint cannot be met by distribution, they propose to use encryption. This scheme supports the efficient execution of queries to attributes whose values are not encrypted. Queries containing encrypted attribute values, however, are not supported efficiently.

The field of secure multi-party computation (MPC) from cryptography deals with a problem that relates closely to ours: Given a set of parties, how can they securely perform a joint function evaluation without any of the parties learning the input of the others? For example, how can two millionaires—without openly discussing their wealth—determine who is the richer of the two? There is a rich body of research, most prominently [6] and [3]. However, most results in MPC don't yield practical techniques applicable to cloud computing scenarios.

## 3 Methods

We combine several techniques in our scheme. The goal is to create a middleware that offers a transparent database encryption with an acceptable tradeoff between security and performance. The following section gives an overview of the employed techniques.

### 3.1 Separation of Duties

In [5], we describe how a *Separation of Duties* can enhance the security of services. We distinguish between serial and parallel separation and show that separating a service can—in combination with cryptographic methods—increase the security of services while maintaining efficiency. Figure 1 shows a parallel separation of a service. The overall service is provided by the adapter and the separated parts.

### 3.2 Encryption

We call a tuple (Gen, Enc, Dec) of a key-generation function, an encryption process and a decryption function an encryption scheme. Unless otherwise noted, we use a randomized symmetric encryption scheme, for example the AES block cipher [1] in CBC mode. The key-generation function Gen outputs a single key of adequate length that is used for encryption as well as decryption. The encryption process Enc and the decryption function Dec transform bit strings of any length to bit strings of the same length. If needed, values are padded before the encryption occurs.

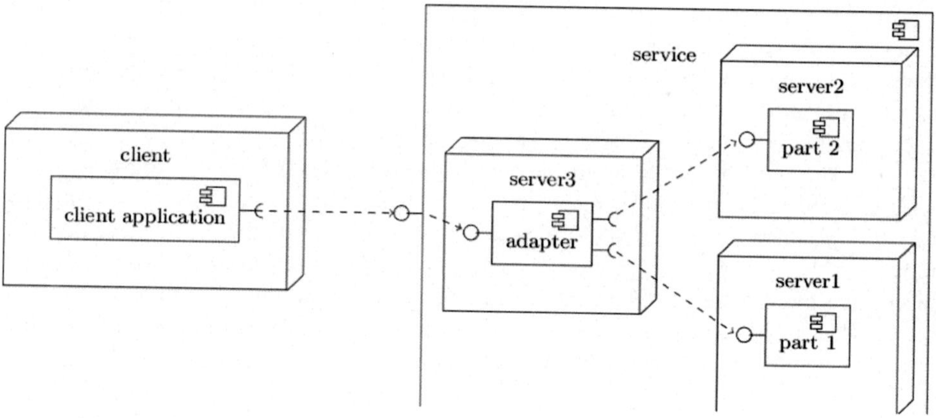


**Fig. 1.** System and deployment view of a parallel separation of a service, and an adapter.

### 3.3 Secure Hardware

With the introduction of cryptographic techniques the problem of key management arises. A simple solution could be the integration of the used key in the client software or on the client's hard drive. In case of theft, all stored data would be exposed, including cryptographic keys used for database encryption. A better solution is to store keys on an external device.

There are specialized hardware tokens which have encryption capabilities. Keys are stored on the tokens themselves and never leave the device. Usually, these devices are tamper-proofed, which makes it infeasibly hard for an adversary to extract the cryptographic keys from the device. To further protect the data, access to the token can be password protected.

## 4 Database Scheme

A database can be stored securely on a remote server by encrypting the whole database beforehand. This approach, however, does not support efficient execution of queries: In oder to perform a query, one needs to download the whole database. To increase performance while retaining security, a more elaborate method must be used. In Section 2 we describe different appproaches that support efficient execution of some queries. They, however, do not suport efficient execution of all classes of queries.

The scheme we propose supports the efficient execution of exact match, range, as well as substring queries while hiding the relations of the attribute values in the database. Our database scheme employs three entities: The database adapter itself, the datastore and an index store. The database adapter provides an SQL interface to the user and accesses the storage backend. The datastore is the encrypted database itself. In the remainder of this section, we explain how to create the index tables and the data table with a simple example.

Consider a database as depicted in Figure 2.

| row | name | surname |
|---|---|---|
| 1 | Alice | Smith |
| 2 | Bob | Smith |
| 3 | Alice | Jones |
| ⋮ | ⋮ | ⋮ |

**Fig. 2.** A simple example of a database: base.

The index tables index name and index surname are created from the base table by creating a table for every attribute containing every attribute value in the first colum, and a list of its occurences in the second column. Then the lists in the second column are encrypted. Figure 3 depicts the index tables for the table depicted in Figure 2.

(a)

| name | rows |
|---|---|
| Alice | Enc(1, 3) |
| Bob | Enc(2) |
| ⋮ | ⋮ |

(b)

| surname | rows |
|---|---|
| Smith | Enc(1, 2) |
| Jones | Enc(3) |
| ⋮ | ⋮ |

**Fig. 3.** The index tables for the table depicted in Figure 2 index_name and index_surname.

The data table is created by encrypting the base table tuple-wise (cf. Figure 4).

| row | $\mathsf{Enc}_{\mathsf{prob}}$(name, surname) |
|---|---|
| 1 | Enc(Alice, Smith) |
| 2 | Enc(Bob, Smith) |
| 3 | Enc(Alice, Jones) |
| ⋮ | ⋮ |

**Fig. 4.** The encrypted data table of the table depicted in Figure 2.

Note that it is still possible to execute queries on this database efficiently. Consider for example the query

```
SELECT * from data WHERE name="Jones" AND
surname="Alice";
```

In order to execute this query, the adapter first queries the name table index name with the query:

```
SELECT * from index_name WHERE name = "Jones";
```

Then the adapter queries the table index surname with:

```
SELECT * from index_surname WHERE surname = "Alice";
```

The adapter decrypts the contents of the rows column in the results and queries the table data with the intersection:

```
SELECT * from data WHERE row = "3";
```

Since the index tables support search for attribute values in sublinear time, queries still can be executed efficiently. The data that has been fetched in this manner is then processed locally to generate a result.

## 5 Implementation

We implemented the adapter described in Section 4 in Java. The prototype uses JDBC to connect to its backend databases. The adapter is transparent to the client application. It transforms queries internally.

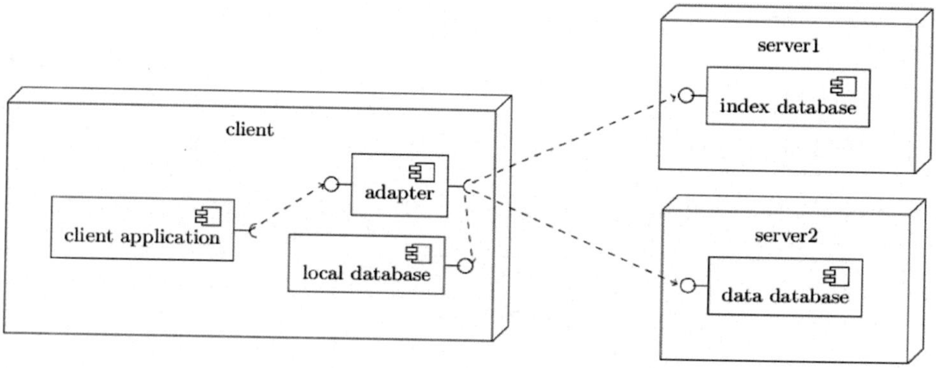


**Fig. 5.** Architecture of our Database. An adapter deployed on the clients' machine provides a standard SQL interface to the client application.

Currently, the prototype supports SELECT and INSERT INTO statements. Query results are determined as follows. The adapter parses issued queries using ZQL (`http://www.gibello.com/code/zql/`) and splits them into subqueries as described in the previous section. The subqueries are issued to the backend databases. The adapter also stores cryptographic keys. After decryption, GROUP BY or aggregate functions are performed locally on the preliminary results before the final result are returned. For local query execution we use a local database. Figure 5 depicts the architecture of our prototype comprising the index database, the data database, the adapter, and the local database. Since we use a standard SQL database for the local database, we automatically support the better part of SQL. Though not implemented in our prototype, the index tables can be distributed on individual servers in order to increase security even further. Since indices can be queries in parallel, this increases query execution compared to the current implementation.

## 6 Discussion

In this paper, we introduce the MimoSecco project and the challenges it addresses. We present the concepts and the methods we implemented in our database encryption prototype. Intuitively, our database scheme hides relations of an outsourced database while supporting efficient evaluation of a large number of queries.

There are numerous challenges we still face and will address in the future. First experiences with our prototype show good performance properties. However, the impact of the database transformation on performance must be measured and evaluated. Also, the security the adapter provides must be analyzed. Therefore, we need to formalize our notion of security and prove that our scheme fulfills this notion. We are currently working on extending our prototype to fully support the SQL standard. We will incorporate tamper-proof hardware in the prototype, for example for storing the encryption keys or handling en- and decryption. This promises to improve the security significantly. We also seek to embed our database solution in larger services and apply our separations of duties approach to them. An example we are currently working on are secure social networks.

## References

[1] Advanced encryption standard (AES). Tech. rep., NIST (2001)

[2] Aggarwal, G., Bawa, M., Ganesan, P., Garcia-Molina, H., Kenthapadi, K., Motwani, R., Srivastava, U., Thomas, D., Xu, Y.: Two can keep a secret: A distributed architecture for secure database services. CIDR 2005 http://ilpubs.stanford.edu:8090/659/

[3] Goldreich, O., Micali, S., Wigderson, A.: How to play any mental game or a completeness theorem for protocols with honest majority. In: Proceedings of the nineteenth annual ACM symposium on Theory of computing. pp. 218–229. ACM, New York, NY, USA (1987)

[4] Hacigümüs, H., Iyer, B., Li, C., Mehrotra, S.: Executing SQL over encrypted data in the database-service-provider model. In: Proceedings of the 2002 ACM SIGMOD international conference on Management of data. pp. 216–227. ACM (2002)

[5] Huber, M.: Towards secure services in an untrusted environment. In: Bühnová, B., Reussner, R.H., Szyperski, C., Weck, W. (eds.) Proceedings of the Fifteenth International Workshop on Component-Oriented Programming (WCOP) 2010. Interne Berichte, vol. 2010-14, pp. 39–46. Karlsruhe Institue of Technology, Faculty of Informatics, Karlsruhe, Germany (June 2010), http://digbib.ubka.uni-karlsruhe.de/volltexte/1000018464

[6] Yao, A.C.: Protocols for secure computations. In: Proceedings of the 23rd Annual IEEE Symposium on Foundations of Computer Science, FOCS'82. pp. 160–164. IEEE Computer Society (1982)

# Implementation of CE in Software Development

Kamran Ahmed Malik [a], Irfan Anjum Manarvi[b,1]

[a]Department of Engineering Management, CASE, Islamabad Pakistan.
[b]Department of Management Sciences, Iqra University, Islamabad, Pakistan.

**Abstract.** This paper presents research on implementing Concurrent Engineering (CE) principles in software development environment. International Turnkey Systems – Pakistan design and development group was selected for experimental research. It was focused on the product architecture and team structure of the Telco design & development group responsible for software development. During implementation a number of barriers were observed. Efforts were made to overcome these barriers using best practices of CE. The requirements of concurrent engineering principles in software development environment were identified; implemented and significant improvements were observed in the organization.

**Keywords.** Concurrent Engineering, Team Structure, Team work, Implementation Barriers, Software development.

## 1 Introduction

Software industry in Pakistan is at a point of momentous change. Service Providers have to rapidly deliver new and profitable Services. They aspire to cut costs and become efficient by modernizing infrastructure and products. Therefore, such organizations are moving towards enterprise applications that adhere to Open Standards of Modularity concepts. The strategy for achievement of high Industry Standardization is opted by the competent software development concerns that are familiar with the swiftly changing requirements of global community. In recent past, software industry in Pakistan has also seen major shifts in context of Integration, product complexity and life cycle costs. Product life cycle cost is based on numerous factors starting from conception through disposal. All software development practices have its own distinct segments of life cycle in terms of Analysis, Design, Development and Maintenance [1]. Out of these factors, Structure of the Team or the Structure of organization is one major aspect that might create hindrance in the way of true performance of the software development concerns. Current organizational structures and team hierarchies demand focused attention for transformation towards efficient systems. Traditionally applied sequential approaches to software product development are getting obsolete [2]. Documentation has also become a core activity of software development process and takes lot of time. It needs to be executed in parallel with all other phases starting

[1]Professor, Department of Management Sciences, Iqra University, Islamabad, Pakistan Tel: +92 336 5025 073; Email: irfanmanarvi@yahoo.com; http://www.iqraisb.edu.pk

D. D. Frey et al. (eds.), *Improving Complex Systems Today*,
Advanced Concurrent Engineering, DOI: 10.1007/ 978-0-85729-799-0_21,
© Springer-Verlag London Limited 2011

from mission definition to maintenance. The information about the software product documents the marketing requirements and plans, project plans, requirements specifications, high level system design, detailed system design, maintenance and support plans [3]. In software industry, the concept of empowerment is also taking its roots for continuous business process improvement [4]. Unfortunately the true teamwork is not observed in Pakistani software industry environment. The people working in different groups on a single product are said to be working in one team. But the individuals are more focused on their personal motives, targets and goals; that leads to uncommon interests towards the target. Knowledge sharing, mutual understanding and coordination activities may be considered as the true concept of Teamwork.

The present research was therefore focused on evaluating the possibilities of implementing concurrent engineering concepts in Product Design and Development Group of International Turnkey Systems – Pakistan. As a first step the key attributes of CE were established [5-9] as follows:

- Parallel Execution of Activities
- Teamwork
- Knowledge Sharing
- Coordination & Communication
- Decentralized Design Approach
- Empowerment
- Re-Structuring
- Employee Participation
- Customer Participation
- Supplier's Involvement
- Barrier's removal among departments
- Documentation with Product Development
- Early Decision Making
- Decision Making by Consensus Building
- Movement as Single Unit
- Continuous Improvement Mechanism
- Team Based Rewards
- Sense of Ownership
- Strong Focus on Definition Phase

## 2 Overview of International Turnkey Systems

Established in 1981, the company is a provider of integrated information technology solutions to a wide range of industries and government organizations in 17 countries with more than 50 implementations in multiple business segments. It has over 2000 professionals and cover various industries and financial institutions including Banks, Telecommunications, and Higher Education. It is in Pakistan operating for last 5 years and is mainly focusing on Telecom Sector. Warid Telecom is the major client, where implementation of Telecom Billing System (TABS Billing Suite) is going on. Its core technical Groups include International

Turnkey Systems - Pakistan maintains three Core Technical Teams besides support departments.

- Design and Development Group.
- Product Support Group.
- Product Implementation Group.

The organization structure of Telco design and development group represents a Flat team structure. Design Manager is leading the Group while keeping a Project Owner / Architect under him. Below the Project Owner / Architect, all the resources including Software Engineers, Developers and Tester / Integrators are working. There is no hierarchy inside the resources. The group is responsible for providing the software solutions for telecom sector worldwide. This group is working on Telecom Billing Solution and was selected for the implementation of Concurrent Engineering practices in present research.

TABS (Telecom Advanced Billing Solution) – Billing Suite is responsible for gathering details of all items that are to be charged against provided products and services. It applies any needed manipulation depending on the company's business rules such as discounting or sometimes taxation, granting bonuses and rewards, and generating a final bill (or invoice) to convey the debt to the customer. Telecom TABS Telecom Billing software is integrated with following modules and are shown in Figure 1. TABS Billing Suite was selected for implementation of Concurrent Engineering practices.

- TABS Billing Suite
- TABS Partner Management
- TABS Collection
- TABS Real Time Convergent Charging
- TABS Incentive Engine
- TABS Customer Relationship Management

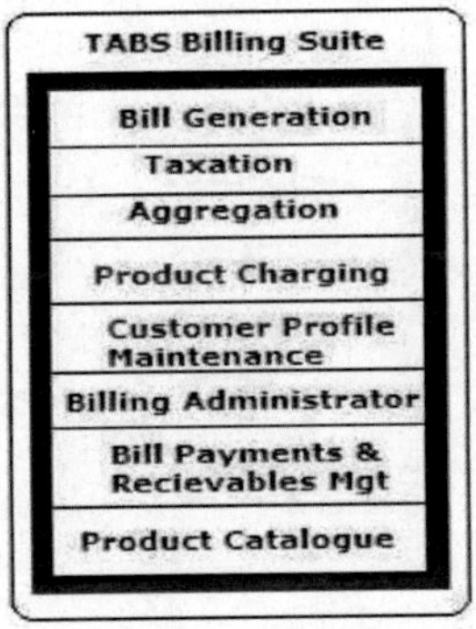


Figure 1 TABS Billing suite

## 3 Existing Software Development Environment

The present software development environment is based on structured approach. It is the sequential process, in which all phases are executed one by one as shown in Figure 1. In first phase, vision is defined. In 2nd phase requirements analysis is carried out. After requirements definition, there is process of review to verify and validate these requirements specifications. If review is successful then next phase begins otherwise changes are suggested for making in requirements. Next phase is proposed solution. After proposed solution, again review is conducted; if changes are suggested then again this phase is repeated otherwise next phase arrives. After proposed solution, the phase of Development arrives. After completion of development, testing phase arrives. If bugs are found, these are sent to development for debugging. If no bugs are found then finally Task is closed. Remaining activities are not in the domain of Telco Design and Development.

## 4 Undesirables in Present Environment

There are some concerns in the existing practices of International Turnkey Systems – Pakistan. These concerns were expected to be eliminated through application of CE. As the existing design approach is discussed, it is traditional sequential process, in which one phase can be initiated after the completion of other. It takes long lead time to complete. This approach is getting obsolete in modern competitive era.

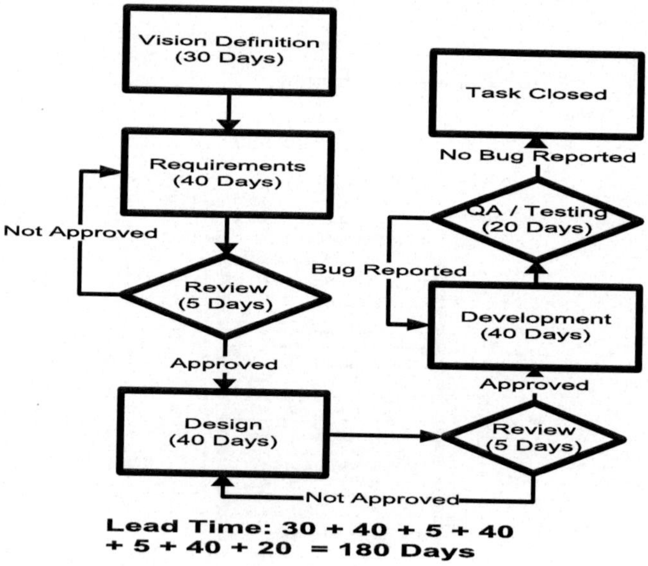


Figure 2 Existing software development process

Currently the TABS Billing Suite, on which Telco Design & Development Group is working, is built on traditional structured design principles. The tool used for development is getting obsolete. As shown in figure 1, team structure is Flat in which all resources are reporting to the Project Owner / Architect. The Architect is reporting to Design Manager. However, there is very less communication and coordination mechanism among them. There is no awareness of concept of empowerment and Centralized decision making mechanism is in place and team member are to just follow it.

Before the execution of Concurrent Engineering principles, a roadmap was drawn which proved to be very beneficial guideline throughout the implementation process. CE implementation requires the full support of Top Management. It is not sufficient, just to get approval of the project manager. All stake holders need to be given importance and involved. Some team members are selected to lead the team and to perform supervision responsibilities. It is also very important to create awareness and value among the team which is going to be affected. For this purpose, comprehensive discussion sessions are arranged on regular intervals. It is however not important to execute all the phases one by one. The most critical element is to gauge either the implementation is on successful path or not. For such purpose, a mechanism for regular assessment is opted and reviews are conducted at every milestone. At the end of implementation lessons learnt during the implementation are discussed. Based on the positive and negative implications of these lessons, decisions are made, either to carry on or drop the specific practice.

Implementation of concurrent engineering principles was not an easy task. Although people had shown their commitment and supported towards the implementation process, but still there were certain factors which need consideration for resolution. Which included lack of knowledge about CE, Cultural constraints and resistance from team members

## 5 Implementation Process

Concurrent Engineering principles which were required to be implemented in existing environment were identified and implemented in various activities considering the inherent modularity in the entire process as shown in Figure 3. The aspect of modularity in System Architecture was given consideration so that changes do not affect the system design in wider perspective. Component based system design approach is adopted. On the basis of architectural composition, the whole system usability and functionality is categorized into TBL Batched business, API/Web Service and TBL Setup, Rules & Configurations segments. The decomposition of the system into 'independent' parts (sections, modules) facilitates the understanding and change of the software. The Client Layer deals with user Input and provides the web interface to the application. The Business Domain

Layer incorporates the business logic by using the data access layer. Specifically, it provides access to the entity beans and business calculations etc. The Data Access Layer (Resources Layer) is responsible to access the enterprise information system (databases or other sources of information). The Data Source Layer is responsible for the storage of data. New team structure was proposed as shown in the figure 4.

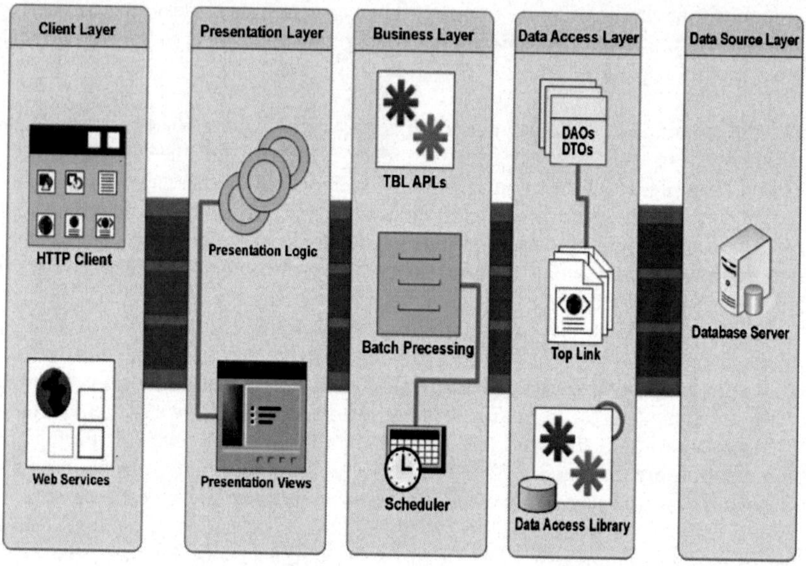


Figure 3 Modular Product Architecture

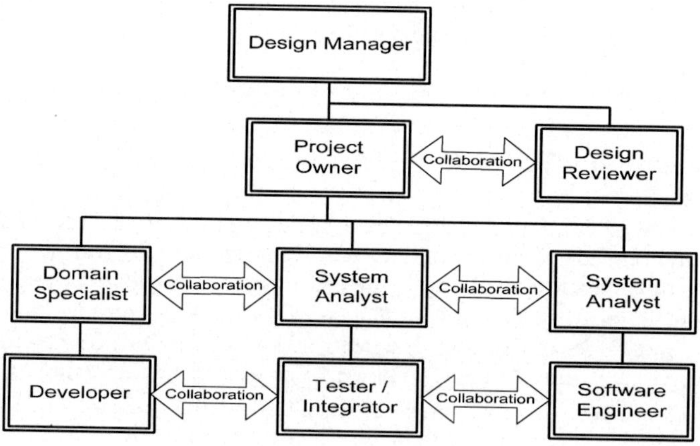


Figure 4 Proposed Team Structure

As shown in Figure 5, the traditional sequential design approach is replaced with object oriented design pattern where activities can be executed in parallel and save product development lead time.

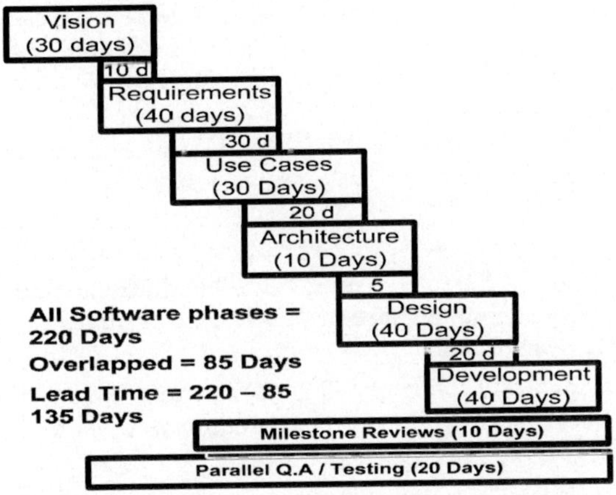


Figure 5 Parallel Software Development Process

# 6 Findings

By implementing concurrent engineering principles, improvements were seen in multiple areas. The activities and processes were observed to have become parallel and started executing simultaneously. The major advantages achieved through implementation of CE were in the following areas.

- Comprehensive orientation and knowledge sharing created sense of ownership.
- In 3 months, team achieved prominent status due to concurrent engineering.
- Processes of system design and development and documentation bindings improved the quality across the Company worldwide.
- Introduction of Object Oriented software development methodology resulted in reduction of software development time.
- Application of CE approach saved 45 days in software development project.

## 7 Conclusions

Implementation of concurrent engineering in ITS Design and Development group was not a simple task. Before initiating the process of implementation, a plan was set up while keeping in view the barriers and cultural constraints. The practices of Concurrent Engineering proved to be beneficial for overall business and resulted in considerable improvements in terms of lead time reduction, improved product quality, comprehensive set of documentation, business returns and customer satisfaction index of the company

## 8 References

[1] Samuel J, Keene and Karen C. Keene, 1993, " Reducing the Life Cycle Cost of software Through Concurrent Engineering", *1993 Proceedings Annual Reliability and Maintainability Symposium.*

[2] R.A. Sprague and K.J. Singh, 1996, "Concurrent Engineering in Product Development", *Proceedings of the IEEE International Conference on Industrial Technology, 1996.*

[3] Lani Hajagos, 1996, "The Role of Documentation in a Concurrent Engineering Environment", *IEEE, 1996.*

[4] M Winchester, 1995, "Enterprise Processes: Organization and Teams", *1995 The Institute of Electrical Engineers.*

[5] Farshad Rafi and Sam Perkins, 1995 "Internationalization software with Concurrent Engineering", *IEEE September 1995.*

[6] F.A Wilson and J.N Wilson, 1993, "Concurrent Engineering Information Systems" *1993 IEEE.*

[7] J. Daniel Sharman, 2004, "Optimal Modes and Levels of Integration, and the Identification of Cross Functional Coordination Deficiencies in concurrent Engineering", *2004 IEEE.*

[8] Jacky Estublier and Sergio Garcia, 2006, "Concurrent engineering support in Software engineering", *21st IEEE International Conference on Automated Software Engineering (ASE'06) 2006 IEEE.*

[9] Barry Brooks, 1996, "Realizing the benefit of concurrent engineering through BPR", *Computing and Control Engineering Journal, April 1996*

# Concurrent Planning Using Semantics-driven Reconciliation

Vitaly Semenov[1], Sergey Morozov, Oleg Tarlapan and Vladislav Zolotov

Institute for System Programming RAS, Moscow, Russia

**Abstract.** Concurrent planning, scheduling, and monitoring are challenging problems relating to effective communication and collaboration among key project participants like managers, planners, designers, cost analysts, and production engineers. Commercially available planning systems provide for a wide range of traditional network techniques such as Gantt chart, Critical Path Method (CPM), and the Program Evaluation and Review Technique (PERT), but have significant limitations for large-scale projects which have to be managed concurrently due to complexity and short deadlines. Being oriented on synchronously working users, such systems usually adopt client-server architectures, centralized data stores and pessimistic transaction models that create well-known performance, availability and productivity shortcomings. The research presented follows alternative approach to collaborative planning based on optimistic replication of project data and use of all-embracing, semantics-driven reconciliation method (SRM). The method enables to identify and to resolve conflicts in concurrently elaborated plans and bring them to semantically correct representation. Being implemented as a data exchange bridge between project management and 4D modelling systems, it supports effective collaboration through synchronizing the project plans and consolidating the results of individual work.

**Keywords.** Concurrent engineering environments, project management systems, planning and scheduling, optimistic replication, reconciliation

## 1 Introduction

Concurrent planning, scheduling, and monitoring are challenging problems relating to effective communication and collaboration among key project participants like managers, planners, designers, cost analysts, production engineers. These stakeholders own individual tasks, plans, and trade expertise, and yet still need to organize each activity in a way that delivers a shared program goal [2]. Tools to aid in project planning and scheduling based on activity durations, precedence relationships, timing constraints and resource allocations have existed for some time. Such tools include traditional network techniques, Gantt charts, Critical Path

[1] Professor (Computer Science), Institute for System Programming of the Russian Academy of Sciences; 25, Alexander Solzhenitsyn st., Moscow, 109004, Russia; Tel: +7 (495) 9125317; Fax: +7 (495) 9121524; Email: sem@ispras.ru; http://www.ispras.ru

D. D. Frey et al. (eds.), *Improving Complex Systems Today*,
Advanced Concurrent Engineering, DOI: 10.1007/ 978-0-85729-799-0_22,
© Springer-Verlag London Limited 2011

Method (CPM), and the Program Evaluation and Review Technique (PERT) [5] and are incorporated in most popular project management systems such as Microsoft Project, Primavera Project Management, Asta Powerproject and Synchro.

But these systems have significant limitations for large-scale industrial projects which have to be managed concurrently due to complexity and short deadlines. Being oriented on synchronously working users, they usually adopt client-server architectures, centralized data stores and pessimistic transaction models that create well-known performance, availability and productivity shortcomings [1]. In practice, to prepare and to analyze multiple alternative plans and schedules, participants usually prefer working in relative isolation on the optimistic assumption that all potential problems and occasional conflicts can be resolved after they have emerged. Taking into consideration that most users has own preferences in planning systems and tools, and integration capabilities are usually restricted by simple file exchange, the use of optimistic replication technologies looks very promising.

Recently the technologies have been used in such domains as collaborative workspaces, concurrent engineering environments, software configuration solutions, mobile databases, version control systems and distributed file services [7]. Sacrificing underlying ACID (atomicity, consistency, isolation, and durability) principles, which the traditional database management systems rely on, optimistic replication enables data to be accessed without *a priori* blocking, but with additional care to permanently maintain the data consistency. However, the mentioned above factor remains a central problem crucial for the applications proceeding with complex model data and semantically rich operations.

The paper presents an approach to collaborative planning based on optimistic replication of project data and use of all-embracing, semantics-driven reconciliation method (SRM) to consolidate the results of individual work. The method enables to identify and to resolve conflicts in divergent data replicas and bring them finally to semantically consistent and functionally meaningful representation. It avoids combinatorial explosion peculiar to many other methods. Due to formalization achieved the method can be implemented within various collaborative environments. Initially the method was developed for collaborative CASE applications and concurrently elaborated UML artifacts [8, 9]. In further research the method was successfully proved for achieving concurrency and consistency in project schedules under various timing constraints [10]. The presented paper focuses on concurrent planning problems to be effectively resolved by the SRM method.

In Section 2 we state the planning problem by specifying information model and semantic constraints imposed upon its underlying data elements. Section 3 briefly introduces the SRM method and provides useful references on its formal description and application details. A meaningful example of reconciliation of concurrently elaborated, divergent plans is given in Section 4. In Conclusions we summarise advantages achieved and outline promising directions for future investigations.

## 2 Problem statement

In the research presented focus is placed on traditional network techniques — Gantt charts and CPM method often used in planning and scheduling projects [5].

Gantt charts were introduced exactly 100 years ago by Henry Laurence Gantt, an American mechanical engineer designed a visual notation to supervise complex, repetitive manufacturing projects like making steam locomotives. Following this notation the project is visually represented as a bar chart depicting activities, precedence relations among them, and durations. The greatest advantage of such charts is seen at the planning stage. Here the user is required to think through a project logically and with sufficient details to establish clear project objectives, activities and the order in which they are to be carried out.

CPM was originally developed in the 1950's by the DuPont Company and Remington Rand Univac for managing plant maintenance and construction work. In the scheduling stage, CPM provides a realistic and disciplined method for determining how to attain the project objectives against time constraints. It enables ultimately to compute start and finish times for each activity as well as to determine the amount of leeway or "float" corresponding to each activity and to each its relationship. Moreover, it will identify critical path activities most sensitive and determinant to overall project duration.

Let us consider the planning problem from the information modelling point of view. Following to object-oriented methodology [6] let a project plan $X$ is a tuple $X = \{A, L, R\}$, where $A$, $L$, $R$ are finite sets of activities, links (precedence relations), and resources consumed by the project activities. Project start is defined by timing attribute of the plan *X.start*. Each activity $a \in A$ has its own attributes *a.UID*, *a.name*, *a.start*, *a.duration* meaning unique user identifier, name, start time and duration. To limit a timing interval the activity must start or finish on, optional attributes *a.low*, *a.high* are introduced with additional specificator *a.constraint* enumerating all typical constraints arisen in the planning applications: *start on*, *start after*, *start before*, *start between*, *finish on*, *finish after*, *finish before*, *finish between*, and *no constraint*.

The model presented admits that activities may aggregate child activities so that an activity containing at least one child can be interpreted as a work breakdown structure (WBS). For this purpose, an optional multiple aggregation *a.children* is used. Hereinafter a set of all the children aggregated by the activity $a \in A$ forms a transitive closure on the aggregation relation designated as $a.children^{+} \subseteq A$ (as opposed to *a.children* that is a set of own child activities). The optional attribute *a.parent* defines an inverse relation between the activity $a$ and its parent WBS.

Each link $l \in L$ interconnects associated preceding and succeeding activities *l.predecessor* and *l.successor* by evaluating the timing lag *l.lag* between them. Additional specificator *l.type* takes one of the predefined values: *finish-start*, *finish-finish*, *start-start*, and *start-finish* pointing out which ends of the activities are connected and what the timing interval for the lag is implied.

Each resource $r \in R$ has its own unique user identificator *r.UID*, name *r.name*, type *r.type* and may have graphic representation *r.representation*. Type of a resource takes one of the predefined values corresponding to *equipment*, *labour*, *material* or *space*. To be assigned to activities, the resource uses bi-directional multiple association *r.activities* with the inverse relation in the target activity *a.resources*.

The described data model can be specified at information modeling languages like EXPRESS [3] in a formal way. We omit such specifications, suggesting that in addition to the basic concepts introduced above the model defines semantic rules. The rules are constraints imposed upon data to satisfy necessary consistency and correctness requirements and to interpret the data in some common way. Violation of any semantic rule would result in the loss of data sense for further elaborations. For brevity we provide only formal definitions below.

**Definition 1.** A plan representation $X = \{A, L, R\}$ is called correct only if the following semantic rules are satisfied:

(S1) $\exists X.start \wedge X.A \neq \varnothing$

(S2) $\forall a \in A \Rightarrow \exists a.UID \wedge \exists a.name \wedge \exists a.duration \wedge a.duration \geq 0$

(S3) $\forall a_1, a_2 \in A \Rightarrow a_1.UID \neq a_2.UID$

(S4)

$$\forall a \in A: \begin{cases} a.constraint = start\,/\,finish\ on \Rightarrow \exists a.low \wedge \exists a.high \wedge a.low = a.high \\ a.constraint = start\,/\,finish\ after \Rightarrow \exists a.low \wedge \overline{\exists}\, a.high \\ a.constraint = start\,/\,finish\ before \Rightarrow \overline{\exists} a.low \wedge \exists a.high \\ a.constraint = start\,/\,finish\ between \Rightarrow \exists a.low \wedge \exists a.high \wedge a.low < a.high \\ a.constraint = no\ constraint \Rightarrow \overline{\exists} a.low \wedge \overline{\exists}\, a.high \end{cases}$$

(S5) $\forall a_1, a_2 \in A : a_1.parent = a_2 \Rightarrow a_1 \in a_2.children \wedge a_2 \notin a_1 \cup a_1.children^{+}$

(S6) $\forall l \in L \Rightarrow \exists l.lag \wedge \exists l.type \wedge \exists l.predecessor \in A \wedge \exists l.successor \in A$

(S7) $\forall l \in L : l.predecessor = p, l.successor = s \Rightarrow p \notin s \cup s.children^{+} \wedge$
$s \notin p \cup p.children^{+}$

(S8) $\forall l_1, l_2 \in L \Rightarrow l_1.predecessor \neq l_2.predecessor \vee l_1.successor \neq l_2.successor$

(S9) $\forall l \in L : l.successor.children \neq \varnothing \Rightarrow l.type = SS \vee l.type = FS$

(S10) $\forall r \in R \Rightarrow \exists r.UID \wedge \exists r.name \wedge \exists r.type$

(S11) $\forall r_1, r_2 \in R \Rightarrow r_1.UID \neq r_2.UID$

(S12) $\forall r \in R, \forall a \in A : a \in r.activities \Rightarrow r \in a.resources$

By satisfying the semantic rules (*S1*)–(*S12*), the project plan representation is suitable for the CPM analysis to be applied and for the visualization using Gantt chart. The maintaining data correctness is especially important for the discussed collaboration scenarios.

## 3 Semantics-driven reconciliation method

The presented research follows a trustworthy "three-way merge" technique that many researches have to address [4]. It generally serves to combine the changes made to a versioned data set in two parallel transactions. It requires a common ancestor version $X$ to be identified, and works by computing deltas $\Delta' = Delta(X',$

$X$), $\Delta'' = Delta(X'', X)$ from the ancestor $X$ to derived versions $X'$, $X''$ elaborated concurrently. The deltas are structurally represented as sets of chunks $\Delta' = \{\delta'\}$, $\Delta'' = \{\delta''\}$ each of which corresponds to a basic transaction operation. In conformity to widely acknowledged object-oriented methodology these operations are to create an object *new(obj)*, to modify its attribute *mod(obj.attr)* and to delete it *del(obj)*. The delta chunks are then combined to produce a merge delta $\Delta^* = Merge(\Delta', \Delta'')$, which being applied to the common ancestor $X$, yields the desired result: $X^* = Apply(X, \Delta^*)$. To obtain valuable results, it would be interesting to form a merge function so that chunks from both deltas are consolidated and maximum degree $|Merge(\Delta', \Delta'')|$ is reached on the sets $\Delta'$, $\Delta''$. In the ideal situation we could accept the merge function as $Merge(\Delta', \Delta'') \equiv \Delta' \cup \Delta''$ to achieve that. Nevertheless, separate chunks in the deltas may conflict with each other preventing correct mutual execution and disturbing the correctness of the final representation.

The SRM method can be successfully applied satisfying both a natural principle of consolidation of concurrent transactions and necessary requirements of correctness. The method allows the formation of a correct final solution $X^* = Apply(X, Merge(Delta(X', X), Delta(X'', X)))$ on the assumption that original and derived versions $X$, $X'$, $X''$ were correct. The assumption above is a rather reasonable hypothesis motivated by trustworthiness of the applications proceeding on local data replicas and running on local machines in isolated mode. In such consideration we try to narrow the explored domain of deltas guaranteeing the correctness of the result. The introduced concept of coherent deltas is defined as follows.

***Definition 2.*** A delta $\Delta^*$ is called coherent to the version $X$ only if all the operations $\delta \in \Delta^*$ can be evaluated correctly and the resulting version $X^* = Apply(X, \Delta^*)$ satisfies all the semantic constraints defined by the information model.

This concept implies that if original and derived versions $X$, $X^*$ are correct, and the delta $\Delta^* = Delta(X^*, X)$ is computed in proper way so that $X^* = Apply(X, \Delta^*)$, then the delta $\Delta^*$ is coherent to the version $X$. The SRM method reinterprets the reconciliation problem as a problem of forming a merge delta $\Delta^* \subseteq \Delta' \cup \Delta''$ coherent to $X$ from available coherent deltas $\Delta'$, $\Delta''$. For that purpose, formal analysis of delta chunks is carried out based on the underlying information model. Covering both data structures and semantic constraints imposed upon them, the analysis results in dependence, precedence and composition relations established among the delta chunks.

The dependence relations are defined using mathematical logics as follows. With each chunk $\delta \in \Delta$, where $\Delta = \Delta' \cup \Delta''$, we connect a corresponding logical variable $\delta^\sigma$. The variable takes true if the chunk must be accepted in producing a final merge delta $\Delta^*$ and false if the chunk must be removed from the delta by some reasons. In the notation applied the values 1 and 0 for negotiation status $\sigma$ correspond to the cases when the chunks are accepted and rejected. In such consideration a vector of logical variables $\{\delta^\sigma\}$ defines a resulting representation for the final delta as a subset of chunks for which negotiation status takes 1. Dependence relations are set among chunks that may belong to the same delta and to concurrent branch deltas. Simple example is an implication relation $\delta_1 \rightarrow \delta_2$

connecting a pair of chunks $\delta_1$, $\delta_2$. The relation implies that a merge delta $\Delta$ must contain $\delta_2$ on condition that it contains $\delta_1$ or in another way $\forall \delta_1, \delta_2 \in \Delta$, $\delta_1 \rightarrow \delta_2$: $\delta_1 \in \Delta^* \rightarrow \delta_2 \in \Delta^*$. Other sorts of binary relations are induced by known logic operations like equivalence, exclusive disjunction, stroke operation, and Peirce function [11]. Dependences, both binary and multiple ones, need to be proceed in a general way. To represent dependence relations, a reconciliation function is defined using normal disjunctive form:

$$D(\Delta) = \bigvee_{\bar{\sigma}} \delta_{i1}^{\sigma_{i1}} \delta_{i2}^{\sigma_{i2}} \delta_{i3}^{\sigma_{i3}} ..., \textit{ where } \delta_{i1}, \delta_{i2}, \delta_{i3} ... \in \Delta$$

The reconciliation function of multiple logic variables $D(\Delta)$ takes false if the corresponding chunk combination $\{\delta\} \subseteq \Delta$ is admissible and true if the combination is prohibited. Here the conjunctive terms go on all the sets of negotiation statuses $\sigma_{i1}$, $\sigma_{i2}$, $\sigma_{i3}$,..., for which particular combinations of chunks violate the imposed dependency relation.

Once the relations are defined, logic deduction methods, in particular, polysyllogistic methods, can be applied to form the merge delta $\Delta^* \subseteq \Delta' \cup \Delta''$ coherent to $X$. It can be done automatically in accordance with a chosen reconciliation policy or interactively giving the user an opportunity to take subsequent decisions and to control obtained results.

An important advantage of the applied method is that it allows the formalization. Using specifications of the underlying data model and its semantics, a system of the relations among delta chunks (more exactly, reconciliation function) can be automatically formed and logically analyzed. Formal analysis covers such peculiar semantic constraints as value domains for primitive and object data types, cardinalities, size of collections, uniqueness and ordering of elements in collections, optional and mandatory attributes, multiplicity of associations, cycles of associations, *etc.* Thereby, all-embracing SRM method has a mathematically solid foundation and allows effective implementations for solving similar problems arising in different application domains. For more details see the works [8–10].

## 4 Concurrent planning

Consider an example of applying the SRM method to the concurrent planning problem.

Figure 1 presents replicas of the concurrently elaborated project plan — common ancestor version $X$ (Figure 1a) and derived versions $X'$ (Figure 1b) and $X''$ (Figure 1c) obtained in the project management and 4D systems respectively. In the version $X'$ a new link *l3* between the activities *a4* and *a3* has been added as well as the activity *a5* has been deleted. So the delta $\Delta'$ takes the form {*new'(l3)*, *del'(a5)*}. In the version $X''$ resources with 3D representations *r1–r3* have been added and assigned to the activities *a1–a3* correspondingly, the activities *a4* and *a5* have been made children of the activity *a3* and interconnected by a new link *l4*. The delta $\Delta''$ takes the form {*new''(r1)*, *new''(r2)*, *new''(r3)*, *mod''(a1.resources)*, *mod''(a2.resources)*, *mod''(a3.resources)*, *mod''(a3.children)*, *mod''(a4.parent)*, *mod''(a5.parent)*, *new''(l4)*}.

Taking into account the semantic rules (*S1*)–(*S12*) the following logic dependencies can be established: *mod″(a1.resources)→new″(r1)*, *mod″(a2.resources)→new″(r2)*, *mod″(a3.resources)→new″(r3)*, *mod″(a3.children)~mod″(a4.parent)∧mod″(a5.parent)*, *del″(a5)⊕new″(l4)*, *new″(l3)⊕mod″(a4.parent)*. The semantic analysis identifies two conflicts between the concurrent deltas. The first conflict appears due to impossibility to set precedence relation for deleted activity that violates the rule (*S6*). The second conflict originates from the semantic rule (*S7*) that inhibits setting links between parent and child activities. Resolving the first conflict by taking the changes of the second transaction and the second conflict by taking the changes of the first transaction, we obtain the reconciled plan given by Figure 1d. The representation is correct satisfying the semantic rules (*S1*)–(*S12*).

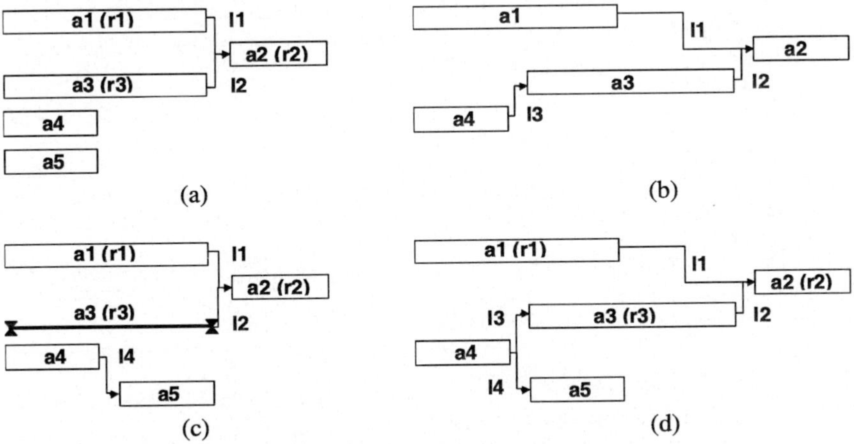


Figure 1. The original project plan replicas and reconciliation result

Essentially that all the meaningful results above have been obtained in formal ways using semantic analysis of concurrent deltas and applying logic deduction methods. This circumstance allows applying the SRM method to more sophisticated situations on the condition that concurrently elaborated project plans share a common information model and obey all its semantic rules.

The described approach to concurrent planning has been implemented as a data exchange bridge between two commercial software systems — Primavera Project Management (P6) system and Synchro 4D modelling system often reclaimed jointly in large industrial programmes. Although both products provide for own collaboration solutions, there is a need to synchronize project plans elaborated in different systems separately. Importantly that the systems adopt similar data models targeted on general-purpose project planning functions like Gantt charts and CPM computations. It enabled to specify the data model in formal way and to apply the presented SRM method. In addition to traditional export/import operations, the implemented software bridge provides for synchronization functions to reconcile the changes concurrently appeared in both systems and to consolidate the results of individual work carried out in relative isolation.

## 5. Conclusions

Thus, the approach to concurrent planning has been proposed. Being based on optimistic replication of project data and use of all-embracing, semantics-driven reconciliation method, it enables to identify and to resolve conflicts in concurrently elaborated plans and bring them to semantically correct and consolidated representation. The approach has been validated and showed significant simplification and formalization of the general problem. It inspires us to propagate the approach into new industrial applications connected with emerging information standards and models.

## 6. References


[1] Anderson T, Breitbart Y, Korth HA. Wool, replication, consistency, and practicality: are these mutually exclusive? In: Proceedings ACM SIGMOD, Seattle, WA, USA, 1998; 484–495.

[2] Boskers N, AbouRizk S. Modeling scheduling uncertainty in capital construction projects. In: Proceedings of the 37th conference on Winter simulation, Orlando, FL, USA, 2005; 1500–1507.

[3] ISO 10303-11: 1994, Industrial automation systems and integration — Product data representation and exchange — Part 11: Description methods: The EXPRESS language reference manual.

[4] Lindholm T. XML three-way merge as a reconciliation engine for mobile data. In: Proceedings of the 3d ACM international workshop on data engineering for wireless and mobile access, San Diego, CA, USA, 2003; 93–97.

[5] Moder J, Phillips C, Davis E. Project management with CPM, PERT and precedence diagramming, 3d edn. Van Nostrand Reinhold Company, New York, 1983.

[6] Rumbaugh J, Blaha M, Premerlani W, Eddy F, Lorensen W. Object-Oriented Modeling and Design. Prentice Hall, New Jersey, 1991.

[7] Saito Y, Shapiro M. Optimistic Replication. ACM Computing Surveys 2005; 37(1): 42–81.

[8] Semenov V. Collaborative software engineering using metamodel-driven approach. In: Proceedings of 16th IEEE International Workshops on Enabling Technologies: Infrastructure for Collaborative Enterprises, Paris, France, 2007; 178–179.

[9] Semenov V. Semantics-based reconciliation of divergent replicas in advanced concurrent engineering environments. In: Complex Systems Concurrent Engineering: Collaboration, Technology Innovation and Sustainability. Springer-Verlag, London, 2007; 557–564.

[10] Semenov V, Morozov S, Jones H, Semenova A. Consistent reconciliation of divergent project schedules under semantic & functional constraints. In: Product and Process Modeling: eWork and eBusiness in Architecture, Engineering and Construction. Taylor &Francis Group, London, 2008; 307–316.

[11] Zakrevskij A. Recognition logic. Editorial Press, Moscow, 2003.

# Research on Service Encapsulation of Engineering Software Based on Jini

Lingjun Kong[a,1], Wensheng Xu[a], Nan Li[b], Jianzhong Cha[a] and Jun Ji[a]

[a]Beijing Jiaotong University, Beijing, China.
[b]Beijing Technology and Business University, Beijing, China.

**Abstract.** In order to realize the effective sharing of resources of engineering software, a Jini-based method of service encapsulation of engineering software is proposed. The encapsulation templates including code template, configuration template and CLI (Command-Line Interface) template are constructed, and the process of service encapsulation of engineering software is discussed. A web-based system is developed to implement the proposed method. Engineers can encapsulate the engineering software as Jini-based engineering software services without the need of programming and can access the services through a standard Web browser. The sharing of resources of engineering software can be achieved. This method can also be used in the sharing of other resources on the dynamic network.

**Keywords.** Service encapsulation, Jini, engineering software, resource sharing

## 1 Introduction

To encapsulate engineering software tools as services is a very useful approach to realize the effective sharing of engineering software among different distributed departments in the enterprises. It can facilitate easy and full exploitation of the software resources such as CAD, CAE, CAPP, etc. which are normally expensive and run on high performance computers. After being encapsulated and provisioned as services through a service Web portal, engineering software tools can be shared and accessed simply through a standard web browser by remote laptop or desktop computers.

Jini is one of the network technologies that are suitable for building adaptive network systems that are scalable, evolvable and flexible [1]. The Jini technology which aims to support the rapid configuration of devices and software within a distributed computing environment can provide a fundamental underlying environment for the resource sharing of engineering software, and the software tools can be made available to remote clients as Jini services on the dynamic network. Based on Jini, a method of service encapsulation of engineering software

[1] Ph.D student, School of Mechanical, Electronic and Control Engineering, Beijing Jiaotong University, Beijing; Email: 08116294@bjtu.edu.cn.

D. D. Frey et al. (eds.), *Improving Complex Systems Today*,
Advanced Concurrent Engineering, DOI: 10.1007/ 978-0-85729-799-0_23,
© Springer-Verlag London Limited 2011

is proposed in this paper. With this method, engineers can encapsulate the engineering software as Jini-based engineering software services without the need of programming and users can access the services through a standard Web browser. The effective sharing of engineering resources can thus be achieved.

In this paper, the service encapsulation of engineering software is first analyzed (Section 2), then the encapsulation templates and process of service encapsulation of engineering software are discussed (Section 3), and an application is introduced (Section 4), and finally the conclusions are given (Section 5).

## 2 Analysis on Service Encapsulation of Engineering Software

In order to encapsulate the engineering software as services, the open interface of the software is needed to be used as the encapsulation interface of the service. Generally speaking, there are two kinds of open interfaces: one is the Application Programming Interface (API); the other is the Command-Line Interface (CLI) [2]. The encapsulation of engineering software based on API usually focuses on how to encapsulate some features of the software into a general service or how to encapsulate the engineering software for a specific engineering application [3-6], and it usually requires further development when new software or new features of the software need to be encapsulated. Because the API of different software varies and does not have a common pattern or a unified development language, it is not easy to construct a unified encapsulation interface for different engineering software. The CLI, which most engineering software supports, has not been fully explored yet in the encapsulation of engineering software. Actually it can be used as the encapsulation interface and it is suitable for constructing a unified encapsulation interface for different engineering software (see Section 3.1.3). This can achieve the automatic encapsulation of the engineering software without the need of programming. Because the services are provisioned on the network, the dynamic nature of the network [7] needs to be considered in the service encapsulation of engineering software.

## 3 Service Encapsulation of Engineering Software

To realize the service encapsulation of engineering software, the CLI of engineering software is used as the encapsulation interface of different engineering software, and Jini is used as the underlying service environment for the sharing of engineering software to adapt the dynamic nature of network.

### 3.1 Encapsulation Templates

Three different types of encapsulation templates are constructed to encapsulate the engineering software as Jini-based engineering software services. They are code template, configuration template and CLI template.

*3.1.1 Code Template*

The code template is for generating Jini-based services of engineering software. Typically, a Jini-based service program consists of three parts: service interface, service provider code and service requestor code. The service interface is to define the service contract between the service provider and service requestor. The service provider code has two main functions: one is to implement the service interface to provide real functions of the service; the other is to register the service and maintain the lease of the service in Jini lookup service (LUS) [1]. The service requestor code is used for invoking the service by dynamically downloading code from a code server, like Webster service in the RIO project [8].

Based on the characteristics of Jini service, different code templates have been constructed for generating the code jars of Jini-based engineering software service. The atomic operational strings which will be replaced when generating the jars of service provider and service requestor are defined by the Velocity Template Language (VTL) [9], and some of them are shown in Table 1.

**Table 1.** Operational strings of code template

| **Operational string:** | **Description:** |
|---|---|
| ${ServiceInterface} | Name of service interface |
| ${ExecuteMethod} | Name of service method in the templates of provider code and requestor code |
| ${ServiceProvider} | Name of Java class file of service provider |
| ${ServiceRequestor} | Name of Java class file of service requestor |

*3.1.2 Configuration Template*

There are three different kinds of configuration information in the Jini-based engineering software services: (1) Jini information: the addresses of lookup locators, names of lookup groups, communication protocol, lease time, etc.; (2) Software information: software name, software version, CLI template, software path, software description, etc.; (3) Service information: service name, service instance No., service owner, service description and etc.

Jini has a dynamic configuration mechanism which supports runtime configuration of Jini service, and there is a spectrum of choices between simple values and a full programming language [10]. Based on this mechanism, the configuration template which acts as an information carrier of the Jini-based engineering software service is constructed by VTL, and some of them are shown in Table 2. This template will be dynamically injected with the three kinds of information including Jini information, software information and service information. In the configuration file, every piece of the three kinds of information will be encapsulated into a Jini Entry object, the service provider will load the entries (Jini Entry objects) when the service starts and register the entries in the LUS, then the service requestor can dynamically query a certain service by defined entries.

**Table 2.** Operational strings of configuration template

| Operational string: | Description: |
| --- | --- |
| ${CommandLineTemplate} | Command-line interface template of engineering software |
| ${BatInputFileType} | Input file type in command-line interface |
| ${ServiceInterface} | Name of service interface |
| ${InstanceNo} | Supported service instance number |
| ${SoftwarePath} | Path of command-line executor of engineering software |
| ${ServiceID} | ID of service template of engineering software |
| ${ServiceImplID} | ID of the encapsulated service of engineering software |

*3.1.3 CLI Template*

Most engineering software tools, especially the commonly used larger-scale and commercial engineering software tools, have the command-line interface with different command-line pattern and a set of pre-defined parameters. Through our past experiences [11, 12], we synthesize the CLIs of different engineering software and have defined a set of atomic operational strings of the CLIs to generate different CLI templates for engineering software in order to realize the automatic encapsulation of engineering software.

**Table 3.** Atomic operational strings for generating CLI template

| Operational string: | Description: |
| --- | --- |
| ${InputFile} | Input file name in command-line interface |
| ${OutputFile} | Output file name in command-line interface |
| ${SoftwarePath} | Path of command-line executor of engineering software |
| ${WhiteSpace} | Represents a space in command-line interface |
| ${Quote} | Represents a quote, path of engineering software executor which contains space needs to be enclosed in quote marks |
| ${Other} | For further extending of command-line template |

Using the atomic operational strings in Table 3, the CLI templates of ANSYS, Pro/ENGINEER and HyperMesh software can be constructed as shown in Table 4. The atomic operational strings can also be used to generate the CLI templates for other engineering software.

**Table 4.** CLI templates of the ANSYS, Pro/ENGINEER, HyperMesh software

| Engineering software: | Script template: |
| --- | --- |
| ANSYS | ${Quote}${SoftwarePath}${Quote}${WhiteSpace}-b${WhiteSpace}-i${WhiteSpace}${InputFile}${WhiteSpace}-o${WhiteSpace}${OutputFile} |
| Pro/ENGINEER | ${Quote}${SoftwarePath}${Quote}${WhiteSpace}-g:no_graphics${WhiteSpace}${InputFile} |
| HyperMesh | ${Quote}${Quote}${SoftwarePath}${Quote}${Quote}${WhiteSpace}-c${InputFile}${WhiteSpace}-continue |

### 3.2 The Process of Service Encapsulation of Engineering Software

The process of service encapsulation of engineering software (Figure 1) is actually the process of filling up the encapsulation templates by relevant service template information and service information. The basic service template which consists of a set of code templates and a configuration template is used for generating different service templates. And the service template refers to the template for a certain type of engineering software with the standard service interface, CLI template, function, etc. For instance, the service template of the ANSYS software is used for the ANSYS software and can be used to generate different customized ANSYS services. The engineering software service is the Jini-based engineering software service after the encapsulation process.

In the first stage encapsulation, the relevant service template information will be encapsulated into the basic service template, such as interface name, method name, CLI template, service template description, operator name and etc. In the second stage encapsulation, the relevant service information including service name, service instance No., service description, software name, software path, software description and etc. will be encapsulated into the service template to generate engineering software service. After the encapsulation process, the engineering software service mainly has three components: provider jars, requestor jars and configuration file. The provider jars and configuration file are used by service provider to register the service in the LUS and encapsulate the CLI of the engineering software to provide real functions of the service. The requestor jars are used by service requestors to invoke the engineering software service.

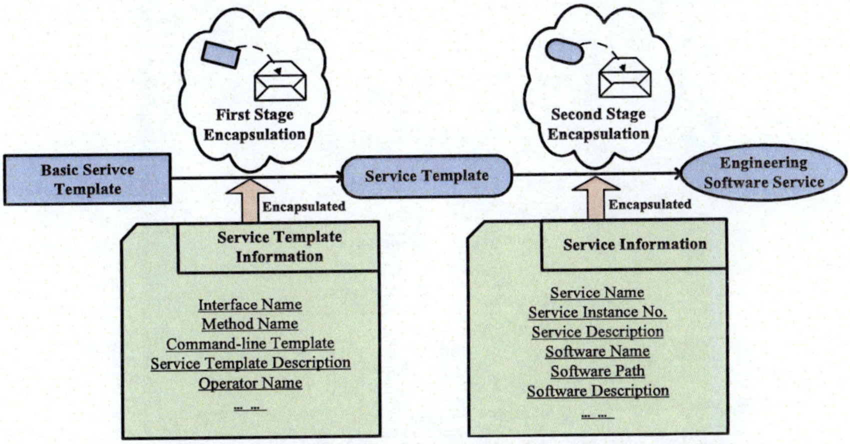


**Figure 1.** The process of service encapsulation of engineering software

## 4 Application

A web-based system is developed to implement the proposed method. Engineers can encapsulate the engineering software as Jini-based engineering software services without the need of programming and access the services through a standard Web browser.

Take the HyperMesh software as an example. In the first stage encapsulation (Figure 2), the engineer chooses the basic service template and fills up the relevant service template information, then generates the service template of the HyperMesh software which can be used to generate customized HyperMesh services of different HyperMesh software. In the second stage encapsulation (Figure 3), the engineer encapsulates the HyperMesh software as an engineering software service through the service template of the HyperMesh software by filling up the relevant engineering software information and deploys the encapsulated service in a selected host. Then the HyperMesh software is exposed as a Jini-based engineering software service on the network. Other engineers can access the service conveniently through a standard Web browser, create task by selected service, upload input task files, and retrieve task results (Figure 4).

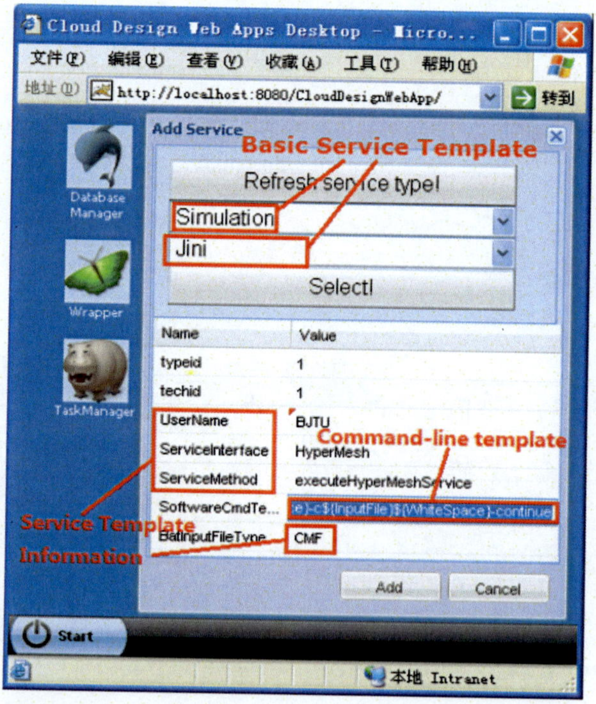


**Figure 2.** Generate service template of the HyperMesh software

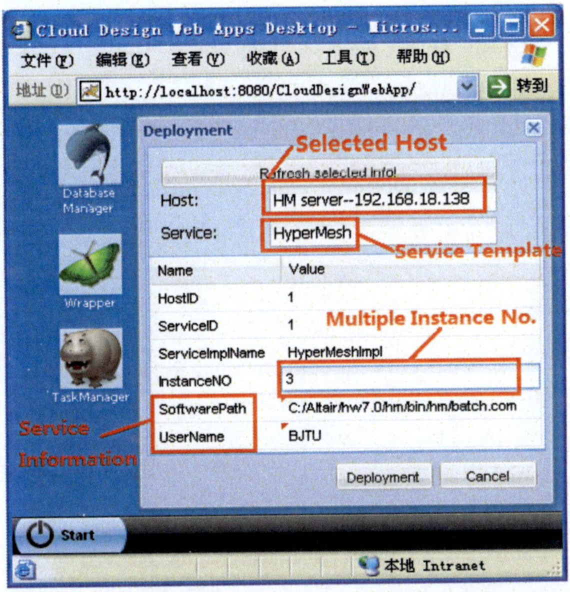


**Figure 3.** Encapsulate the HyperMesh software as a HyperMesh service

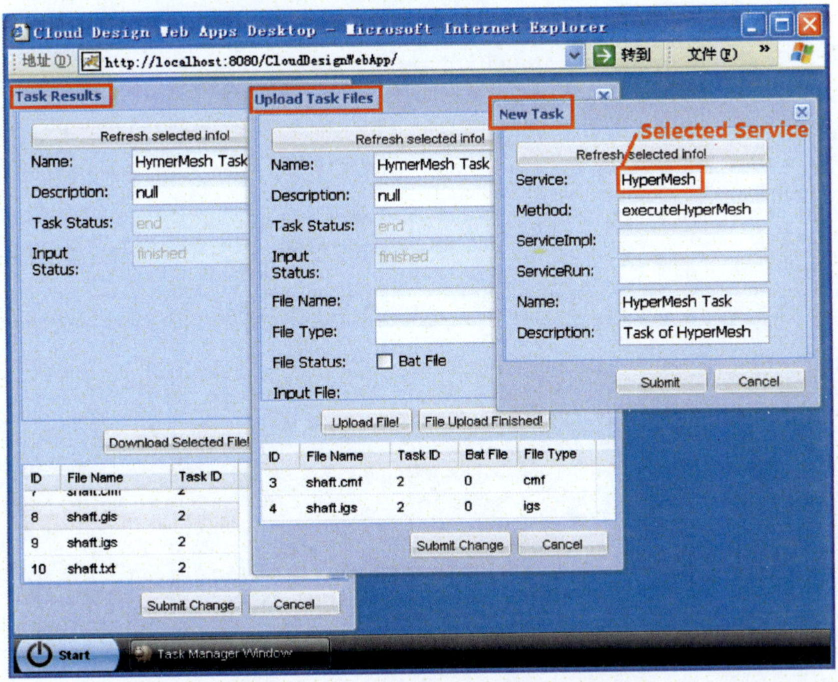


**Figure 4.** Submit task by the selected service

## 5 Conclusions

A Jini-based method of service encapsulation of engineering software is proposed in this paper. With this method, engineering software can be conveniently encapsulated as Jini-based engineering software services without the need of programming on the network, and the services can be accessed conveniently by engineers through a standard Web browser. The sharing of resources of engineering software can be achieved. This method can also be used in the sharing of other resources on the dynamic network.

## Acknowledgment

This research is supported by the Fundamental Research Funds for the Central Universities from the Chinese Education Ministry (2009JBM084).

## References

[1] Jini. Available at: <http://www.jini.org/>. Accessed on: Jan. 18$^{th}$, 2011.

[2] Command-line interface. Available at: <http://en.wikipedia.org/wiki/Command-line_interface>. Accessed on: Jan. 18$^{th}$, 2011.

[3] Liya Fan, Yunli Wang, Gang Chen, Yongbin Han, Tianyuan Xiao. CAD software encapsulation system based on CORBA. Computer Engineering and Applications 2003;39(9):61–62 (in Chinese).

[4] Bo Li, Rongqiao Wang. Distributed part parametric design system based on COM/DCOM. Journal of Beijing University of Aeronautics and Astronautics 2004;30(12):1195–1199 (in Chinese).

[5] Guanghui Zhou, Pingyu Jiang. Encapsulation and integration for networked manufacturing resource based on mobile Agents. Computer Integrated Manufacturing Systems 2002;8(9):728–732 (in Chinese).

[6] Guohai Zhang, Pingyu Jiang, Guanghui Zhou. Portalet-based encapsulation for CAx software. Computer Integrated Manufacturing Systems 2006;12(7):1134–1140 (in Chinese).

[7] Fallacies of Distributed Computing. Available at: <http://en.wikipedia.org/wiki/Fallacies_of_Distributed_Computing>. Accessed on: Jan. 18$^{th}$, 2011.

[8] The RIO project. Available at: <http://www.rio-project.org/>. Accessed on: Jan. 18$^{th}$, 2011.

[9] The Apache Velocity Project. Available at: <http://velocity.apache.org/>. Accessed on: Jan. 18$^{th}$, 2011.

[10] Jan Newmarch. Foundations of Jini 2 Programming. Apress, USA, 2006.

[11] Wensheng Xu, Jianzhong Cha. A manufacturing grid architecture based on Jini and SORCER. In: 16th ISPE International Conference on Concurrent Engineering. Springer, London, 2009;855–864.

[12] Jiaqing Yu, Jianzhong Cha, Yiping Lu, Wensheng Xu, M. Sobolewski. A CAE-integrated distributed collaborative design system for finite element analysis of complex product based on SOOA. Advances in Engineering Software 2010;41(4):590–603.

# Ontology Development for the Integration of CAD Models in a Collaborative Environment.

Samer Abdul-Ghafour[a], Parisa Ghodous[a], Behzad Shariat[a] and Eliane Perna[a,1]

[a]LIRIS Research Center, Claude Bernard Lyon I University, FRANCE.

**Abstract.** A major issue in product development is the exchange and sharing of product knowledge among many actors. This knowledge includes many concepts such as design history, component structure, features, parameters, and more. Thus, to efficiently share design information, the design intent should be persistently captured and the semantics of the modeling terms should be processed both by design collaborators and intelligent systems. Our research investigates the use of Semantic Web technologies for the exchange of "intelligent" CAD models, while maintaining the original relations among entities of the model. Thus, we have proposed an ontological approach based on the construction of a common design features ontology, used as an Interlingua for the exchange of product data. This ontology is represented formally with OWL DL. The ontologies integration process relies basically on reasoning capabilities provided by description logics in order to recognize automatically additional mappings among ontologies entities.

**Keywords.** CAD, Feature-based design, Semantic Web, Ontology, SWRL, Reasoning.

## 1 Introduction

The increasing number of product development tools entails effective collaboration among different partners. CAD data exchange is a critical point because CAD models represent the core result of product development, namely the product's shape. Actually, most of the current CAD systems provide feature-based design for the construction of solid models. Firstly, geometric modelers were regarded as the appropriate representation of product information, but more recently the feature approach has been proposed to enhance the capabilities of solid modelers [1]. Thus, efficient data exchange implies the possibility to integrate all relevant engineering data into a compound product model in a reusable form.

However, existing solutions for exchanging product information are limited to the process of geometrical data, where semantics assigned to product model are

[1] LIRIS (Lyon Research Center for Images and Intelligent Information Systems), Claude Bernard Lyon I University; 43 Bd. Du 11 novembre 1918, 69622 Villeurbanne, France. Tel: +33 (0)4 72 44 83 09 ; Fax: +33 (0)4 72 44 83 64; Authors e-mails adresses: sbatrouny@hotmail.com, {ghodous, behzad.shariat, eliane.perna}@liris.cnrs.fr; http://liris.cnrs.fr/

D. D. Frey et al. (eds.), *Improving Complex Systems Today*, Advanced Concurrent Engineering, DOI: 10.1007/ 978-0-85729-799-0_24,
© Springer-Verlag London Limited 2011

completely lost during the translation process. Current standards, such as STEP (STandard for the Exchange of Product model data) [2] have attempted to solve this problem, but they define only syntactic data representation so that semantic data integration is not possible [3]. Moreover, STEP does not provide a sound basis to reason with knowledge.

Our research investigates the use of Semantic Web technologies, such as ontologies and semantic web rule languages for the exchange of product data semantics among CAD systems. Ontologies have been proposed as an important and natural means of representing real world knowledge for the development of database designs. Furthermore, Ontology offers an additional benefit for exchanging CAD models by way of its reasoning ability. Hence, we propose in our paper a method of sharing CAD models based on the creation of a sharable "Common Design Features Ontology", called CDFO, to enhance the interoperability of CAD systems. Implicit facts and constraints are explicitly represented using OWL and SWRL (Semantic Web Rule Language). These technologies will facilitate collaborative engineering and improve product data management. However, it is not the intention of this article to introduce a complete ontology for features or shapes, but to illustrate the beneficial synergy between feature technology and ontology-based approach.

The rest of our paper is composed as follows: In section 2, we present an overview of related work for CAD models interoperability. In section 3, we present briefly our methodology for sharing CAD models. A generic view of our ontology is presented in section 4. Integration methodology is described in section 5, and examples are provided in section 6. Conclusions are drawn at the end of this paper.

## 2 Related Work

CAD data exchange problem is highly addressed by the international standard STEP which provides a system independent format for the transmission of data, in computer interpretable form. STEP, as various neutral formats, has been proven successful in the exchange of product geometry on a high quality level. Nevertheless, the problem consists in the ability of editing the outset model in the target system. Indeed, design intent including construction history, parameters, constraints, and features are potentially lost [4].

The problem of defining an appropriate and comprehensive feature taxonomy has been recognized as one of the central problems in order to share feature-based CAD models. Choi *et al.*, [4] define a set of neutral modeling commands describing the history of construction and modeling features. The data exchange is realized by using an XML file containing the history of commands used during the design process. Although this method allows the mapping between different terminologies, having the same meaning but different syntax, the mapping is performed only syntactically but not semantically. To consider the semantic aspect, Seo *et al.,* [3] extend this approach by developing an ontology based on the macro-parametric approach to achieve semantic interoperability.

The project PSL (Process Specification Language) [5], developed at NIST (National Institute of Standards and Technology) proposes an ontology, as neutral format, for the representation of information in the domain of manufacturing. The ontology is represented with the language KIF (Knowledge Interchange Format) [6]. Although PSL focuses on the representation of data relevant to the manufacturing process, efforts were made to match concepts of design and manufacturing. However, these efforts have been limited to the mapping of geometry and related information from design to manufacturing [7].

Dartigues *et al.,* [8,9] elaborated the drawbacks of STEP for integrating features between two phases of product development: design and process planning, and thereby proposed an ontological approach using the language KIF.

Patil *et al.,* [7] theorized an ontology-based framework to enable exchange of product data semantics across different application domains. The authors proposed as a tool the PSRL (Product Semantic Representation Language). In the prototypal implementation, descriptions and axioms are defined from a comparative study between two feature-based CAD systems: Solidworks and Unigraphics.

## 3 Ontology-based Approach

Our approach uses Semantic Web technologies for exchanging feature-based CAD models by considering semantics assigned to product data. This will enable data analysis, as well as manage and discover implicit relationships among product data based on semantic modeling and reasoning. As described in our previous work [10,11], we consider the interoperability issues at two levels of heterogeneities: syntactic and semantic. Treating the former, through homogeneization process, has led to the development of specific CAD applications ontologies, and a common ontology, CDFO, serving as a neutral formal (*cf.* section 4). On the other hand, we tackle the semantic heterogeneities by developing a methodology to explicitly state and exchange meaning of terminologies by associating them with their context. Thus, axioms and mapping rules are defined to achieve the semantic integration between the applications ontologies and the common ontology (*cf.* section 5).

## 4 Development of CDFO Ontology

In this section, we propose a feature-based design ontology, with a special attention to part design features. The aim is to share CAD models as instances of this ontology, enabling access to design knowledge for collaborative designers using semantic queries. A generic scheme of our ontology is illustrated in Figure 1. In the following, we describe key concepts defined in our ontology:

*Model_3D* represents the 3D CAD model, including among other data the product structure, parameters, and geometric representations. A model is designed and assembled to fill a functional need. It consists of either a *PartDocument* defining a manufactured component, or a *ProductDocument* as an assembly of components brought together under specific conditions to form a product and perform functions. Some properties could be defined for a model, *e.g.*, its version and material. In addition, a part is composed of a *BodyList*, an ordered set of *Body*.

A body could be either solid *Solidbody* or surface *SurfaceBody*. A solid body is in its turn composed of an ordered set of solid components *SolidBodyComponentList*, where a component may refer to a *Sketch*, or a *Feature* characterizing the shape of the body.

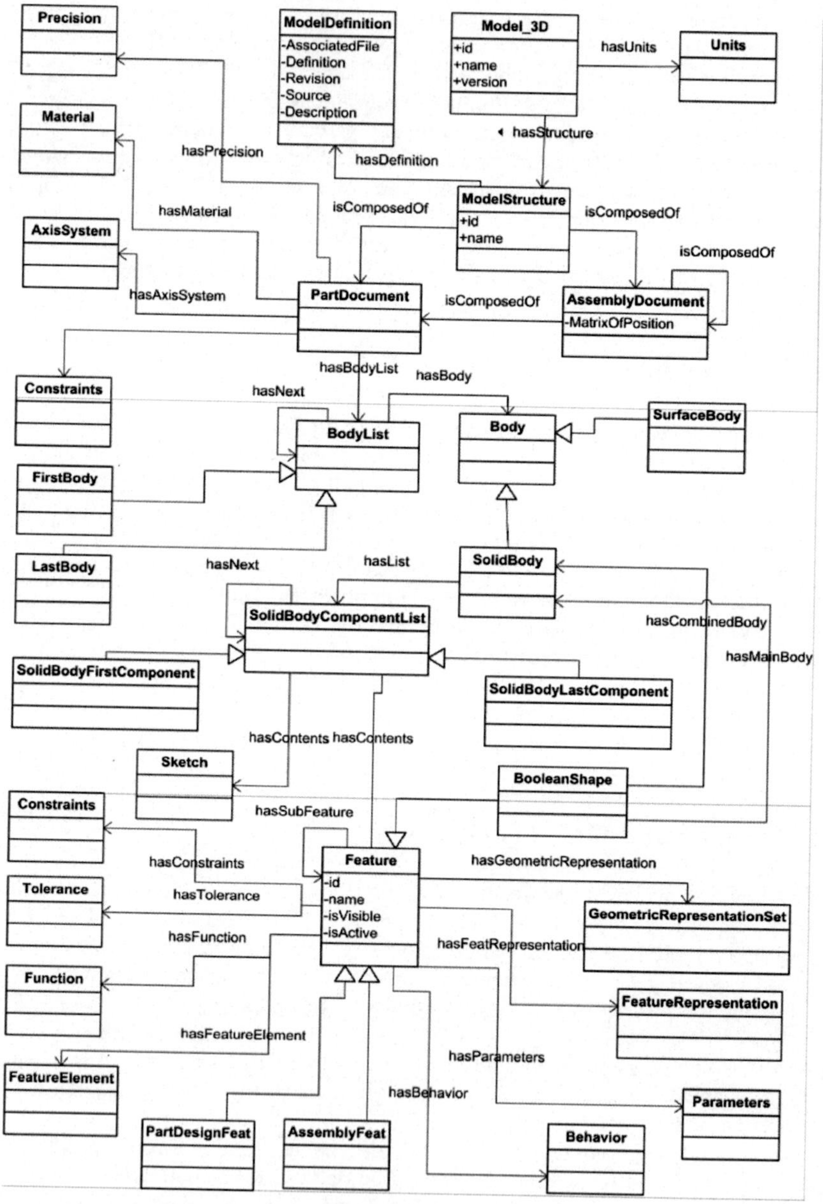


**Figure 1. A generic Schema of Common Design Features Ontology**

*Feature* is a main class of our ontology, which constitutes a subset of the form of an object that has some function assigned to it. It may refer to different specifications, such as design features *PartDesignFeat*, assembly features *AssemblyFeat, etc.* Compound features can be generated from primitive features and are related to their sub-features with the property *hasSubFeature. Function* represents what the feature is supposed to do. The feature satisfies the engineering requirements largely through its function. Moreover, there is a need to handle geometrical entities that constitute the feature, a created face for instance. For this reason, a *GeometricRepresentationSet* is added to our ontology to describe the way the feature is represented, such as its B-Rep (Boundary Representation). Other aspects related to a feature are constraints, parameters, tolerance, and behavior.

# 5 Semantic Integration using Reasoning

Our integration method is accomplished by defining axioms and rules which enable terms to be reasoned as being equivalent semantically, for instance, even though they are using different terminologies. OWL expressivity allows the definition of logical classes (intersection, union and complement operators), which enables automatic classification of product components. Thus, we create axioms and rules to map different entities. These axioms and rules constitue a basic element to perform reasoning operations. Consequently, we use ontologies reasoning ability to recognize automatically additional mappings between entities.

## 5.1 Axiomatization and Defining Rules

Axiomatization process aims at enriching ontologies semantics by the definition of axioms and rules between different entities. It's processed manually by experts of the domain. The axiomatization can be applied between entities of the same ontology, *intra-ontology*, or belonging to various ontologies, *inter-ontology*. Moreover, axioms can be defined for concepts and properties. However, axiomatization is realized thanks the high level expressiveness of OWL constructs, besides the use of SWRL to define formally more general relationships flexibly.

## 5.2 Semantic Mapping Recognition

Semantic mapping recognition aims at inferring automatically additional mappings between entities by using reasoning mechanisms provided by the inference services of a reasoner, *e.g.*, Pellet [12]. To integrate two ontologies, mappings are created explicitily to define correspondences between ontologies entities. These mappings, called "*Anchor Axioms*", are stored in a third *ad hoc* ontology, called mapping ontology. Thus, to perform the semantic mapping recognition, the mapping ontology is loaded into the inference engine in order to recognize automatically additional mappings between ontologies entities. Generated mappings are defined as a set of triplets $<E_1, Corresp, E_2>$ representing a correspondence between two entities $E_1$ and $E_2$, belonging respectively to ontologies $O_1$ and $O_2$. *Corresp* determines the type of relationship, *e.g.*, equivalence or subsumption. The main operations executed by the reasoner are based on the test of subsumption between two entities, or on the test of satisfiability of a concept. Thus, the reasoner performs a mutual subsumption test between two entities $E_1$ and $E_2$. For example, *Corresp* may refer to *Subsumption* if a sumbusption is inferred in only one direction, or to *Equivalence* if inferred in both directions.

Once the inferred mappings validated, they can be added to the mapping ontology as "*Anchor Axioms*" for further reasoning operations. Semantic mapping recognition could be applied at two levels: terminological and factual. The former enables recognition for descriptions of concepts and properties. The latter allows inferring the types of instances representing the CAD model in the target ontology.

## 6 Use case

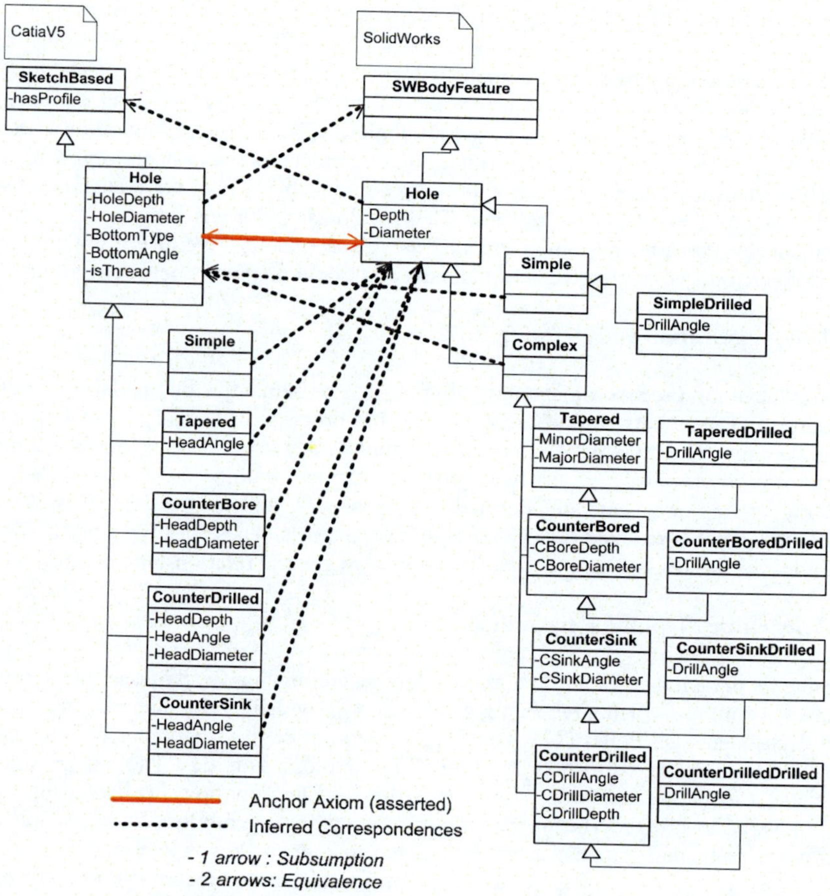


**Figure 2. Mappings on the feature Hole between CatiaV5 and SolidWorks**

Consider the integration of application ontologies of two CAD systems: CatiaV5 and SolidWorks. We focus on processing the feature Hole defined in both systems.

Figure 2 illustrates the classification of the feature *Hole* in both systems. It is defined in application ontologies as a primitive concept. Its subclasses represent different subtypes of holes, *e.g.*, complex hole. In this sample, we created an anchor axiom, illustrated with a solid line in Figure 2, to specify an equivalence between *Hole* concepts of the two systems. Once the inference engine is loaded, additional mappings have been inferred, illustrated with dash lines in Figure 2. These mappings represent subsumption correspondences between entities of both ontologies. For instance, all the CatiaV5 *Hole* subtypes are inferred as a subclass of *Hole* in SolidWorks. Furthermore, *Hole* in SolidWorks is inferred as a sketch-based feature in CatiaV5.

Subsequently, we considered the impact of creating mapping axioms on object properties defined in the applications ontologies of both systems. Consider the object properties defining the bottom angle of a hole in CatiaV5 and SolidWorks respectively: *Catia:hasBottomAngle* and *SW:hasDrillAngle*.

Descriptions of these two properties in the application ontologies are represented respectively as following:

*Catia:hasBottomAngle (Catia:Hole, Angle)*

*SW:hasDrillAngle (SW:ComplexHole, Angle)*

These two axioms define the domain and range for each object property. In fact, we added to the mapping ontology an anchor axiom defining an equivalence correspondence between these two properties, that's to say:

*Catia:hasBottomAngle* ≡ *SW:hasDrillAngle.*

This anchor axiom has led to the inference of a set of additional semantic mappings between entities of these ontologies. The list of inferred mappings is as follows:

*SW:SimpleDrilled* ⊑ *Catia:Hole*

*SW:TaperedDrilled* ⊑ *Catia:Hole*

*SW:CounterBoreDrilled* ⊑ *Catia:Hole*

*SW:CounterSinkDrilled* ⊑ *Catia:Hole*

*SW:CounterDrilledDrilled* ⊑ *Catia:Hole*

## 6 Conclusion

This paper presents a neutral format ontology of feature-based design modeling. The main objective of developing our ontology is the enhancement of collaboration among designers by considering product semantics. The exchange of Product shape is not sufficient. Other design engineering data should be considered such as parameters, constraints, features, design history, assemblies information, functionality, tolerances, material, etc. This could be carried out by providing the ability to access design knowledge and retrieve needed information not only asserted by designers, but also implicit facts inferred by reasoning engines.

Thus, this research takes full advantages of Semantic Web technologies to represent complicated relations that are scattered among form features and parts. We use a descriptive logic language, namely OWL DL to define formally concepts, their properties and their potential interactions in our ontology. More complex rules are created using SWRL. We focus on integrating various ontologies by means of ontology reasoning and matching issues to support more effective sharing of product semantics across heterogeneous life cycle processes.

# 7 References


[1] K Case, J X Gao, N N Z Gindy. The implementation of a feature-based component representation for CAD/CAM integration. Proceedings of the Institution of Mechanical Engineers. Part B. Journal of engineering manufacture ISSN 0954-4054 208 (1994): 71-80,.

[2] ISO. Industrial Automation Systems and Integration Product data Representation and Exchange, ISO 10303 - International Organization for Standardization (ISO), 1994.

[3] Tae-Sul Seo, Yoonsook Lee, Sang-Uk Cheon, Soonhung Han, Lalit Patil, Debasish Dutta. Sharing CAD models based on feature ontology of commands history. International Journal of CAD/CAM (2005).

[4] Guk Heon Choi, Dunhwan Mun, Soonhung Han. Exchange of CAD Part models Based on the Macro-parametric Approach. Int. Journal of CAD/CAM vol. 2, No.1, (2002).

[5] Michael Gruninger, Christopher Menzel. The process specification language (PSL) theory and applications. AI magazine 24 (2003): 63.

[6] M R Genesereth, R E Fikes. Knowledge Interchange Format (KIF) Version 3.0 Reference Manual, Logic-92-1. Computer Science Department, Stanford University, 1992.

[7] Lalit Patil , Debasish Dutta, Ram D Sriram. Ontology-based exchange of product data semantics. IEEE transactions on automation science and engineering ISSN 1545-5955 (IEEE transactions on automation science and engineering ISSN 1545-5955) 2, n° 3 (2005): 213--225,.

[8] Christel Dartigues. Product data exchange in a collaborative environment. Lyon: PhD Thesis, University of Claude Bernard - Lyon, 2003.

[9] Christel Dartigues, Parisa Ghodous, Michael Gruninger, Ram Sriram. CAD/CAPP Integration using Feature Ontology, Concurrent Engineering: Research and Applications 15(2):237-249, SAGE publications, ISSN 1063-293X, 2007.

[10] Samer Abdul Ghafour, Parisa Ghodous, Behzad Shariat, and Eliane Perna. Towards an Intelligent CAD Models Sharing Based on Semantic Web Technologies. 15th ISPE International Conference on Concurrent Engineering. Belfast: Springer, 2008 - b.

[11] Samer Abdul Ghafour, Parisa Ghodous, Behzad Shariat, and Eliane Perna. Using the Semantic Web for the Integration of Feature-based CAD Models Information. 2008 International Conference on Industrial Informatics and Systems Engineering (IISE 2008), System and Information Sciences Notes ed. Glasgow, UK. Communications of SIWN . ISSN 1757-4439. 2008.

[12] Bijan Parsia, Evren Sirin. Pellet: An owl dl reasoner. MINDSWAP Research Group - University of Maryland, College Park, MD. 2004.

# Ontological Modelling of the Aerospace Composite Manufacturing Domain

Wim J.C. Verhagen[a], Richard Curran[b,1]

[a] PhD candidate, Air Transport & Operations, Delft University of Technology
[b] Chairholder, Air Transport & Operations, Delft University of Technology

**Abstract.** This paper focuses on the development of an ontology for the aerospace composite manufacturing domain. It uses an amalgam of ontology development methodologies to arrive at a semi-formally declared ontology, which is tested for practical validity in a case study focusing on the construction of an architecture for a knowledge-based engineering solution for an aerospace Original Equipment Manufacturer (OEM).

**Keywords.** Ontology, Knowledge Management, Knowledge-Based Engineering

## 1 Introduction

In recent years, headlines in aerospace engineering news have frequently focused on the large-scale adoption of composite materials in new aircraft programs, as evidenced by the development of the Boeing B787 and the Airbus A350 XWB (Extra Wide Body). The development of the B787 and the A350 XWB represents a milestone as the major aerospace producers now switch from predominantly aluminium to predominantly composite structures. One critical aspect of this change is that it imposes the need to create and update existing knowledge bases to adapt industry to this new demand. Converted and newly developed knowledge bases must be supported by the right knowledge structure to ensure that the maximum benefit from the included knowledge is reached; in particular, the capture, sharing, consistent exploitation and re-use of knowledge throughout its life-cycle are important to ensure. To create such a knowledge structure, it is critical to have a solid understanding of the concepts of the attendant domain, as well as the interrelationships and the underlying concept attributes. Ontologies are widely used to model these aspects, resulting in reliable, verifiable and computer-interpretable mappings of a domain.

---

[1] Corresponding author. Chairholder, Air Transport & Operations, Faculty of Aerospace Engineering, Kluyverweg 1, 2629 HS Delft, The Netherlands. Tel: +31 (0)15 27 81513. Fax: +31 (0)15 278 2062. E-mail: R.Curran@tudelft.nl

D. D. Frey et al. (eds.), *Improving Complex Systems Today*,
Advanced Concurrent Engineering, DOI: 10.1007/ 978-0-85729-799-0_25,
© Springer-Verlag London Limited 2011

In this paper, a semi-formal ontology is developed which specifically addresses the aerospace composite manufacturing domain; the reported ontology is work in progress. Targeted ontological modeling of the aerospace composite manufacturing domain has not been performed in earlier research (as discussed in Section 2), but would contribute greatly to initiatives for sharing, management and re-use of knowledge in this particular domain. Furthermore, this paper briefly illustrates a case study in which the ontology is used to construct an architecture for a knowledge-based engineering (KBE) solution. This KBE architecture supports knowledge use and re-use in an aerospace OEM, thus validating the ontology approach in practice.

To enable the construction of the aerospace composite manufacturing domain ontology, existing perspectives on ontologies and supporting development methodologies are discussed in Section 2. Following this, the ontology construction process is illustrated in Section 3, which includes amongst others the elicitation of concepts, hierarchical structuring, relationship modeling via predicates and implementation in Protégé-OWL. The resulting ontology has been converted and implemented in a knowledge management application, which is elaborated in Section 4. Finally, conclusions, limitations and recommendations for further research are presented.

## 2 Ontologies: A Theoretical Perspective

Ontologies are defined as 'explicit (formal) specifications of a conceptualization' [1]. Four elements in this definition need further clarification. As Uschold [2] notes, a conceptualization can be seen as 'a world view, a way of thinking about a domain that is typically conceived and/or expressed as a set of concepts, their definitions and their inter-relationships'. An ontology necessarily includes a vocabulary of terms and a specification of their meaning [2]. Without specification, the set of ontology concepts would be variously interpretable by different sets of users. An ontology is explicit when it is or can be articulated, coded and stored in certain media, and readily transmitted to others. Finally, the formality of the ontology indicates the level of expression in an artificial, formally defined language, which extends to the possible ontology property of being machine-interpretable. Ontologies can be expressed along a range of formality degrees; this is one of the three key dimensions along which ontologies vary, as mentioned by Uschold [2]:

- **Formality:** the degree of formality by which a vocabulary is created and meaning is specified. Uschold [2] posits a formality continuum that moves from highly informal to rigorously formal (meticulous definition of terms with formal semantics, theorems and proofs of properties such as soundness and completeness). The ontology developed in this paper can be characterised as semi-formal (expressed in an artificial formally defined language).
- **Purpose:** Uschold and Gruninger [4] identify three main categories of use for ontologies: communication, interoperability and achieving system

engineering benefits. Both the first and the last are relevant for the aerospace composite manufacturing ontology, as substantiated in Section 4..

- **Subject matter:** Uschold [2] identifies three main categories, namely domain ontologies, task/problem solving ontologies, and meta-ontologies. The latter are also called foundational ontologies [5]. The aerospace composite manufacturing ontology is unequivocally a domain ontology, which is defined as 'focusing specifically on one particular subject domain (e.g. medicine, geology, engineering) or sub-domain' [2].

Though ontologies have been developed for the manufacturing domain [5,6], the aerospace domain [7, 8], the composite material and manufacturing domains [9, 10, 12] and the aerospace manufacturing domain [11], an ontology for the combined aerospace composites manufacturing field has not been developed and presented in literature yet. Given the developments in this field (e.g. the aforementioned B787 and A350 XWB, and other development programs), this constitutes a significant research gap. This paper addresses this specific research gap by proposing an aerospace composite manufacturing ontology and highlighting its first development steps.

To develop an ontology, a number of ontology construction methodologies are available. Examples include the methodologies by Uschold [2], Noy & McGuinness [3], Uschold & Gruninger [4] and the METHONTOLOGY methodology [13]. All these methodologies share common steps, though the exact representations may vary from methodology to methodology. The common steps have been summarized by Pinto & Martins [14]:

1. **Specification:** identification of the purpose and scope of the ontology.
2. **Conceptualization:** identification of the domain concepts and the relationships between concepts.
3. **Formalization:** organizing the concepts into class hierarchies and subsequent construction of axioms to formally model the relationships between concepts.
4. **Implementation:** codification of the class hierarchies and axioms into a formal knowledge representation language.
5. **Maintenance:** updating and correcting the implemented ontology.

For the construction of the aerospace composite manufacturing ontology, the first four steps of the five mentioned above have been carried out, while respecting the aforementioned supporting activities. The specification, conceptualization and formalization steps are performed in Section 3: Ontology Development for the Aerospace Composite Manufacturing Domain. Implementation in the context of this paper concerns both codification into a formal knowledge representation language (see Section 3.4) and practical implementation of the ontology as a backbone for a knowledge-based engineering application (Section 4).

# 3 Ontology Development for the Aerospace Composite Manufacturing Domain

The development of the aerospace composite manufacturing ontology development follows the steps identified by Pinto & Martins [14], which is reflected in the structure of this section.

## 3.1 Specification

The purpose of the aerospace composite manufacturing ontology is two-fold. In a wide sense, the purpose of this ontology development effort is to fill a research gap by introducing a new domain ontology. In a narrow sense, the purpose of the ontology is to support a knowledge management application by providing a knowledge structure, allowing the use and re-use of composite manufacturing knowledge in several business processes. The scope of the ontology is limited to its domain, aerospace composite manufacturing. Within this scope, the emphasis lies on the concepts and relationships that are employed in the business process (as shown in Section 4).

## 3.2 Conceptualization

To elicitate the applicable concepts and relationships for the aerospace composite manufacturing ontology, various sources have been employed. First of all, a small number (N = 4) of experts from a large aerospace manufacturer and integrator company have been interviewed. The results have been augmented by analysis of company sources (including product and process specifications), as well as analysis of general literature (e.g. [15]). Furthermore, a number of existing ontologies [5, 16, 17] have been studied to enable re-use of previous development efforts and to check for compliance of the aerospace composite manufacturing ontology with meta-ontologies.

The source analysis has resulted in a library of concepts and relationships. The ontology top-level concepts and their interrelationships have been expressed in Figure 1.

The Activity and Resource classes from the Knowledge Management (KM) ontology [17], the Activity, Constraint and Entities classes from the informal model of MOKA (Methodology and tools Oriented to Knowledge-based engineering Applications [16]) and the Product and Process plan classes from ADACOR (ADAptive holonic COntrol aRchitecture for distributed manufacturing systems [5]) have been incorporated and where necessary adapted to enable ontological interoperability.

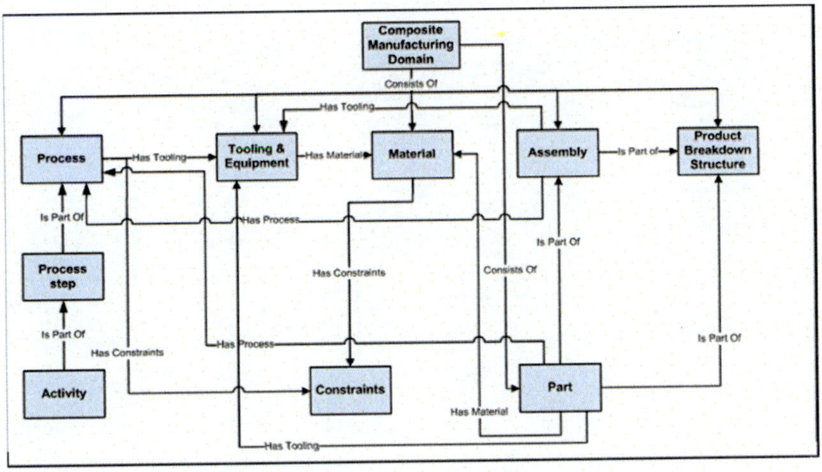


**Figure 1: Top-level ontology structure**

### 3.3 Formalization

The top-level classes have evolved into class hierarchies. A partial overview of the top-level classes and underlying class hierarchies is given in Figure 2.

During the development of the classes, the appropriate properties have been assigned. For example, the top-level Process class has Process_Cycle_Time and Process_Lead_Time as properties. Many potential properties have been identified; so far, just 80 datatype properties have been implemented as the focus of the ontology development effort lies more on class and relationship development.
The relationships (including the top-level relationships shown in Figure 1) have been formalized into the ontology as predicates. Examples include:

- HasPart(x,y): assembly x has part y.
- HasConstraint(x,y): part or process x has constraint y.

With classes and predicates in place, the elements for declaring axioms are available. The top-level concepts and their relationships have been modeled in axioms using the predicates. However, the aerospace composite manufacturing ontology is currently still in the first steps of expression into axioms; most sub-classes in the various hierarchies have not been expressed in axioms yet. Consequently, the ontology is still semi-formal. There are some examples of axioms available, but as these are based on literature and the input of a few experts and have not been validated in a large expert group, no examples will be given at this point in time.

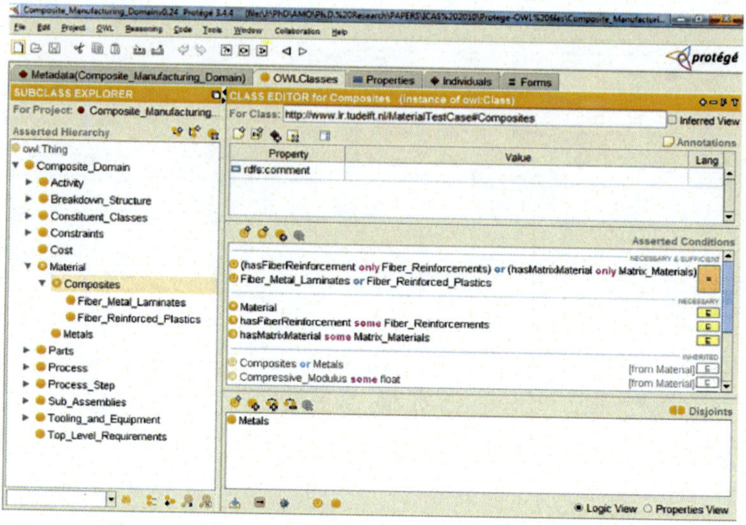


**Figure 2: Ontology implementation in Protégé-OWL**

### 3.4 Implementation

The class hierarchies, attendant properties and predicates have been implemented in Protégé-OWL. An example that also shows some of the class properties and axioms is shown in Figure 2.

The current iteration of the Protégé implementation has 188 classes with 152 restrictions. 46 predicates along with the inverse versions account for 92 predicates (object properties in the Protégé-OWL nomenclature). Finally, 80 datatype properties have been included.

## 4 Ontology Application as Knowledge Management (KM) Architecture

To validate the completeness and usability of the ontology, it has been used to generate the main knowledge architecture to support a knowledge-based engineering (KBE) application. This KBE application supports three business tasks of an aerospace OEM: the generation of composite manufacturing brochures, support of composite manufacturing cost modeling in the conceptual design phase and supporting composite ply continuity optimization for manufacturing constraints.

The knowledge architecture has been implemented in Ardans Knowledge Maker (AKM) [19]. AKM does not directly support importing ontologies, but it instead offers the possibility to implement class hierarchies directly. Furthermore, it has a facility to create knowledge models. These can be tied directly to the class hierarchy. Furthermore, these knowledge models can be used to define class properties and their slots (e.g. value ranges, value types). The knowledge models can also be used to prescribe the allowed relations between knowledge elements. Finally, the knowledge models can be associated automatically within the implemented class hierarchies. These provisions make it possible to 're-create' the ontology in AKM in a roundabout manner.

When instantiating a knowledge model, the resulting knowledge article is a class instantiation with the required class properties. It is possible to directly create so-called neighbouring knowledge articles, which are instantiations of a different class and automatically share a pre-defined relationship with the original knowledge article. In this manner, the ontology relationship predicates are implemented in practice.

The class hierarchies, class properties and relationships have been implemented in AKM. Together, these provide a knowledge structure which is subsequently used to support business tasks. More information about the practical implementation and the support for knowledge-based cost modeling can be found in literature [18, 20].

## 5 Conclusions

A semi-formal ontology for the aerospace composite manufacturing domain has been presented. This ontology can be used to support business tasks in this domain, either directly or through other applications. In this paper, it has been shown that the ontology can be used in a knowledge management environment to support engineering tasks.

The aerospace composite manufacturing domain ontology is subject to a number of limitations. First, the ontology is currently in a semi-formal format and possesses a low level of axiomatic expression. Furthermore, the concepts, their definitions and the concept relationships have been verified by a small user group only. Consequently, recommendations for future research are to establish an independent expert group that can serve to confirm or refuse ontology elements. By user group interaction, the informal expressions of the concepts and relationships of the current ontology can be tested, improved and validated. Subsequent expression of the ontology into an axiomatic form will enable the use of automatic inference capabilities, such as consistency and completeness checking and handling. Finally, further research will focus on expansion of ontology implementation to support business tasks.

## Acknowledgements

This work was conducted using the Protégé resource, which is supported by grant LM007885 from the United States National Library of Medicine.

# 9 References


[1] Gruber TR. A Translation Approach to Portable Ontology Specifications. *Knowledge Acquisition*, Vol. 5, No. 2, pp.199-220, 1993.

[2] Uschold M. Building ontologies: Towards a unified methodology. *Proceedings of 16th Annual Conference of the British Computer Society Specialist Group on Expert Systems*, 1996.

[3] Noy NF and McGuinness DL. *Ontology development 101: a guide to creating your first ontology.* Stanford University, 2009.

[4] Gruninger M and Uschold M. Ontologies: principles, methods and applications. *The Knowledge Engineering Review*, 11, pp 93-136, 1996.

[5] Borgo S and Leitão P. Foundations for a core ontology of manufacturing. In *Ontologies: A Handbook of Principles, Concepts and Applications in Information Systems*, Springer, pp. 751-776, 2006.

[6] Zhang L, Yuan W and Wang W. Towards a framework for automatic service composition in manufacturing grid, *Grid and Cooperative Computing - GCC 2005*, pp. 238 – 243, 2005. DOI: 10.1007/11590354_35.

[7] Malin JT and Throop DR, Basic Concepts and Distinctions for an Aerospace Ontology of Functions, Entities and Problems. *IEEE Aerospace Conference*, paper #1357, 2007.

[8] Lanfranchi V, Bhagdev R, Chapman S, Ciravegna F and Petrelli D. Extracting and Searching Knowledge for the Aerospace Industry. *ESTC*, 2007.

[9] Lukibanov O, Martinez I, Lenz T, McDowell J, Radcliffe C and Sticklen J. Socharis: The instantiation of a strategy for conceptual manufacturing planning. *AI EDAM*, vol. 14, pp 323-335, 2000. doi:10.1017/S0890060400144063.

[10] Susca L, Mandorli F, Rizzi C and Cugini U. Racing car design using knowledge aided engineering. *AI EDAM*, vol. 14, pp 235-249, 2000. doi:10.1017/S0890060400143057.

[11] Hale P, Scanlan J and Bru C. Design and Prototyping of Knowledge Management Software for Aerospace Manufacturing. In *Proceedings of the ISPE International Conference on Concurrent engineering (CE 2003)*, Madeira Island, Portugal, pp. 1083–1090, 26–30 July 2003.

[12] Wei H and Ma Y. A UML-based describing and organizing approach for complex domain knowledge and its usage in composite materials manufacture. *International Conference on Artificial Intelligence and Computational Intelligence*, Shanghai, vol. 3, pp.277-281, 2009.

[13] Fernándes López M, Gómez-Pérez A and Juristo N. Methontology: From ontological art towards ontological engineering. *Symposium on Ontological Engineering of AAAI*, 1997.

[14] Pinto HS and Martins JP. Ontologies: How can they be built? *Knowledge and Information Systems*, Vol. 6, No. 4, pp. 441–464, 2004.

[15] Mazumdar, SK. *Composite manufacturing: materials, product, and process engineering*, CRC Press, 2002.

[16] Stokes M (ed.), *Managing Engineering Knowledge: MOKA - Methodology for Knowledge Based Engineering Applications.* London, UK: Professional Engineering and Publishing Limited. 2002.

[17] Holsapple CW and Joshi KD. A Formal Knowledge Management Ontology: Conduct, Activities, Resources and Influences, *Journal of the American Society for Information Science and Technology.* Vol. 55, No. 7, pp. 593-612, 2004. DOI: 10.1002/asi.20007.

[18] Verhagen, W.J.C. et al. Knowledge-Based Cost Modelling of Composite Wing Structures. *International Journal of Computer Integrated Manufacturing*, Special issue on 'Managing Costs through Life'. DOI: 10.1080/0951192X.2010.531292

[19] Mariot P et al. Méthode, modèle et outil Ardans de capitalisation des connaissances. In: Cepadues, ed. *Revues des Nouvelles Technologies de l'Information, RNTI E-12,7èmes journées francophones Extraction et Gestion des Connaissances*, Namur, Belgium, pp. 187-207, 2007.

[20] Verhagen, W.J.C. et al., Knowledge-based optimization of ply stacking sequences for conceptual composite wing design , *Advanced Engineering Informatics*, Submitted.

# Part IV
# Knowledge Engineering

# A Knowledge-Based Approach to Design Automation of Wire and Pipe Routing through Complex Aerospace Structures

Christian Van der Velden[a1], Cees Bil[a] and Xinghuo Yu[b]

[a]*School of Aerospace Mechanical and Manufacturing Engineering, RMIT University, Melbourne, Australia*
[b]*School of Electrical and Computer Engineering, RMIT University, Melbourne, Australia*

**Abstract.** The automation of engineering processes through Knowledge Based Engineering methods and technologies can provide significant savings in project scheduling and cost, increasing competitiveness in a changing aerospace market. In this paper we present outcomes of a research project aimed at improving engineering automation capability through development of a tool for automatic rule based path-finding for the complex engineering task of aircraft electrical harness and pipe routing.

**Keywords.** Knowledge-Based Engineering, Design Automation, Wire Routing

## 1 Introduction

The current aerospace engineering environment is very different from that of two to three decades ago which saw high activity in the military sector in the years surrounding the cold war, and in the civil sector with high levels of tourism. A significant reduction in the number of major civil and military aerospace programs in recent years has led to highly cyclic work patterns of high then low levels of activity, often in response to the current economical and political climate. Accordingly, for engineering companies to remain competitive in this dynamic environment it is vital to exploit opportunities and respond to customer needs rapidly to gain a share of engineering workload resulting from major projects.

The automation of engineering processes through Knowledge Based Engineering (KBE) methods and technologies can provide capability to respond quickly to needs faced in the engineering market, and can deliver significant savings in project scheduling and cost [1]. Development of knowledge based software applications for process automation provides structure to product development processes and captures engineering and product knowledge for later reuse. In this paper we present outcomes of a research project aimed at improving

[1] Postgraduate research candidate, School of Aerospace, Mechanical and Manufacturing Engineering, RMIT University, PO Box 71, Bundoora, VIC 3083, Australia.



D. D. Frey et al. (eds.), *Improving Complex Systems Today*,
Advanced Concurrent Engineering, DOI: 10.1007/ 978-0-85729-799-0_26,
© Springer-Verlag London Limited 2011

engineering automation capability through development of a software tool for automatic rule based path-finding for aircraft electrical harness and pipe routing.

## 2 Problem Background

The increasing complexity of aircraft electrical systems has led to an increase in the number and size of electrical harnesses required to connect subsystems and equipment throughout the airframe. Electrical harness routing is a complex and largely manual task with numerous governing rules and best practices. Wiring looms typically comprise hundreds of harnesses all of which are manually routed, with engineers using personal knowledge and experience of the problem domain.

Also, adding to the problem size and complexity, subsystem design (including the wiring system) is often conducted in parallel with principle structural design in large scale projects. Therefore changes in structure and subsystem layout occurring over the development phase can impact wiring looms, requiring time-consuming and expensive rework. This can lead to lengthy delays in the aircraft development, as has been seen recently with the Airbus A380 wiring systems [2].

Major aerospace companies often have proprietary standards and practices for harness routing, which can vary between aircraft development programs depending on requirements. The generic process for harness routing involves manually creating a set of points in the CAD structural model from which the harness will be clamped to the main structure. Following this, the spine of the harness is passed through these points; ensuring sufficient clearance is given from structure, particular subsystems, certain types of harnesses, moving parts, and areas of high heat. The process can be largely trial and error, and often the only way to determine whether sufficient clearance has been allowed, is to make manual measurements in the CAD model which can be time consuming. These characteristics make the routing task a prime candidate for process automation.

## 3 Approach

The routing problem is encountered in numerous fields including electronics (e.g. microprocessors and printed circuit boards), navigation systems, and Artificial Intelligence (AI). Many algorithms have been developed in these fields, but few have been tailored and applied to the aerospace domain.

Algorithms for microprocessor design employ powerful path-finding algorithms including maze routers, channel and switchbox routers, and line routers [3]. These algorithms are driven by technology improvements in component manufacture including the ever-reducing size of interconnecting wires and number of two dimensional layers available for routing layout. Commercial software packages for multilayered printed circuit board design are also available, employing similar techniques.

Improving AI in computer games is another area which continues to drive intelligent path-finding development. The way in which non-player characters (NPCs) move and react in the game environment largely determines the realism of the gaming experience. To this end, numerous algorithms have been developed, incorporating various behavioural techniques including shortest path, stealthy movement for avoiding detection, cautious movement using cover to avoid damage, etc.

Developments in these fields have provided an excellent base of knowledge which can be utilized for engineering automation of aircraft electrical harnesses and pipe layout. The tool developed incorporates path-finding techniques from microprocessor routing and game AI domains, together with knowledge modeling techniques for capturing design rulcs, enabling the model to be constrained in such a way as to produce a path which accurately satisfies requirements, minimizing the amount of manual work required. The system layout is shown below in figure 1.

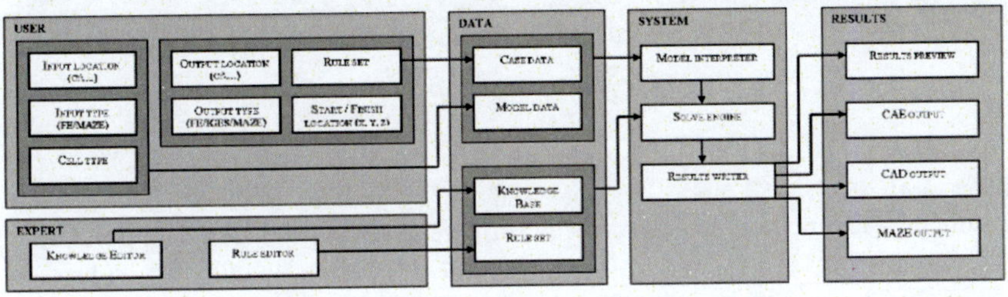


**Figure 1.** Automated router system structure

## 4 Test Model

System functionality will be described with reference to a test case consisting of a simplified model of an internal weapon storage area, typical of new generation fighter aircraft such as F-22 Raptor and F-35 Lightning II. Weapon bays are complicated and densely populated assemblies consisting of complex metallic structure, payload envelope, and various subsystems and equipment with hundreds of interconnecting electrical harnesses and pipes. Figure 2 shows one of the weapon bays of the F-35 Lightning II [4]. It shows a complex weave of harnesses and pipes throughout the length of the bay. As discussed above, the route for each harness is determined manually on CAD workstations. It does not take a lot of imagination to see that design of such a complicated loom is a very time consuming and resource intensive task.

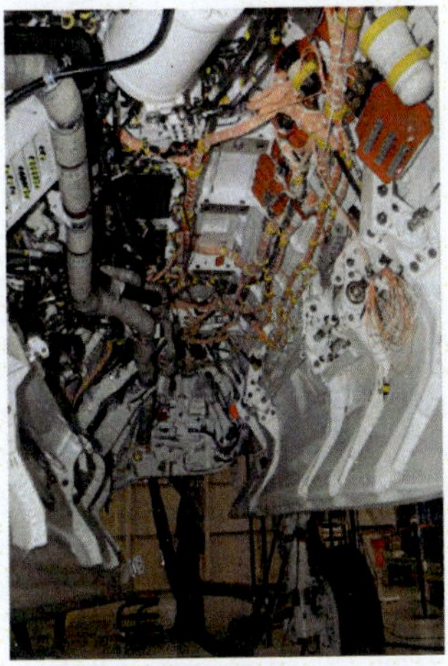

**Figure 2.** Weapons bay of F-35 Lightning II [4]

The test case for the routing tool is a simplified version of the assembly shown in figure 2, based on observations of the CAD model of the F-35 weapon bay structure and subsystems.

The test geometry consisting of principle structure (excluding bay doors), arbitrary subsystems, and sample payload was modelled using CAD software and is shown in figure 3.

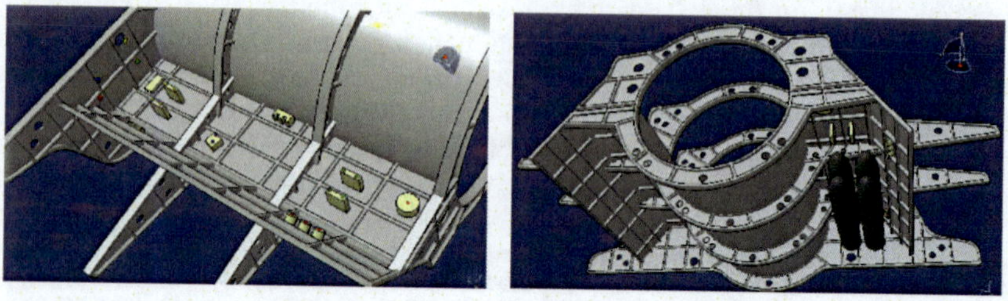

**Figure 3.** Test case geometry in CAD representation

Geometry is represented internally by the tool as a discrete, grid-based maze object with each grid node characterised by an integer address (in Cartesian coordinates) and node type (e.g. principle structure, subsystem, payload, pipe/cable type, searchable space, etc.). Categorisation of nodes within the solver is significant as it allows various obstacle types to be treated independently by rules implemented in the path-finding algorithm, such as following certain types and avoiding others.

## 5 Knowledge Modelling

Representation and implementation of knowledge by KBE applications is one of the most critical tasks in developing automated solutions. Domain knowledge generally exists in a number of forms which range in complexity. In cases where all rules are explicitly documented, modeling knowledge can be a relatively straight forward matter of quantifying and categorising rules. Engineering processes requiring tacit knowledge stored within experienced engineers can present difficulties in knowledge capture, and remains one of the main challenges in KBE. The routing system described in this paper supports path routing from numerous domains including electrical harnesses, hydraulic and pneumatic pipes, and fuel lines. Engineering knowledge of each particular routing domain must be accurately represented in the system for results to be valid. Effective knowledge capture processes are therefore important to obtain an accurate and complete coverage of all rules applicable.

The automated routing system includes a rule editor for modeling domain rules. The editor consists of a simple form containing a set of controls for defining rule types, conditions for validity, area of influence, and action to be taken. Families of rules for different routing domains and path types are stored in separate libraries in Comma Separated Variable (CSV) format.

Rules supported by the system currently include: bend radius, path profile, attract and repel rules causing paths to favour certain obstacle types and avoid others, turn penalty rules restricting amount of bends in the path, restrictions on movement (single, two and three dimensional movement), and clamping rules. Implementation of these rules within the solver is discussed below.

Solver settings, import and export options and hard-coded rules are defined within a database termed the "knowledge base" which can be customised for different routing applications. These are accessed through a knowledge editor.

## 6 Path Discovery

The algorithm implemented in the automated routing system extends the popular A* search algorithm used in shortest path problems in game AI [5]. In the basic A* algorithm, path movement is determined by evaluating a cost function for each node interrogated (Equation 1). Node cost, *f(n)*, is equal to the sum of the shortest distance from the source node, *g(n)*, and the estimated cost to the goal using a heuristic function, *h(n)*. At each iteration of the search, the lowest cost node is selected as the new node to expand. Provided the heuristic which calculates the remaining distance does not overestimate the distance to the target, the algorithm is both optimal and complete [5]. Example heuristic functions are discussed in reference 6.

$$f(n) = g(n) + h(n) \tag{1}$$

To tailor the algorithm to the complex rule-based domain of electrical harness design, the cost function is extended to include additional terms representing rules in domain libraries (Equation 2). The *g(n)* and *h(n)* terms are determined in the same way as normal A*. The *i(n)* term modifies total node cost, depending on proximity to given obstacle types, causing the path to favour obstacles of a particular type and avoid others. The magnitude of the *i(n)* term is determined by performing a local search within a given radius for each node interrogated in the main search. A separate *i(n)* term is added for each proximity-type rule in the library, *k*.

$$f(n) = g(n) + h(n) + \sum_{N=1}^{k} i_N(n) + T(n) \tag{2}$$

For example, a rule may specify that harnesses must be routed close to principle structure for clamping purposes. Such a rule would reduce the cost for nodes falling close to structural obstacles, resulting in a lower cost solution closer to structure. Other rules may state that harnesses must have a certain clearance from heat zones, moving parts and cables with particular electromagnetic sensitivities. Such rules increase node cost for nodes near these obstacle types, resulting in a lower cost solution away from these areas.

A turn penalty term, *T(n)*, is also introduced for models with complex geometry, encouraging the path to minimise unnecessary deflections. The contribution of the term is determined by the turn deflection angle. Mild turns are penalised less than harsh or backwards turns.

For the test case, 16 paths were routed, separated into four categories with four harnesses in each. The rule library implemented included a rule for harnesses to follow structure as closely as possible, minimising the length of unsupported runs for clamping purposes. Rules specifying each harness category to follow previously routed paths of the same type were also implemented.

## 7 Results Output

The system delivers output paths in a number of ways. Firstly, a simple three view path diagram referenced against input geometry is given in the software application itself. This gives the user a quick indication of the quality of the routing job in terms of whether paths were placed in the desired region of the solution space. For example, if a model features a number of cut-outs and lightening holes which are not properly constrained, the path may follow the outside of the structure rather than the inside. The simple preview can reduce analysis time for results which clearly require refinement. Several statistics relating to path quality are also displayed, including path length, number of turns, solution time, and number of nodes searched.

Secondly resultant path spines are output as CAD-readable wireframe IGES models consisting of straight segments connected with a user defined bend radius. Path models can be imported and integrated into the existing CAD geometry assembly, and detail added as necessary for a complete digital representation of the routed part using CAD software tools (e.g. thickness, detail to terminals, etc.). The output for the test case described above is given below in figure 4. The results clearly indicate successful implementation of the rule set including the rule for following structure and previously routed cable types.

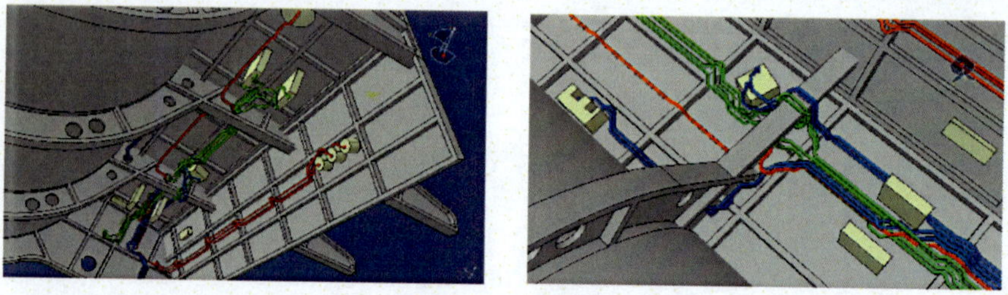

**Figure 4.** CAD output representation

The third output method is a Finite Element (FE) model comprising several component layers including input obstacles, routed paths and knowledge components describing the automated routing process. The purpose of this output method is to assist users in understanding results and implementation of rules by the solver. Obstacles encountered in the search space are included for referencing rule implementation against input geometry. Obstacles are represented by hexahedral elements colour coded according to category (e.g. primary structure, subsystems, cable category, etc.). Routed path spines are represented in wireframe using bar elements. Additional layers represent rules and search characteristics used in determining path placement, communicating design justification to the user. These layers include 3D maps showing locations where individual rules were implemented in the search space, allowing users to determine regions where particular rules are significant for any manual touching up required. A map of all nodes interrogated by the algorithm is also given, allowing users to determine whether the algorithm is considering the correct area in the search space, or whether additional constraints are necessary. The FE output for the test case is given in figure 5.

**Figure 5.** FE output representation

Various components in the solution can be activated independently to view interaction of rules using visualisation options available in the third party FE pre-processor used for viewing results. The two images above highlight different characteristics of the solution. The left image shows all nodes interrogated by the solver, referenced against input geometry. The image on the right shows all points on the reference geometry where the rule specifying paths to follow structure was implemented. As expected, the majority of nodes searched were in the vicinity of points where this rule was implemented, as it was a lower cost solution for points close to primary structure. Similar outputs are available for points where the various path-attract rules were implemented.

After a number of test runs of the system with various rule combinations were conducted, it was found that quality of resultant paths is closely coupled with the weight factor applied to individual rules. Thus a process for tuning the rule library for individual models is required to produce optimal results, which can be a time consuming process. Despite this, the system works very well as a proof of concept application. The solution time is sufficiently small that a number of solutions can be generated for various combinations of rules and weightings in a relatively short time compared to manual routing practices, allowing a large number of results to be assessed for suitability relatively quickly.

Further improvements to the system would focus on implementation of an optimisation process for tuning rule weights, leading to higher quality results without the manual trial and error process for applying various rule combinations. Such work would also require development of more effective methods for quantitatively assessing path quality.

## 8 Conclusion

The automatic routing tool outlined in this paper successfully implements knowledge based methods and techniques together with path-finding principles from microprocessor routing and game AI domains, to produce routed paths satisfying user defined design rules and constraints. Knowledge and rule editors simplify the knowledge capture process, extending capability to domain experts to implement new routing methods and rules for application to new domains (e.g. water pipes, air conditioning ducts, etc.). The benefits of using such a tool in industry are clear, with solution times of a few minutes per path for detailed, high resolution models (depending on the number of rules applied, and the degree to which the model is constrained), in comparison to the manual process which can take two to three days.

As opposed to many current "black-box" type of automated solutions where little if any detail is provided about the methods used in the automated processes, the automated routing tool outputs detailed information of rules and methods implemented within the search space for analysis of the solution. In cases where the automated solution does not produce paths of acceptable quality, users can use the rule output visualization as decision support for modifying sections of the path, maintaining validity against the relevant rule base.

Ultimately, project success requires rapid mobilization of resources to respond quickly to problems faced by both customers, and internally on the engineering floor. Implementation of knowledge based methods and technologies to develop automated engineering solutions can provide this capability, delivering time and cost savings, and improving competitiveness in the global engineering market.

## 9 References

[1] Smith, A. L., and Bardell, N. S. "A Driving Need for Design Automation within Aerospace Engineering". *11th Australian International Aerospace Congress*, Melbourne, Australia, 2005.

[2] "Airbus confirms further A380 delay and launches company restructuring plan", *Airbus Website*, URL: http://www.airbus.com/en/presscentre/pressreleases/pressreleases_items/06_10_03_a380_delays_company_restructuring_plan.html [cited 3 October 2006].

[3] Groeneveld, P., "Electronic Design Automation, Part 1". Technische Universiteit Eindhoven, The Netherlands, 2005.

[4] "F-35A Manufacturing", F-35 Joint Strike Fighter Program Website, URL: http://www.jsf.mil/gallery/ gal_photo_sdd_f35amanf.htm [cited 3 September 2007].

[5] Russel, R., and Norvig, P., *Artificial Intelligence A Modern Approach,* 2nd ed., Prentice Hall, New Jersey, 2003, Chap. 4.

[6] Haslum, P., and Gener, H., "Admissible Heuristics for Optimal Planning" *Proceeding of the 5th International Conference on AI Planning and Scheduling*, AIAA Press, 2000, pp. 140-149.

# The Cluster Grouping Approach of NM Forecasting Model for Book Publishing

Kenji Tanaka[a1] , Yukihiro Miyamura[a2] and Jing Zhang [b]

[a] Graduate School of Engineering, The University of Tokyo, Japan.
[b] Global Research Ltd

**Abstract**: Sales of new products peak after their release, then suddenly decline after the sales peak. This trend of sales has resulted in much waste and opportunity loss. In this study, the NM forecasting model is modified with the clustering grouping approach in addition to the original expert knowledge-based approach. As a case study, the proposed model is applied to a Japanese publishing sales case and showed better performance. This method was verified to reduce the return goods ratio through the real-time simulation of Japanese publishers' cases.

**Keywords**: forecasting, knowledge utilization, book retail, decision support

## 1 Introduction

Demand forecast decisively influences the stock disposal rate. As supply becomes excessive, the cost of final disposal will reduce revenue, and as supply becomes insufficient, in-store stock shortage will result in opportunity loss. The decision about the number of products to be produced and the production timing in such industries is extremely important for effective of revenues. The precision of demand forecast in planning decisions had significant influence on the final revenues. Since existing established methods of forecast are limited to a linear forecasting method that is only applicable to the sales trend of long-selling products, the forecast of new products immediately after their release when their behavior is unstable and nonlinear is extremely difficult. For this reason, sales forecasting whose goal is practical management of revenues has not been introduced despite its importance, due to the difficulty and low precision of forecasting. There have been studies on a forecasting method using the neural network approach, but since this method lacks stable forecasting precision and requires much preliminary preparation, it has not yet been applied to products with

[1] Assistant Professor, Department of Systems Innovation, Graduate School of Engineering, The University of Tokyo, Japan. E-mail: kenji_tanaka@sys.t.u-tokyo.ac.jp, +81-3-5841-6522

D. D. Frey et al. (eds.), *Improving Complex Systems Today*,
Advanced Concurrent Engineering, DOI: 10.1007/ 978-0-85729-799-0_27,
© Springer-Verlag London Limited 2011

many items (Ching-Chin et al [1], Decker et al [2], Kumar et al [3], Trappey et al [4]).

The subject of this study, book products, is a typical short life cycle product that peaks within one month after the release and then suddenly peaks out. Therefore, the book industry is hovering at an average rate of books returned of around 40%, which is a serious business problem for publishers. Further, due to the environmental changes in the book industry, the sales of publications have been decreasing since 1996 when they peaked, and in 2009, the sales of books and magazines were less than two trillion yen for the first time in 21 years, resulting in the changing or closing down of publishing companies. Improvement of the rate of books returned is an immediate business challenge for the industry (Research Institute for Publications [5]).

In order to improve the situation, since 2005, the University of Tokyo, book wholesale companies, and engineering companies have jointly promoted the BBI (Book Business Innovation) project, with the objective of data sharing among players of the industry and improving the rate of books returned through sales forecasting based on sufficient product demand and an efficient supply system. In this project, Tanaka et al. [6] proposed a book forecasting method using the tendency that sales of book products of the same group have certain common characteristics, validated it against existing forecasting methods, and confirmed that its forecasting precision was higher than the existing methods. This study was conducted as part of the BBI project; it aims to improve the NM forecasting method using the clustering approach for products that require high-precision forecasting, and to implement the method into a system for supporting publishers in reprinting.

## 2 Improvement of sales forecasting method

### 2.1 NM forecasting method

Tanaka et al. [6], focusing instead on the accumulated daily or weekly sales, proposed the NM method that enables sales forecasting of new products with a short life cycle, which was extremely difficult using the existing linear method, and it obtained a practicable level of forecasting precision. While this cannot be used for the forecasting of a single short life cycle product, it uses the characteristics that the accumulated sales trends within a group of similar products resemble one another and there is high correlation in the accumulated sales on any given two days (Nth day and Mth day). The NM forecasting method forecasts the accumulated sales in the future (until the Mth day) based on the actual accumulated sales at present (until the Nth day).

Since the NM performs its prediction using the empirical rule of similar group sales, the method depends on how the reference NM sales trend of the target product is associated with a similar product group, and whether the target product can be categorized in the appropriate NM group immediately after the release when the sales trend is not yet known. Tanaka [6] obtained a practical level of precision below the rate of books returned by adopting a knowledge-based grouping before release, using book classification by experts at book agents. However, the classification before release did not contribute to sufficiently improved precision. Therefore, in the current study, improvement has been made to the forecasting precision by using the filtering approach and clustering approach. In the filtering approach, peculiar titles that are included in a knowledge-based group are removed from the group. The clustering approach is a method of creating NM groups by clustering similar sales trends using the K-means method.

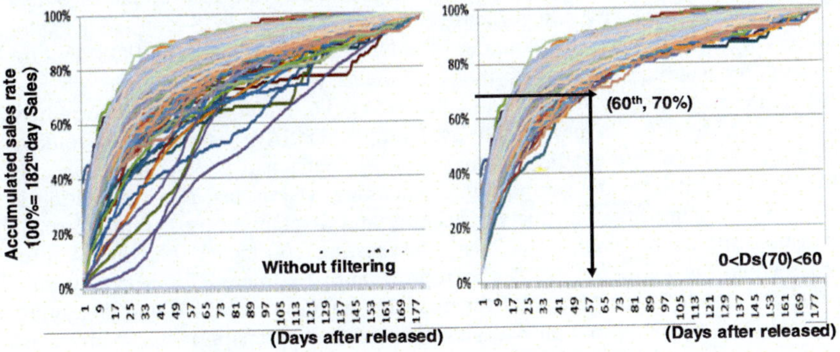


**Figure 1** Example of filtering (paperback book)

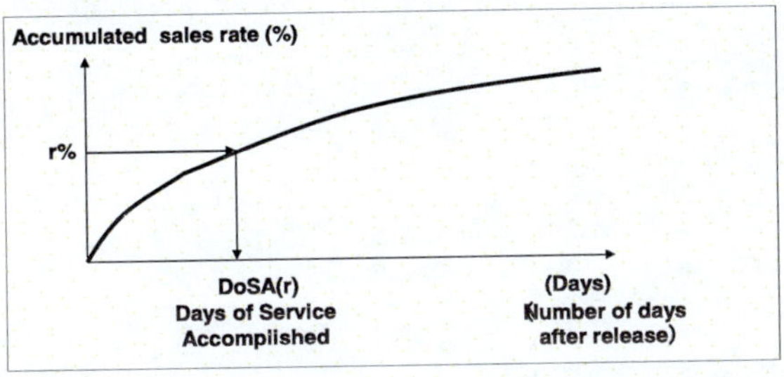


**Figure 2** Definition of Day of Sales Accomplished (DoSA)

## 2.2 Filtering method

Knowledge-based grouping before release may include a few titles (abnormal titles) with different sales trends due to human error in categorization, etc., causing reduce precision (Figure 1). This is mainly because titles with originally different sales trends are categorized in the same group.

Therefore, precision is improved by removing these peculiar titles from the NM group. For identification of peculiar titles, the Day of Sales Accomplished (DoSA) until the target sales rate is defined as an index representing sales trends. DoSA[days] indicates the number of days required to accomplish r% of target sales against the total sales during the period (Figure 2). By indicating the upper limit of the threshold of number of days as Dmax, and the lower limit as Dmin, it is possible to perform filtering by Dmin < DoSA(r) < Dmax.

Figure 1 shows an example of filtering at 0 < DoSA(70) < 60 in a certain group in the paperback book category. The horizontal axis shows the number of days after release, and the vertical axis shows the rate of sales accomplished, with the accumulated sales for 182 days being 100%. In this example, the title group that peaks after release and then decreases is separated from peculiar titles that are constantly sold, and the latter is removed by filtering. Figure 3 shows the flow until the determination of the NM coefficient using optimal parameters.

First, the NM group that requires improved precision is selected. Next, title candidates to be included in the NM group are extracted from this sample POS data based on knowledge-based categorization. Here, the sample POS data are the sales data of sample titles used in calculating the NM coefficient, and the POS data for verification mean the sales data of titles for forecast precision verification.

Further, parameter r for filtering and thresholds $D_{max}$ and $D_{min}$ are determined. Using these, filtering is performed by Dmin < DoSA(p) < Dmax to remove peculiar titles from title candidates, and the NM coefficient is determined and the NM table created from the sample titles remaining. Using this, NM forecasting is performed by using title data for verification, and assessment by absolute error is performed. In the combination (N, M) in forecasting, (N) means the Nth day after release when the publisher decides to reprint, and (M) means the Mth day after release for which forecasting is performed. By changing target r% and thresholds $D_{max}$ and $D_{min}$, the combination that yields the highest precision is discovered, and each value is determined.

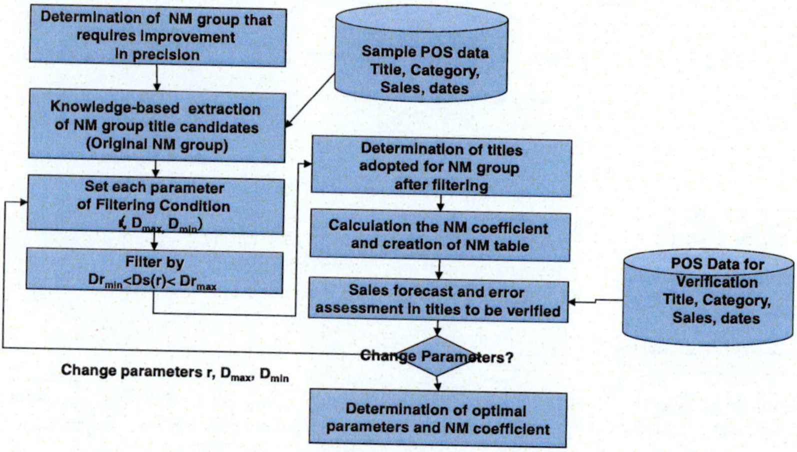


**Figure 3** Improvement of NM group by filtering

## 2.3 Improvement by clustering

Refinement using the clustering method aims to improve NM forecasting precision by categorizing a certain NM group into a cluster with similar sales trend. This improves the situation where there are titles that should originally have been categorized in separate groups due to too rough classification, and the precision of forecasting is reduced due to multiple sales types within the same category. The flow of determination of NM forecasting coefficients based on the clustering method is following. First, the target NM group is selected, and the titles that belong to the group after filtering are extracted. Next, the accumulated sales accomplished r(%) and the number of clusters k, which are variables, are determined. DoSA(r) is calculated based on r(%), and using these as elements, they are categorized into k clusters using the K-means method. On the basis of this categorization, the NM coefficient and NM table of each cluster are calculated. Figure 4 shows an example of categorizing title groups of the NM group of a certain book into three clusters on the basis of their sales trends.

In knowledge-based grouping by experts, the NM group to which the target title belongs was known before release. However, in grouping based on sales trends, the NM group to which the target title should belong is not known before release, and a daily estimate needs to be made on the basis of the sales record after release. At first, as Figure 5 shows, two days (p, q) before the day of forecasting (Nth day) should be set. Next, from their combinations, the NM forecast on the qth day is calculated for all clusters using the records of the pth day. By comparing the forecast value based on the NM coefficient of each cluster with the records as of the qth day, the NM coefficient of the cluster with the smallest error is used for forecasting as the coefficient of the target title on the Nth day. The setting pattern of (p, q) is determined as required such as {(p, q) = (N-14, N)}, indicating determination by comparing two weeks ago and the present on every Nth day.

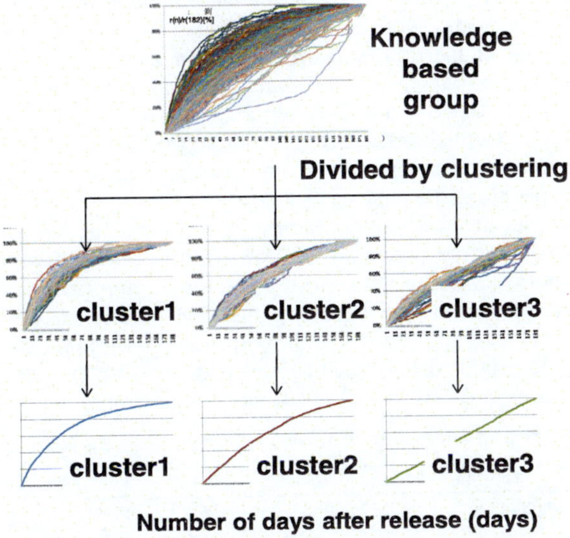


**Figure 4** Transition of accumulated sales accomplished and clustering based on sales trend (upper figure: target original NM group, middle figure: each cluster by 0 < DoSA(70) < 100 (k=3), lower figure: representative line of each cluster)

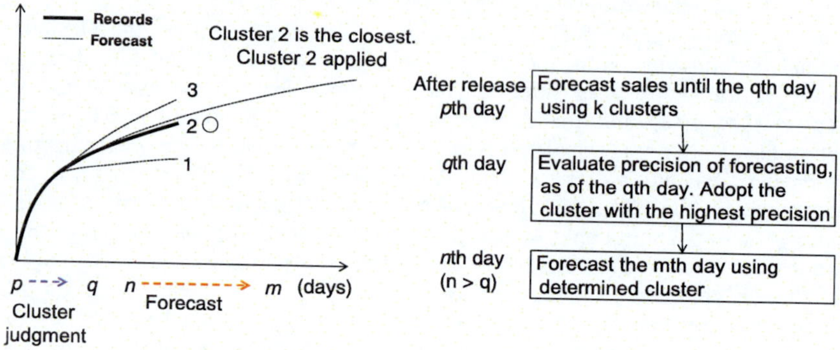


**Figure 5** Forecasting flows using sales trend clustering

# 3 Result

## 3.1 Comparative verification with existing forecasting

The forecasting method developed in the current study is verified using actual data from a major book agent in Japan. This company maintains a share of approximately 40% of the domestic market. Using 7,661 titles in the paperback book category released in Japan from March 2003 to September 2006 as sample data, 1,021 titles released from October 2006 to March 2007 were verified. Assessment was performed of the ratio of titles so that the absolute error is within ±30% as the index of error assessment. This was adopted based on the results of interviews with people in the publishing industry.

The forecasting method using filtering is verified. In the paperback book category, by changing the parameter of DoSA(r) to perform filtering of various patterns, the optimal filtering with the highest forecasting precision was sought. Table 1 shows the results. By obtaining the NM coefficient after removal of peculiar titles by filtering, the forecasting precision of the entire group was improved. In this group, the forecasting precision of $14^{th}$day after release was improved by 60% from 19.9% to 80.1% or more at maximum.

**Table 1** Improvement of forecasting presision by filtering

| (Nth-Mth) | 14→182 | 30→182 | 60→182 | Ave. |
|---|---|---|---|---|
| Knowledge based (original) | 19% | 3% | 16% | 13% |
| Filtering method | 81% | 94% | 97% | 90% |
| **Improvement** | **61%** | **90%** | **81%** | **77%** |

### 3.2 Method using clustering

Next, the forecasting method using clustering is verified. In the existing method, the large classification and middle classification by experts and the revised NM method [6] based on feedback for these classifications are prepared for grouping. In the method using clustering, the sales trend described in 2.3, the number of first-run copies, and the clustering method using both of these are prepared.
Table 2 and 3 show the comparison of forecasting precision between the existing method with the highest forecasting precision and the method using clustering with the highest forecasting precision. Here, Group 1 and Group 2 indicate publisher A and publisher B, respectively. The precision of forecasting improved by 11% and 6% respectively, for each publisher, thus increasing the number of titles that error was reduced less than 30%, which indicates the effectiveness of the forecasting method developed in the current study.

Table 2 Improvement of forecasting precision by clustering (publisher A)

| Group 1 | 14→182 | 30→182 | 60→182 | Ave. |
|---|---|---|---|---|
| Knowledge based (original) | 32% | 46% | 67% | 48% |
| Clustering Method | 45% | 55% | 79% | 60% |
| Improvement | 13% | 9% | 12% | 11% |

Table 3 Improvement of forecasting precision by clustering (publisher B)

| Group2 | 14→182 | 30→182 | 60→182 | Ave. |
|---|---|---|---|---|
| Knowledge based (original) | 39% | 52% | 79% | 57% |
| Clustering Method | 47% | 55% | 87% | 63% |
| Improvement | 8% | 3% | 8% | 6% |

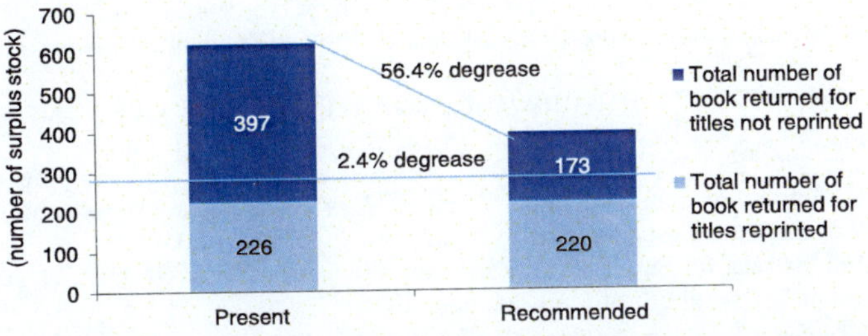


**Figure 6** Effect of reduction in the number of books returned by application of the system

### 3.2 Verification of publishing support system

Sales forecasting was performed using sales trend clustering shown in the current study, and reprinting simulation was performed for the numbers shown in the publishing recommendation. The target is all titles in the business category published from 2006 to 2009 by two medium-sized publishers specializing in business publications. Using the accumulated recommended printing records and sales records, the effect is verified through reproductive simulation of temporal progress. The printing lead time was set to two weeks. Scenarios related to reprinting recommendation can be established taking into consideration the lead time and forecasting precision. As a result of the reproductive simulation of the publishing support system, the number of books returned was reduced by 35% or 218,267 copies as a whole. Figure 14 shows the effect of reducing the number of books returned. As far as the titles reprinted, the number of books returned was reduced by 56%

## 3 Result

The conclusion of the current study is as follows

1. By developing a method of data filtering and clustering, the problem with the existing NM forecasting method was resolved and forecasting precision was improved by 6–11%.
2. Using the forecasting method developed, a publication management system for publishing companies was constructed. The results of simulation using actual data showed the possibility of reducing excess stock by 35%.
3. A demonstration simulation was carried out at publishing companies. The results showed that the system could be used for practical purposes.

## References

[1] Ching-Chin, C., Ieng A.I.K., Ling-Ling, W. et al. (2010) 'Designing a decision-support system for new product sales forecasting', Expert Systems With Applications, Vol. 37, No. 2, pp. 1654–1665.

[2] Decker R. and Gnibba-Yukawa K. (2010) 'Sales forecasting in high-technology markets: A utility-based approach', Journal of Product Innovation Management, Vol. 27, No. 1, pp. 115–129.

[3] Kumar, M. and Patel, N.R. (2010) 'Using clustering to improve sales forecasts in retail merchandising', Annals of Operations Research, Vol. 174, No. 1, pp. 33–46

[4] Trappey, V.C. and Wu, H.-Y. (2008) 'An evaluation of the time-varying extended logistic, simple logistic and Gompertz models for forecasting short product lifecycles', *Advanced Engineering Informatics*, Vol. 22, No. 4, pp. 421–430.

[5] Research Institute for Publications, Publication Indicator 2010, All Japan Magazine and Book Publisher's and Editor's Association (2010)

[6] Kenji Tanaka, A Sales Forecasting Model for New-released and Non-linear Sales Trend Products, the Journal of Expert System with Applications, 2010, ESWA4632, Volume 37, Issue 11, pp. 7387-7393

# A Prosthesis Design Based on Genetic Algorithms inthe Concurrent Engineering Context

Osiris Canciglieri Junior[a,1] , Marcelo Rudek[a,1] and Thiago Greboge[b,2]

[a]Professor in PPGEPS - Pontifical Catholic University of Paraná (PUCPR), Brazil.
[b]Mechatronic Engineer .

**Abstract.**Concurrent engineering is an important philosophy in achieving better time to market in new product development. While the use of design teams is achieving some success there is a need for modern software tools which support the design process to be radically improved. In this context, the research explores the use of three-dimensional images reconstructed from computed tomography that provides support for prosthesis modeling in CAD (Computer-Aided Design) system. This article presents a method for using a genetic algorithm (GA) that improves the virtual modeling applied to prosthesis conception. In the context, the GA was used to fit an ellipse in order to fill completely a defect in a skull. The bone piece that was missed in skull could be virtually created using the arcs extracted from the adjusted ellipses for each tomographic slice and it can generates profiles from the cloud of points in order to build a 3D geometric model using a CAD system. A synthetic image of the missing bone piece was built using the GA to fill its defect in the skull.

**Keywords.**Prosthesis Modeling, Imaging Processing, 3D Reconstruction, Product Development, CAD/CAM.

## 1 Introduction

Prototyping processes that involve image data of tomography can be seen in several applications levels, as in 3D virtual modelling of bone structure to prosthesis production. The prosthesis conception based on 3D reconstructed images is an important research area. In this case, a CAD (Computer-Aided Design) system can be used to modelling of bone shape and to generate the profiles used in machining preparing.

[1]Professor in the Department of Production Engineering at Pontifical Catholic University of Paraná (PUCPR), RuaImaculadaConceição, 1155, Prado Velho, Curitiba, CEP 80215-901, PR, Brazil; Tel: +55 (0) 32711304; Fax: +55 (0) 32711345; Email: marcelo.rudek@pucpr.br, osiris.canciglieri@pucpr.br; http://www.pucpr.br/cursos/programas/ppgeps/corpo_docente.php

[2]Master Degree of Production Engineering at Pontifical Catholic University of Paraná (PUCPR), RuaImaculadaConceição, 1155, Prado Velho, Curitiba, CEP 80215-901, PR, Brazil; Tel: +55 (0) 32711304; Fax: +55 (0) 32711345; Email: thiago.greboge@pucpr.br

D. D. Frey et al. (eds.), *Improving Complex Systems Today*,
Advanced Concurrent Engineering, DOI: 10.1007/ 978-0-85729-799-0_28,
© Springer-Verlag London Limited 2011

Basically, all the modelling process works from 3D reconstructed images from tomography scans. These images can be rotated and viewed in any plane, allowing acquiring of the morphologic features of the bone structures as in [1]. Three-dimensional figure generated from computed tomographic (CT) images can aid in understanding the region of complex anatomy, because the evaluation of this region is difficult [2]. Then the 3D surface representation must bring guaranties of real bone's measures. Some solutions to representation of skull curvature have been developed in the last years. Some of them are proposed based on the symmetry, as the bilaterally symmetric closed Fourier curves developed by [3]. In this case, we have some guaranties of the shape is very similar in the left or right side of human body as described by [4].

This idea can be applied in the skull problems, because the existing good bone area of one side can be mirrored to repair a missing area in opposite side. However it's true only when the local of defect is symmetric from the both sides how shows [5]. In several cases, the skull defect is not ever symmetric. In this case, a missing bone piece must be created from a criterion that guarantees their curvature. For this problem [6] proposes a method based on sub-cube pixel technique to obtain a surface model of a loss part of a defective skull. His work operates the 3D modelling by triangulations techniques.

In a different way in this research, we are proposing a manner to adjust an ellipse with parameters based on the skull border curvature to fill a missing area. The problem is that several ellipses can be created with similar shape of the bone border, differing themselves only by small values. Then, the question is to decide what the best solution is. In this context, optimization approaches can be useful, such as genetic algorithms (GAs) that can be used together to image processing in many ways, as presented in [7],[8] and [9]. Once defined the best ellipse for each CT slice, the missing bone area (hole) in a skull can be filled by a logic subtraction from existing border in original image. The volume can be reconstructed using the profiles exported to a CAD/CAM(Computer-Aided Design/Computer-Aided Manufacturing) system.The result is a 3D model of the virtual bone piece that was missing. The work covers the main steps of the proposed methodology and presents the progress of research in prosthesis modelling through a case study of a skull problem.

## 2 Proposed Method

The method proposed here is an extension of traditional methodology described in [10] that presents a conceptual model to prosthesis design. The figure 1 shows the context of process. Basically exits two fundamental parts. There is a part concerned with the medical requirements about the orthopedics problems and surgery process. Despite it is the basis of research this theme is not focused here. Also, there is a second one that deals about engineering of product development. This second aspect is the main objective of this research. The focus are: (i) The software requirements to image handling, (ii) the mathematical models to 3D reconstruction, (iii) the expert algorithms to image processing, and (iv) the tools used to virtual modeling in CAD. The interest here is to joint all involved areas in preliminary conception of product development, because there are some important questions to be answered before the virtual machining.

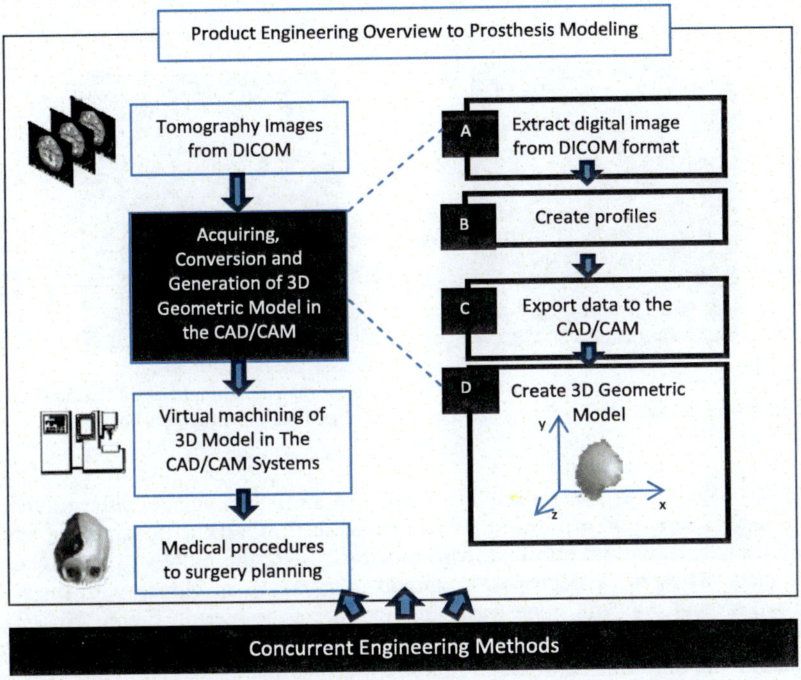


**Figure 1.** The Concurrent Engineering to support in Prosthesis Design.

According figure 1, the step (A) is the image acquiring from DICOM file and its respective bone's segmentation. This part deals how to extract the bone edge from each slice of tomography. These are the problems well known and they are documented in several researches as [1] and [11]. However, some problems can occur in this stage when the bone has a defect. The main problem in this case is if the edge of bone has a missing part (for example a hole in a skull can do this in the tomography slice). Then, in the step (B), the profiles creation is based on an automatic fill process. In this moment a correction is executed using a GA technique to create a virtual bone piece that not exists. This is the key point in this research. Once it's solved, the next stage (C) is to create the 3D cloud of points to be used in 3D reconstruction process (step D). It is also well documented in the literature and recently used in [12].

All stages of the presented method depends that we have a corrected representation of bone virtually processed. Then, the part of method of interesting here is focused in how to build a piece of bone missing in the image. In this research the GA deals about of the creation of an ellipse fitted into existing bone area in the image. An arc from the adjusted ellipse can be obtained for each CT slice, and so to create a virtual missing bone piece.

## 3 Ellipse Adjustment Processing

If a border at a slice has not a closed contour, a piece of correcting edge must be virtually built on the opening. The figure 2 shows the main idea of the process.

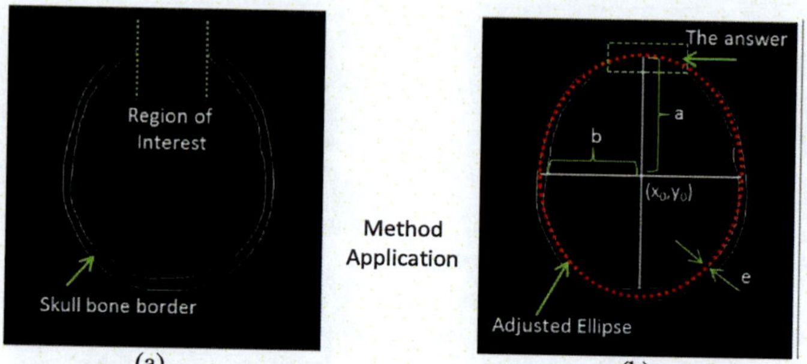


(a) (b)

**Figure 2.** The expected result after method application: (a) An example of a CT image slice with no closed border. (b) The virtual ellipse adjusted and parameters identification.

In the Figure 2(a) we can see an example of a CT slice image with no closed skull bone. The region of interest (ROI) can be seed in a frontal position of skull slice. There is a gap with no pixels information. Then, the problem is to create a bone piece adjusted. The proposed method is to create an ellipse with the size adjusted around the skull bone for each slice. It can be seed in Figure 2(b). The answer searched is an arc formed by a piece of the adjusted ellipse. The parameters to find are the Cartesian coordinates (x0,y0) of center, the major and minor axes ‘a’ and ‘b’ and also, the thickness ‘e’ of the adjusted ellipse. In this context, the GA is implemented to search these parameters. It starts with a set of solution formed by binary-encoded individuals (chromosomes) with the form described in Figure 3.

| Individual: | a | b | $x_0$ | $y_0$ | e |
|---|---|---|---|---|---|
| Size bits: | 8 | 8 | 9 | 9 | 4 |
| Lower Value: | 80 | 100 | 230 | 230 | 1 |
| Higher Value: | 200 | 200 | 270 | 270 | 8 |

**Figure 3**.The GA parameters.The individual format, the size of bits and the range of values.

In figure 3 it is shown the size in bits of entire chromosome and the range of values that can be assumed for each one. The limited range restricts the points in the initial population by specifying the lower and upper bounds. By the analysis of the image in tomography, a lower and a higher value can be defined. The unit of measurement is in pixels.

The main problem is to know the best solution, because a lot of ellipses can be created with different adjustments. In a genetic algorithm, values of fitness are the basis of the selection process. The selection mechanism searches the improvement of the next population quality. During selection, statistically only the most fitted individuals are chosen in order to allow them to take part in creation of the next population [12][13].It is used a fitness function that will find the best set of ellipse parameters among a population of ellipses. The fitness function is given by the equation 1.

$$f(x) = \frac{\sum_{i=1}^{l} \sum_{j=1}^{c} x(i,j) * E(i,j)}{\sum_{i=1}^{l} \sum_{j=1}^{c} x(i,j)} \tag{1}$$

where, the $l$ and $c$ are respectively the total of lines and columns in the image $x$. The $x(i,j)$ is the position of one pixel in the slice image. And, the $E(i,j)$ is the position of a pixel in the generated ellipse. The ellipse point $(x(t),y(t))$ can be obtained by,

$$x(t) = x_0 + a * \cos(t)$$
$$y(t) = y_0 + \mathrm{b} * \sin(t) \tag{2}$$

The equation 2 is the parametric form of ellipse equation. The $t$ is a variable with values from 0 to $2\pi$, $x_0$ and $y_0$ are the centers, and '$a$' and '$b$' are the axis. Here it is considerate the center is not in the origin of system. The parameters of center and axes can be evaluated by the GA that generates the best solution. After these values were obtained, the thickness must be still calculated.

The implementation of a problem in GA always starts from the parameter encoding. In this study, the solution is formed by binary-encoded individuals (chromosomes) with the form described in Figure 3. Then, a lot of ellipses can be drawn due to given range of possible parameters. We are looking for one solution in the search space which will be the best among all others. A possible basic form of a GA was presented by [13], [14] and [15]is:

<u>Step 1</u>. *Create an initial population filled with individuals generated randomly using a uniform distribution that are obtained from initial image within the minimum and maximum limits of the control variables and it is chosen as a parent population.*

<u>Step 2</u>. *Each individual in the current population is evaluated using the fitness measure according to the equation 1.*

<u>Step 3</u>. *Create a new population formed by applying the genetic operators to these chromosomes. The better individuals get higher chance to be selected in the next generation.*

<u>Step 4</u>. *Return until Step 2 to next generation. The loop starts from Step 2 and it is repeated while the stopping criterion is not satisfied.*

The adjusted ellipse has the values of '$a$' and '$b$' known, but it has a unitary thickness yet. It is useful to eliminate the main border of image, and it can be used to define an arc only along the gap. Then, this first ellipse marks the position of curvature. Once it defined, the GA must to operate with the thickness ('$e$' parameter).

# 4 The Case Study

It is proposed here a case study based on a hollow human skull. Here, it is simulated a hypothetic case when a piece of bone are missing. The hole in the bone was open of synthetic form. The analysis was made using some TC slices. The GA script was programed in MATLAB. Figure 4(a) shows the bone's pixels for a sampled slice where there is a defect. This is one that contains the initial set of values (pixels positions).

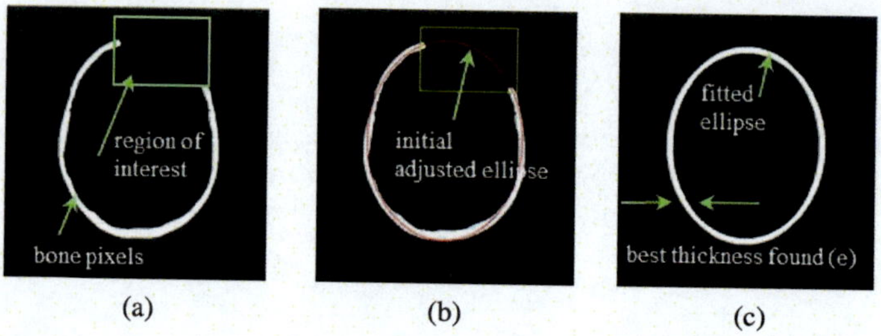


**Figure 4.** This is a sample of a slice with a defect. (a) The original pixels in CT image. (b) The initial adjusted ellipse. (c) The final ellipse with adjusted thickness.

In the figure 4(b) it is presented an initial ellipse adjusted after the genetic algorithm run. In this step, the center and axes of ellipse were obtained. The found partial solution has a unitary thickness yet. It is useful to eliminate the main border of image, and it can be used to define an arc only along the gap. Then, this first ellipse marks the position of curvature. Once it defined, the process must to operate with the thickness improvement. Despite these two phases apparently have existed, the all parameters values of ellipse were obtained together for the same fitness function in the same time. Figure 4(c) shows an example for the best fitness in the search space. Thus it is possible to known the best ellipse that can be used to form a curvature of bone shape.After the process finished, it is obtained the final solution highlighted with 80% in figure 5.

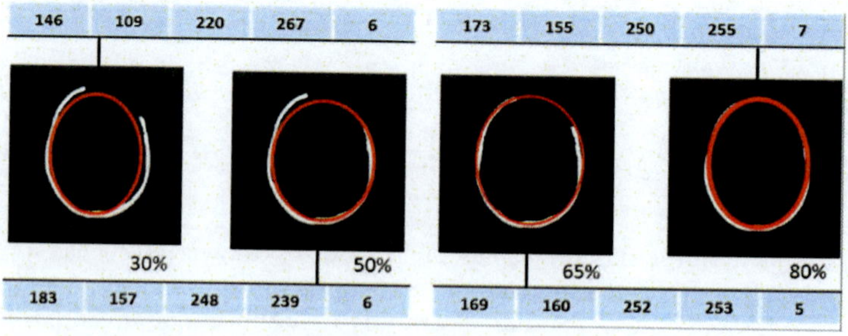


**Figure 5**. Some ellipses generated and the possible solution.

The fitness function gives the best parameters combination as seed in Figure 5. In this case, the percentage value of 80% permits to select the ellipse that is the solution. This solution contains: ($a$ = 173; $b$=155; $x_0$=250; $y_0$=255; $e$=7).Then it is still necessary to evaluate this result. Figure 6 presents a grafted piece forming the missing bone region.

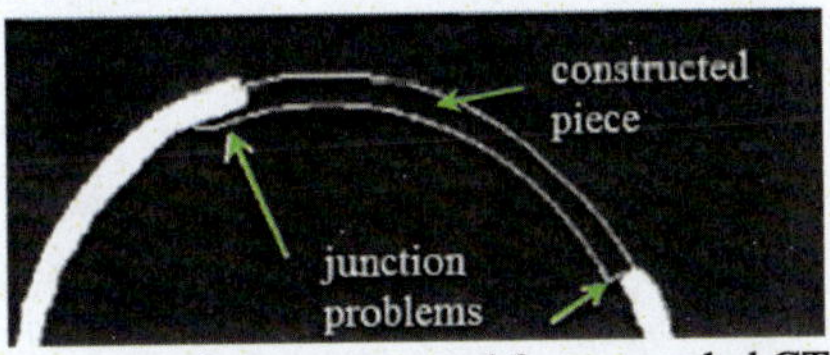


**Figure 6.** The best piece constructed for a sampled CT slice.

The best ellipsefound is not necessarily a useful solution. Some junction problem it can be seen in Figure 6. In this presented case, the both inner and outer borders are not completely aligned with the bone circumference. Then, by human viewpoint the solution is not good enough. For all simulations the result was very similar, and a few cases the adjusted ellipse was better. The good or weak result depends of the slice which was processed.Despite the problem exists, it is not focused in this research at the moment.According the proposal, here we are concerned with the process to generate a CAD model.

For the all computed slices, this process generates various arcs that can be superposed to make a 3D shape. The surfaceis modeled inSolidWorks system and the virtual area was created in the position of hole. The final solution gotten is the complete surface model as can be seen in the Figure 7.

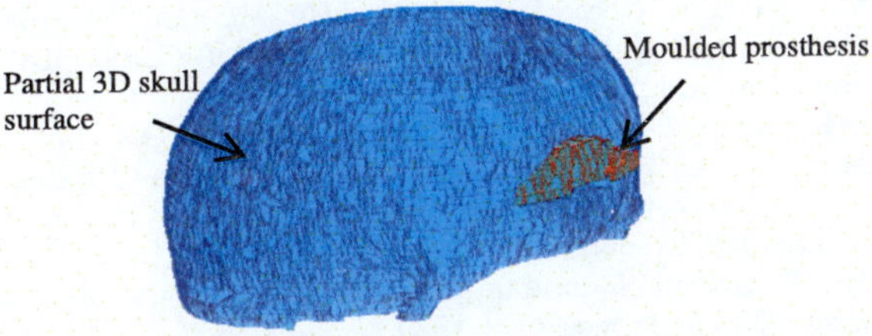


**Figure 7.**The 3D skull surface reconstructed with adjusted prosthesis.

## 5 Conclusion

In this paper was presented a method to build a 3D virtual bone piece to be used as prosthesis model. The problem of a hole in a skull was presented as study case. The suggested method was based in the creation of ellipses that were capable to perform a self-adjustment at bone curvature through a GA. The obtained results show that there are some problems yet to be solved. The union problems between bone segment and reconstructed piece are an open question. The main importance of this work is to know, how it is possible to modeling a piece of inexistent information.Thenit was verified a manner to make prosthesis modeling automatically.As the GAs have been used as optimization techniques and they are become very useful in many recent researches in engineering problems as a resource to aid the subjective human intervention.Here we showthat the proposed method is a promising technique to prosthesis modeling of this kind of problem.

## 6 References


[1] Rudek M, CanciglieriJr O, Catani F. A Method for Displacement Measurement of Femoral Head and Acetabulum Based on Tomography Images. Proceedings of the 20th IASTED International Congress in Computer Graphics and Imaging (CGIM 2011), Innsbruck, Austria, 2011;47-52.

[2] Fatterpekar GM, Doshi AH, Dugar M, Delman BN, Naidich TP, Som PM. Role of 3D CT in the evaluation of the temporal bone. Department of Radiology, Mount Sinai Medical Center, New York, USA, 2006.

[3] Sengupta DP, Sengupta D, Ghosh P. Bilaterally Symmetric Fourier Approximations of the Skull Outlines of Temnospondyl Amphibians and their Bearing on Shape Comparison. Journal of Biocience. 2005; 30:3:377–390.

[4] Li H, Xie Z, Ruan S, Wang H. The Measurement and Analyses of Symmetry Characteristic of Human Skull Based on CT images. Biomechanics and Vehicle Safety Engineering Centre, Tianjin, China, 2009; 26:1:34–37.

[5] Greboge T, CanciglieriJr O, Rudek M.Modelagem 3D de PrótesesBaseadaemImagens de TomografiaComputadorizada. XVII Simpósio de Engenharia de Produção – SIMPEP, Bauru, SP, Brazil, 2010.

[6] You F, Hu Q, Yao Y, Lu Q. A New Modeling Method on Skull Defect Repair. IEEE International Conference on Measuring Technology and Mechatronics Automation, Shanghai, 2009; 1:568–572.

[7] Hashemi S, Kiani S, Noroozi N, Moghaddam ME. An Image Enhancement Method Based on Genetic Algorithm. IEEE International Conference on Digital Image Processing, New York, NY, USA, 2009; 167–171.

[8] Jennane R, Almhdie-Imjabber A, Hambli R, Ucan ON, Benhamou CL. Genetic Algorithm and Image Processing for Osteoporosis Diagnosis. IEEE Annual International Conference of Engineering in Medicine and Biology, BuenosAires, Argentina, 2010; 5597–5600.

[9] Goldberg DE. Genetic Algorithm in Search, Optimization and Machine Learning. 1st Ed., Addison-Wesley Professional, 1989.

[10] Canciglieri, O. Rudek, M. Francesconi, T. de Souza, T.M.Geometric modeling from the medical images in the CAD system to support prosthesis design.40th International Conference on Computers and Industrial Engineering (CIE), 2010.

[11] Francesconi, T. PropostaMetodológicaparaModelagemGeométrica a Partir de ImagensMédicas. Master Dissertation, Curitiba, PPGEPS- PUCPR, 2008.

[12] Lodygowski T, Szajek K, Wierszycki M. Optimization of Dental Implant Using Genetic Algorithm. Journal of Theoretical and Applied Mechanics, 2009;47:3:573–598.

[13] Renner G, Ekárt A. Genetic Algorithm in Computer Aided Design. Computer-Aided Design, 2003;35:8:709–726.

[14] Kanungo P, Nanda PK, Samal UC. Image Segmentation Using Thresholding and Genetic Algorithm. Image Processing and Computer Vision Lab, 18 dec. 2010.

[15] Joshi R C, Tapaswi S. Image Similarity: A Genetic Algorithm Based Approach. World Academy of Science, Engineering and Technology, 2007; 327–331.

# Conceptual reasoning model for supporting strategic planning of dental implant surgical process

Anderson Luis Szejka[a,1] , Marcelo Rudek[b,1], José Mauricio Perussolo[c] and Osiris Canciglieri Junior[b,1]

[a]Automation Engineering, Curitiba - Brazil.
[b]Professor, Pontifical Catholic University of Paraná - PUCPR - Brazil.
[c]Dentist Periodontist - Curitiba - Brazil.

**Abstract.** The technological evolution over the last decades brought the necessity of integration between different areas of knowledge; an example of this is the integration between the Odontology and Engineering in order to find new solutions to improve the surgical process of dental implants. This work proposes a conceptual reasoning model for supporting strategic planning of dental implant which offers support to the dentist through a three-dimensional geometric modeling of the dental arch (bone, nerves, existing teeth) allowing the realization in advance of the strategic planning of the surgical process. In this geometric modeling the reasoning system interacts with the dentist presenting the bones density, nerve location, among others features, becoming, in this way, a tool that will assist the dentist in the decision-making process. The proposed model creates an auxiliary mask that will work as a guide for locating and positioning the implants making the procedure quicker and less painful. The article's main contributions are: i) conception and development of a computational reasoning tool that supports the process of dental implantation; ii) the interactivity in the development of surgical planning through a three-dimensional geometric model of the dental arch; iii) the reduction of surgical time and the patient's recovery time.

**Keywords.** Implant, Imaging Process, Strategic Planning, Product Development, Reasoning Systems.

## 1 Introduction

Since ancient times mankind has searched the improvement of its well-being and longevity through the development of means and techniques which could help in daily life and especially in the treatment of the diseases. As consequence areas of knowledge had emerged (engineering, medicine, dentistry, etc.), in order to improve the people's quality of life through new inventions even though working individually, detached from each other.

Over the years came the necessity for integration between the areas in order to

[1]Master Degree of Production Engineering at Pontifical Catholic University of Paraná (PUCPR), Rua Imaculada Conceição, 1155, Prado Velho, Curitiba, CEP 80215-901, PR, Brazil; Email: anderson.szejka@pucpr.br

[2]Professor in the Department of Production Engineering at Pontifical Catholic University of Parana (PUCPR), Rua Imaculada Conceição, 1155, Prado Velho, Curitiba, CEP 80215-901, PR, Brazil; Tel: +55 (0) 32711304; Fax: +55 (0) 32711345; E-mail: osiris.canciglieri@pucpr.br

[2]Dentist Periodontist - Curitiba - Brazil, Rua Pe Anchieta, 1846 cj 402 Bigorrilho - CEP: 80.730-000PR - Brazil.

D. D. Frey et al. (eds.), *Improving Complex Systems Today*,
Advanced Concurrent Engineering, DOI: 10.1007/ 978-0-85729-799-0_29,
© Springer-Verlag London Limited 2011

search more efficient solutions. This search has been promoting ideological changes in the processes of conception and development of products and processes and can be considered, nowadays, as the advent of the globalization where the search for the integration and unification of the correlated areas of knowledge [7]. As a result, it has appeared the opportunity of application of the areas such as medicine, dentistry, mechanical, electrical, electronics and software engineering in the search of new results for the conception and development of products and processes [1]. In this scenario it can be highlighted the evolutionary technological process in the conception and design of dental implants. This process is characterized by the integration of different fields of knowledge through a secure planning of the whole surgical procedures [6]. However, this process requires a prior evaluation of some characteristics of arch bone that will receive the implant, such as the bone structure, the position of nerves, the dimensions of the teeth, among others. This assessment is extremely important since it can avoid errors in the surgical process that would bring harm to the patient, i.e., facial paralysis, the immediate loss of sensitivity, among others [11]. In this way, the research reported in this paper proposes a conceptual reasoning model for strategic planning that can offer support to the dental implant surgery. This conceptual model becomes an important tool assisting the dentist, in a virtual way, in the planning of the implant surgical procedures. This support happens through the 3D geometric visualization of the part of the jaw bone, the geometrical location of the nerves, the bone density, the recommended dimensions of the implants elements, the implant placement in the bone part of the jaw and mainly the planning of the surgical tools that will be used.

## 2 Research Methodology

This research is considered practical since the studied knowledge were applied to solve a specific problem. It has a qualitative approach because it sought a deep understanding of a specific phenomenon through descriptions, comparisons and interpretations; it is exploratory as it provided greater familiarity with the problem in order to make it explicit, and finally, relating to the technical procedures it can be considered bibliographical and experimental. It is a bibliographical research because it was elaborated from already published material, consisting mainly of books, journal articles and currently available material on the Internet and experimental research in the determination of the object of study, in the selection of variables that would be capable of influencing so, in defining the ways of control and in the observation of the effects that the variables would produce in the object of study. Figure 1 illustrates the steps of the research process whose main objective was to propose conceptually a software tool that would offer support to the planning of the dental implant surgical process. Detail "A" represents the necessary knowledge regarding to the existing implant techniques and the new technologies that are being used in this process; the detail "B" shows the propose of a conceptual reasoning model to support the implant surgical process; the detail "C" presents the implementation process of the proposed conceptual model through case studies. The model was developed using the C++ platform where it can develop a system that is dedicated to the image processing and files exporting. Finally, the detail "D" emphasizes the analysis of the preliminary results obtained from the comparison between the proposed model and the expected results from the experiments.

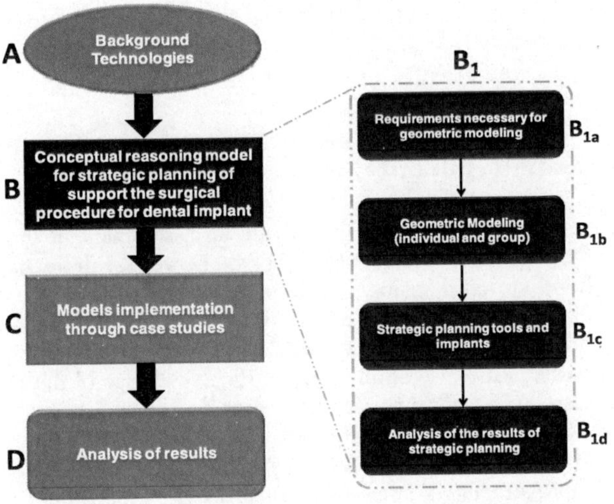


Figure 1 – Methodological Sequence of the Research.

## 3 Background Technologies

One reason for the increase in the life expectancy of the world population is the technological developments in various areas of science. Thus, it can be verified that the boundaries between different areas of knowledge became increasingly tenuous allowing, in this way, a greater synergy in cooperative actions where the results are significant improvements in the people's quality of life [1]. This phenomenon occurs, also, in dentistry and more specifically in the areas of implant surgery and prosthesis. In this manner, the dental implants came in order to replace missing teeth caused by old age, illness or accidents. This replacement, therefore, may have an aesthetic or clinic character and have been widely accepted by dentists and patients because of the positive results [8, 13, and 14]. However, the traditional surgical procedures demand great experience and skills from the dentist in order to choose since the implant type and its respective dimensions until its location in the mandible of the patients. This lack of an appropriate strategic planning can lead to high rates of functional and aesthetic failure in surgery. This framework has been transformed using techniques such as computed tomography and magnetic resonance imaging in the precise determination of the diagnosis regarding the bone condition [9]. Therefore the use of computed tomography and magnetic resonance imaging in the dental implant process made the procedure safer likewise other areas of medicine that are already using the files obtained in these techniques in the three-dimensional geometric modelling process (3D) for the cranial reconstruction [5]. This is possible through the file export in the DICOM (Digital Imaging and Communications in Medicine) pattern with the aim of standardize and make possible the communication of the information related to the diagnosis, to the images and data of any kind [10]. With this, the dentist can analyse with precision the images, after that will be able of planning accurately the procedures as well as determining the surgical tools that will be necessary in the implant surgical process [12].

## 4 Conceptual Reasoning Model for Supporting Strategic Planning of Dental Implant Surgical Procedure

In the traditional surgical procedure it is made an assessment of the arch of the patient before the procedure which determines the nerves location and mechanical properties of bone structure using magnetic resonance imaging and computed tomography in a way that the dentist can define the geometric boundaries that will be used in the implant surgical process. However, this procedure is not precise and the whole process is at the discretion solely of the experience of the dentist. Based on this, this research proposes a conceptual reasoning model for strategic planning that is able to offer support to the dental implant surgical procedure as illustrated in detail "B1" of the Figure. This model aims to assist the dentist during surgery through the improvement of the surgical process (procedures and surgical instruments) and the improvement of the post-surgery patient's recovery. So that, the idea proposed by the reasoning model is to realize the whole strategic planning of the tools that are necessary for the dental implant surgery, offering, in this way, the support to the dentist in the decision-making process of their surgical procedures. The proposed reasoning model consists of four steps, which are:

- Requirements necessary for the geometric modelling (Detail $B_{1a}$ – Fig. 1);
- Geometric modelling (individual and set - Detail $B_{1b}$ – Fig. 1);
- Strategic Planning of the surgical tools/instruments and the implants (Detail $B_{1c}$ – Fig. 1);
- Analysis of the results of the strategic planning (Detail $B_{1d}$ – Fig. 1).

### 4.1 Requirements Necessary for the Geometric Modelling

The geometric modelling must be done through tomographic cut in DICOM format files containing digital image. The DICOM files contain the information of the digital image of the transverse cuts, axial and panoramic views of the coordinates "X,Y,Z", of the cuts and colour tone of each cut for the appliance of the Hunsnfield Scale. The DICOM pattern is different from others image formats as JPEG,TIFF,GIF because it allows that the information related to the patients be keep together with the image in a structured way [4]. This protocol represents the standardization of a structured data form in order to interchange digital images between medicine and engineering [2], as illustrate in the right side of the figure 3. The figure presents three images in DICOM format (axial, panoramic and transverse cut views). For the reconstruction of the missing teeth it is necessary to use the re-engineering techniques by the 3D digitalization (Rapid Prototyping) for the generation of a geometric model in CAD (Computer Aided Design) [1] through real models or manual modelling.

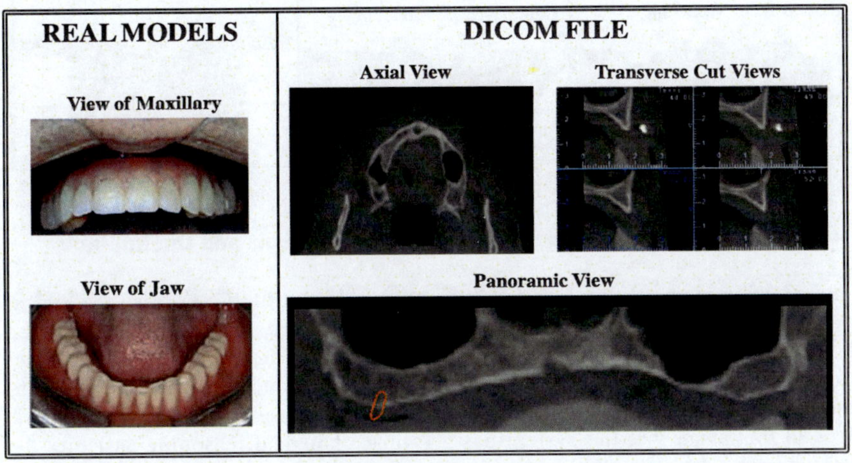


Figure 2 – Requirements for geometric modelling.

**4.2 Geometric Modelling**

Based on the information obtained in the previous phase, now it is done the modelling of the bone arch of the mandible/maxilla through tomographic cuts or magnetic resonance imaging. However, the dental lack will be modelled individually using rapid prototyping techniques in a re-engineering process of the real geometric model (figure 4). This process is fundamental since virtually it is possible to plan the precise positioning of the implant ant its accessories. In this geometric modelling it is used algorithms of image processing and artificial intelligence for the three-dimensional reconstruction through the cloud of points in order to generate the 3D models in CAD [3], as illustrated in the "individual modelling part" and "assembly modelling" of figure 5.

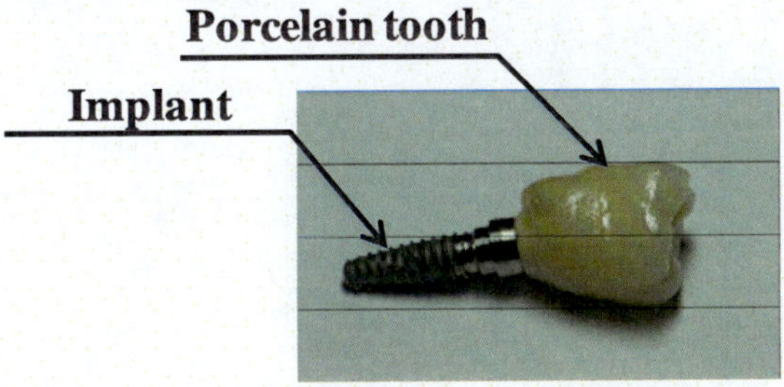


Figure 3 – Dental implant and the porcelain tooth.

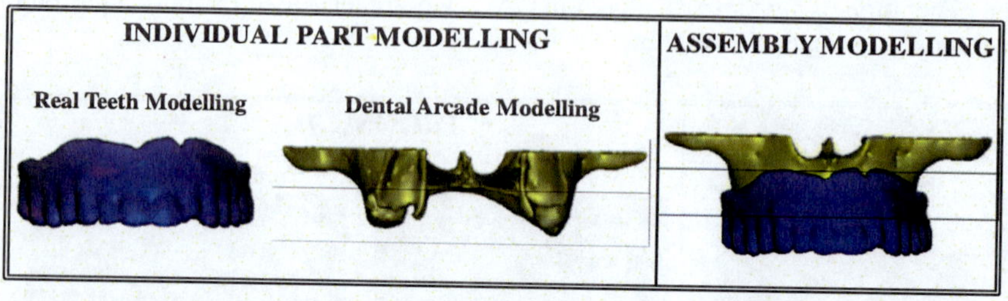


Figure 4– Three-dimensional geometric modelling.

### 4.3 Strategic Planning of the Surgical Tools/Instruments and the Implants

The strategic planning can be performed on two fronts: the first presents a logical order of the execution of the implants process activities, that is, the study of the bone structure of the arch and the nerves location, the stages of pre-drilling, drilling and threaded opening for insertion of the implant; the second is related to the planning of all surgical material/tools that will be used in the surgical process. It is in this stage that the system has a format of logical reasoning assisting the dentist in the decision-making process for the realization of the surgical planning. The Figure 6 presents the logical reasoning of the system which will help the dentist in the decision making process predicting in advance the possible problems that may occur during the surgical process.

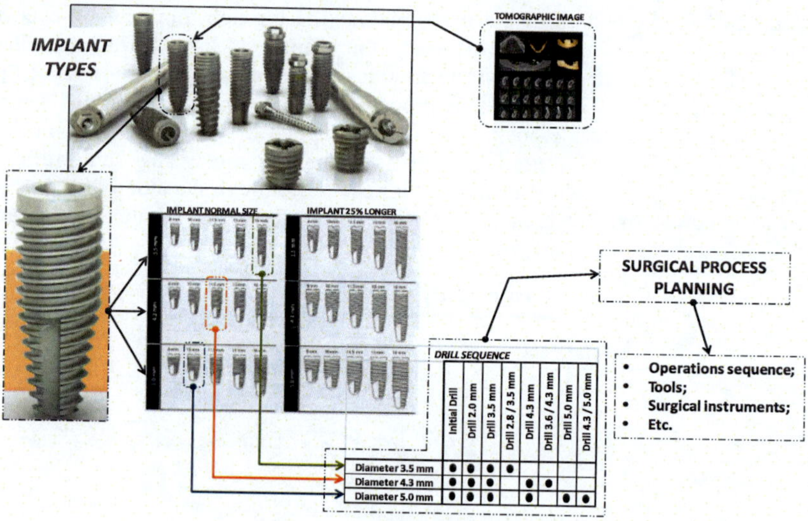


Figure 5 – The system logic.

### 4.4 Analysis of the Results of the Strategic Planning

The final process of the reasoning model is executed in two parts, which are: the emission of a report showing chronologically and in detail each step of the surgical procedures with its respective tools that will be used in the process; virtual simulation of the surgical process through 3D visualization of the bone arch, geometric position of the implant and of the auxiliary mask that will be used as a guide in the processes of drilling and threading. This mask will be created virtually based in the planning generated and after can be export in a STL file to be manufactures in rapid prototyping (figure 7).

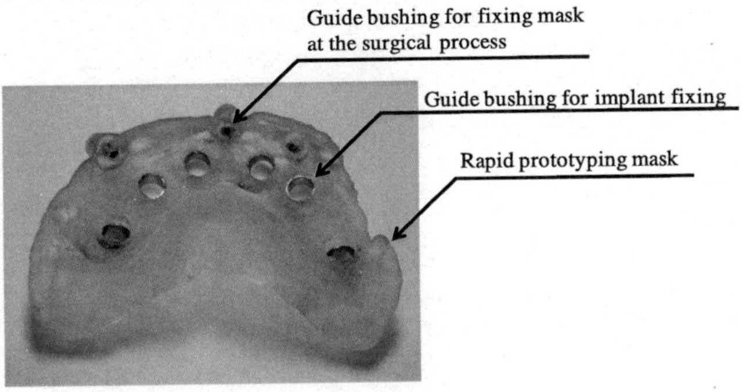


Figure 6 – Mask model (Rapid prototyping).

## 5 Analysis of the Results

The conceptual model presents an evolution in the dental implant process as: i) reduction in the surgery total time; ii) faster recovery of patients; iii) the surgical procedure will be better planned and precise; and iv) better quality of the aesthetic results for the patients. Based in tomographic images the implants are guided through the models and has lower risk of being out of its position and especially determines the depth of the hole so avoinding reaching the nerve.This makes possible to re-create precisely the patient's dental arch allowing a more accurate visualization of the mouth details through the data obtained by the vectorial digitalization.Thus it is possible to conceive the auxiliary mask (Figure 6)that will guide the dentist at the time of the implant incision. However, the model still has some limitations that should be objects of future scientific explorations.

## 6 Conclusion

This article presentes a conceptual reasoning model for supporting the strategic planning of surgical dental implant procedure which provides images and characteristics of the patient dental arch aloowing the virtual visualization of the nerves in order to define precisely the dimensions of the holes for the implant insertion reduncing drastically the risk of partial or total facial paralisys. It also presented the development phases of the conceptual model illustrating its singularities in the research context. Despite the results still preliminaries it is already possible visualize the method potenciality by the results obtained in the sirgery planning.

The authors believe that the model can be deeply explored in order to offer more autonomy to the system in the decision- making process. Therefore, several future researches themes can be proposed, for instance:

- The use of genetic algorithms in order to generate systems rules searching more autonomy for the surgical planning;
- Development of a expertise system for selection of implant types and surgical tools;
- Development of new tools and accessories.

# 7 References

[1] Bertol LS. Contribuição ao estudo da prototipagem rápida, digitalização tridimensional e seleção de materiais no design de implantes personalizados. Master Dissertation, Porto Alegre, PPGEM (Programa de Pós-Graduação em Engenharia de Minas, Metalúrgica e de Materiais), UFRGS, 2008.

[2] Bidgood WD, Horil SC, Prior FW, Van Syckle DE. Understanding and using DICOM, the data interchange standard for biomedical imaging. J Am Med Inform Assoc 1997; 4: 199 212.

[3] Canciglieri Junior O, Greboje T, Rudek M. Modelagem 3D de próteses baseada em imagens de tomografia computadorizada. XVII Simpósio de Engenharia de Produção, Bauru, 2010.

[4] Carita, EC, Matos ALM, Azevedo-Marques PM. Ferramentas para visualização de imagens médicas em hospital universitário. Radiologia Brasileira 2004;37,6:437-440.

[5] Grauer D, Cevidanes LSH, Proffit WR. Working with DICOM craniofacial images. American journal of orthodontics and dentofacial orthopedics : official publication of the American Association of Orthodontists 2009;136:460-470.

[6] Jivraj S, Chee W. Rational for dental implants. British Dental Journal 2006; 200:661-665.

[7] Lastres H, Cassiolato J, Lemos C, et al. Globalização e inovação localizada. IE/UFRJ, 1998. Available at: <http://www.ie.ufrj.br/redesist/P1/texto/NT01.PDF>. Acess on: Jan. 15th 2011.

[8] LI T, HU K, Cheng Libo, Ding Y, et al. Optimum selection of the dental mplant diameter and length in the posterior mandible with poor bone quality – a 3D finite element analysis. Applied Mathematical Modelling 2011;35:446-456.

[9] Menezes PDF, Sarmento V, Lamberti P. The use of rapid prototyping in implantology. Innovations implant Journal – Biomaterials and esthetics 2008; 3:39-44.

[10] Mildenberger P, Eichelberg M, Martin E. Introduction to the DICOM standard. European Radiology 2002; 12:920-927.

[11] Pye AD, Lockhart DEA, Dawson MP, et al. A review of dental implants and infection. The journal of hospital infection 2009; 72:104-110.

[12] Stanford CM. Dental Implants – A role in geriatric dentistry for the general practice? J Am Dent Assoc 2007;138:34S-40S.

[13] Wijaya, SK, Oka H, Saratani K, Sumikawa T, Kawazoe T. Development of implant movement checker for determining dental implant stability. Medical Engineering and Physics 2004; 26:513-522.

[14] Zarb GA, Albrektsson T. Consensus report: toward optimized treatment outcomes for dental implants. Int J Prosthodontics 1998;11:389.

# Conceptual Designing Supported by Automated Checking of Design Requirements and Constraints

Wojciech Palacz[a,1], EwaGrabska[a] and SzymonGajek[a]

[a]Faculty of Physics, Astronomy and Applied Computer Science, JU

**Abstract.** This paper aims at contributing to a better understanding of essential concepts of supporting system for initial visual design. Towards this end, we first outline the system architecture corresponding to our model of conceptual design aided by computer, in which designer's drawings created on the monitor screen are automatically transformed into elements of a graph-based data structure and next into the first-order logic formulas. Then, we describe particular modules of the proposed model paying attention to the role of a graph-based data structure gathering information on which design knowledge is based. Finally, the approach is illustrated on examples of designing floor layouts where fire code regulations and ranges of sensors are checked on the base of the proposed design knowledge reasoning module.

**Keywords.**CAD system, visual language, graph-based data structure, design knowledge, first-order logic.

D. D. Frey et al. (eds.), *Improving Complex Systems Today*,
Advanced Concurrent Engineering, DOI: 10.1007/ 978-0-85729-799-0_30,
© Springer-Verlag London Limited 2011

## 1 Introduction

Any design process must start with an understanding of what is required by the design and end with a design containing valid solution of the design task. There are several levels of formulating and implementing ideas and methods during the design process. Sometimes, the first part of the process, consisting of understanding the requirements, goes together with the visualization of early design solutions. On this conceptual level, designers seek a fresh approach to the problem. Visual exploration of the preliminary ideas can be done with help of pencil sketches or their computer equivalents [1]. Recent studies [2] state that increased sketching at the beginning of the project correlates with better design outcome. In the Internet age the designer often creates early design drawingson the screen by means of appropriate CAD tools [3]. Contemporary CAD tools could be treated as useful assistants of designers, reminding them about requirements and constraints during design process and suggesting modifications of design solutions. Project constraint reasoning is one of the desired techniques that is missing in Knowledge-Based Engineering [4]. Suggesting modifications of design solutions could also be achieved, if process of design knowledge reuse, which is rather common among designers [5], was partially automated.

This paper proposes a conceptual CAD model in which designer's drawings are automatically transformed into appropriate elements of a graph-based data structure and then these elements are translated to formulas of first-order logic, which constitute design knowledge. This approach has allowed one to implement a prototype CAD system called Hypergraph System Supporting Design and Reasoning (HSSDR), which constantly verifies and validates the solution being created. Being designer's active assistant distinguishes this system from other proposed approaches which validate finished projects [6, 7].

[1] Faculty of Physics, Astronomy and Applied Computer Science, JagiellonianUniversity, ul.Reymonta 4, 30-059 Krakow, Poland; Tel: +48 12 6635547; Fax: +48 126337086; Email: wojciech.palacz@uj.edu.pl; http://www.fais.uj.edu.pl/

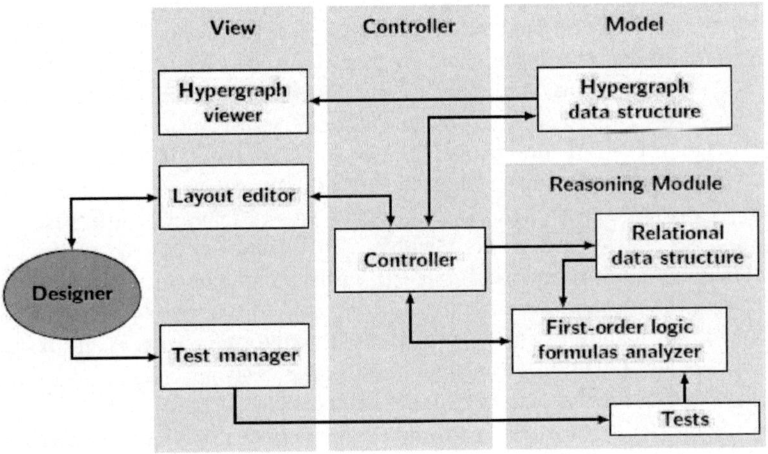


Fig. 1. HSSDR System Architecture

HSSDR was implemented in Java, as a tool for designing floor layouts. Its internal architecture is displayed in Fig. 1. The designer interacts with layout diagrams, but all his/her actions are internally translated into appropriate modifications applied to the hypergraph model. These modifications are then reflected back as changes in the diagrams.

Design requirements and constraints are specified as first-order logic sentences and stored in external files. The designer can select a set of such files in the test manager. Selected constraints are loaded by the reasoning module. In order to decide if they are fulfilled or not, the module requires a relational structure corresponding to the current state of the hypergraph. This structure provides an interpretation for the logic sentences [8].

To illustrate our method a collection of user tests related to floor-layout designing has been created using HSSDR. We distinguish two groups of tests. The first group concerns fire code regulations and verifies if evacuation route constraints are fulfilled for all areas. The second one deals with sensors and their ranges as an example of abstract artifacts which can also violate design constraints [9]. Test results are presented for example projects created with HSSDR.

It is worth noticing that the proposed approach has also allowed one to use extracted knowledge to generate navigation maps for robots [10].

## 2 Visualization of Design Solutions

A CAD system needs to communicate with its users. It has to respond to their input by appropriately modifying the design and by displaying results of these modifications. Majority of CAD systems use some kind of graphical user interface and visualize design solutions as diagrams. These diagrams constitute a specific visual language.

The language of UML diagrams, widely used in software engineering, is an example of a visual language. The language of architectural drawings is another one. It is worth noticing that there are many dialects of this language-different countries use different symbols to represent the same elements, etc.

Authors of CAD tools must define their visual languages carefully. On the one hand,such a language should be complete (i.e. be able to express all com-ponents and properties ofa design solution), unambiguous, and intuitive. On the other hand, it should be alsoextensible because CAD systems sometimes expand in scope - this usually introduces newgraphical shapes which have to be expressible in the visual language.

As an example, let us consider the visual language of HSSDR. It originally provided means of describing room layouts and door locations. Layouts were constrained by an underlying square grid - walls had to follow grid lines, doors had to be placed on grid cell borders. Later HSSDR was used to design floor-layouts where certain areas should be trackable by sensing devices such as cameras, motion sensors, etc. The consequence of such design task was adding a new graphical shape representing a camera.

Not all parts of a diagram must be created in response to designer's actions. A CAD system can provide its own information. For example, HSSDR knows what the range of cameras is and when drawing design solutions it uses this knowledge. Each camera is represented by a gray "pie slice". Its origin and direction are determined by the designer and treated as camera attributes by the system, but its size is determined by HSSDR. In other words, a created diagram is a result of the designer-system cooperation.

Another problem related to diagram visualization concerns constraint checking and error reporting. There is a class of requirements which can be violated by a presence of a particular room. HSSDR highlights offending rooms in red. This highlight is a part of our visual language, but does not, strictly speaking, correspond to any element or attribute of a design solution.

## 3 Graph-Based Data Structure

HSSDR uses hierarchical hypergraphs as its internal data structures (detailed description of hypergraphs can be found in [11] and [12]). They consist of hyperedges and nodes. There are two types of hyperedges: layout component hyperedges and relation hyperedges. The former are labeled by names of components and the latter by names of relations, respectively. Component hyperedges represent elements of a layout, i.e. areas and rooms. Nodes attached to a component hyperedge represent fragments of this component, i.e. walls delimiting this room or area. A component hyperedge and its nodes are treated as a single unit. They can be nested inside another component hyperedge. This provides for representing hierarchical organization between areas and rooms. Relation hyperedges represent relations between fragments of layout components, e.g. between walls of two rooms.

In HSSDR top-down philosophy of design is assumed. Therefore hierarchical graphs are used as internal representations of diagrams. A designer starts with a diagram depicting one big empty area corresponding to the whole floor, and then recursively divides it into smaller areas, stopping when they reach the level of single rooms. A hierarchical hypergraph model, representing created solution, has one hyperedge at the top level (the whole floor) with additional hyperedges recursively nested inside. These hyperedges are arranged in a tree structure with hyperedges representing rooms as its leaves.

Let$\sum_G^C$ and $\sum_G^R$ be an alphabet of component hyperedge labels and an alphabet of relation hyperedge labels, respectively. Let $\sum_G^V$ be an alphabet of node labels.Let $\sum_G^E = \sum_G^C \cup \sum_G^R$ . Let $At_G$be a set of attributes (i.e. mappings which associate values with hyperedges and/or nodes). Let [i] denote the interval [1, i] of natural numbers.

By an attributed hierarchical hypergraph over Σ we mean a system

$$G=(E_G,V_G,s_G,t_G,lb_G,att_G,ext_G,ch_G),$$

where

- $E_G$is a nonempty and finite set of hyperedges being a disjoint sum ofrepresentations of layout components ($E_G^C$) and relations ($E_G^R$), i.e., $E_G = E_G^C \cup E_G^R, E_G^C \cap E_G^R = \emptyset$;
- $V_G$is a nonempty finite set of nodes;
- $s_G: E_G \to (V_G)^*$ and $t_G: E_G \to (V_G)^*$are two mappings assigning to each hyperedge sequences of source and targets nodes, respectively, in such a way that $\forall e \in E_G^C: s_G(e) = t_G(e)$;
- $lb_G = lb_G^E \cup lb_G^V$, where $lb_G^E: E_G \to \sum_G^E$ is a hyperedge labeling function suchthat $\forall e \in E_G^C\ lb_G(e) \in \sum_G^C$ , $\forall e \in E_G^R\ lb_G(e) \in \sum_G^R$ , and $lb_G^V: V_G \to \sum_G^V$ is a node labeling function;
- $att_G = att_G^E \cup att_G^V$, where $att_G^E: E_G \to P(At_G)$is a hyperedge attributing function, and$att_G^V: V_G \to P(At_G)$ is a node attributing function;

- $ext_G: [n] \rightarrow V_G$specifies a sequence of external nodes;
- $ch_G: E_G \rightarrow P(E_G \cup V_G)$is a child nesting function which is acyclic (i.e., hyperedge cannot be its own descendant), a hyperedge or a node cannot be nested in two different parents, a component hyperedge and its source and target nodes must have the same parent.

Attributes are used to store additional information associated with hyperedges and nodes. The following attributes belong to the set $At_G$ of attributes:

- $area_G: E_G^C \rightarrow \mathbb{R}$- surface area of room or area represented by thiscomponent hyperedge;
- $dir_G: V_G \rightarrow$ {"N", "S", "W", "E"} - orientation of the wall represented by this node;
- $length_G: V_G \rightarrow \mathbb{R}$- length of the wall;
- $coord1_G: V_G \rightarrow \mathbb{R} \times \mathbb{R}$- coordinates of the lower-left corner of the wall;
- $coord2_G: V_G \rightarrow \mathbb{R} \times \mathbb{R}$- coordinates of the upper-right corner of the wall.

## 4 Design Knowledge and Reasoning

Design constraints are specified as first-order logic formulas. They can refer to elements of the layout (rooms, walls, doors and cameras) and to core relations and functions provided by HSSDR. An example constraint, specified below, uses the type function to query the role assigned to a room. This set of core relations and function constitutes a vocabulary.

```
exists r in Rooms: type(r) = "Bedroom"
```

The reasoning module is tasked with evaluating design constraints and reporting any failures. To do that, it must provide meaning to the vocabulary symbols. Instead of getting this knowledge directly from the graph model, it constructs an auxiliary model - the relational structure [8].

Relational structure $\mathcal{A}$ consists of a nonempty set of individuals (known as the domain, $dom(\mathcal{A})$) and a set of functions and relations defined on $dom(\mathcal{A})$(corresponding to those specified in the vocabulary). In HSSDR's case, the domain contains objects representing elements of the current layout, as well as labels and numbers. The set of relations and functions includes, among others:

- $Room(x)$ - unary relation, true if x is an object representing a room;
- $accessible(x, y)$ - binary relation, true if both x and y represent rooms, and these rooms are connected by a door;
- $label(x)$ - unary function, returns the label of object x;
- $doorsDist(x, y)$ - binary function, if x and y represent doors returns their distance in meters.

Existence of the relational structure separated from the graph model allows for easy implementation of user-defined relations. In HSSDR, a test can contain definitions of auxiliary relations, specified in terms of already available relations and functions. Examples presented in the following pages contain such user-provided definitions.

## 5 Examples of Design Knowledge Reasoning

As it has been mentioned, two example projects created with HSSDR with test results will be presented. The first one demonstrates a sensor-related constraint, while the second one is related to fire safety regulations.

Let us consider an example floor layout of single-family home presented in Fig. 2a. Two areas in this project are highlighted, indicating violation of a rule. One requires that all paths leading from external doors to important areas (office rooms, etc.) be monitored. This condition is formed in the language of first-order logic (predicate isPathWatched checks if a path between doors in the same room is in range of a sensor, while failure_msg defines a message that is shown to the user if a violation is detected):

```
SecuredArea(r) <=> Rooms(r) and (type(r) = "Office"
or type(r) = "Living_Room" or type(r) = "Garage");
UnwatchedPath(d1,d2) <=> d1 != d2 and Doors(d1) and Doors(d2)
and (exists r in Rooms: doorsInRoom(d1,r) and
doorsInRoom(d2,r) and not isPathWatched(d1,d2));
UnwatchedPathInduction(d1,d2,n) <=>
 (n=1 and UnwatchedPath(d1,d2)) or
(n<>1 and UnwatchedPathInduction(d1, d2, n-1)) or
 (n<>1 and exists d in Doors:
UnwatchedPathInduction(d1,d,n-1) and UnwatchedPath(d,d2));
UnwatchedPathPlus(d1,d2) <=>UnwatchedPathInduction(d1,d2,99);

failure_msg "a path leading to secured area is not monitored"
not exists d1, d2 in Doors: exists r in Rooms:
SecuredArea(r) and doorsInRoom(d2,r) and
UnwatchedPathPlus(d1,d2);
```

Placement of an additional sensor in the living room fixes the problem and violation is suppressed, see Fig. 2b. If cameras are used instead of infrared

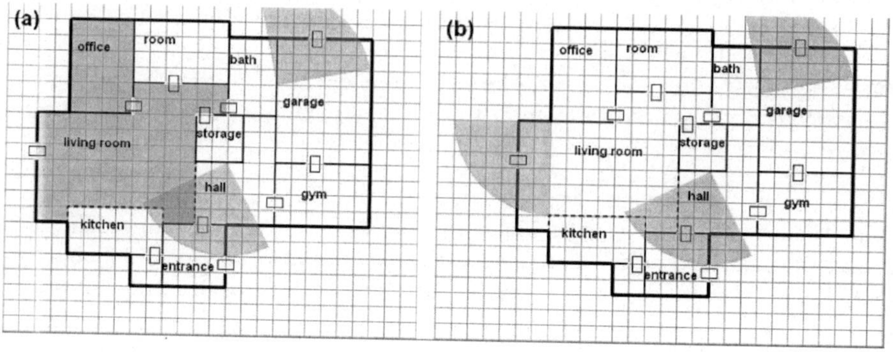


Fig. 2. Design of a floor of a single-family home violating security rule (a) and a correct design (b)

motion detectors, an additional constraint may be needed: private areas should not be in cameras' range.

The second use case deals with office buildings. For such buildings strict fire safety regulations are specified. Two following constraints are taken from the Polish regulations:

- an evacuation route leading to a common stairs or a final exit should measure less than 30 meters;
- from any point in the building, an access path to an evacuation route should pass through no more than three rooms.

For these rules an appropriate test was coded, however due to its size (42 lines) it is not presented in this paper. Fig. 3a shows floor layout which violates first of the two rules given above (the cells of the grid underlying the layout diagram measure 1 by 1 meter). Designer, warned by the system, creates additional exit and obtains a valid layout displayed in Fig. 3b.

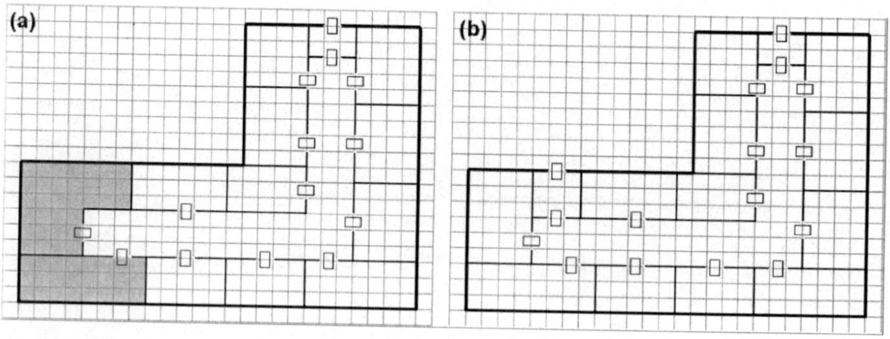


Fig. 3. Design of a floor of an office building violating fire safety rule (a) and a correct design (b)

## 6 Conclusions

This paper is the next step in developing graph-based data structures to support the first conceptual phase of design. The modification of drawings representing early design solutions are reflected by operations performed on their graph representations. The presented prototype system has been tested on functional-oriented designing floor-layouts. Succeeding test will be concerned with form-oriented designing.

## References

[1]Goldschmidt G. The Dialectic of Sketching.Creativity Research Journal vol 4,1991.

[2]Yang M. Observations on concept generation and sketching in engineering design. Research in Engineering Design 2009; 20:1-11.

[3]Tang HH, Gero JS. Cognition-based CAAD. In: Vries B, Leeuwen JP, AchtenHH (eds) CAADFutures 01. Eindhoven, The Netherlands, 2001; 521-531.

[4]Baxter D, Gao J, Case K, Harding J, Young B, Cochrane S, Dani S. An engineering design knowledge reuse methodology using process modelling. Research in Engineering Design 2007; 18:37-48.

[5]Demian P, Fruchter R. An ethnographic study of design knowledge reuse in the architecture, engineering, and construction industry. Research in Engineering Design 2006; 16:184-195.

[6]Yurchyshyna A. Modélisation du contrôle de conformité en construction: uneapprocheontologique. PhD thesis, Universite de Nice-Sophia Antipolis, 2009.

[7]Eastman C, Lee J, Jeong Y, Lee J. Automatic rule-based checking of building designs. In: Automation in Construction vol 18, 2009; 1011-1033.

[8]Fagin R, Halpern JY, Moses Y, Vardi MY. Reasoning About Knowledge. MIT Press, 2003.

[9]Bhatt M, Freksa C. Spatial computing for design: An artificial intelligenceperspective. In: Visual and Spatial Reasoning for Design Creativity (SDC '10),2010.

[10] Grabska E, Borkowski A, Palacz W, GajekSz. Hypergraph System Supporting Design and Reasoning. In: Wolfgang Huhnt (ed) Computing in Engineering EG-ICE Conference 2009. Shaker Verlag, 2009; 134-141.

[11] Grabska E, Lembas J, Łachwa A, Ślusarczyk G, Grzesiak-Kopeć K. Hierarchical layout hypergraph operations and diagrammatic reasoning. Machine Graphics & Vision 2008; 16(1/2):23-38.

[12] Palacz W. Algebraic Hierarchical Graph Transformation. In: Journal of Computer and System Sciences vol 68, 2004; 497-520.

# Extended KBE in Mechanical Engineering - discussion of concepts

Maciej Gil[1], Jerzy Pokojski[2] and Karol Szustakiewicz[3]

**Abstract.** The paper deals with the computer aided process of developing Knowledge Based Engineering (KBE) applications. When creating KBE applications, both the product and its design process are modeled. The main problem of building an advanced, comprehensive model of a design process is the selection of appropriate knowledge representations and implemented tools. In this work the specification of conditions affecting the form and function of so called Extended KBE templates are presented. The case study is presented in the [24].

**Keywords.** CAD, Knowledge Based Engineering.

## 1 Introduction

The paper deals with the computer aided process of developing Knowledge Based Engineering (KBE) applications [2, 3, 12, 16, 25, 26, 27, 29].

Classic Knowledge Based Engineering applications are built in the environments offered by CAD/CAE systems producers. These environments have available knowledge representation structures which allow to model engineering knowledge elements and to integrate these elements with detailed geometric product models. At the stage of knowledge articulation and modeling, the suitable geometric models or at least their parts should already exist. Applications built in CAD/CAE systems environments are often integrated with external modules which support engineering activities [4, 16, 18]. As a result product model data are obtained.

In the KBE modules of CAD/CAE systems, engineering knowledge elements can be modeled in the form of rules, formulas and others. The process of knowledge modeling is realized in the environment of a CAD/CAE system in the available representations. By this the KBE application in the form of geometric models and the knowledge in the available representations form one deeply integrated whole [4, 16, 18].

[1] Warsaw University of Technology, maciej.gil@gheell.com
[2] Warsaw University of Technology, Institute of Machine Design Fundamentals, Faculty of Automotive and Construction Machinery Engineering, Narbutta 84, 02-524 Warszawa, tel. 048 22 2348286; jerzy.pokojski@simr.pw.edu.pl;
[3] Warsaw University of Technology, ka.szustakiewicz@gmail.com



D. D. Frey et al. (eds.), *Improving Complex Systems Today*,
Advanced Concurrent Engineering, DOI: 10.1007/ 978-0-85729-799-0_31,
© Springer-Verlag London Limited 2011

Applications built that way are characterized by a high efficiency of functioning. They allow a fast creation of geometric models which are differentiated parametrically and structurally.

Generally, developing KBE applications is a labour-intensive task. Appropriate geometric models must be built and suitable knowledge elements must be integrated with them [12, 16, 25].

When creating KBE applications, both the product and its design process are modeled. Usually, product modeling in KBE applications is carried out very carefully and precisely. Modeling the design process in KBE applications, however, is much more difficult. As a consequence it is often performed in a simplified way.

In many cases the process of the KBE application development is concentrated on the perfect geometric model making while at the same time its knowledge layer is reduced to the modeling of design recommendations concerning the values of particular parameters and their mutual relationships. The knowledge captured and modeled this way – in most cases in forms of rules and formulas – does not establish a real model of the inferencing performed by the designer in the design process. The knowledge only reflects conclusions accompanying a real design process [1, 3, ,5]. A real design process is often more complex and more complicated, and more difficult for modeling [3, 13, 28].

If a KBE application is made on the level of design recommendations the process of its further development is not so easy. It is difficult to determine which recommendations are still valid, which of them should be modified and how contradictions between them can be eliminated. Thus we come to the point where the analysis and verification of the modeled knowledge is really needed. Often on their basis a new model of design recommendations for the next version of the KBE application must be built.

The effort of building a new KBE application is especially high when the modeled knowledge consists of recommendations delivered by different knowledge suppliers and when it concerns issues typical for design and manufacture. In practice, cases are known where after a longer period of time a knowledge authorship is not clear.

An approach based on the concept of design recommendations is one extreme. Building a detailed model of a design process, which considers all key inferences and decisions, is a second one. An advanced and complex model of the design process, however, can be used in designing and it can be further developed as well.

## 2 Concepts of Extended KBE

The main problem with building an advanced, comprehensive model of a design process is the selection of appropriate knowledge representations and implemented tools. For that purpose tools available in the KBE modules of CAD/CAE systems and also strict software programming tools (both algorithmic and declarative approach) can be applied. The software programming tools offer enormous possibilities when creating models for the design process, but their application requires a lot of work and time [7, 8].

### 2.1 Basic Concepts

Structures available in the KBE modules as well as those in programming languages refer primarily to a formal, mathematical and rather elementary way to capture design knowledge elements. When building a KBE application the issue of selecting an integrated development environment (IDE) together with its available structures depends largely on the knowledge and preferences of its developers. In spite of the different software realizations of many solutions similar features are possible.

Engineering design issues have also become objects of research. According to Pahl and Beitz's theory [13] the engineer may experience the following phases in the design process: clarification, conceptual design, embodiment and detailed design. Each of these phases has its own characteristics and can be modeled in a particular way, using different classes of computer tools.

The classification of the types of design problems proposed by D. Ullman [28] can also lead to concrete product and design process models [18, 22].

Table 1 presents various groups of KBE applications and their implementation specifics. A large variety of design processes comes out but it is difficult to judge the impact of individual design process characteristics on the form of the software application (see also [4]).

The analysis of the above material and the opinions of people involved in the KBE application development lead to the conclusion that this task is often accomplished with the use of certain patterns [18, 23]. In their conceptions these patterns refer to strict programming models and knowledge as well as to structures typical for engineering design theories [13, 28].

These structures may be templates which correct or generate geometric design models. There may be templates for modeling the designer's structure of preferences and other templates may analyze the relationships between objects. If the template resources are large enough, the process of building the KBE application means the selection and integration of particular templates. The authors call this possibility Extended KBE. So Extended KBE is understood as tools with the same functionality as in classical KBE, e.g. in knowledge modules of UGS, Catia or others, yet enriched by an appropriate template library. In this work the specification of conditions affecting the form and function of Extended KBE templates are presented.

When the set of the previously developed KBE applications is large, the whole approach is often reduced to the selection of one of the available, implemented design process models (which reminds the method of CBR [11, 14, 15]).

### 2.2 Knowledge modeling, templates

There is a specific knowledge behind every model of a computer design process [3, 9, 15]. It is the designer's very own knowledge which she/he takes advantage of while working. Often parts of this knowledge are formally expressed. This formally represented knowledge is used for building complex models of different stages of the design processes. Each of this models can be built in of various techniques and formalisms. Of course, it is possible to model this articulated design knowledge in representations typical for KBE modules. But it is also possible to build a formal representation of that knowledge in a specific programming language and later integrate this module with parametric models built in a CAD system. In this case the set of the available structures/representations is quite elementary and not large.

Usually, the most labor-consuming applications are those where not only the detailed design stage is modeled but also the selection of the product structure or the conceptual design stage is applied [13, 28]. This kind of modeling is based on very abstract ideas about product functioning. Solutions to that application are presented in Table 1.

**Table 1.** Characteristics of KBE applications

| Author | Domain | Characteristics of product and design process models | Knowledge representation | Tools | Design theory characteristics |
|---|---|---|---|---|---|
| [21] | car gear box design | classic calculations of gears -> optimization -> storage of calculated cases -> KBE geometric modeling of gears | procedural programming (MS VB, MS C/C++), rules, formulas (KBE module) | CATIA, MS VB, MS C/C++, MS Access | Parametric design, optimization, single structure. |
| [22] | car gear box design | classic calculations of gears -> KBE geometric modeling of gears | rules, formulas (KBE module) | CATIA | Parametric design, limited set of structures. |
| [19, 20] | truck gear boxes design | classic calculations of gears -> optimization (decomposed structure)-> storage of calculated cases | procedural programming (MS VB), | MS VB | Parametric design, optimization. |
| [30] | speed reducer design | the structure of geometric models modeling -> geometric model generation | procedural programming (MS VB), rules, formulas (KBE module) | MS VB, CATIA | Parametric design, large set of structures. |
| [17, 18, 23] | spiral stairs design | the structure of geometric models modeling ->optimization -> geometric model generation | Procedural/object oriented programming (MS VB), CAD system | MS VB, AutoCAD | Large set of structures, parametric design, optimization, geometric model generation. |

The above comparison illustrates essential concepts of various formalisms for different parts and for different stages of the design processes. The analysis of the above examples leads to the conclusion that in each case specific templates which are useful in the process of building and servicing new applications can be separated.

Table 2 presents selected conceptual characteristics of the solution templates applied in the application [17, 18, 23]] developed by the authors of the paper. The listed templates can be used for building other applications. The presented solutions are relatively universal. The proposal offers a number of concept solutions for particular stages of the design processes, visualization standards, the interaction level with the user, integration with another software and others.

**Table 2.** Templates derived from developed application

| Design phase | Design problem | Solution | Template name | Template category |
|---|---|---|---|---|
| 2. Conceptual | Complex task of selection of parameters values concerning set of step stairs in a tier | Generation of multiple solutions concerning fulfilment of a tier by step stairs | Generative Solution Development | Design activities |
| 5. Detailed/ Embodiment | • Gathering information from engineer about selected design solution types and parameter values;<br>• Structuring and formalization of identified routine design knowledge (sequences of design activities) | • Object model of engineer preferences;<br>• Object model of data-driven inference mechanism;<br>• Distinction of the information gathering, the inference model and the detailed model objects creation activities | Detailed Model Wizard (complex template containing preferences, UI and object generation models templates) | Design activities |
| 10. Detailed | • Appearance of specific part/assembly in many places in the model;<br>• counting of particular parts/assemblies;<br>• spatial orientation of parts/assemblies in the higher level assembly;<br>• management of parts and assemblies numbering series | Model of occurrences of parts/assemblies; similar to idea of Usage Class from CPM of NIST | Multi-Part/ Assembly Instantiation | Design activities |
| 12. Output | Laborious, by hand creation of the BOM | • Automatic searching of the model with parts/ assemblies recognition and BOM mechanism;<br>• Dynamic DB creation populating all parts/ assemblies;<br>• performing DB query (SQL) for final BOM | Design Report | Manage-ment |
| 19. All phases | • System maintenance, versioning;<br>• Connection of the model, control and visualization | Application of the MVC software design pattern for product modeling and dynamic visualization (event–driven approach) | Design Product MVC | Manage-ment |
| ... | ... | ... | ... | ... |

The literature survey in this field brings out that similar programming solutions are present in the works of other authors [4, 6, 10, 25, 29]. People who build KBE applications normally try to generate sets of their own templates for their own purposes which are to improve their work efficiency. But it cannot be observed that these achievements are made public.

In industry, the costs of the applied programming solutions as well as the developer's professional background and experience strongly influence their form, complexity and the level of advancement. A model which merely follows the "design recommendations" functions widely outside any computer system because a lot of the designer's knowledge is not represented in the process. As a result the model is difficult to improve or to develop.

## 2.3 Diversity of Templates

As a rule, every design team dealing with a particular group of products and the classes of design processes connected with it has its specific way of tackling the problem [3, 15]. This becomes obvious in the knowledge resources they use, the channels of knowledge development, the concepts of knowledge modeling and re-using and the ways of information processing in general.

Probably each group of designers of any domain possesses a set of ideas and metaphors which arises from their knowledge.

These ideas and metaphors may have their appropriate variables and data structures in the existing computer tools. But rarely these ideas and metaphors find their one-to-one representation in commercial computer tools. Consequently, the way commercial computer tools are used by a particular team strongly depends on the specificity of the model which is created. The situation when ideas and metaphors used by particular team in their design processes fit to the ideas and metaphors available in commercial computer tools happens relatively rare. However more often ideas and metaphors used by particular design team contain many components which are specific for this domain, for this team, for individual designers. Then the way how computer tools are used by particular team can be characterized by very strong modeling specificity.

## 2.4 Modeling Diversity of KBE Applications

In the following, the circumstances resulting from the factors enumerated in the previous chapter are characterized further. It is assumed that the considered class of KBE tools is suitable for supporting real industrial design problems.

Design problems performed by industry nowadays are very differentiated from the point of view the levels of the applied models. Some models are very precise, but others unfortunately base on earlier experiments or historical/outdated knowledge. According to Pahl and Beitz theory, it is possible to meet elements typical for conceptual design and typical for detailed design side by side in one design process. Different parts of the design process can also be realized by falling back on historical knowledge. Because of this, in many design offices and processes the structure of the links between activities is deeply considered and well tested.

Similar observations can be made with individual designers. They also exploit their own specific, personal, professional knowledge. If we map this phenomenon onto the concept of a more advanced set of computer tools, then this set should be strongly linked with the elements enumerated above.

The presented concept aims at the customization of universal software for issues of specific domains. One important feature of this approach is the integration of conceptual design components with detailed design components into one sensible unit, so that we obtain one single product/process model. The approach is based on classic engineering knowledge and the knowledge which is created during the realization of design processes.

## 3 Conclusion

The importance of the early phases of a design process should not be underestimated. The decisions made at the beginning are fundamental. All subsequent stages evolve from the first design decisions. However, in most cases the early, conceptual phases are difficult to automate.
The reasoning performed in human minds is often aided by paper and pencil. The concept and design rationale emerging in this phase is hard to capture and to transfer to electronic media. On the other hand, computer environment offers quick and easy access to various simulations and analyses which often can be done ad hoc.

The case study is described in the [24].

## 4. References

[1] Badke-Schaub P, Frankeberger E. Analysis of design projects. Design Studies 1999; 20; 465-480.
[2] Baxter D, et al. An engineering design knowledge reuse methodology using process modelling. Res. Eng. Design 2007; 18; 37-48.
[3] Clarkson J, Eckert C. Design Process Improvement, A review of current practice, Springer-Verlag, 2005.
[4] Curran R, Verhagen WJC, van Tooren MJL, van der Laan TH. A multidisciplinary implementation methodology for knowledge based engineering: KNOMAD. Expert Systems with Applications 2010; 37; 7336–7350.
[5] Dorner, D. Approaching design thinking research. Design Studies 1999; 20; 407-415.
[6] Fan I, Bermell-Garcia P. International Standard Development for Knowledge Based Engineering Services for Product Lifecycle Management. CERA 2008; Vol. 16; 271-277.
[7] Fenves, S. et al., CPM: A Core Model for Product Data. Manufacturing Systems, Integration, Division, National Institute of Standards and Technology, Gaithersburg, USA Available at: http://www.mel.nist.gov/msidstaff/sriram. Accessed on: March 10th 2007.
[8] IDEF0: Function modeling method, Knowledge Based Systems, Inc., 2004. Available at: www.idef.com. Accessed on: April 20th 2009.
[9] Kok Sim S, Duffy AHB. Towards an ontology of generic engineering design activities. Res. Eng. Design 2003; 14; 200-223.
[10] Kuhn. O, Ghodous P., Dusch T., Collet P.: Collaboration for Knowledge-based Engineering Templates Update. In: Pokojski J, Fukuda S, Salwiński J, New World Situation: New Directions in Concurrent Engineering: Proceedings of the 17th ISPE International Conference on Concurrent Engineering (Advanced Concurrent Engineering), 2010; 249-256.
[11] Maher M.L, Pu P. Issues and Applications of Case-Based Reasoning in Design. Lawrence Erlbaum Associates, Publishers, 1997.

[12] Managing Engineering Knowledge, MOKA – project, Professional Engineering Publishing Limited, London, 2001.
[13] Pahl G, Beitz W, and al. Engineering Design: A Systematic Approach, Springer-Verlag, 2007.
[14] Pokojski, J., (ed.) Application of Case-Based Reasoning in Engineering Design, WNT, Warsaw, 2003. (in Polish)
[15] Pokojski J. IPA (Intelligent Personal Assistant) – Concepts and Applications in Engineering, Springer-Verlag, London, 2004.
[16] Pokojski J. Expert Systems in Engineering Design, WNT, Warsaw, 2005. (in Polish)
[17] Pokojski J, Gil M, Skotnicki S, Szustakiewicz K, Szustakiewicz K, Bekasiewicz T. Application supporting design process of spiral stairs, integrated with AutoCAD 2006-project developed for POLIMEX- MOSTOSTAL Siedlce S.A., 2006-2008.
[18] Pokojski J, Gil M, Szustakiewicz K. Engineering Knowledge Modeling in Design. In: New World Situation: New Directions in Concurrent Engineering: Proceedings of the 17th ISPE International Conference on Concurrent Engineering (Advanced Concurrent Engineering), Editors: Pokojski J, Fukuda S, Salwiński J, Springer-Verlag, 2010; 257-266.
[19] Pokojski J, Niedziółka K. Transmission system design – intelligent personal assistant and multi-criteria support. In: Next Generation Concurrent Engineering – CE 2005, ed. by M. Sobolewski, P. Ghodous, Int. Society of Productivity Enhancement, NY; 2005; 455-460.
[20] Pokojski J, Niedziółka K. Transmission System Design – Decisions, Multi-Criteria Optimization in Distributed Environment. In: Leading the WEB in Concurrent Engineering, Next Generation Concurrent Engineering, IOS Press, Frontiers in Artificial Intelligence and Applications, 2006; Vol. 143; 595-602.
[21] Pokojski J, Okapiec M, Witkowski G. Knowledge based engineering, design history storage, and case based reasoning on the basis of car gear box design. In: AI-Meth-Proceedings, Gliwice 2002; 337-340.
[22] Pokojski J, Skotnicki S, Bieliński M, Kozłowski R, Młyński Z, Wyszyński K. Object Oriented Models and Knowledge Based Engineering in Machine Design. Mechanik, nr 1/2010, CD, 1-10. (in Polish)
[23] Pokojski J, Szustakiewicz K. Towards an aided process of building a Knowledge Based Engineering application. Machine Dynamics Problems 2009; Vol. 33, No 3; 91-100.
[24] Pokojski J, Gil M, Szustakiewicz K. Extended KBE in Mechanical Engineering-discussion of solutions. (Delivered for CE 2011).
[25] Skarka, W., Methodology of Knowledge Based Engineering, Wydawnictwo Politechniki Śląskiej, Gliwice, 2007. (in Polish)
[26] Sriram R.: Intelligent Systems for Engineering. A Knowledge Based Approach, Springer-Verlag, 1997.
[27] Sriram R., Tong C. Artificial Intelligence in Engineering Design. Academic Press, 1992.
[28] Ullman DG. The Mechanical Design Process, McGraw-Hill (Third Edition), 2002.
[29] Verhagen WJC, Curran R. Knowledge-Based Engineering Review: Conceptual Foundations and Research Issues, In: Pokojski J, Fukuda S, Salwiński J. New World Situation: New Directions in Concurrent Engineering: Proceedings of the 17th ISPE International Conference on Concurrent Engineering (Advanced Concurrent Engineering); 239-248.
[30] Woźnicki Ł. Building an application supporting the process of designing the structure of single-stage reducer with KBE system. Diploma, Institute of Machine Design Fundamentals, Warsaw University of Technology, Warsaw, 2011.

# Extended KBE in Mechanical Engineering - discussion of solutions

Maciej Gil[1], Jerzy Pokojski[2] and Karol Szustakiewicz[3]

**Abstract.** The paper deals with the computer aided process of developing Knowledge Based Engineering (KBE) applications [7]. When creating KBE applications, both the product and its design process are modeled. The worked out concepts are presented in the paper. They base on an real world example developed by the authors [4, 5, 6].

**Keywords.** CAD, Knowledge Based Engineering.

## 1 Introduction

The solutions presented in the paper [7] can be realized with the help of different tools: from very elementary to very advanced. But in each case we should remember that the knowledge which defines the background of each solution is developing. Therefore KBE applications should also offer the possibility to capture knowledge development without significant costs.

The worked out concepts are presented in the following chapter. They base on an example developed by the authors [4, 5, 6].

## 2 KBE in LaScala- case study

Staircase design has special characteristics [4, 5, 6]. There is a wide range of activities with many small but different design problems. In the early phase, the process focuses on developing a preliminary list of functional requirements and a conceptualization of the staircase form. Throughout the process, many detailed design problems must be solved, such as site, structure, form, interoperability, standard parts materials, manufacturing, and costs. The process can be split into single phases, as presented in the Figure 1.

---

[1] Warsaw University of Technology, maciej.gil@gheell.com
[2] Warsaw University of Technology, Institute of Machine Design Fundamentals, Faculty of Automotive and Construction Machinery Engineering, Narbutta 84, 02-524 Warszawa, tel. 048 22 2348286; jerzy.pokojski@simr.pw.edu.pl;
[3] Warsaw University of Technology, ka.szustakiewicz@gmail.com



D. D. Frey et al. (eds.), *Improving Complex Systems Today*,
Advanced Concurrent Engineering, DOI: 10.1007/ 978-0-85729-799-0_32,
© Springer-Verlag London Limited 2011

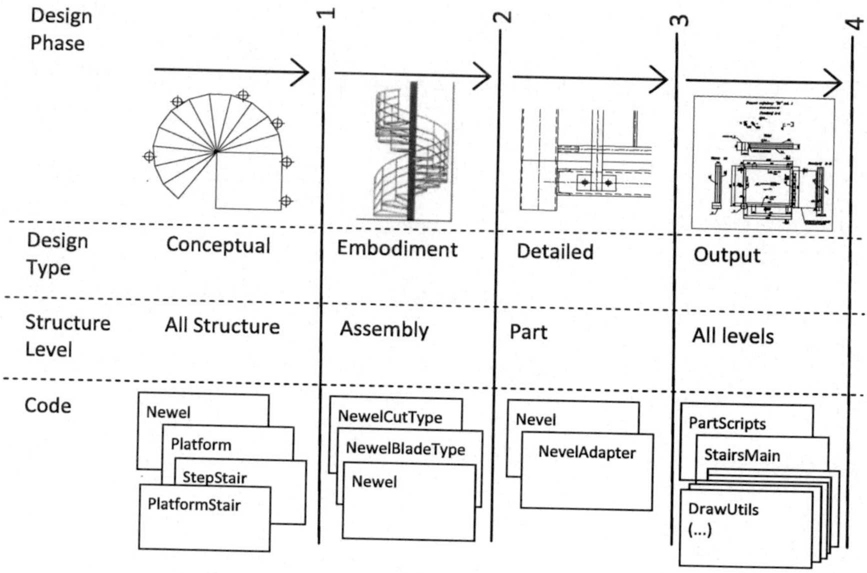


**Figure 1.** LaScala support for staircase design process

Each phase has its own characteristics and scope, each phase focuses on a different part of the staircase and each phase is represented in a different form. Given this diversity in the field, KBE should be specific to the kind of problems being addressed and the solutions expected.

At various stages of the design process and at various levels of the system abstraction, different design activities are carried out. At each level we are free to choose the most appropriate KBE. There are many examples in the LaScala system where the various KBE techniques were introduced. One of the most evident examples is present in the early phase, in which the concept of the main form of the staircase is developed.

## 2.1 Concept generation

A concept is an idea that is sufficiently developed to evaluate the physical principles that govern its behavior. This understanding is also encapsuled in the following statement [8]: “If you generate one idea, it is probably a poor one. If you generate twenty ideas, you may have a good one. Who spends too much time developing a single concept realizes only that concept.” Therefore at the conceptual design stage of the design process the LaScala system has been equipped with a software module featuring the generation of the possible layout variants that fulfill the generic design requirements (like step collection continuum, step-by-step height increment, step maximum standard height etc.) All the knowledge gathered from the designers was categorized into axiomatic and not obvious. With the help of the axioms a generation engine was created, with the freedom to find specific solution can be narrowed down using certain filtering criteria, where the criteria can be applied to key design variables (applicable to design scope). Figure 2 represents two windows; the background window is the generator output, while the top one is filtering for the best variant solution criteria definition.

The layout solution is immediately available in various graphical forms available for the designer's convenience, like 2D schemas (also referenced there as "lemons") or 3D simplified models. There are also several checkpoints at this stage; for example the height checker that controls if there is enough space between the step and the corresponding step on the next level. All these utilities make certain that the proposed concept will operate as anticipated and that, with reasonable further development (embodiment), it will meet the targets set.

## 2.2.Template example

In this chapter we discuss in details one of the templates mentioned in the paper [7] (in the Table 2) - Generative Solutions Development template. It concerns the design problem initially depicted in the section 2.1. Following is the more detailed description.

### *2.2.1. Design problem: Selection of the set of step stairs for the particular tier.*

One of the most important and yet most difficult tasks with which the engineer has to deal at the beginning of the design is suitable composition of the staircase structure, taking into account the constraints imposed. Basic, initial restrictions in this case are the total height of the stairs; starting and ending levels of individual tiers; the relative angular location of tiers, exits directions and staircase starting direction; the rotation (clockwise or counterclockwise); and also very often shape and size of the platforms. As additional restrictions, preferences for internal angle and height of single steps in the particular tier may be applied or possibly landing-steps along with the parameters describing them, etc. All these parameters must of course be within the ranges established in the standards. Traditionally, engineer performed this task "manually" leaning on calculations made on a sheet of paper or using a calculator, and sketching top views called "lemons" (these sketches resemble the cut lemon) on paper or in AutoCAD.

A major problem was to select height and inner angle of step-stairs and theirs number in such a way that the vertical distance and the range of the angular difference between adjacent tier levels have been completed with step-stairs without error, i.e. without any gap. It often happens that in the face of initial constraints and requirements the task is not feasible in terms of height, angle, or both. In some cases it is possible to use some alternative solutions e.g. with different first step-stair height (of course, in the admissible range) or with addition of the landing-step, that is the step-stair with the inner angle larger than the others. Sometimes any manipulation of the starting and ending edges of the platforms, landing-steps or stairs entrance helps. This activity is called "rotating". While in the single-tier staircases such operations are not too complicated, in the multiple-tier problems may occur, especially with the angular inaccuracy. Platforms are often attached to certain parts of the environment which are strictly positioned. For this reason, it is impossible to tamper with the setting unlimited, and thus the angular space between levels. Engineer in order to choose suitable solution has to check the many possibilities before it hits the right. With more tiers, besides additional constraints, the exponentially growing number of possibilities is another difficulty.

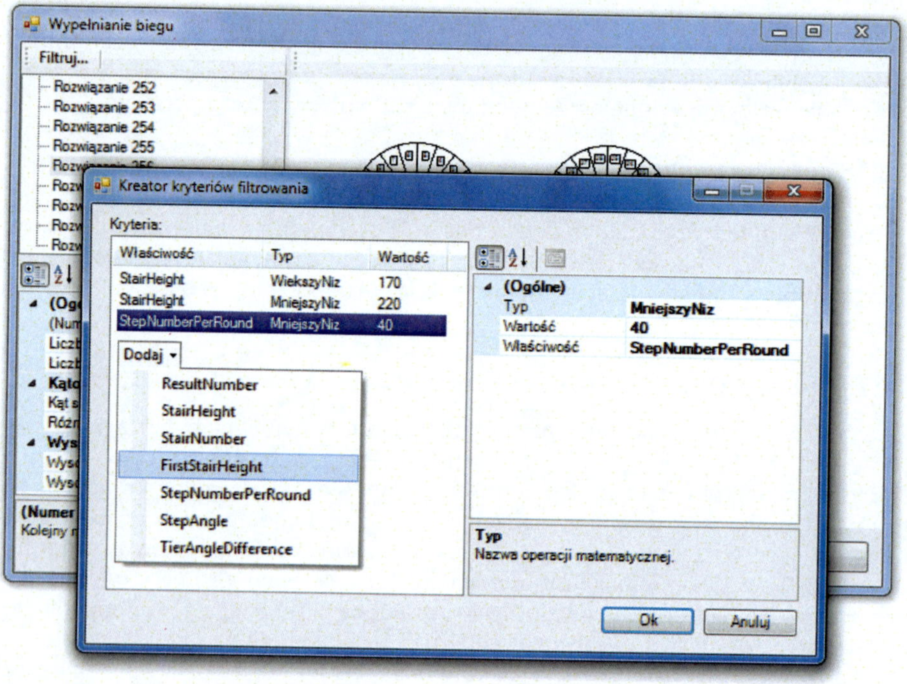


**Figure 2.** Generic design knowledge provides hundreds of possible variants. Specific problem at hand forms the selection criteria for the best solution

### *2.2.2. Simplification of the problem.*

The above description shows how difficult and arduous this design task can be, if done by the engineer in the classical way, by hand. Not to mention the effort and time-consuming activities of creating sketches and calculations.

The problem has been simplified by dividing the task of the structure selection for the whole staircase into smaller tasks of the structure selection for the particular tiers. Additionally, generated solutions consist of step-stairs only, excluding potential landing-steps (taking into account multiple landing-steps with different inner angles in this case greatly complicate the task by numerous additional solutions). However, angularly inaccurate solutions are acceptable. Engineer opting for such implicitly intends to make necessary amendments e.g. by changing one step-stair angle manually and making it a landing-step.

*2.2.3. Working scenario for the generational step-stairs set solution creation module.*

In order to proceed to the activity of the step-stairs set selection for the tier the earlier pre-preparation of a conceptual model is necessary. This involves defining the basic, global constraints of the staircase (starting level, starting angle, direction of rotation ,...) and adding the tier, together with a platform with defined position in the global structure. When starting the module for the selected tier, all the necessary information is retrieved: the conceptual model parameters defining the space to fill; the application settings parameters concerning standards, such as the minimum allowable height of a step-stair. This information is transmitted to the newly created instance of the *TierCounter* class. This is the main class of the described module model. Other classes are the *Result* and the *ResultCollection.* The first describes the properties of the single solution, while the second allows to store-and-manage the collection of feasible solutions. After loading necessary information the mechanism of combinatorial generating of acceptable solutions is run (method *TierCount* of *TierCounter* class). They are stored in the instance of the *ResultCollection* class. And then are displayed in the module user interface (Figure 2). It is a Windows Form with three main areas: the list of generated solutions, the chosen solution properties and the area of the graphical presentation. Behind each of these areas an object model is hidden. Presentation of the solutions takes place in a dynamic way. Appropriate mechanisms carry out mappings of the solutions model to the views models. The appearance of the interface corresponds to the appearance of the main application view. The area of the graphical presentation is based on the same model used in the view of the main application window ("Lemons"). Fragment marked with a red line in the graphical view presents a aforementioned angular "uncertainty" of the solution.

Engineer to select the most appropriate solution is able to use the functionality of filtering the list of solutions according to appropriate criteria. This can definitely reduce the number of potential solutions and accelerate the procedure of selection. The task of filtering is described by means of a specific object model attached to the main model of the module.

*2.2.4. Generative Solutions Development template.*

On the basis of worked out detailed solution the template was abstracted. It consists of the schema of tasks performed in implementing similar problems and the overall structure of classes (Figure 3).

Implementation schema:

- construction of a well-defined product model - prepared for a corrective development of the problem solution;
- definition of a *Solution* class together with its detailed description and the *SolutionCollection* class;

- definition of a mechanism for combinatorial generating of the set of solutions;
- construction of Filtering and/or Sorting and/or Optimization tasks models;
- construction of a solution model visual representations;
- user interface design.

The Generative Solutions Development template class diagram is illustrated in Figure 3. It consists of a set of classes grouped into packages. These packages are: Management, Functionality, Visualization, Solution and Application Model. The latter represents classes of the application main model (e.g. product model) participating in the activity of generating solutions.

The main template class is *SolutionGenerator.* This is the class from the Management package that represents solutions generation module. All actions performed within the framework of the module are called from its level. Generated solutions are collected in it. The module communicates with Application Model through this class receiving some input parameters and sending results of the generation. It transforms the Solution Model into the form acceptable by the Application Model.

The *Solution* class of the *Solution* package is a description for the solutions generated by the module. This description can be extended in particular case by adding other classes to this package. The *Solution* class is associated with the main class – *SolutionGenerator* - by three relationships. Fields *_solutions* and *_narrowedSolutions* represent a collection of all generated solutions and a collection narrowed by filtration, optimization and possibly sorted respectively. *SelectedSolution* property represents one solution selected for the use in the Application Model.

The Functionality package gathers classes representing possible actions of the module. Class *ISolutionsCounter* with *Count* function represents a mechanism for performing calculations and generating a set of potential solutions. *ISolutionsFilter* class is a filter which, based on the criteria represented by instances of a class *FilterCriterium*, narrows the set of feasible solutions. This package can be further expanded with the classes corresponding to the optimization and sorting functionalities. Classes in this package are interfaces, which means that these are only definitions of functionality. It allows you to use many different ways to implement each of the functionality without worrying about compatibility with other classes of the module.

In addition to the *SolutionsGenerator* class, the Management package still contains two classes: *SolutionVisualizer* and *ModelViewerBinder*. These three classes form the MVC programming pattern (ModelViewController). The first one is a model of the module. The second acts as a view representing the user interface and being a container for detailed views presenting solutions in various forms. The third, in turn, is a controller class, which allows the integration of separated model and view classes. It functions as a mapper between the solutions generator model and the views models while handling events emerged from both the model of the module and the user.

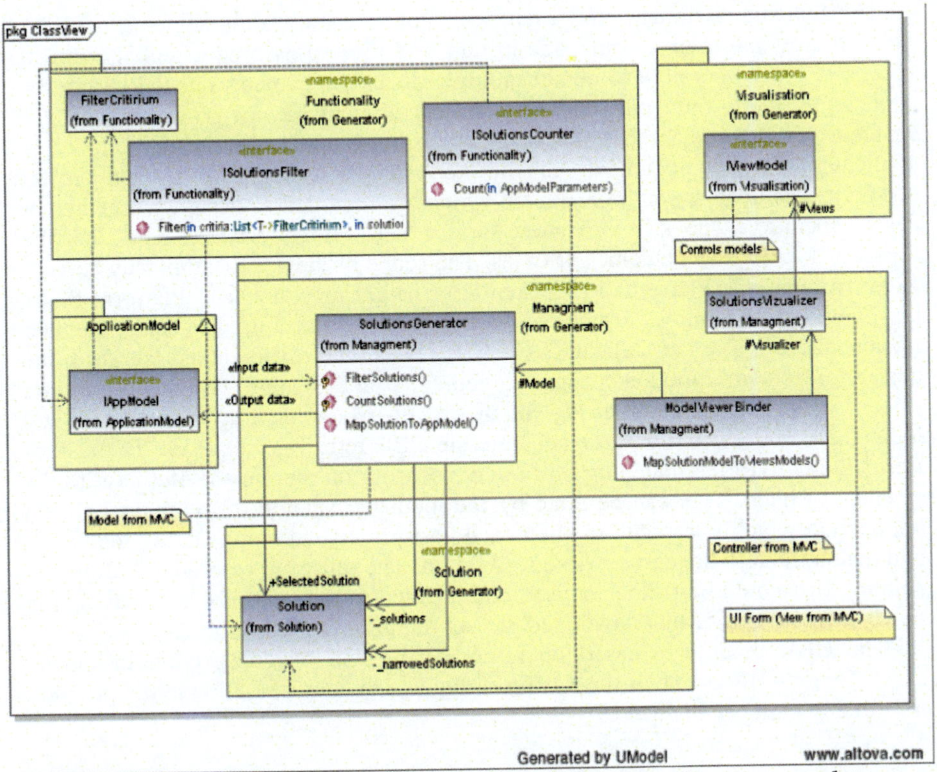


**Figure 3.** Class diagram of the Generative Solutions Development template

The Visualization package contains the *IViewModel* interface that is the abstract definition of every specific view model, such as graphical representation model. Similar to Functionality package classes there is a possibility to add different views models, to which the Solution Model can be mapped .

## 2.3 Case Based Reasoning

The LaScala system provides software solutions for the initial phase of a design process when the designer is looking for design solutions [2, 3]. The system is based on the assumption that designers store and later examine and compare previous design episodes. This stored knowledge may be used as a starting point when setting about a new design task. By recalling previous project files which are formatted as cases and reasoning with them, the designer can find solutions similar to past situations. Hereby the Case Based Reasoning module can be regarded as another form of the KBE. The CBR module does not consist of generalized knowledge rules. It rather functions as a memory of stored cases of specific designs from the past. The collection of cases could be the designer's own staircase LaScala projects or a wider range of design precedents. LaScala is able to provide assistance in the early phases of the design process.

The LaScala document model has been implemented using singleton patterns. This ensures that there is only one instance of itself providing a global point of access. It is then possible to obtain multiple document-projects opened within the application, while only one, activated project is available for processing. It also ensures that the project has not been opened for update before, by any other application. Project documents can be stored at each stage. From each project certain key indicators can be extracted which are regarded as design precedents or design memory. The key indicators form a selection of features that describe staircase solutions as outcomes of cases. The scope of these indicators may be very wide: from simple Platform Type, Step-Stair height or Number of Platform-Stairs per Tier to more complex structures, like the set of criteria used to select the Tier layout solution. The key indicators can also cover other different aspects from the design precedent: functional aspects, geometry and shapes, ornaments, texture finish, materials, manufacturing methods, environment etc. The more key indicators are defined, the more accurate the CBR retrieval process is. The task of retrieval involves the matching of a query focusing on the most useful case to the problem at hand. This can be done by the modified version of the main LaScala application which is capable to open each past project, examine all key indicators and match them to the current design situation. The retrieved case would provide a solution that could be further revised and adapted to the new problem. Saving the newly adapted and further developed project retains a new case for later use.

The CBR module is based on remembering and encoding previous design cases. Using CBR as one of the KBE techniques provides a solution to the problem of utilizing memory in design.

## 3 Conclusion

KBE integrates and works with (not: instead-of) human designers. Supporting the whole design process of the staircase using single KBE application was not acceptable. Thus the process was split into the phases and the tasks, so that it was possible to target less complicated problems using a selection of KBE approaches. Consequently, Extended KBE with its tools based on the developed library of templates should support the development of KBE applications. Nevertheless, there is still a need for extending KBE, so that it will not only support a single task but the whole or partial design process [1].

## 4. References


[1] Kodiyalam S, Sobieszczanski-Sobieski J. Multidisciplinary Design Optimization – Some Formal Methods, Framework Requirements, and Application to Vehicle Design. Int. Journal Vehicle Design 2001; 3-22.

[2] Maher M.L, Pu P. Issues and Applications of Case-Based Reasoning in Design. Lawrence Erlbaum Associates, Publishers, 1997.

[3] Pokojski, J., (ed.) Application of Case-Based Reasoning in Engineering Design, WNT, Warsaw, 2003. (in Polish)

[4] Pokojski J, Gil M, Skotnicki S, Szustakiewicz K, Szustakiewicz K, Bekasiewicz T. Application supporting design process of spiral stairs, integrated with AutoCAD 2006-project developed for POLIMEX- MOSTOSTAL Siedlce S.A., 2006-2008.

[5] Pokojski J, Gil M, Szustakiewicz K. Engineering Knowledge Modeling in Design. In: New World Situation: New Directions in Concurrent Engineering: Proceedings of the 17th ISPE International Conference on Concurrent Engineering (Advanced Concurrent Engineering), Editors: Pokojski J, Fukuda S, Salwiński J, Springer-Verlag, 2010; 257-266.

[6] Pokojski J, Szustakiewicz K. Towards an aided process of building a Knowledge Based Engineering application. Machine Dynamics Problems 2009; Vol. 33, No 3; 91-100.

[7] Pokojski J, Gil M, Szustakiewicz K. Extended KBE in Mechanical Engineering-discussion of concepts. (Delivered for CE 2011).

[8] Ullman DG. The Mechanical Design Process, McGraw-Hill (Third Edition), 2002.

# Towards a Semantic Knowledge Life Cycle Approach for Aerospace Design Engineering

Isaac Sanya [a], Essam Shehab [a,1], Dave Lowe [b], Maksim Maksimovic [a] and Ahmed Al-Ashaab [a]

[a] Decision Engineering Centre, Manufacturing Department, Cranfield University, College Road, Cranfield, Bedfordshire, MK43 0AL United Kingdom
[b] Rolls-Royce Plc, Moore Lane, Derby, DE24 8BJ United Kingdom

**Abstract.** The efficient and effective management of knowledge is becoming increasingly important within the aerospace design engineering sector due to the complexity of product development. Semantic technology is becoming mainstream technology and is being applied by many disciplines for the management of complex knowledge. However, there is a lack of a semantic knowledge life cycle to support the semantic knowledge management discipline. This paper presents a systematic knowledge life cycle (KLC) for supporting the semantic knowledge management discipline with a particular emphasis on the importance of structuring knowledge. The semantic KLC comprises eight stages namely: (1) Understand the domain (2) Structure (3) Enrich vocabulary (4) Capture (5) Represent (6) Interpret the 'know how' (7) Share (8) KBE system. This research project adopts a qualitative approach and a five-phased research methodology. An illustrative scenario within the aerospace engineering industry for producing gas turbine systems is used to demonstrate the practicality and applicability of the proposed approach. The semantic KLC supports a shared agreement of meaning and understanding between design and manufacturing engineers.

**Keywords.** Design engineering, semantic knowledge life cycle, ontology, shared agreement of meaning.

## 1 Introduction

Aerospace engineering is considered to be one of the most advanced and complex branches of engineering. The complexity of designing and manufacturing flight vehicles requires careful understanding and balance between technological advancements, design, management and costs. Thus, it has become imperative to manage and maintain the appropriate capture, structure and dissemination of product and process knowledge within this sector in order to maintain competitive

[1] Cranfield University (Bedford), Building 50, Cranfield Bedfordshire, MK43 0AL, UK; Tel: +44 (0) 1234 750111 x5241; Fax: +44 (0) 1234 754605; Email: e.shehab@cranfield.ac.uk; http://www.cranfield.ac.uk/sas/aboutus/staff/shehabe.html



D. D. Frey et al. (eds.), *Improving Complex Systems Today*,
Advanced Concurrent Engineering, DOI: 10.1007/ 978-0-85729-799-0_33,
© Springer-Verlag London Limited 2011

advantage and retain both design and manufacturing engineering experience built up over many decades. Figure 1 illustrates the main phases of the product development life cycle, which is described by Whitaker [12] as the product-creation process, this starts with a concept phase that is transformed into a detailed set of instructions for manufacture and assembly.

Due to the subjective and domain-dependent nature of knowledge, it has been identified that one of the major issues in traditional knowledge management is the complexity of establishing a shared agreement of meaning between people, processes and technology [5]. Consequently, miscommunication is a major barrier that exists between both design and manufacturing engineers. In regards to Information Technology, this barrier is usually the result of lack of computer supported open-source tools [8] to enable engineers within a specific domain to collaboratively share and reuse knowledge. However, recent research suggests that the barrier of miscommunication within the aerospace industry is a people issue rather than an IT issue [10]. This is inherently due to the diversity of individuals, different perspectives and inconsistent use of vocabulary. This has made it more difficult to develop a shared understanding of a given domain between a group of users. The semantic knowledge management discipline aims to address this issue.

Davies et al [4] describes semantic knowledge management as a set of practices that seek to classify content so that the required knowledge it contains may be immediately accessed and transformed for delivery to the desired audience in the required format. In this research, Semantic KM is not only about the technologies and platforms used to support such a practice. Semantic KM can be defined as a systematic process that aims to enrich and integrate both domain and operational forms of knowledge in order to ensure a shared agreement of meaning between domain experts and end users.

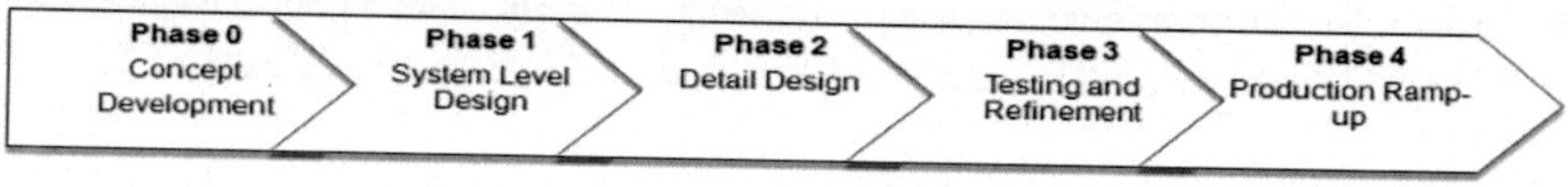


**Figure 1.** Phases of Product Development Life Cycle

Recently, some research work has been reported that use ontologies in conjunction with semantic web technologies within the aerospace sector. These are promising success stories that demonstrate the capability and benefits of semantic technology. However, there is lack of a knowledge life cycle to support the semantic knowledge management discipline. Consequently, many of the current ontological methodologies for semantic knowledge management are generic philosophical guidelines rather than explicitly defined activities.

This paper presents a systematic knowledge life cycle for semantic knowledge management using concepts from the soft system methodology (SSM) in particular rich pictures, software engineering (object oriented paradigm) and semantic web (ontologies) in order to enhance a shared understanding of a given domain. Emphasis is on the knowledge structure stage, which has often been neglected in traditional knowledge life cycles. An illustrative scenario within the aerospace industry for producing gas turbine systems is used to demonstrate the practicality of the proposed stages within the semantic KLC approach.

## 2 Related Work

The past decade has seen increasing interest in the field of knowledge management and ontological engineering for semantic web technologies as more sectors apply these disciplines. This section details some of the well-established knowledge life cycles within knowledge management.

Firestone and McElroy [5] suggested, "Knowledge management is about managing the KLC". In 1995, Nonaka and Takeuchi pioneered the term knowledge life cycle (KLC) and proposed the SECI (Socialisation, Externalisation, Combination, Internalisation) model. This is a model that has been described as the knowledge creating process and represents various stages of knowledge conversion. Bukowitz and Williams [1] approach of the KLC is divided into two dimensions. These are tactical and strategic. The tactical stages of the KLC are to acquire, use, learn and contribute whilst the subsequent strategic stages are to assess, build, sustain and divest. The McElroy's [7] KLC consists of two major processes, namely knowledge production and knowledge integration. The focus of this proposed KLC is on knowledge production, which formulates the following stages: individual and group learning, knowledge formulation, information acquisition and knowledge validation. The originality of this approach is in the single and double loop leaning processes. The five stages of the Jashapara [6] KLC are as follows: discover knowledge, generate knowledge, evaluate knowledge, share knowledge and leverage knowledge. The three stages of the Dalkir [3] KLC are as follows: knowledge capture and/or creation, knowledge sharing & dissemination and knowledge acquisition & application. Knowledge is assessed between stages one and two. However, no approach is mentioned as how this is achieved. This KLC is strongly attributed to the generation of new knowledge, which emphasise a cultural change in organisation learning. The methodology for knowledge based engineering applications (KBE) (MOKA) [11] is a generic KLC for the KBE domain. The 6 stages of MOKA KLC are as follows: identify, justify, capture, formalise, package and analyse. The main focus of the MOKA KLC is the capture and formalise stages. The Rodriguez and Al-Ashaab [9] KLC approach is considered to be distinguishable from other KLCs. This approach is also used within the knowledge based engineering (KBE) discipline. The KLC [9] proposed the use of a collaborative knowledge based system to support product development and manufacturing activities performed in dispersed locations. The stages of the KLC are as follows: identify, capture and standardize, represent, implement and use. The Chao et al [2] Semantic Web Life Cycle consists of 6 stages, which are as follows: representation, interconnection, reasoning, retrieving, validation and integration. Many of the KLCs reviewed suggest the importance of learning and understanding as an important step in achieving effective KM. However, there is no consistent vocabulary between some of the proposed KLC stages. For example, do identify, discover, and learn signify the same activity? The authors' perspective has a significant part to play. It was discovered that there is a lack of clear consistent meaning between capture and represent. Something more clear and concise is needed at a stage before capture and represent, whereby knowledge is structured using appropriate vocabulary. This particular stage has commonly been neglected and has not been identified as a stage of its own in the reviewed KLCs.

## 3 Semantic Knowledge Life Cycle Approach for Aerospace Design Engineering

Figure 2 illustrates the various stages of the proposed semantic knowledge life cycle (KLC). The semantic KLC comprises of eight stages which are: (1) Understand the domain (2) Structure (3) Enrich Vocabulary (4) Capture (5) Represent (6) Interpret the 'know-how' (7) Share (8) and KBE system. Stages (2), (3) and (6) reinforce the semantic knowledge management KLC.

Understanding the domain involves definition of the scope as well as identifying knowledge sources. The next stage is to develop an initial structure/construct of domain knowledge. The structure stage is comprised of the modularisation of knowledge into different chunks. It is not enough to structure knowledge; it has to use an appropriate vocabulary that is agreed upon by both domain experts and users. Iteration may be required between stages (2) and (3) and it is possible to refine the vocabulary until a more universal and agreed vocabulary is reached by a group of users. The next stage is to capture knowledge; if new knowledge is captured, it is imperative to feed back to stage (2) and restructure/add to the domain knowledge construct. Once knowledge has been captured, it is then represented. Visual representation is always appealing and is deemed as one of the best ways of eliciting knowledge from experts.

Stage (6) allows for the declaration and interpretation of 'know how' rules. Rules are interpreted from the construct of the domain knowledge in stage (2) and these are considered as operational knowledge. Knowledge is only beneficial when it is used. Stage (7) allows for the sharing and validation of both domain and operational forms of knowledge. Stages (2 - 7) demonstrate the process used to integrate both domain and operational forms of knowledge. These stages also illustrate the knowledge creation process. In addition, extensive validation should occur within these stages (2 – 7). The last stage (8) of the semantic KLC process involves development of knowledge-based systems using a semantic or object oriented approach. Iteration is primiarily between stages 2-3, 2-4 and 2-7.

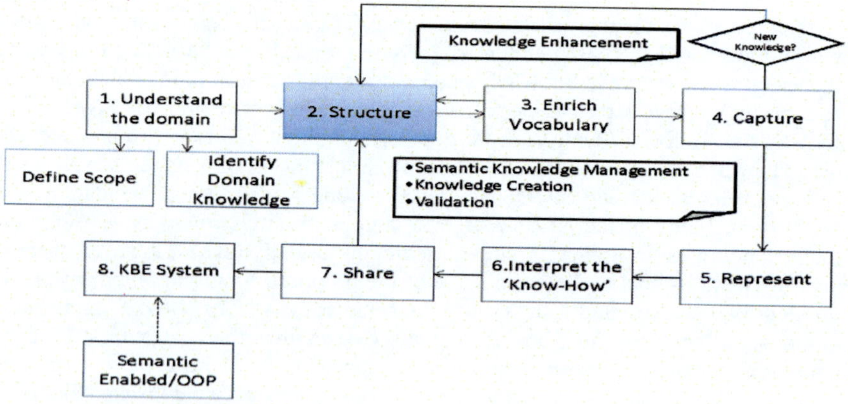


**Figure 2.** Semantic Knowledge Life Cycle

### 3.1 Understand the Domain

It is imperative to agree the scope of the problem domain before commencing with the semantic framework development. In the context of this research project, the manufacture of holes on a combustor wall (part of a gas turbine engine) has been selected and process-mapping activities have been used to help understand the domain. This has been achieved by mapping various design activities as well as identifying the various skill types, data inputs/outputs feeding into each activity, the roles involved in providing and consuming data, decision points as well as minor and major iteration loops. The process map has captured the preliminary design stage of the combustor product development process, which has helped in identifying key experts, end users, key documents, etc. IDEF0 functional modelling has been applied to support identification of the domain knowledge.

### 3.2 Structure

Having understood the engineering domain, it is imperative to begin structuring and categorizing knowledge into various segments in order to enhance knowledge systematisation. Figure 3 illustrates the proposed ontology framework. There are many similarities between ontological engineering and the object-oriented paradigm (OOP). The main three aspects of the object-oriented paradigm are: Inheritance, Encapsulation and Polymorphism. To incorporate the OOP way of thinking into the ontology development, the right questions must be addressed as illustrated in Figure 3. Fourteen concepts have been developed within the aerospace domain through the use of a suitable taxonomy and consistent vocabulary. The ontology can be readily extended as new knowledge is captured.

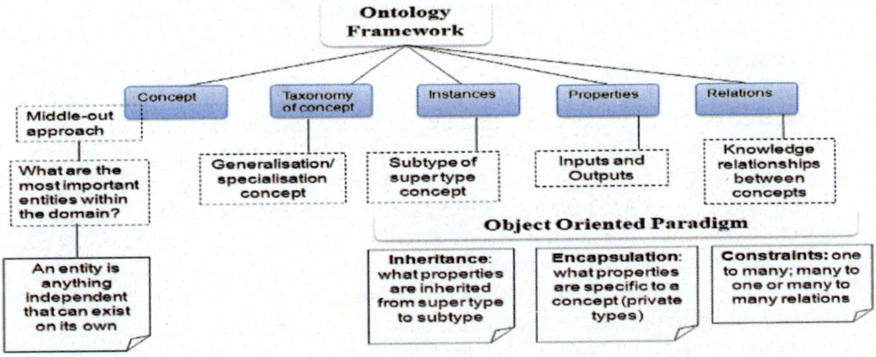


**Figure 3.** Object Oriented Paradigm way of thinking into Ontology Development

### 3.3 Enrich Vocabulary

The next stage of the semantic KLC is to enrich the vocabulary of the domain structure that was determined in the previous stage. This is a crucial aspect of the semantic KLC. Enriching the vocabulary includes ensuring a universal shared agreement of terminologies between domain experts. This includes identifying key terminologies within the ontology and quantifying their meaning.

After the development of the ontology, experienced design and manufacturing engineers contributed towards enhancing the vocabulary of the domain ontology with a view to develop a common understanding of meaning. To illustrate a scenario, one of the concepts within the domain ontology was defined as 'Tool', it was identified that the term 'Tool' in manufacturing engineering is interpreted as a physical manufacturing device (e.g. chipless machining). However, the term 'Tool' in design engineering is often thought to be computer software, excel spreadsheet with macros or even a design method. Due to the variety of meanings and context, it was identified that the term 'Tool' was not a suitable name for a concept within the domain ontology and this term was changed to the term 'Software' which was agreed and shared between both the design and manufacturing engineers. One of the enabling tools of this stage has been the use of rich pictures to display aspects of the ontology. This has proven to be effective in eliciting knowledge. The real value of this technique is in the way it encourages the creator to think deeply about the problem/scenario and understand it well enough to represent it pictorially. By having the domain experts contribute towards the creation of rich pictures, they helped develop a shared understanding of both design and manufacturing domain.

### 3.4 Capture

New knowledge will always be produced and captured within a domain as a result of new experience. Each time new knowledge is captured, it is essential to loop back to stage (2) and restructure/add to the domain knowledge construct (i.e. the domain ontology). This process is considered as the knowledge enhancement stage because the domain knowledge is enriched through this process. Through the capture and storage of new knowledge, the domain knowledge is enhanced.

### 3.5 Represent

It is important to represent knowledge in a manner that can be easily understood and interpreted in a consistent way. There is also a need to adopt a visual notation for the representation of the domain ontology. The Unified Modeling Language (UML) for the Object Oriented Paradigm (OOP) has proven to be an effective means of eliciting and representing knowledge within the software engineering discipline. Many researchers have suggested the use of UML as an effective way of representing ontologies, although there is still a need for a better notation for ontological representation. The use of the UML class diagrams can be used to represent the relationships and properties between various concepts.

### 3.6 Interpret the 'know-how'

Stage (6) of the semantic KLC involves interpreting a set of 'know-how' rules from the domain construct ontology developed in stage (2). Rules defined and interpreted from the domain ontology are known as the operational ontology. The operational ontology involves using the domain construct ontology (i.e. concepts, taxonomy of concepts, instances, properties, and relations) to define rules with particular emphasis on the vocabulary used to interpret such rules.

Figure 4 illustrates a design-centric user case to illustrate the interpretation of a 'know how' rule defined from the domain ontology. The advantage of this stage is in the reusability of the elements within the domain ontology construct.

IF (ROLE has *skillType* of "Aerothermal") AND (COMPONENT has FEATURE with *featureType* of "Cooling Hole" with *holeDimension* '**x-y cm**) AND (ANALYSIS TYPE required for COMPONENT has *analysisType* "CFD") AND (SOFTWARE has *softwareType* of "Code Z") AND (PILM Process is stage "**1**")

{
*modelType:* **3D CFD Model**
***cfdModel3DanalysisGeometry***
}

Concept
Property
Instance

**Figure 4.** Design-Centric User Case Scenario

### 3.7 Share

Both domain and operational ontology is disseminated to all experts in order to ensure the validation of the ontology. Workshops and one to one feedback sessions involving design and manufacturing engineers have been used to validate the ontology. Stages 2 to 7 demonstrate the semantic knowledge management process, which solidifies the integration of domain and operational knowledge.

### 3.8 KBE System

The final stage is the development of a KBE system, which will integrate and demonstrate both domain and operational ontology. Due to the ontological and object oriented nature of the semantic KLC, it is vital to adopt an object oriented or semantic enabled platform to demonstrate the KBE system.

## 4. Validation

The developed semantic KLC has been presented to key experts within the aerospace engineering domain. This includes participation of an engineering process lead, design and manufacture engineers and a knowledge management specialist. The semantic KLC has been deemed as a useful process with potential for delivering significant benefits if applied correctly. However, there is still need for extensive validation requiring a larger group of stakeholders. All of the interviewed experts suggested the importance of structuring and standardising knowledge using appropriate vocabulary and there is a strong focus on this in the semantic KLC.

## 5. Conclusions and Future Work

A semantic KLC to support the semantic knowledge management discipline within the aerospace industry has been presented in this paper. Practical scenarios alongside experts' validation have been used to demonstrate the applicability of the proposed approach. The semantic KLC is used to support a shared agreement of meaning between design and manufacturing engineers. Stages 2, 3 and 6 are significant stages emphasising the semantic knowledge management discipline.

Future work involves further demonstration of the semantic KLC and the development of a semantic KBE system to demonstrate the proof of concept. Extensive validation requiring a larger audience base will also be conducted. Lastly, the generic nature of the research will be quantified in the applicability of the semantic KLC to other sectors (e.g. medical, pharmaceutical, automotive, etc).

## 6. Acknowledgements

The authors would like to thank the Engineering and Physical Sciences Research Council (EPSRC), and the industrial sponsor (Rolls-Royce Plc) for funding this research project.

## 7. Reference Lists

[1] Bukowitz WR, Williams RL. Knowledge Management Fieldbook. London: Prentice-Hall, 1999.

[2] Chao L, Xiaomi A, Liugi Y. Knowledge Life Cycle on Semantic Web. In: Int Forum on Information Technology and Applications. Web Manning China: Chengdu, 2009; 369 – 371.

[3] Dalkir K. Knowledge Management in Theory and Practice. Massachusetts: Butterworth, Heinemann, 2005.

[4] Davies J, Fensel D, Harmelen F. Towards the Semantic Web: Ontology-driven Knowledge Management. New York: John Wiley & Sons, 2008.

[5] Firestone JM, McElroy MW. Key issues in the new knowledge management. Massachusetts: Butterworth-Heinemann, 2003.

[6] Jashapara A. Knowledge Management: An Integrated Approach. London: Pearson Education, 2004.

[7] McElroy MW. The New Knowledge Management. Massachusetts: Butterworth-Heinemann, 2003.

[8] McMahon C, Lowe A, Culley S. Knowledge Management in Engineering Design: Personification and Codification. Int J of Engineering Design 2004; 15: 307-325

[9] Rodriguez K, Al-Ashaab A. Knowledge Web-Based System to Support E-Manufacturing of Injection Moulded Products. Int J of Manufacturing Technology. 2007; 56: 125- 140.

[10] Sanya I, Shehab EM, Roy R, Houseman O, Jonik M. Semantic Based Approach for Product Maintenance Support. In: CIRP proceedings of the 20th design conference, France: Nantes, 2010; 90 - 97.

[11] Stokes M. Managing Engineering Knowledge: MOKA: methodology for knowledge based engineering applications. London: Professional Engineering, 2001.

[12] Whitaker R B. Value Stream Mapping and Earned Value Management: Two Perspectives on Value in Product Development. MSc Thesis, Cambridge MIT, 2005.

# Regeneration of 3D profile line using a combination of photo images and target markers

Teruaki Ito[a,1], Masayoshi Taniguchi[b] and Tetsuo Ichikawa[c]

[a]Institute of Technology and Science, The University of Tokushima, Japan.
[b]Department of Mechanical Engineering, The University of Tokushima, Japan.
[c]Institute of Health Biosciences, The University of Tokushima, Japan.

**Abstract.** Photogrammetry technique is regarded as one of the promising tools to reconstruct 3D shape from several photo images. The feasibility of the technique is studies in various areas such as, reconstruction of architectural objects, building constructions, industrial products, etc. However, applications of this technique have been conducted for rather large sized objects. This study focuses on this technique and proposes a method of 3D model reconstruction based on the combination of digital camera images and several target markers in order to apply for rather small sized target objects. First, this paper shows the conventional reverse modelling technique. Then, the overview of photogrammetry-based reverse modelling technique is presented to compare it with the conventional method. In order to measure the accuracy of photogrammetry-based reverse modelling, several experiments were conducted, of which results are presented. Some reverse modelling examples based on photogrammetry is presented, including the contour line and surface for freeform surface of a target object.

**Keywords.** Photogrammetry, reverse modelling, contour line and surface

## 1 Introduction

Thanks to the development of digital engineering techniques, shape design approaches based on the combination of CAD modeling and reverse modeling are becoming popular today. It is most often the case that a typical measuring devices, such as 3D scanner is used in the reverse modeling technique to capture the 3D surface data of the target object. 3D modelling data, or CAD model is created based on the captured surface data, which means that the model could be physically manufactured by CAM machining or rapid prototyping. Therefore, a designer today often makes a clay model for complicated surface shapes which are hard to design by CAD software, and then the models are converts to CAD data by

[1] Institute of Technology and Science, The University of Tokushima, 2-1 Minami-Josanjima, Tokushima 770-8506, Japan; Tel: +81 (88) 656 2150; Email: ito@me.tokushima-u.ac.jp

D. D. Frey et al. (eds.), *Improving Complex Systems Today*,
Advanced Concurrent Engineering, DOI: 10.1007/ 978-0-85729-799-0_34,
© Springer-Verlag London Limited 2011

reverse modeling [1]. In this way, even complicated, sophisticated, or aestheric shapes could be designed in CAD software. However, some special measurement devices, such as a 3D scanner [2], or special measuring environments are required [3] in the reverse modeling.

In dental applications, reverse modeling technique is used for oral impression, is an accurate representation of part or all of a person's dentition and other areas of the mouth. The dental impression forms an imprint of those teeth and gums, which can then be used to make a cast and used for the fabrication of dentures, crowns or other prostheses. Today, a handy 3D scanner could be used for making digital impression, which then can be used to make a cast in the same manner as the conventional method mentioned above. However, the 3D scanners are still in the categories of very expensive devices and limited to some special purposes. It is not always possible to use them in ordinary use.

This research focuses on the photogrammetry technique, which makes it possible to reconstruct 3D shape objects based on a combination of several 2D photo images which could be taken by any of the regular digital cameras. This technique is often applied to regenerate relatively large objects such as buildings or constructions. If the technique is applied to small objects, accuracy of the generated object may be an issue. However, a primary dental impression for initial processing does not require so much high accuracy because the accurate impression could be prepared in the secondary phase of treatment. Therefore, the photogrammetry technique may be used for reverse modelling of primary dental impression.

First, this paper shows the conventional reverse modelling technique. Then, the overview of photogrammetry-based reverse modelling technique is presented to compare it with the conventional one. In order to measure the accuracy of photogrammetry-based reverse modelling, some experiments were conducted, of which results are presented. Some reverse modelling examples based on photogrammetry is presented, followed by some concluding remarks.

## 2 Reverse modeling by a 3D scanner

This section shows the overview of a typical reverse modeling method using a 3D scanner. A 3D scanner is a non-destructive that takes photo images of a real-world object placed on the measurement stage, analyzes and collects data on its shape and possibly its appearance, and regenerates its 3D shape as digital data. The scanner makes it possible to regenerate any complicated shape as long as it is place on the scanning stage. However, there are some drawbacks in reverse modeling using a 3D scanner. For exmple, the scanner cannot take care of the black color or blind spots of the object. Measurement error often occurs in merging several photos. The target object is required to be placed on the measuring stage. Considering these drwabacks, some counter measures are taken. For example, measurement error could be covered by a post image processing. The use of a handy-type 3D scanner makes it possible to scan a large object which cannot be placed in the measreument stage. However, 3D scanners are special devices which are suitable for specific purposes and not for general use.

Figure 1 shows an example of 3D shape digital model of oral impression generated by the reverse modeling using a 3D scanner. The figure shows that the regenerated 3D shape of the lower mouse is almost identical to the original model. Since this is a digital model representing 3D shape of the original model, its physical model can be manufactured by rapid prototyping such as a 3D printer using plaster materials, of which result would be similar to the conventional oral impression.

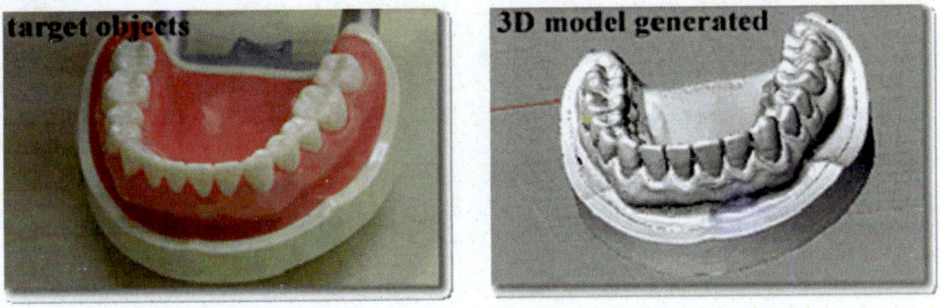


**Figure 1.** 3D digital model created by conventional reverse modelling

## 3 Photogrammetry for reverse modeling

Photogrammetry is a method to obtain the geometrical features of the target object based on the photo images, of which origin goes back to the middle of 19 th century. One of the typical examples of Photogrammetry is to measure the distance between the points which are located on the two parallel planes each. If the scale of the photo is known, the real distance can be calculated based on the combination of photo scale and the photo measurement data. As for an example of Photogrammetry application, stereophotogrammetry involves estimating the three-dimensional coordinates of points on an object. These are determined by measurements made in two or more photographic images taken from different positions. Common points are identified on each image. A line of sight (or ray) can be constructed from the camera location to the point on the object. More sophisticated algorithms can exploit other information about the scene that is known a priori, for example symmetries, in some cases allowing regeneration of 3D coordinates from only one camera position.

The photogrammetry of this research combines several photos taken by a calibrated digital camera [4], to measure the 3D distance of the target object. This method enables the reconstruction of 3D shape from the several digital camera photos. Calibration means the measurement of specific parameters to the camera, such as focal length and resolution, and initialization of the camera for the photogrammetry. The basic idea of photogrammetry is to calculate the depth distance, which cannot be obtained from a single photo image, by the combination of another photo which was taken from a different angle. 3D shape reconstruction can be possible from this idea. However, the actual size of the target object cannot be measured because the distance of the camera from the target object is unknown.

Therefore, one actual distance in the real space is measured on the target object and the distance is used in the calculation of the model. In this way, the distance between the camera and the target object can be calculated. Then all of the other remaining part of the object can be calculated. The photogrammetry enables the reconstruction of the 3D shape of the target object only from the combination of several 2D photo images.

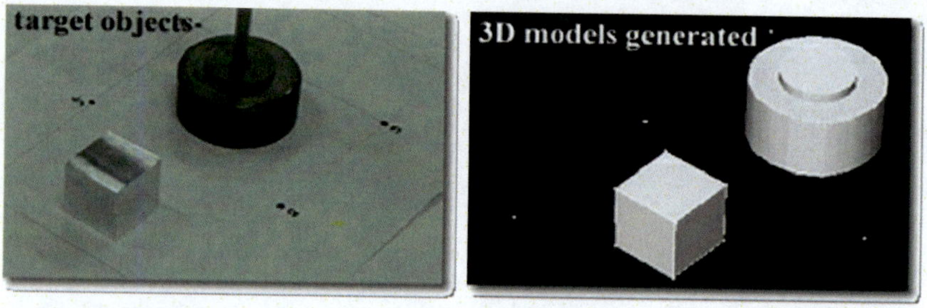


**Figure 2.** 3D digital models created by photogrammetry-based reverse modelling

Figure 2 shows example of the regenerated 3D digital models on the right as opposed to the original objects on the left. The corners and edges of the objects are used to regenerate the digital model. Therefore, no additional markers are required to deal with these types of geometric shapes.

## 4 Experiments and discussions for 3D reconstruction of the target object by photogrammetry

In order to measure the accuracy of photogrammetry-based reverse modelling, measurement experiments were conducted, of which details are presented in this section.

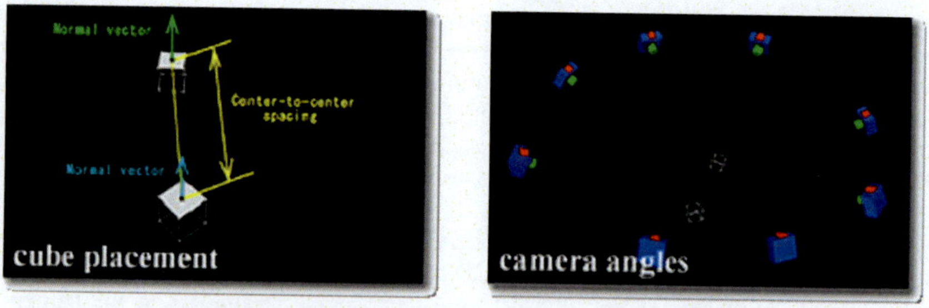


**Figure 3.** Overview of cube placement and camera agnles

Photogrammetry requires several photo images taken from different angles. For example, if the angle difference is 90 degree, then four photos may be taken. If the difference is 45 degree, then the number of photos may be eight. Some additional photos may be required to cover the blind spots from upper angles or lower angles.

From a point of view of experience, many numbers of photos are not necessarily but the minimum number of photos is obviously required. The relationship between the number of photos and the measurement accuracy are not clear for the application of this research. Therefore, this experiment was conducted to clarify the relationship. Moreover, this experiment also studied how the distance between camera lens and the target object affect the measurement accuracy.

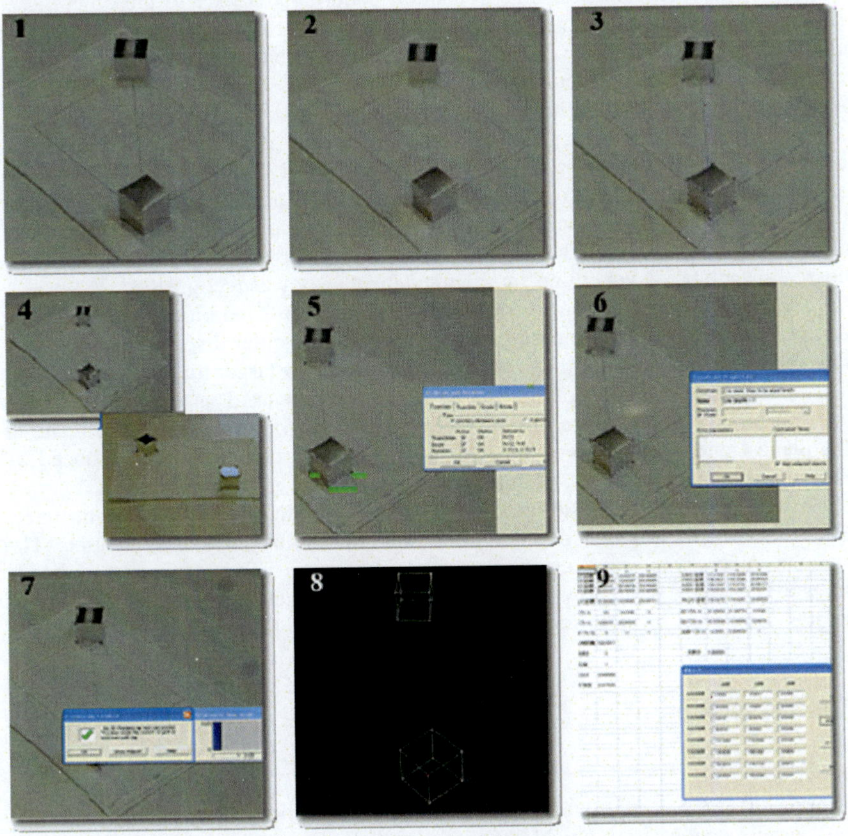


**Figure 4.** Process flow of photogrammetry-based reverse modelling

Three types of cubic aluminium objects with the accurate sizes of 10mm, 20mm and 30mm sides were used in this experiment. A pair of the same size cubes was placed on the horizontal test plate so that the upper surfaces of the pair, or plane I and plane II, should be in parallel to the horizontal plane, and the distance of the centre of plane I and Plane II should be either 100mm or 200mm. The number of photos to be taken was 4, 8, and 13. Figure 3 shows the overview of this experiment, showing the definition of center-to-center distance and the eight camera angles to take photos.

Figure 4 shows the process flow of reverse modelling by photogrammetry in this experiment and how the measurement was made using this model.

As shown in Figure 4, the experiment was conducted based on the following nine procedures. (1) Photo shots: Photos of the target object are taken by a digital camera, (2) Image loading: Photo images are loaded to a PC. (3) Tracing corners/edges: Tracing of the corners and edges are conducted in each photo. (4) Mapping photos: The corners and edges are mapped among the photos. (5) Axis setting: The basic edge is determined on the object and its actual length is measured. (6) Constraint setting: Constraint conditions are set to the objects. (7) 3D analysis and modelling: Digital model of the target object is generated by 3D analysis and modelling. (8) 3D measurement: Measurement process is conducted using the model. (9) Calculation: Distance and angle are calculated using the measured data for the experiment.

Three sets of aluminium blocks were used in the experiment, with the inter-centre distance of 100mm and 200mm. The number of photos was 4, 8 and 13 in each. From the experiment results, the maximum error of distance was 2.28mm for the inter-centre 200mm blocks, and 1.25mm for 100mm blocks, of which average error was between 0.88 and 1.17 mm. Even though the error was fluctuated some degree in each block set but no significant dependency was recognized.

The measurement of centre coordinate was made in each block set. As for the error of centre coordinate measurement, 100mm inter-center distance block set has the smallest error. Overall, the error regarding plane I is larger than that of plane II. This is because that one of the corner of plane I is used as the origin of the coordinate axes. Inter-centre distance was calculated from the centre coordinate of the plane I and II. Therefore, the error of inter-centre coordinate was reduced by the counterbalance of the co-ordinate errors.

As for the error of angle measurement, the maximum error of 200mm inter-center block set was 2.5 degree, while that of 100mm was 0.87 degree. The average error was between 0.59 and 1.31 degree. The accuracy of angle measurement was somehow related to the size of target object or the number of photos. However, no significant dependency was recognised in these parameters.

## 5 Regeneration of 3D profile line and profile surface for freeform shape objects

Experiments in section 4 shows that 3D model generation is possible if the edges and corners of the target object are recognized. However, freeform surface cannot be obtained by this approach because those corners and edges are not recognised. This section shows an approach of a combination of photo images and target markers to regenerate the freefrom surface. The target markers include physical targets and projections targets.

Figure 5 shows the result of freeform shape generation by the combination of 50 target markers affixed on the target object and 7 photo images. Even though the generated shape is not accurate as opposed to the one generated by a 3D scanner, a contour surface was obtained. However, attachment of 50 targets is not always available. Therefore, three-target attachment experiment was conducted to obtain a contour line of a target freeform shape, which could be used to generate a contour surface. Figure 6 shows the lower jaw bone plastic model affixed with three target markers. As a result of photogrammetry, a contour line was generated as shown in the middle of Figure 6. When the line is superimposed on the digital model of the object, it shows a good match as shown in the figure. The similar results were obtained when three spots of laser projection were used as the target markers.

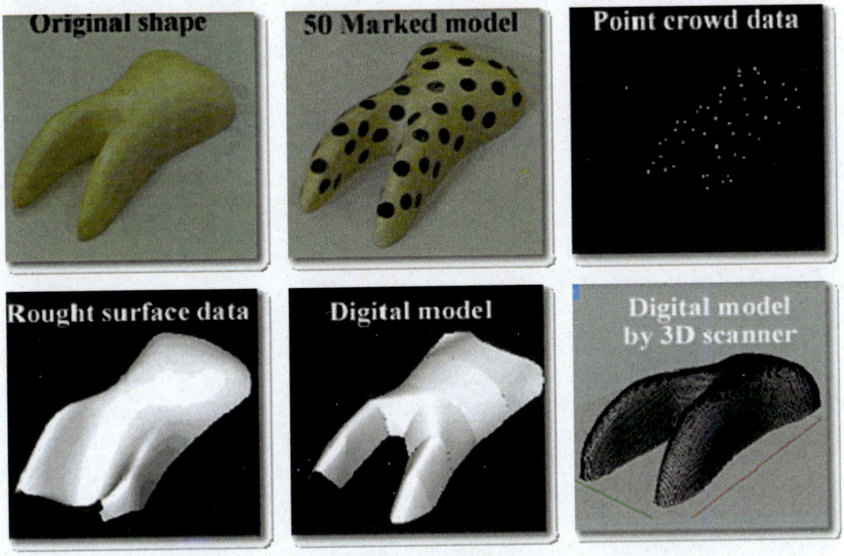


**Figure 5.** Flow of surface regeneration vs. its digital model

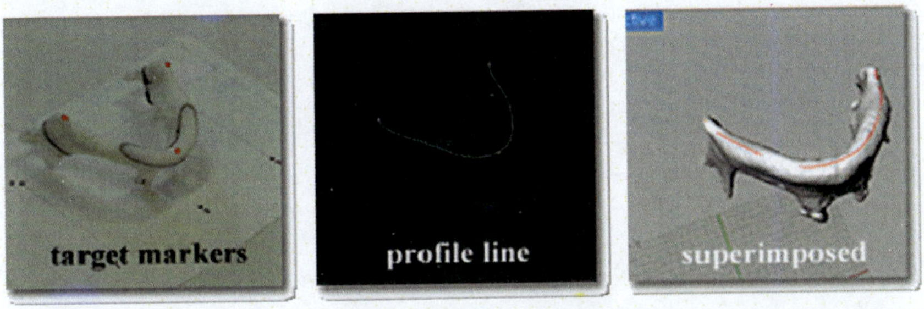


**Figure 6.** Regeneration of profile line and its superimposed image on the original data

Figure 7 shows the three contour lines obtained using three spots of laser projection in each. Figure 7 also shows a freeform surface generated by the combination of these three lines. Apart from the accuracy of generated surface, the experiment shows that the contour surface of the target object was obtained from the combination of three contour lines, which required a set of three photos from different angles and three spots of laser projection target markers for each line.

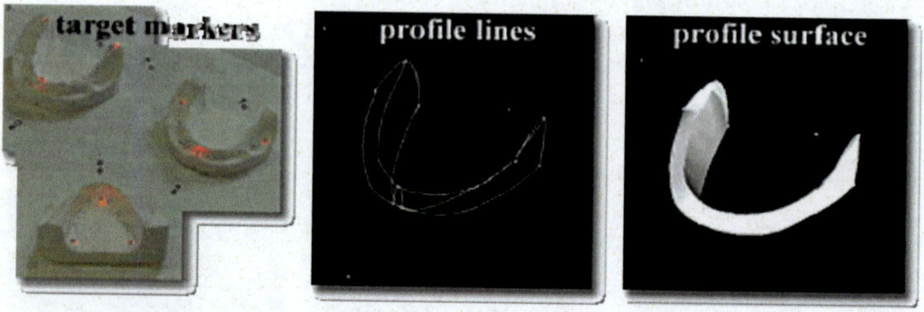


**Figure 7.** 3D profile surface reconstructed by the combination of three profile lines.

## 6 Concluding remarks

This paper presented the photogrammetry-based reverse modelling technique in comparison with the conventional one in order to regenerate a freeform surface shape such as an oral model for dental applications. It is shown that a combination of several photos taken by one of the general digital cameras enables to regenerate the 3D digital model of the target object and to make measurement on the digital model. It is expected that this method will make it easier to perform reverse modelling with the use of common digital cameras. On the other hand, the accuracy of measurement may become an issue. Therefore, this study reviews the measurement accuracy on this method. As a result, the accuracy was independent of the size of target object or the number of photos. On the other hand, the accuracy cannot be improved even the more number of photos or the bigger size of the target object. However, it is expected that his digital camera-based reverse modelling method could be used in such an application where high accuracy is not required, such as the primary preparation of dental impression. For future study, further experiments are schedule to be conducted to clarify mode detailed conditions for feasibility of this approach.

## 8 Citations and References

[1] Shibasaki, R., Laser scanner and 3D measurement, Journal of the Japan Society of Photogrammetry and Remote Sensing. 2000. Vol.39. No.2; 6-7. (in Japanese)

[2] Yamanouchi, H., Imazato, M. Measurement for Objects of Complex Shape by 3D Laser Beam Scanner. Transaction of the Visualization Society of Japan. 2008;Vol.28 (Suppl.1); 341–342. (in Japanese)

[3] Furuichi H., Kitagaki R., Kanematsu, M. and Noguchi, T. Characterization of Visual Roughness hi Building Cladding Surfaces by Using Contact-type 3D Scanner, Proceedings of Architectural Institute of Japan. 2003; 47-48. (in Japanese)

[4] Nakano K. Chikaku, H., Exif dependent camera calibration in close range digital photogrammetry. Journal of the Japan society of Photogrammetry and Remote Sensing. 2009. Vol.48. No.3; 180-187. (in Japanese)

# Virtual Reality-Based Lathe Operation Training Based on Brain Activity Assessment Using Functional Near-Infrared Spectroscopy

Keiichi Watanuki[a,b] and Lei Hou[c]

[a] Professor, Graduate School of Science and Engineering, Saitama University, Japan.
[b] Professor, Saitama University Brain Science Institute, Japan.
[c] Doctoral student, Graduate School of Science and Engineering, Saitama University, Japan.

**Abstract.** This paper proposes a new virtual reality-based lathe operation training and human resource development. In our proposed system, the education content is displayed in the immersive virtual environment with haptic information, whereby a trainee may experience work in the virtual site operation. The brain functional activities of the worker under virtual and real skills training are measured using near-infrared spectroscopy, and the characteristics of these activities are analyzed in detail.

**Keywords.** Virtual reality, Skills transfer, Lathe operation, Near-infrared spectroscopy, Brain activity analysis

## 1. Introduction

Despite a general trend toward automated lathe operation with NC lathes and other systems, many types of high-precision machining are difficult to automate and require manual lathe operation by highly skilled machinists with years of experience who can judge the state of the machining operation by sensing chatter vibration, sounds, and even smells generated during lathe turning. This ability is generally referred to as "instinct", or simply "skill". In recent years, however, with the widespread aging of skilled lathe operators, concern is rising that this skill could be lost [1]. In the past, it was learned at the side of a master machinist, who guided the trainee in the actual use of the lathes. Instruction was given chiefly through physical guidance, rather than by verbal explanation. In this on-the-job training (OJT), the operating skill was thus learned in guided experience and the trainee was in this way able to gain the required implicit knowledge, awareness, and sensitivity to potential problems in lathe operation [2]. This learning method, however, is time- and labor-intensive, and it is highly dependent on the capabilities of the trainer in nurturing the requisite skill.

D. D. Frey et al. (eds.), *Improving Complex Systems Today*,
Advanced Concurrent Engineering, DOI: 10.1007/ 978-0-85729-799-0_35,
© Springer-Verlag London Limited 2011

In this light, the authors have proposed a training method which merges OJT with a virtual reality (VR) environment for physical experience in the processing operation [3]. In this method, lathe operation can be readily repeated under given conditions. It also has the advantage of providing safe, virtual training and experience in the occurrence of accidents and their consequences due to errors in operation. In this paper, we describe the construction of a VR system for the transmission of lathe operation skill, and neurological analysis of the effect on the trainee of using the VR system as compared with training on an actual general-purpose lathe, based on near-infrared spectroscopy (NIRS) measurements of brain activity during lathe operation with the VR system and with an actual lathe.

## 2. Measurement of brain activation response by NIRS

During neural activity in the brain, humans transmit and process information and decide upon actions or responses. When neural activity occurs, blood flow and blood quantity increase in the tissue near the active neurons, and the ratio between oxygenated and deoxygenated hemoglobin in the blood changes. Hemoglobin characteristically changes in near-infrared (700~900 nm) absorbance in accordance with its oxygen content, and it is therefore possible to determine changes in the oxygen level in the hemoglobin by measurement of this absorbance by NIRS. In NIRS, the cranial target area of the subject is overlaid with a special holder for optical fibers which are inserted into the holder at selected measurement positions on the scalp, thus enabling non-invasive measurement of NIR incidence and reception. This configuration places little or no restriction on bodily movement of the subject, and the measurements can therefore be made with the subject in a natural state of posture and movement. In the present study, we thus used NIRS to determine the state of the hemoglobin oxygenation at the cerebral surface and on that basis performed real-time color mapping of the state of brain activity (the brain activation response) in the target regions.

Before attempting measurements for the complex actions performed in lathe processing, we performed preliminary trial measurements of brain activation responses for the three basic actions of moving the center of gravity, bending at the waist, and rotating a handle. For each experiment, the subjects were two males in their twenties. The NIRS instrument used for the measurements was the Shimadzu FOIRE-3000.

## 3. Operation of general-purpose lathe

In lathe working, which is a type of machining, a cylindrically shaped workpiece is fixed on the lathe and rotated, and a tool bit is pressed against the workpiece to machine it. Broadly classified, lathe working is performed either on NC lathes, in which the machining is performed automatically, or on general-purpose lathes, in which the machining is performed manually. The latter was used in the present study.

The general-purpose lathe contains a chuck for holding the workpiece, a toolpost which holds the tool bit, a carriage with a handle for longitudinal movement of the workpiece, a crosspiece with a handle for transverse movement of the toolpost, and a lever for actuation of the main spindle. The workflow essentially consists of: (1) attachment of the tool bit to the toolpost, (2) attachment of the workpiece to the chuck, (3) initiation of main spindle rotation, (4) alignment of the zero point, and (5) rotation of the handles to maintain tool bit engagement and thus cut the workpiece to the target configuration. Generally, the cutting speed (the rotational speed of the workpiece), the cutting depth (thus, the transverse position of the tool bit), and the feed rate (the rate of longitudinal movement by the tool bit) are controlled by the machinist in accordance with the workpiece material, its shape, the type of tool bit used, and the nature of the machining design. Together, these elements are referred to as the cutting conditions, and their control is the key factor determining lathe operating skill.

The above process steps (1) to (3) are preparatory steps. It is steps (4) and (5) that require a particularly high level of skill, and which are the subject of the present study. In this study, moreover, the recommended cutting conditions were set in accordance with the ideal cutting conditions based on the cutting target material and diameter.

## 4. VR system for transmission of lathe operating skill

In these trials, the operator stood in front of a screen, wearing a head tracking device enabling detection of the position and orientation of his head, and polarizing glasses. The operator thus had a stereoscopic view of the lathe image which was projected on the screen, and could manipulate the model handles which were located near his hands, as shown in Figure 1, and synchronized with the screen image. One handle was modeled on the carriage handle of an actual lathe for

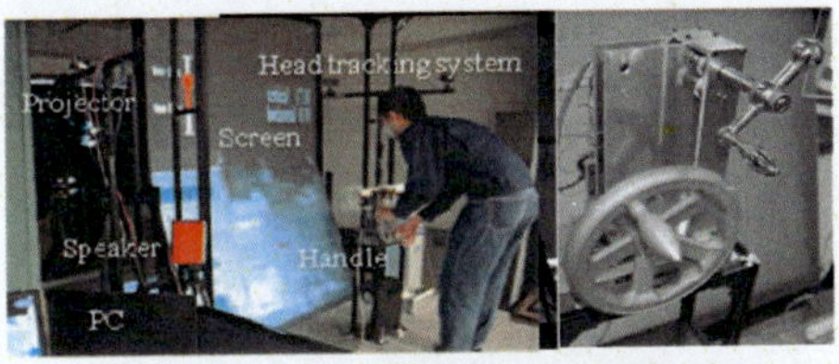


**Figure 1.** VR system for transmission of lathe operating skill

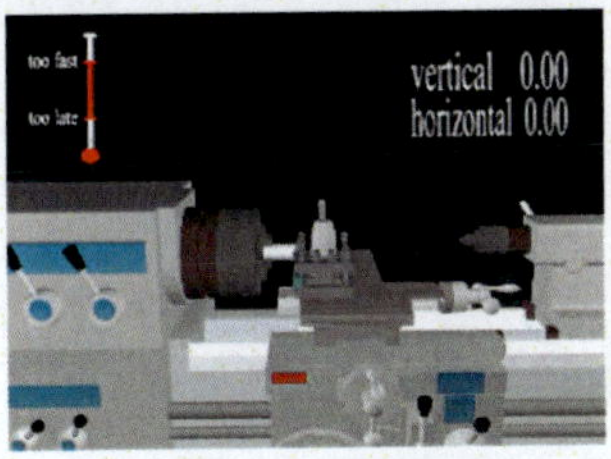


**Figure 2.** VR lathe

longitudinal movement, and the other on a toolpost handle for transverse movement. Each handle unit contained the handle, a magnetic brake, and an encoder. The handle and the magnetic brake were attached coaxially on the same crankshaft, and the encoder was linked to each crankshaft by coupling. The magnetic brakes imparted a sensation of a force on the operator when he rotated the handle. Simultaneously, a signal was sent from the encoder, the information was read by a PIC microcontroller, and the rotational speed was transmitted to a personal computer (PC). From the rotational speed, the number of rotations was determined and the depth of the cut was computed from the positional relationship of the tool bit and the workpiece. Based on this information, an image of the processing state was constructed and projected onto the screen. Figure 2 shows the projected screen image. Scale markings are present on actual lathe handles, but none were provided on the fabricated system handles and they were therefore shown in the image. As an added training function in the virtual system which simulated one aspect of OJT, it was possible to show not only a reproduction image of the actual object but also the feed rate in comparison with the feed rate of the recommended cutting conditions. During the operation, processing sounds were provided by an ITU-R5.1ch speaker system, to present the changes in sound when the tool bit came into contact with the workpiece and the various sounds depending on the cutting state. In this way, it was possible to present haptic, visual, and auditory information to the trainee.

# 5. Brain activation response during lathe operation as measured by NIRS

## 5.1 Measurement of brain activation response by NIRS during lathe operation

In these trials, measurements relating to the subject's brain activation level were performed during lathe operation in the VR- lathe environment and in the actual lathe environment. The subject was a male in his twenties with experience in lathe operation. The trials in the VR environment were performed in alternating segments of observation of another person operating the lathe and operation by the subject himself. The parts of the brain measured by NIRS were the motor cortex and visual cortex in combination, and the frontal lobes and visual cortex in combination. For each combination, the trial was performed four times. The measurement positions are shown in Figure 3. In the actual lathe environment, similarly, the subject observed and performed the lathe operation by turns, each

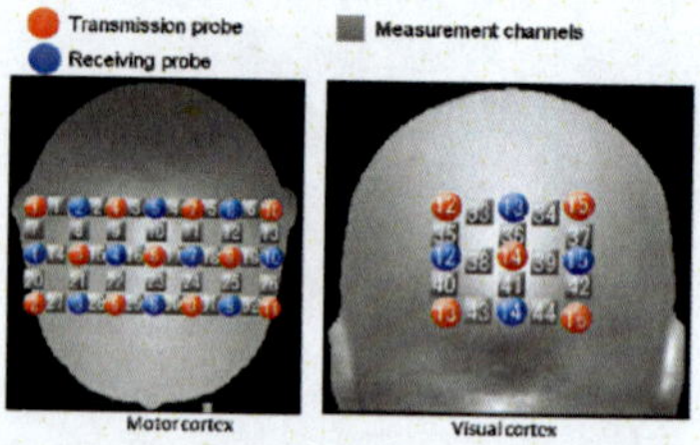


**Figure 3.** Measurement regions during lathe operation

eight times. The workpiece material was aluminum. The cutting conditions, of which the subject was instructed in advance, were: cutting depth, 0.2 mm; cutting length, 20 mm. The time of the operation (the "task time") was 40 s, and the non-operating time (the "rest time") was 80 s.

## 5.2 Analysis of brain activation response measurements by NIRS during lathe operation

### *5.2.1 Trials in the VR-lathe environment*

We measure the changes in oxyHb levels in the motor cortex and visual cortex during the observation of operation in the VR environment by another person. The oxyHb in the visual cortex tended to decline in both the right and the left hemisphere. A decline in oxyHb in the motor cortex was also observed. To investigate whether the decline in oxyHb was attributable to the task observation, a test for a significant difference between oxyHb during the task time and during rest time was performed using the Holm-Bonferroni method. The level of significance was taken as 1%. The results of this test showed a significant difference at almost all of the measurement positions in the visual cortex and motor cortex between the task time and the rest time, thus indicating that the measured decline in oxyHb was actually attributable to the task observation. In regard to the positions in which there was no significant difference, moreover, a test performed with the assumption that the task-time oxyHb increased showed no significant difference from the oxyHb change during the rest time. In summary, the results indicate that, in the case of observing another person perform the lathe operation in the VR-lathe environment, the in the motor cortex and the visual cortex decline or do not change.

Figure 4 shows the results of the motor cortex and visual cortex measurements during operation of the VR- lathe by the subject himself. In this case, oxyHb increased at almost all measurement positions in both the motor cortex and the visual cortex. In the visual sensory area, the oxyHb increase was larger in the upper positions than in the lower positions. Also, the increase in oxyHb was larger in the

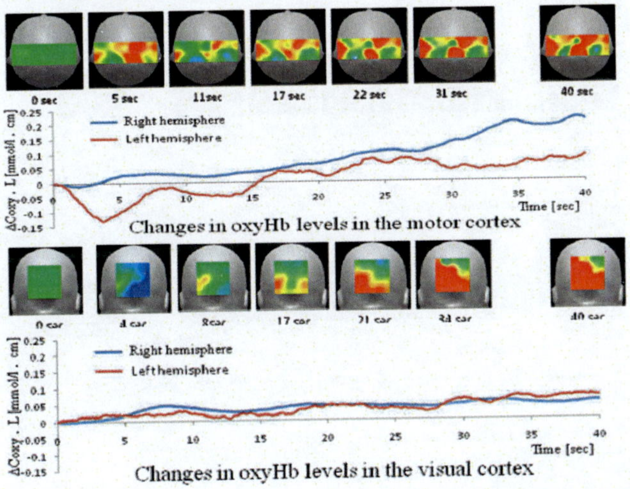


**Figure 4**. Brain activation response during of VR lathe operation performance

right hemisphere than in the left. In these trials, again, with a significance level of 1% as tested by the Holm-Bonferroni method, a significant difference was found between the task time and the rest time in almost all measured positions in the motor cortex and the visual cortex. It can therefore be concluded that the increase in oxyHb as measured by NIRS was attributable to the performance of the lathe operation. In the motor cortex, the positions in which the oxyHb increase was particularly large were in the vicinity of the temporal region which responds to hand movement and the parietal region which responds to trunk movement. In the visual cortex, the response at the lower measurement positions was particularly strong. The increase in oxyHb during performance of the operation was particularly large in comparison with the oxyHb changes during observation of the performance by another person. In the frontal cortex also, the oxyHb increase was clearly larger during operation by the subject than in observation of operation by another person.

Taken together, the results of the trials in the VR environment show that oxyHb tends to decline while watching someone else perform the lathe operation, but increases substantially while performing the operation. In short, they show that brain activity is greater while performing the lathe operation than while seeing it performed.

### *5.2.2 Trials in the real lathe environment*

Figure 5 shows the change in oxyHb level in the motor cortex and the visual cortex of the subject as he observed the operation of the actual lathe by another person. Although the change was small, in this case as in the case with the VR-lathe, a decline in oxyHb was observed in the visual cortex and the motor cortex. The decline was again tested for significance by the Holm-Bonferroni method with a significance level of 1% to investigate whether it was attributable to the task, and the results again showed a significant difference between the task time and the rest time, for the measured positions in the overall visual cortex and nearly all of the motor cortex, thus indicating that the oxyHb decline was attributable to observation of the task. For the positions in which there was no significant difference, no significant difference from the change in hemoglobin during the rest time was

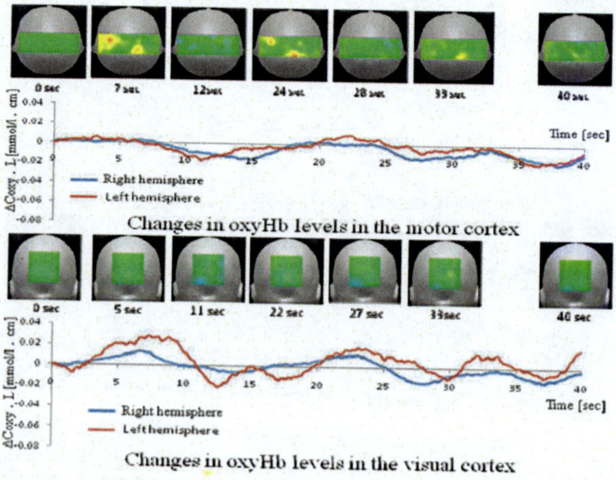


**Figure 5.** Brain activation responses during observation of real lathe operation

found when it was assumed that the oxyHb increased during the task observation. In summary, the results indicate that, during observation of the lathe operation by another person in the real lathe environment, oxyHb in the visual cortex and the motor cortex declined or did not change.

We measure the results for the motor cortex and the visual cortex during real lathe operation by the subject. As before, to investigate whether the increase in oxyHb was attributable to the task, the results were tested by the Holm-Bonferroni method with a significance level of 1%. The test results showed a significant difference between the rest time and the task time at nearly all measured positions in the motor cortex and the visual cortex, thus indicating that the increase in oxyHb was actually caused by the task performance. The measurement positions in the motor cortex that showed particularly large increases in oxyHb were in the temporal cortex. In the visual cortex, the oxyHb increase was particularly large in the lower positions. In the frontal cortex, the increase in oxyHb was larger during performance of the operation than during observation of its performance. The results thus indicate that in the real-lathe operation, as in the VR-lathe operation, the brain activation response was larger during performance of the lathe operation than during observation of its performance by someone else.

### *5.2.3 Brain activation response during performance of VR-lathe and real-lathe operation*

In the real-lathe operation by the subject, oxyHb was found to increase greatly in the lower measurement positions in the visual cortex. This may be attributed to the relative location of the workpiece and the tool bit on the actual lathe. As shown in Figure 6, both were below the usual line of sight of the subject, who therefore had to lower his gaze in their direction during the operation. During VR-lathe operation by the subject, similarly, the oxyHb increased greatly in the lower measurement positions of the visual cortex. As the VR-lathe was presented in the same size and position as the actual lathe, it was similarly necessary for the subject to lower his gaze during the operation, thus presumably resulting in the large increase in oxyHb in those measurement positions in the lower portion of the visual cortex.

In summary, the above results show very close agreement between the trends in frontal cortex, motor cortex, and visual cortex activation responses during VR-lathe and real lathe operation. This shows that essentially the same movements were performed by the operator with both lathes, and thus indicates that the lathe operation in the VR environment very closely mimics that in the real environment.

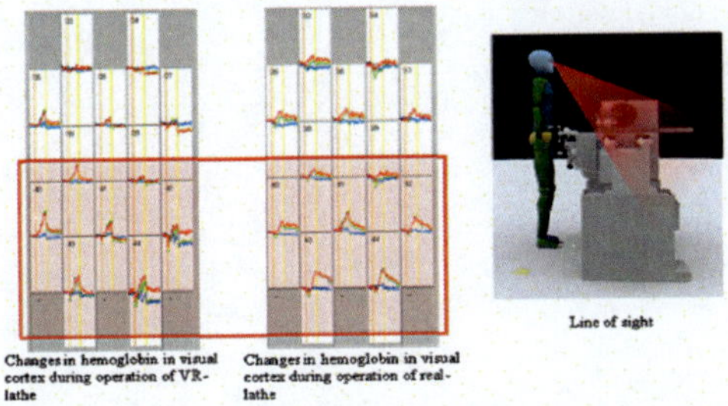


**Figure 6**. Gaze direction and change in hemoglobin

## 6. Conclusion

In this paper, brain activation responses at various NIRS measurement positions during the performance of three basic movements showed that the positions of change in oxyHb were clearly influenced by direction of movement and body posture at the time of the movements, and that the degree of change was largely governed by the level of force which was applied. The results also showed that the change in oxyHb was delayed by several seconds from the time of the actual movement. In the light of these findings, NIRS measurement was performed to investigate brain activation response during manual operation of a VR- lathe and an actual lathe. The results showed qualitative agreement in the trends of oxyHb change for operation in the VR-lathe environment and the real lathe environment. The results also showed that, in both environments, performance of the lathe operation induces greater changes in oxyHb than observation of its performance by another person, and thus that the brain is more highly activated by performance of the operation than by its observation.

The results, taken together, show that brain activation response during the performance of lathe operation in the VR environment is very close to that which occurs during operation in the actual environment, and thus indicate that training in such a VR environment will be effective and useful. In the future, we plan to perform more detailed analyses of movements performed during lathe operation and the relationship between the process of developing lathe operation skills and brain activation responses, and to apply the findings to more effective virtual learning.

## References


[1] The Ministry of Health of Japan, Analysis of Labor Economy, (2005), pp.162-176.

[2] Kimura, Y., Taniguchi, Y., Fujinami, M., Factor to promote OJT in manufacturing, Journal of The Japan Institute for Labor policy and Training, No. 26, (2007)

[3] Watanuki, K., Knowledge Acquisition and Job Training for Fundamental Manufacturing Technologies and Skills by Using Immersive Virtual Environment, Journal of Japanese Society for Artificial Intelligence, Vol. 22, No. 4, (2007), pp.480-490.

# Part V
# Value Engineering

# Addressing complexity in sustainable aircraft design

Marcel Schroijen[1], Michel van Tooren[2,a], Mark Voskuijl[2] and Richard Curran[2]

[1]PhD student, Delft University of Technology, Faculty of Aerosapce Engineering, Department of Systems Engineering and Aircraft Design (SEAD)
[2]Delft University of Technology, Faculty of Aerospace Engineering

**Abstract.** Next generations of civil transport aircraft will need to be evaluated not only against their behaviour as an aircraft system but also as a part of the larger air transport system. The sustainability issues related to noise and emissions, represented by environmental impact, combined with the complex behaviour of the stakeholders in the transport system require a different way of evaluating future aircraft designs. The study case of the box wing type aircraft, i.e. Prandtl Plane, shows the potential of reduced induced drag and direct lift control, when evaluated using the current system level computational design framework (DEE). However, this aircraft type also shows a potential for alternative use, preventing environmental impact prediction.The proposed DEE extension provides an integrated approach to address these complexities in sustainable design and facilitate impact evaluations.

**Keywords.** Environmental impact, Value engineering, Agent based simulation, Stakeholder approach

## 1 Introduction

To sustain our current wealth and health on earth, many societal and technological changes will be required. Science and technology are expected to have a major impact in their contribution to the satisfaction of future human needs for food, health, transport, energy and leisure. Within the transport sector, aviation will be one of the highly challenged segments. To absorb the expected global growth in transport of passengers and freight, all elements of the air transport system will need severe changes[11]Environmental impact needs to be addressed at the global aviation, or system-of-systems, level[6]. To design future aircraft more efficiently, computational frameworks automating repetitive work, Design and Engineering Engines (DEE) have been proposed[15]. However, the challenges posed by the sustainability conundrum require an even more integrated approach to design. The complexities that have to be addressed surpass the system level and require the incorporation of system-of-systems level considerations.

---

[a] Full Professor, Delft University of Technology, Faculty of Aerospace Engineering, Kluyverweg 1, 2629HS Delft, tel: +31(0)152784794, email: M.J.L.vanTooren@TUDelft.nl

D. D. Frey et al. (eds.), *Improving Complex Systems Today*,
Advanced Concurrent Engineering, DOI: 10.1007/ 978-0-85729-799-0_36,
© Springer-Verlag London Limited 2011

This paper addresses the framework proposed to extend the current DEE. Complexity analysis[13] is used identify the issues to be addressed by the DEE to facilitate design for sustainability. As the basis for the extension the qualitative Quality Function Deployment (QFD) method is used. This basis is augmented by Agent Based Modelling and Simulation (ABMS) --- not to be confused with the agents connecting the DEE components ---, Value Engineering (VE) and Multidisciplinary Design Optimization (MDO) using either a single level (SL) or multilevel (ML) approach to address the identified complexities. To capture design knowledge, the Knowledge Based Engineering (KBE) approach is used[15]. To illustrate the complexity of the problem a sample novel aircraft design configuration, the Prandtl plane (Fig. 1) will be used. This aircraft concept has several features not present in the current aircraft fleet making it difficult to discover to what extent these features can contribute to a sustainable future for aviation.

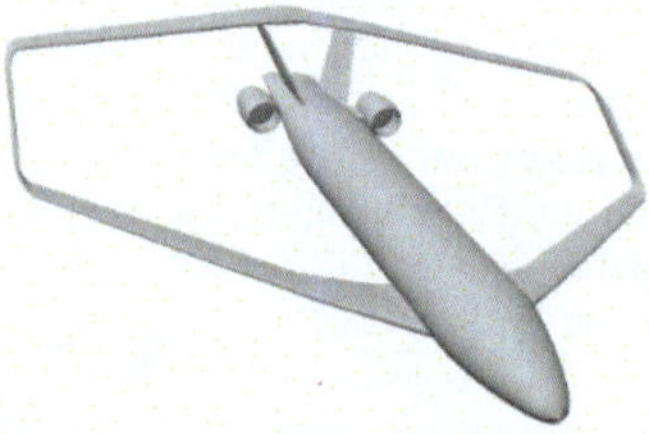

**Fig. 1:** The Prandtl Plane Concept [5].

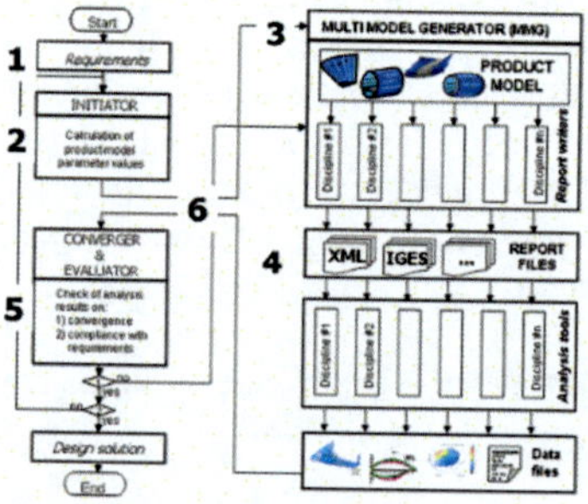


**Fig. 2:** Current Design and Engineering Engine (DEE) framework.

## 2 Research method

In order to support the design of future aircraft Van Tooren and La Rocca proposed the design and engineering engine (DEE) concept[15], intended to automate the repetitive work and free up time and resources for creative and value adding work. The framework is shown in Fig. 2. Its main components are: 1. Requirements / design options specificator; a module that allows the user to specify objective functions, constraints and parametric product models. 2. Initiator; a module that calculates initial values for the parameters in the product model. 3. Multi-Model Generator (MMG); a KBE application that implements the parametric product model including links to the discipline tools. 4. Expert tools covering all the analyses required to derive the behaviour of the

System in the different Life Cycle phases. 5. Converger and Evaluator; a module checking convergence of the behavioral models and a module with the optimizer.

6. An (agent-based) framework; the links between the different components of the DEE.

To be able use the DEE for the design and evaluation of novel and innovative technologies and aircraft configurations, the complexities arising in are addressed by the complexity analysis as defined by Sussman[13]. Sussman identifies four complexities for Complex Large-scale Interconnected Open-Social (CLIOS) systems: *evaluative*, *nested*, *behavioural* and *structural*. The computational approach taken by the DEE framework adds *modelling* complexity.
The complexities which are going to be discussed, the current DEE and its proposed extension are schematically shown in

Fig. 3. First the complexities and their origin in aviation are discussed. Second solutions to address these complexities in the design of novel technologies are proposed.

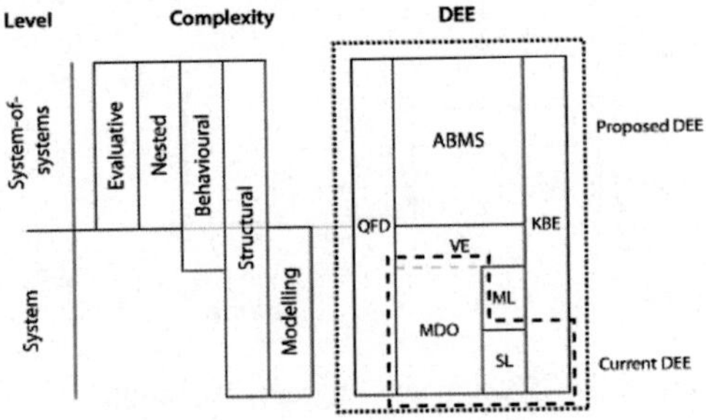


**Fig. 3:** Relation between complexity, level and availabel or proposed tools.

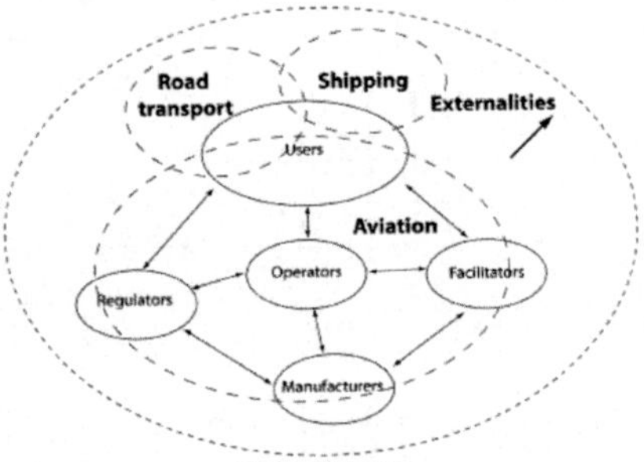


**Fig. 4:** Schematic (subjective) overview of the stakeholders and their interactions in aviation.

## Complexity perspective

**System-of-systems:** To design for sustainability, measured at the aviation, or system-of-systems, level is far from trivial[8]. Aviation consists of many stakeholders of which a subjective set is shown in
Fig. 4. A stakeholder is defined as[10]; ``*A stakeholder for an aircraft development programme (by definition) any group or individual who can affect the achievement of the programme's objectives*". This includes people experiencing the drawbacks, e.g. noise and emissions, of aviation as they will be empowered in the future to make aviation more sustainable. This large number of intertwined stakeholders (*structural*) all portray their own behaviour (*behavioural*) and with it they affect the impact of aviation[9]. Furthermore they all have different wants and needs from the new design (*evaluative*) and often want the new system to be integrated in available infrastructure[2], which is optimized for conventional/ current aircraft (*nested*).

**System:** A system should be designed to fulfill the requirements and comply to the constraints. Due to the indication that conventional means are likely to be insufficient in achieving the Vision 2020 goals[14], additional complexity is introduced into the design process as well, *structural complexity* and *modelling complexity*, where the former arises from the large number of interconnected subsystems compromising the system and the latter arises from the fact that the validity of the results produced by first principle and/or high fidelity tools remains uncertain for unconventional configurations. The modelling complexity directly influences the predicted behaviour of the aspect system and consequently the overall system and contributes to the *behavioural* complexity.

**QFD**

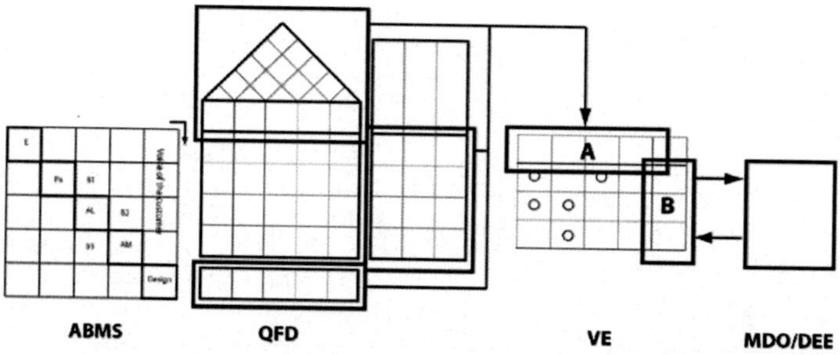


**Fig. 5:** Connection between agent based framework and Design and Engineering Engine.

To address the complexities at the system-of-systems level and the system level, an integrated approach is required at both these levels. Quality function deployment (**QFD**) is chosen for its focus on customer needs and tracing those through the design[3]. The sustainable design approach is based on the inclusion of the needs of *all* stakeholders, and is therefore different from the conventional approach of ``finding the latent pocket of demand". QFD does not include the interactions between the various stakeholder needs. Consequently, the systems-of-systems level complexities arising from these intertwined stakeholder needs are not addressed. Furthermore, QFD is a qualitative approach and the extension of the DEE requires a quantitative computational approach.

To address the structural and behavioural complexity at the system-of-systems level quantitatively an agent based modelling and simulation (**ABMS**) approach is used. The stakeholders and their behaviour are represented by agents and their interactions are studied in an artificial environment. This computational approach has to address the behavioural complexity of human behaviour to evaluate the emerging behaviour in response to a novel technology. This behaviour includes the decision for the acquisition of a novel technology and consequently the nested complexity. The ABMS approach only provides a means of modelling the interacting stakeholders and emerging system-of-systems behaviour, e.g. environmental impact, in response to an existing/fully designed system, e.g. aircraft.

To design a novel technology, the evaluative complexity is addressed by value engineering (**VE**). This approach tries to formulate a value measure defined as the ratio between the benefits, i.e. usefulness of the design, and the costs, i.e. the environmental costs as measured by the ABMS approach. This measure is applied to various design solutions, e.g. Prandtl Plane, Blended-Wing-Body or conventional aircraft. The value engineering approach consequently formulates a quantitative objective function from the often qualitative and abstract needs of the customer[4-5]. In this way the evaluative complexity is addressed at the system-of-systems level. Value engineering formulates for a given design solution a optimization problem,

$$\min_{\mathbf{x}} \quad \phi(\mathbf{x}, \mathbf{y})$$

$$s.t. \mathbf{h}(\mathbf{x}) = \mathbf{0}, \mathbf{g}(\mathbf{x}) \leq \mathbf{0}$$

This mathematical formulation can be addressed by the current DEE, addressing the system level behavioural, structural and modelling complexity. Nevertheless, a more suitable multidisciplinary design optimization (**MDO**) approach might be a multilevel (ML) approach, like Bi-Level Integrated System Synthesis[12], instead of the current single level (SL) optimization approach. This BLISS approach not only decomposes the analysis of the system into various expert tools, but also decomposes the optimization problem into a system level optimization and multiple decomposed blackbox optimizations. The previous discussion is schematically shown in Fig. 5.

## 3 Proposed Study Case For the Evaluation Framework

The Prandtl plane was introduced earlier in this paper as the case study. This aircraft concept aims at minimization of the induced drag of aircraft. The systems optimization level has been set-up and tested already. The systems level DEE has been used to optimize the location and size of the control surfaces[7]. In this study, use was made of a first order, commercial of the shelf, panel code with viscous boundary layer integration[1] for the aerodynamic analysis. The aerodynamic results were used in an in-house developed flight mechanics toolbox to analyze the overall behaviour of the aircraft. An optimization study was done to minimize the total control surface area, whilst keeping an adequate level of control power in all axes. The resulting architecture is presented in Fig. 6. It can be observed that control surfaces are present on both the front and rear wing. The inboard controls are mainly intended for longitudinal control and the outboard controls for lateral control. Each control however can have multiple functionalities. This control architecture provides two clear opportunities;

1. a pure moment can be generated by deflecting the front- and rear-wing controls in a differential manner,
2. direct lift control can be applied by deflecting the surfaces on front- and rear-wing simultaneously in one direction.

A mix of these two distinct control strategies can also be applied. If a conventional aircraft pitches up, then first a down force has to be created on the horizontal tail plane. The aircraft will initially descend slightly and once it has rotated, it will start to climb. This behaviour is typically designated as non-minimum phase behaviour. On the Prandtl plane, this behaviour does not have to be present. The advantage of the control layout is twofold. First, airfield performance can be improved. Second, aerial refuelling and precision height control becomes significantly easier due to the direct lift capability. The aircraft response to a height command is presented in Fig. 7. One can see that the height is rapidly captured without any overshoot or non-minimum phase behaviour. The pitch attitude is continuously kept within 0.1 degrees from the trim value. This demonstrates the direct lift capability. In principle a flight control system can be designed that creates a response type somewhere between the two extremes (direct lift, pure moment) presented here. The next step would be to apply and evaluate the PP concept for a typical air transport segment to evaluate the proposed framework. However, it is already clear that the substitution of a conventional aircraft with a PP --- *ceteris paribus* --- ignores the additional potential originating from possible alternative uses.

To evaluate the impact of the Prandtl Plane at the system-of-systems level, the stakeholder behaviour in response to the novel features has to be accounted for. A selection of the stakeholders the PP encounters in its operational life is shown in Fig. 8. Each of these stakeholders affects the desirability of the direct lift capability provided by the PP. Consider for example the potential alternative take-off procedures. Airlines might not see any benefit from this capability if airports and air traffic control refuse to adapt their procedures. Only if these alternative trajectories are accepted, the noise burden on the community might be reduced resulting from the PP. As a consequence of the reduced noise burden, more flights might be allowed on a noise constrained airport, resulting in a benefit for the airlines and passengers from increased service level. This is merely an illustration of the intertwined needs of stakeholders and their effect on the potential of novel technologies. These conflicting needs should be balanced by the QFD/VE approach in order to correctly design the PP for environmental impact reduction. This applies to other concepts, e.g. blended wing body, as well. In addition to this, it is clear that technology design limited to system level evaluations is insufficient for environmental impact reduction. However the proposed system-of-system level approach is likely to be iterative: formulate system level problem, design a solution, evaluate the true impact, reformulate the system level problem.

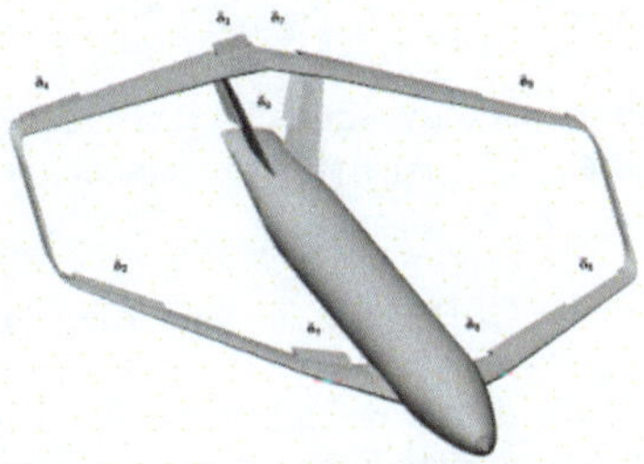

**Fig. 6:** Control surface layout.

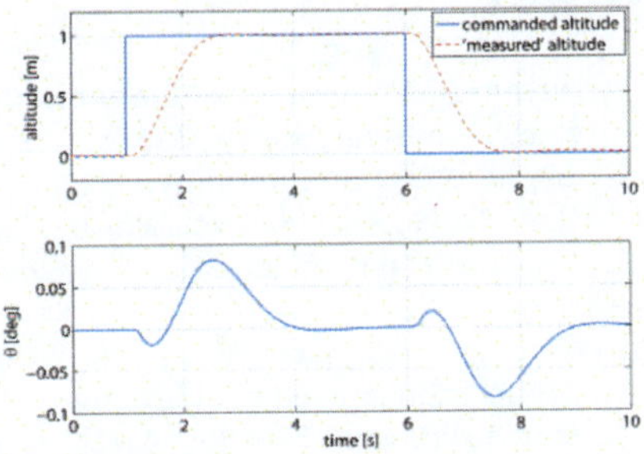


**Fig. 7:** Response to height command.

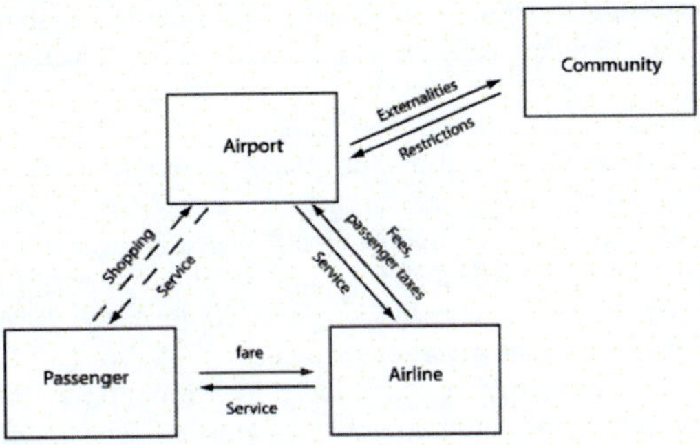


**Fig. 8:** A selection from the stakeholders encountered in the operational life of the PP.

## 4 Conclusion

This article proposes an architecture for a framework to include system-of-systems considerations in the DEE. In addition it proposes a test case for the evaluation of the framework itself. The identified complexities, evaluative, nested, behavioural, structural and modelling are addressed in the extended DEE. The ABMS addresses the nested, behaviour and structural complexity at the system-of-systems level, by providing a framework for the dynamic interaction between the stakeholders. QFD and VE provides a means to translate the qualitative needs into an optimization formulation addressing evaluative complexity. Furthermore VE evaluates the design spaces for their suitability partially addressing modelling complexity. Finally, the MDO framework addresses the behavioural and structural complexity at the system level. The Prandtl plane (PP) concept has features that can be optimized and applied in different ways. Future work on framework and the PP concept should elicit the practicality of the proposed framework. However, the need for the system-of-system level evaluation capabilities is clear from the identified uncertainty in stakeholder behaviour in response to the PP direct lift capabilities.

## References


[1] Anonymous. VSAero, A code for calculating nonlinear aerodynamic characteristics of arbitrary configurations, User's manual Version 7.1. Analytical methods,Inc., Redmond, Washington, U.S.A., 2005.

[2] J. Clark. Buying the big jets. Ashgate, 2001.

[3] L. Cohen. Quality function deployment: how to make QFD work for you. Addison-Wesley, 1995.

[4] R. Curran, T. Abu-Kias, M. Repco, Y. Sprengers, P. van der Zwet, and W. Beelearts. A value operations methodology for value driven design: medium range passenger airliner validation. In 48th AIAA aerospace sciences meeting including the new horizons forum and aerospace exposition, 4-7 January 2010, Orlando, Florida, 2010; AIAA 2010-0847.

[5] R. Curran, S. Raghunathan, and M. Price. Review of aerospace engineering cost modelling: the genetic causal approach. Progress in aerospace sciences, 2004; vol 40:487–534.

[6] D. DeLaurentis. Understanding transportation as system of systems design problem. In AIAA Aerospace Sciences Meeting and Exhibit 10-13 January, Reno Nevada, 2005; AIAA 2005-0123.

[7] D. v. Ginneken, M. Voskuijl, M. van Tooren, and A. Frediani. Automated control surface design and sizing for the prandtl plane. In 6th AIAA MultidisciplinaryDesign Optimization Specialist Conference, Orlando, Florida, 2010; AIAA 2010-3060.

[8] E. Han and D. DeLaurentis. A network theory-based approach for modelling a system-of-systems. In 11th AIAA/ISSMO Multidisciplinary Analysis and optimization conference 6-8 september, Portsmouth Virginia, 2006; AIAA 2006-6989.

[9] G. Hardin. The tragedy of the commons. Science, 1968; vol 162:1243–1248.

[10] R. Mitchell. Toward a theory of stakeholder identification and salience: defining the principle of who and what really counts. Academy of management review, 1997; vol 22(4):853–886.

[11] Report of the group of personalities. European aeronautics; a vision for 2020, 2001.

[12] J. Sobieszczanski-Sobieski, J. Agte, and R. Sandusky, Jr. Bi-Level Integrated System Synthesis (BLISS). Technical Report NASA/TM-1998-208715, NASA Langley Research Center, Hampton, Virginia, August 1998.

[13] J. Sussman. The new transportation faculty: The evolution to engineering systems. Transportation Quarterly, 1999.

[14] E. Torenbeek and H. Deconinck, editors. Innovative configurations and advanced concepts for future civil aircraft. Von Karman Institute, 2006.

[15] M. van Tooren, G. L. Rocca, and T. Chiciudean. Further steps towards quantitative conceptual aircraft design, chapter 27. Springer optimization and its applications. Springer New York, 2009.

# Balancing Value in Complex Systems Development

Márcio Silva Alves Branco[a,1], Geilson Loureiro[b] and Luís Gonzaga Trabasso[c]

[a]Technologist, National Institute of Space Research (INPE), Brazil.
[b]Technologist, Laboratory of Integration and Testing, (LIT), Brazil.
[c]Associate Professor, Aeronautics Institute of Technology (ITA), Brazil.

**Abstract.** The analysis of alternatives during the concept exploration must support the transformation of a need into a balanced design, must be able to reconcile the differences in the present set of physical and functional requirements and must evaluate the operational scenarios in terms of several attributes. However, the analysis of alternatives in the early stages of complex systems development is poorly structured and characterized by not sufficient detail for the assessment of the initially identified needs. The understanding of the relationship between the preferences of stakeholders and potential solutions for the systemic analysis of alternative designs is one of the most important activities in pursuit of information that can distinguish good from bad solutions. The evaluation of the system properties during the project requires the ability to analyze and map the architecture value space to find the best solutions. Thus, decisions made early in the development of a program should be supported by systems engineering analysis, involving teams of users, purchasers and others involved in the project, namely the stakeholders. In this paper are discussed several value dimensions and implications for the systems development to achieve and balancing value in complex system development.

**Keywords.** Complex systems, value, development, stakeholders, analysis of alternatives

## 1 Introduction

The analysis of alternatives during the exploration of concepts must support the transformation of a necessity in a balanced design, be able to reconcile the differences in the set of physical and functional requirements and evaluate the operational scenarios in terms of several attributes. In this way, a major part of the system engineer's role is to provide information that the system manager can use to make the right decisions.

According to Keeney [11] individuals are concerned with values. Therefore, these values should direct efforts in decision making, i.e., they must be part of the effort and time that is spent on decisions. But this is not the normal route followed. Instead, decision making, often focusing on the choice between alternatives. In fact, it is common to characterize a decision problem by the alternatives available.

[1] Technologist (Systems Engineer), National Institute of Space Research (INPE), Av dos Astronautas, 1758 – Jardim da Granja, CEP 12227-010, São José dos Campos, São Paulo, Brasil; Tel: +55 12 3208 6593; Fax: +55 12 3941 1890; Email: marcio@dss.inpe.br.

D. D. Frey et al. (eds.), *Improving Complex Systems Today*,
Advanced Concurrent Engineering, DOI: 10.1007/ 978-0-85729-799-0_37,
© Springer-Verlag London Limited 2011

Thus, understanding how value is identified, planned and delivered is vital for developing systems for success.

In this paper are discussed several value dimensions and implications for the systems development to achieve the balancing value in complex system development.

## 2 Value

According to Keeney [11], Dominguez [5], INCOSE [8] companies considered leaders began to emphasize the perceived value, believing it to be, instead of consumer satisfaction, the driver of customer loyalty. So, the first step to understand the implications and dimensions of perceived value to the systems development is to have a clear understanding of concepts related to value.

In literature, there are many definitions of value. Johansson *et al.* [10] propose that the value can be quantified in terms of product quality, $Q$; service, $S$; selling price, $SP$, and lead time, $LT$; according to Equation 1.

$$Value = \frac{Q * S}{SP * LT} \qquad (1)$$

Park [16] proposes that the value can be based on product functionality ($F$) and cost ($C$):

$$Value = \frac{F}{C} \qquad (2)$$

To Downen [6] value is a term often only loosely defined in vague terms by the ratio between benefits and costs, and thus is problematic to make it operational in practical applications.

Mandelbaum and Reed [14] emphasize that the relationship between the value or utility of an item and its real cost represents his concept of value. The highest value is represented by a product with an essential quality with the lowest total cost and that will perform reliably the required function in the desired time and place.

In this context, the product value is expressed by a relationship between its utility ($U$) and cost ($C$), as modeled by Equation 3.

$$Value = \frac{U}{C} \qquad (3)$$

## 3 Perception of Value

Briggs, Reinig and Vreede [3] proposed a causal theory that was written originally to explain the mechanisms that could give rise to responses of satisfaction and dissatisfaction of the product. The authors argue that the cognitive mechanisms that give rise to satisfaction responses may also give rise to the perception of value for an object, and that the satisfaction response is integral to perceptions of value.

Like satisfaction, perceptions of value may arise from the mechanisms of mind that relate to goal attainment. A goal is a state or outcome that an individual desires to attain [13].

Zeithaml [18] finds that the perceived value is the total consumer evaluation of the utility of a product based on perceptions of what is received (benefits) and what is given (sacrifices).

Despite variations, the various authors cited above seem to converge on the concept that perceived customer value is linked to the use or utility of the product or service. Thus, the perceived value is related to customer perception and involves the notion of exchange of benefits for sacrifices.

Dominguez [5] adds that some aspects of the analysis of perceived value are:

i. Timing dimension, i.e., it can vary with the time of evaluation;

ii. External an internal view to the company. According to Zeithaml [18], it may have differences between the expectations of customers regarding the value of the product attributes and perceptions of the company about those expectations.

iii. Nature of the market. Customers may define value differently depending on the nature belong to different markets.

iv. Personal dimension. The purchase decision always involves people who may have different views of value, according to their own perceptions

v. Supply chain coverage. Companies must operate in the entire value chain, seeking partnerships with suppliers, dealers and distributors, to maximize the value delivered by the chain to the customer.

According to these literature definitions, the perception of value $\upsilon$ can be represented by the ratio between the benefits received by the customer $i$ in the evaluation time ($\beta_i(t)$) and the sacrifices given by the customer $i$ in the evaluation time ($S_i(t)$), as written in the Equation 4.

$$\upsilon = \frac{\sum_{i=1}^{n} \beta_i(t)}{\sum_{i=1}^{n} S_i(t)} \qquad (4)$$

## 4 System Value in the Development

Chase [4] notes it is difficult to quantify value, particularly within the context of product development, because there are many perspectives on value. These perspectives depict the complexity of value, which is seen differently by the business customer, end user, shareholder, employee, etc. Each of these will typically have a different perspective on what is valuable.

To Downen [6] despite to seemingly quantitative nature of value definitions, all of them involve qualitative parameters such as quality, function, benefit need, and levels of satisfaction..

Hallander and Stanke [7] consider that the value of a system includes the evaluation of multiple dimensions and implications throughout its life cycle. These value dimensions include aspects such as performance, cost and time. Another concern that arises in developing new systems is large technological uncertainties. From this account, the risk, therefore, becomes another dimension of value. Chase [4] adds that these dimensions should be considered as evaluation criteria for solutions effectiveness.

If the value, as a function of the attributes of the system, was estimated, the value of a proposed solution can be calculated from the variables related to each of the attributes that characterize it.

In this line, Shinko [17] states that a measure of effectiveness of a system describes the systemic objectives achievement in a quantitative way. Thus, each system has its own measures of effectiveness.

Based on all information collected by the various authors referred to here, Equation 5 for the system would be:

$$\upsilon_{system} = \sum_{i=1}^{n} \upsilon_i(\chi_d) + \sum_{i=1}^{n} \upsilon_i(\chi_c) + \sum_{i=1}^{n} \upsilon_i(\chi_r) + \sum_{i=1}^{n} \upsilon_i(\chi_s) \qquad (5)$$

where, $\upsilon_{system}$ is the system value, $\upsilon_i(\chi_d)$ is the value of the performance attribute $\chi_d$ related to the valuer *i*, $\upsilon_i(\chi_c)$ is the value of the cost attribute $\chi_c$ related to the valuer , $\upsilon_i(\chi_r)$ is the value of the risk attribute $\chi_r$ related to the valuer *i*, and $\upsilon_i(\chi_s)$ is the value of the schedule attribute $\chi_s$ related to the valuer *i*. All of the value assessment evaluated at a determinate moment to *n* valuers.

Therefore, the value of system architecture is strongly associated with the personal value related to the attributes that characterize the architecture of a system. Thus, the system value regarding the attributes should be considered as a measure of effectiveness in assessing the value of the solutions of the problem presented.

## 5 The Value Context in Complex System Development

The INCOSE [8] postulates that the main challenge for the industry in the XXI Century involves identifying and delivering value to all stakeholders. This challenge is compounded by geographical and philosophical distance between all stakeholders in the development of a system, despite the need for collaboration between them.

In this way, when considering the perspective of stakeholders during the development of new systems can be identified a variety of interests. These interests are usually related to performance, development time and delivery, system cost or price and, therefore, risks of development.

Lemon Bowitz, Burn and Hackney [12] suggest that the successful development of a system is strongly related to perceptions of stakeholders in the value of the project and their relationships with the development team.

Thus, in a systemic way, measures of effectiveness should be established based on perceived value, i.e. the relationship between the attributes of a system and the needs of stakeholders in its development.

Keeney [11] states that the value-centered thinking is a philosophy to guide decision makers. It has three main ideas: starting with the establishment of values, use these values to generate better alternatives, and use them to evaluate these alternatives. The author points out a way for the analysis of alternatives that begins with values, rather than starting with alternatives, i.e. start with the objectives of stakeholders and decision makers and use them to generate better alternatives. Finally, the author demand to use values to evaluate the alternatives generated by the technique of multiple objective decision analysis.

On the other hand, Murmam et al. [15] propose a creation value system for the development process with activities, inputs and outputs. Figure 1 illustrates this structure with three distinct stages: the value identification, the value proposition, and the value delivery.

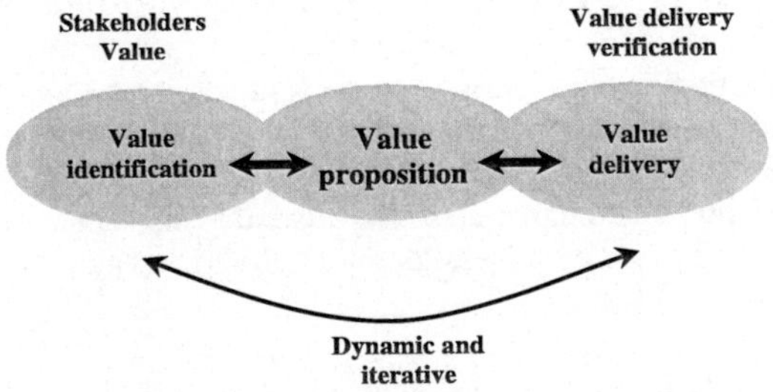


**Figure 1** – Value creation process [15].

Thus, it becomes important to improve the goals, needs, direct or indirect interests, of the technical definitions, and programmatic constraints during the early stages of complex systems development. Thus, if the selection criteria are correct and well defined, those involved in the project can do analysis to better understand the design space and then choose one that best balance the satisfaction with the architecture chosen according to the needs be met.

The assumptions below detail the changes proposed for balancing the perception of stakeholders value in developing a complex system.

**Assumption 1:** A better understanding of the problem to be solved can be accomplished through the study of all internal and external interfaces of the system being developed. Several systemic values come from various areas such as technical, political, social and financial can be incorporated to the development through analysis of stakeholders. Thus, communication and information exchange can be accomplished in a more appropriate way, following the requirements of a new governance in system development.

**Assumption 2:** Most of the value that will be delivered to the stakeholders is defined at early development stage by the architecture that will be selected among the proposals. Thus, the main objective of this premise is to drawn the proposed concepts from the values identified in the systemic stakeholder analysis. For this, all the tasks of proposing requirements, functional analysis, synthesis, design and definition of figures of merit for characterizing the system should be based on stakeholder satisfaction, i.e. the value given by the stakeholders to these characteristics.

**Assumption 3:** The purpose of the third premise is to set the solution analysis of effectiveness based on the evaluation of the importance placed by stakeholders at every critical parameter of the various proposed architectures, namely the valuation of the attributes of the system of special interests identified. The architecture analysis must be performed based on the perception of value, i.e., in evaluating the impact of the attributes of the architecture, measured by the figures of merit of the system on the needs of stakeholders during the development.

## 6 Balancing Value in Complex Systems Development

The central goal of this paradigm shift in systems development at an early stage of exploration of concepts is to balance the satisfaction of stakeholders with respect to the system figures of merit, as opposed to the position of performing a direct analysis of the attributes that characterize the system. This satisfaction is the basis for enhancing the confidence of stakeholders in delivering value in the long run in complex system development.

Figure 2 illustrates the relationships between stakeholders interests, a systemic solution, and their figures of merit that characterizes the perceived value relationship, used as a measure of effectiveness that, consequently, model the evaluation in delivering value decision.

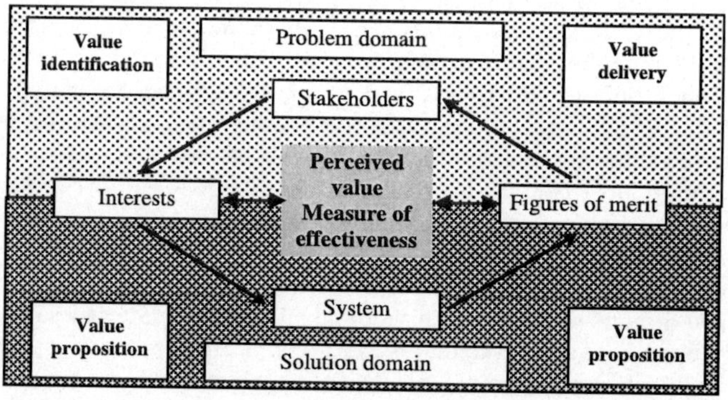


**Figure 2** – Value criation in complex system development [1].

Thus, it is proposed that during the definition of a system is more appropriate, natural and palpable evaluate a solution by the total value perceived by stakeholders, as a practical way they will give financial, technical, political and social support to the development of the system.

Changing the perceived value of the attribute with the value given by the stakeholders for the performance, cost, risk and schedule attributes and introducing a weight for each value factor, the Equation 5 becomes:

$$\begin{aligned} \upsilon_{stk} = & \sum_{i=1}^{n} \omega_{psi} \times \upsilon_{psi}\left(\varphi_p, \eta_{si}\right) + \sum_{i=1}^{n} \omega_{csi} \times \upsilon_{csi}\left(\varphi_c, \eta_{si}\right) \\ & + \sum_{i=1}^{n} \omega_{rsi} \times \upsilon_{rsi}\left(\varphi_r, \eta_{si}\right) + \sum_{i=1}^{n} \omega_{ssi} \times \upsilon_{ssi}\left(\varphi_s, \eta_{si}\right) \end{aligned} \tag{6}$$

where, $\upsilon_{stk}$ is the stakeholders's system value, $\upsilon_{psi}$ is the value of the performance figure of merit $\varphi_p$ related to the $i$ stakeholder interest $\eta_{si}$; $\upsilon_{csi}$ is the value of the cost figure of merit $\varphi_c$ related to the $i$ stakeholder interest $\eta_{si}$; $\upsilon_{rsi}$ is the value of the risk figure of merit $\varphi_r$ related to the $i$ stakeholder interest $\eta_{si}$; $\upsilon_{ssi}$ is the value of the schedule figure of merit $\varphi_s$ related to the $i$ stakeholder interest $\eta_{si}$; $\omega_{psi}$, $\omega_{csi}$, $\omega_{rsi}$, and $\omega_{ssi}$ are the weight factors for each value factor for the Equation 6.

The proposed methodology seeks to highlight the importance placed by stakeholders to the properties of a system early in development, namely through the relationship between attributes and interests, characterizing the perceived value of the stakeholders. In this way, the Equation 6 capture the elements to get a better balancing of the value given by stakeholders of the system for the analysis of alternative architectures in the development of complex systems. Traditional methodologies provide only a partial picture of these elements and their interactions.

Thus, explain to people how the systems produce tangible stakeholder's value benefits should become a major task for all actors involved in the complex systems development.

## 8 Conclusions

One of the most difficult aspects of the product development process is to recognize, understand and manage the development work in a way that maximizes the product's success. This is particularly important for complex systems [2].

The main objective of the analysis of architectures has been to meet the demands associated with high performance in a cost-effective long-term and low-risk, i.e., the common is the optimization of these parameters directly. In other words, the traditional analysis does not focus its efforts on balancing stakeholder satisfaction in developing a system with respect to its attributes, but in balancing the attributes that characterize the system architecture.

Based on results achieved in [1] is indeed possible to balance the stakeholder satisfaction with the attributes of an architecture. This owes to the fact that attributes can be evaluated in a specific manner according to the interests involved, i.e., through the perceived value of each stakeholder with respect to the attribute. Thus, a figure of merit of a system can be evaluated differently according to the perception of who is evaluating.

The approach seems to be closer to the reality lived nowadays regarding the improvement of the quality of decisions made within the various centers of development of complex systems, harmonizing political, budgetary, programmatic and technical decisions, that by definition, are taken at different levels of society.

## 9 References


[1] Branco, M. S. A. Balanceamento da Percepção de Valor dos Stakeholders no Desenvolvimento de Sistemas Complexos. Tese de doutorado – ITA, 2010.

[2] Branco, M. S. A., Loureiro, G. Trabasso, L. G. Stakeholder value analysis of architecture alternatives for sustainable space systems developments. Sixth International Aerospace Congress IAC'09. Daejeon Republic of Korea. October 2009.

[3] Briggs, R. O, Reinig, B. A., E Vreede, G. J. Satisfaction Attainment Theory and its Application to Group Support Systems Meeting Satisfaction. Information Science and Technology Working Paper Series, 1(1), College of Information Science and Technology, University of Nebraska at Omaha, 2003.

[4] Chase, J. Measuring Value in Product Development. The Lean Aerospace Initiative. WP00-05 - Massachusetts Institute of Technology - March 2000.

[5] Dominguez, S. V. Valor Percebido Como Elemento Estratégico para Obter a Lealdade dos Clientes. Ensaio. Caderno de Pesquisas em Administração, São Paulo, V. 07, nr 4, Outubro/Dezembro. 2000.

[6] Downen, T. D. A multi attribute value assessment method for the early product development phase with application to the business airplane industry. PhD Thesis. 2005. Massachusetts Institute of Technology. All rights reserved. Thesis.

[7] Hallander, I. E Stanke, A. Lifecycle Value Framework for Tactical Aircraft Product Development. The Lean Aerospace Initiative. WP01-02. April 2001.

[8] INCOSE. Systems Engineering Handbook - A Guide for System Life Cycle Processes and Activities. INCOSE- TP-2003-002-03, Version 3, 1 June 2004. International Council on Systems Engineering. Copyright © 2006 by INCOSE.

[9] Jiang, J. E Klein, G. Risks to Different aspects of System Success. Information and Management 36(5), 263 – 271- 1999.

[10] Johansson, H. J., et al. Business Process Reengineering: Breikpoint Strategies for Market Dominance. New York: John Wiley & Sons, 1993.

[11] Keeney, R. L. Value-Focused Thinking: A Path to Creative Decision Making. Cambridge, MA: Harvard University Press, 1992.

[12] Lemon, W. F., Bowitz, J., Burn, J., E Hackney, R. Information Systems Project Failure: A Comparative Study for Two Countries. Journal of Global Information Management, 10(2), 28-40, 2002.

[13] Locke, E. A. & Latham, G. P. A Theory of Goal Setting and Task Performance. Englewood Cliffs, NJ: Prentice Hall. 1990.

[14] Mandelbaum, J. E Reed, D. L. Value Engineering Handbook. Institute for Defense Analysis Analyses, 4850 Mark Center Drive, Alexandria, Virginia. September 2006. Approved for public release. IDA Paper P-4114 Log: H 06-000611.

[15] Murman, E. et al. Lean Enterprise Value: Insights from MIT's Lean Aerospace Initiative. Palgrave, 2002.

[16] Park, R. V. Value Engineering: A Plan for Invention. New York: St. Lucie Press, 1988.

[17] Shinko, R. NASA Systems Engineering Handbook. National Aeronautics and Space Administration, 1995.

[18] Zeithaml, V. Consumer Perceptions of price, quality and Value: a Meand-end Model and Syntesis of Evidence. Journal of Marketing, New York, jul 1988. v. 52, n. 3.

# Value Stream Analysis of Manufacturing Engineering New Product Introduction Processes

Malachy Maginness[a], Essam Shehab[b,1] and Chris Beadle[c]

[a, b] Decision Engineering Centre, Manufacturing Department, Cranfield University, College Road, Cranfield, Bedfordshire, MK43 0AL, United Kingdom
[c] Rolls-Royce plc., PO Box 31, Moor Lane, Derby, DE24 8BJ, United Kingdom

**Abstract.** This paper describes the use of a methodology for value stream mapping and analysis of Manufacturing Engineering New Product Introduction processes. The applicability and usefulness of the technique to process improvement in this domain is explored in a case study where the production system for a new component part is planned and proven. This analysis enables an improvement strategy for the Manufacturing Engineering process to be successfully outlined.

**Keywords.** New Product Introduction, Manufacturing Engineering, Aerospace, Lean

## 1 Introduction

Understanding current state conditions is the essential first step taken in any business seeking to improve how it performs its core processes. Successful process improvement strategies rely on acquiring rich, quantitative measures of the current state. The importance of such understanding is demonstrated in the measure and analysis phases of the widely used DMAIC (Define-Measure-Analysis-Improve-Control) model for processes improvement. Value stream mapping and analysis methodologies are well established tools for process improvement in physical manufacturing processes [12, 14]. This paper describes the use of the value stream analysis to a novel area of the enterprise value chain: the transactional processes of Manufacturing Engineering New Product Introduction (NPI). A case study was carried out at a large aerospace manufacturer where Manufacturing Engineering

[1] E.Shehab, Manufacturing Department, Cranfield University, College Road, Cranfield, Bedfordshire, MK43 0AL, United Kingdom; Tel: +44 (0)1234 750111 x524; Email: e.shehab@scranfield.ac.uk

D. D. Frey et al. (eds.), *Improving Complex Systems Today*,
Advanced Concurrent Engineering, DOI: 10.1007/ 978-0-85729-799-0_38,
© Springer-Verlag London Limited 2011

(ME) performs a key role in developing and delivering the production system for the products and components developed by Design Engineering [13]. Value stream analysis methodology has recently been successfully applied to assist improvement efforts in the Design Engineering processes of this manufacturer [5]. The broader extension of the methodology as a standard for value stream analysis in transactional processes is explored here. Manufacturing Engineering can be understood as an information driven, transactional process aimed at creating physical production systems. This case study considers the processes associated with planning the production system for a particular component part. The complex geometry requires multiple manufacturing methods (for confidentiality these are identified as Method X and Method Y). The value proposition of Manufacturing Engineering is defining quality solutions to achieve design intent at required levels of cost and lead time. Lead times for physically creating all parts of the production system are a significant feature of the transactional process. A value orientated Manufacturing Engineering NPI process is one that can arrive at quality definitions of the method more quickly to enable rapid introduction of new products to market.

## 2 Related Literature

Techniques for evaluating value and waste in product development as a critical step toward improvement in information driven processes are emerging in literature [1-2, 4-11]. Parallels are drawn between the information products of product development and physical process products, and lean principles are extended across both domains [6, 10]. The concept of 'value' in product development processes has been matured in a number of applications and remains consistent with user orientated definition of value in physical process domains [2, 6-8, 10, 12, 14]. 'Waste' is also considered. Information that waits in queues for the next processing activity is equated with physical inventory queues in machining systems [11]. The 'aspects' of value are further developed to stipulate those that define the product and production system and eliminate risk to the contrary. Tasks enabling value-add tasks to proceed (documentation) and those that are non-value adding (facilitating communication) are additionally proposed [4, 9].

A move towards a standard method for value stream analysis in the transactional product development process domain is evident [5, 9]. Two main areas of investigation are the metrics that are relevant for describing the transactional process flow, and the approach to visually map or represent that flow. Wasted time is advocated as a key improvement focus for lean product development [11]. To that end, the proportion of wasted time present in the lead (or elapsed) time for an activity is explored by distinguishing cycle (pure processing) time from waiting time (delays and interruptions). Furthermore, cycle time is decomposed into components of manual or automated time to indicate the degree of effort personally required from engineers to complete the process. The description of the value stream is completed by activities attributes which include the system tool used, those responsible and the inputs and output associated with each activity including format (the information flow) [5, 9]. Visual representation and mapping of the process flow is associated with the analysis approach. A standard format that combines all the relevant information for value stream analysis in a singular representation has emerged [5]. The process flow is presented in 'activity boxes' named with a suitable description and accompanied by the attribute details and 'data-boxes' containing the metrics noted above. Information flows between activities are represented by arrows and the 'castle-wall' details respective iteration timelines. Crucially this approach is not yet applied to the Manufacturing Engineering transactional process domain [5].

## 3 Method

Data collection was carried out in three phases. In phase one the case study project was identified with the key stakeholder, the NPI lead in the relevant business area. The problem was bounded to a specific component example from a recent development project. A two hour workshop was conducted with input from a range of cross-functional representatives who were identified by the key stakeholder as relevant participants in the NPI process. The output was a high level map of 25 activities and information flows which was used to identify the boundaries of the end-to-end process. Post-it® notes were adopted as a flexible, interactive means of obtaining the depiction of this flow. This was later distributed among attendees and other stakeholders to obtain correction comments. The second phase consisted of a five day schedule of individual interviews with engineers. The initial map was used as a framework for developing a detailed map of the actual NPI process flow. The map was subject to numerous revisions as the data collection interviews continued. In its final version this map documented 107 activities and the information flows between them in a Role-Activity format [5]. The process map was presented in a one hour cross-functional workshop for validation. The final (five day) phase identified the key value stream for analysis during a workshop review of the detailed map with the main stakeholders. Further interviews were carried out and value stream metrics collected. A value stream map depiction was created and included activities and iterations identified in the detailed process flow map, along with the value stream metrics and time lines to indicate the specific activities that are revised in iteration loops. The final deliverable was the value stream mapping and metrics analysis and was presented to the key stakeholder for review and identification of future state improvement actions. All process flow and value stream map depictions were created in Microsoft Visio graphics software.

The main information source was the engineers who participated in the component NPI processes. These ranged from engineers with direct involvement in the activities to those at a business level of project management. Individual semi-structured interviews conducted by the researcher were of approximately one-hour's length each and were recorded by note taking, Dictaphone and transcripts. The question set was developed in line with the value stream metrics concepts discussed above [5] and piloted to verify comprehension and the recording technique. The first section of the question set defined an Activity Description as experienced by the interviewee (Table 1). This was structured around the Supplier-Input-Process-Output-Customer (SIPOC) model in order to capture the information flow [3]. Capturing the system used in the activity considers the significant role computer aided design and manufacturing (CAD/CAM) now plays in engineering processes. The second section captured data consistent with the value stream metrics. Validation was achieved by corroboration with project planning literature and presentation of findings at workshops. A total of 11 interviews (excluding short sense-making conversations) and three workshops were completed.

**Table 1.** Collected Data Types

| Section 1: Activity Description | Section Two: Value Stream Metrics (Hours) |
|---|---|
| Specialist Role/Activity Owner<br>Activity Name<br>System Used<br>Number of Engineers involved<br>Work Output (including format)<br>Inputs (format) & Suppliers<br>Outputs (format) & Customers | Activity Lead Time (LT)<br>Cycle Time (CT)<br>Waiting/Delay Time (WT)<br>Manual Time (MT)<br>Automated Time (AT)<br>(LT = WT + CT)<br>(CT = AT + MT) |

# 4 Findings

## 4.1 Description of the Value Stream

Manufacturing Method X consists of stages of operations in which a sequence of tools forms simple material geometry into a shape approaching the complexity of the design intent. Manufacturing Method Y finishes the component made in Method X to a state that matches the design intent. Planning for Method X occurs simultaneously with Design Engineering processes and planning for Manufacturing Method Y uses certain of its definitions and physical parts to complete. As an intermediate process, planning of Method X has the potential to delay downstream processes and is compelled to complete within shorter lead times. The planning process uses a number of iterations that are a challenge to reducing lead time. Certain component geometry that is critical to the performance of the complete product is formed in the method. Physical trials (typically a total of three) are used to determine all aspects of tool geometry that influence the creation of a quality part that matches the design intent. It is for these reasons that a lean planning process for Manufacturing Method X is desirable and analysis was applied here.

The value stream map (Figure 1) depicts the particular value stream for planning Method X (including 25 activity steps) and begins with the first release of the component design model. Also recorded are the CAD/CAM systems used in defining the method. These include the company standard system used and a number of specialist alternatives. The final, intermediate and first stages are derived from the design model in a sequence that is the reverse of the production method. These definitions consist of the part shape expected at the end of each stage and the tool geometry to form it. All are created as CAD models. The final shape definition is evaluated for approval by the laboratory authority. Using the

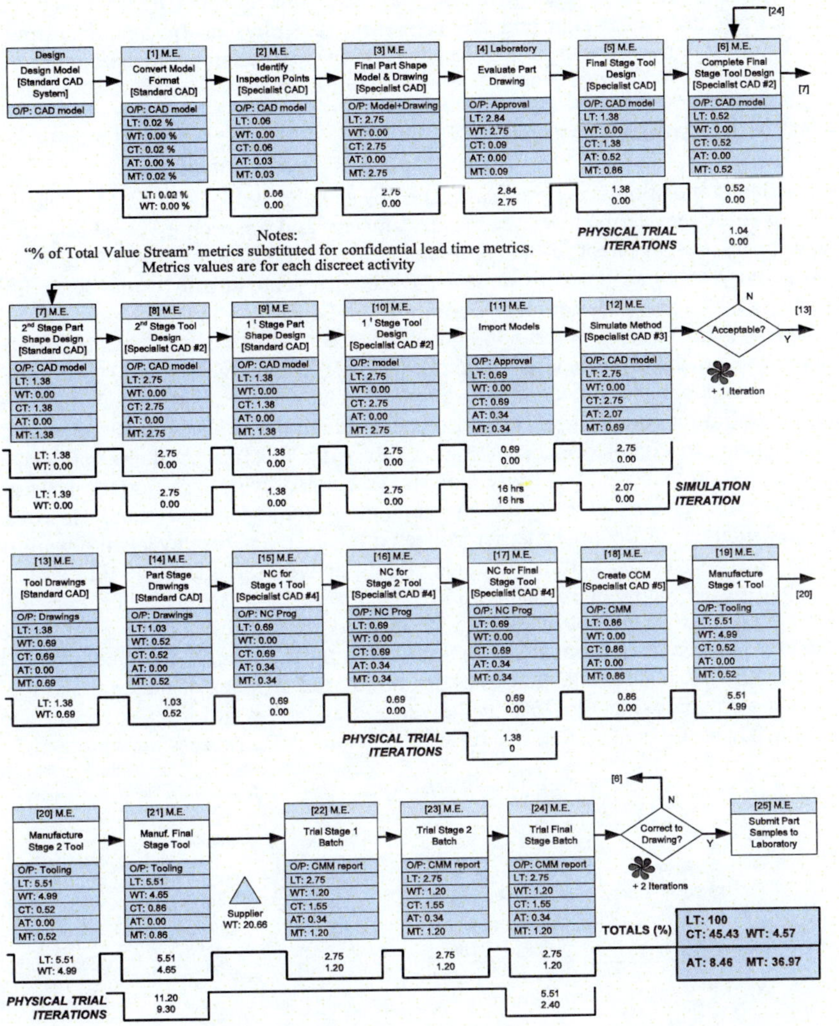


**Figure 1.** Manufacturing Engineering Value Stream for Method X Planning

models, the method is simulated virtually and alteration information fed back. Once acceptable, both part shape and tool geometry is described in technical drawings. Where necessary these are used to order tools and raw material in the external supply chain. Numerical Control (NC) and Coordinate Measurement Machine (CMM) sequences are generated from the models to drive in-house manufacture of tools and the inspection of parts. Upon availability of the material and tools, Method X is trialled, and CMM inspection conducted at each stage. Inspection measurements inform alteration requirements to ensure the part conforms to required geometry. These iteration loops are both transactional (re-engineering the tool design and NC sequence) and physical (re-manufacturing). The value stream ends with submission of parts for inspection by the laboratory.

## 4.2 Value Stream Analysis

The value proposition was agreed with the key stakeholder to be that "a production process may be defined within manufacturing and inspection capability that captures the design intent of the component." At a high level of analysis the value add activity was calculated as almost 94% of the value stream's total lead time (Figure 2). The activities for stage definition (tool and part model creation and NC sequences) are directly value adding, as are the simulation and physical trials. These reduce the risk that the method will not create the required geometry. Activities enabling value adding examples include creating drawings that are formal definition of the planning, or model creation tasks begun in one CAD system prior to completion in another. Converting the format of the models is classed a non-value add activity and is associated with transportation waste although this claims an insignificant amount of the total lead time. However detailed analysis of the time metrics collected for each activity reveals the waiting time hazards that exist within the value adding activities. Total cycle time is less than half of the total value stream while 'waiting' accounts for 55% of the critical path. Waiting is particularly evident in method trials (30% of the total value stream). The reason attributed for this is accessing production equipment that is shared with full scale production. A wait for external supplies of material and tool items is also approximately 20% of the value stream. A wait for approvals is also evident. These are the key findings influencing future state process improvement.

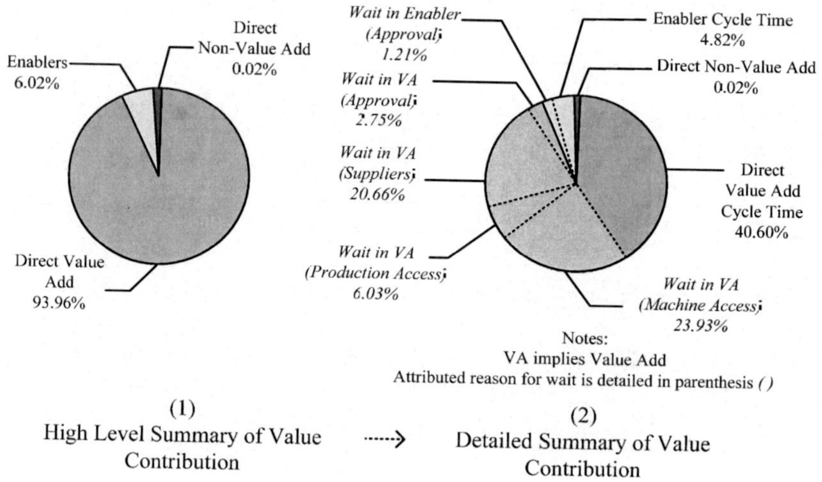


**Figure 2.** Value Stream Contributions in Cases

## 6 Discussion

The usefulness of the value stream analysis and the metrics collected is verified by the ability of stakeholders to identify process improvement opportunities. The value stream mapping presented the process for planning Method X in a manner that is distinguished from wider Manufacturing Engineering process flow and allowed analysis in terms of constituent activities. The value adding, enabler and non-value adding tasks were identified in this value stream. Although made in a manner that was consistent with that established literature [9] and validated with the key stakeholder, this identification remains reliant on the analyst's interpretation. The activity metrics offered a more detailed and quantifiable level of analysis of the value stream that revealed the interaction of wastes with the value adding tasks. In particular the metrics revealed the areas of waiting that occur in the value adding planning process. A rapid process for defining a proven production process enhances the flexibility of NPI. Removing waiting wastes from the total lead time represents the obvious and quantifiable improvement targets. An outline of approaches to address this includes a better upfront planning process for securing both production equipment access and laboratory resource availability for evaluations. More advanced solutions will reduce the dependency on physical iterations with an enhanced virtual simulation capability. Iteration lead times here are notably shorter. An additional Pareto chart, populated by each activity, was considered by the stakeholder to be a powerful representation of the greatest lead time and waiting time contributors in the value stream. In this way the data is used to inform priority improvement strategies

Insights gained from this analysis are dependent on the quality of the original data. A detailed end-to-end map of the process that documents information flow and systems used to complete the identifiable engineering activities was necessary for initial comprehension of how value is added in the process. No such map pre-existed for this case to use. It was created predominantly from the experiences of the engineers elicited from the interviews. Capturing all necessary opinion is important to the integrity of the results. For this, the support of the key stakeholder was crucial. Not only did this elicit the support needed within the business (access to engineers) but it also aided the ultimate verification that was required.

## 7 Conclusions and Future Work

A value stream analysis methodology has been applied in an investigation of the current state transactional processes of Manufacturing Engineering. The planning work associated with a specific production method served as the case study. The success of the approach is measured by the ability of the stakeholders to outline performance improvement targets. The process flow and value stream maps document an accurate end-to-end description of the actual current state process. This effort satisfies the measure and analysis phases of the DMAIC model of process improvement and enabled a number of measurable improvement opportunities to be outlined. Further work is necessary to define an accurate description of the future state. However, this case study has served to explore and illustrate the applicability of value stream analysis to the Manufacturing Engineering domain and this is the contribution made by this work.

## 8 Acknowledgements

The authors would like to thank the sponsors, Rolls-Royce plc, the Decision Engineering Centre in the School of Applied Sciences at Cranfield University, and the Engineering and Physical Sciences Research Council for supporting this work.

## 9 References

[1] Baines T, Lightfoot H, Williams GM, Greenough R. State-of-the-art in lean design engineering: A literature review on white collar lean. Proceedings of the Institution of Mechanical Engineers, Part B: Journal of Engineering Manufacture 2006; vol. 220, no. 9, pp. 1538-1547,

[2] Baines TS, Williams GM, Lightfoot H, Evans S. Beyond theory: An examination of lean new product introduction practices in the UK. Proceedings of the Institution of Mechanical Engineers, Part B: Journal of Engineering Manufacture 2007, vol. 221, no. 11, pp. 1593-1600. 2007

[3] Browning TR, Fricke E, Negele H. Key concepts in modeling product development processes. Systems Engineering 2005; vol. 9, no. 2, pp. 104-128

[4] Chase J. Value Creation in the Product Development Process. Masters thesis, Massachusetts Institute of Technology, Cambridge, 2001.

[5] Darwish M, Haque B, Shehab E, Al-Ashaab A. Value Stream Mapping and Analysis of Product Development (Engineering) Processes. Proceedings of the 8th International Conference on Manufacturing Research (ICMR 2010), Durham University, Sept 2010.

[6] Haque B. Lean engineering in the aerospace industry. Proceedings of the Institution of Mechanical Engineers, Part B: Journal of Engineering Manufacture 2003; vol. 217, no. 10, pp. 1409-1420.

[7] Haque B, Moore MJ. Measures of performance for lean product introduction in the aerospace industry. Proceedings of the Institution of Mechanical Engineers, Part B: Journal of Engineering Manufacture 2004; vol. 218, no. 10, pp. 1387-1398.

[8] Liker JK, Morgan JM. The Toyota Way in Services: The Case of Lean Product Development. Academy of Management Perspectives 2006; vol. 20, no. 2, pp. 5-20.

[9] McManus H. Product Development Value Stream Mapping (PDVSM) Manual. MIT Lean Aerospace Initiative, Cambridge, 2005.

[10] Reinertsen DG. Let it flow. Industrial Engineering (American Institute of Industrial Engineers) 2005; vol. 37, no. 6, pp. 40-45.

[11] Reinertsen DG. The Principles of Product Development Flow: Second Generation Lean Product Development. Celeritas Publishing 2009.

[12] Rother M, Shook J. Learning to See: Value-Stream Mapping to Create Value and Eliminate Waste. Lean Enterprise Institute: Brookline 2003.

[13] Ulrich KT, Eppinger SD. Product Design and Development 4th Ed. McGraw Hill 2008.

[14] Womack JP, Jones D. Lean thinking (2003 Edition), Simon & Schuster 2003.

# Part VI
# Product and Service Development

# Concurrent Product Development - Case Study of LED Linear Tube Lamp

Myongwon Cho[a,1], Shinji Ayao[b], Kenichi Yoshimoto[a], and Hirokazu Noguchi[a]

[a]EVERLUCE Inc., Japan.
[b]MIT Enterprise Forum of Japan.

**Abstract.** Due to the environmental circumstances, for example, the global mercury regulation and the revised energy conservation law, the market potential and demands of LED lamps are increasing to substitute for fluorescent lamps. Component parts of our LED linear tube lamps have been developed concurrently and in parallel with supply chain partners in order to make rapid response to various customer requests regarding luminous performances, power efficiency, and quality. As a result, we have achieved outstanding improvement of luminous output at the same level of power consumption compared to competitors' products and also price reduction - almost double brightness and 40% price reduction achieved. Furthermore, we are planning to use the external resources for rapid prototyping in order to expedite the scheme and process of new product development in the fast changing and growing LED lighting market. This paper describes a case study about the effectiveness of concurrent product development in the case of LED linear tube lamp of EVERLUCE Inc., Japan.

**Keywords.** LED lighting, concurrent product development, supply chain

## 1 Background Circumstances

### 1.1 Global Mercury Regulation

The governing council of United Nations Environment Programme agreed on the need to develop a global legally binding instrument on mercury in February 2009 and the intergovernmental negotiation will be completed before the Global Ministerial Environment Forum in 2013 [1].

[1] General Manager of Technology Strategy, EVERLUCE Inc., South Tower 15F, Yurakucho Denki Bldg, 1-7-1 Yurakucho, Chiyoda-ku, Tokyo, Japan; Tel: +81 (3) 6212 0011; Fax: +81 (3) 6212 0012; Email: davidmw.cho@gmail.com; http://www.cyber-coin.com

D. D. Frey et al. (eds.), *Improving Complex Systems Today*,
Advanced Concurrent Engineering, DOI: 10.1007/ 978-0-85729-799-0_39,
© Springer-Verlag London Limited 2011

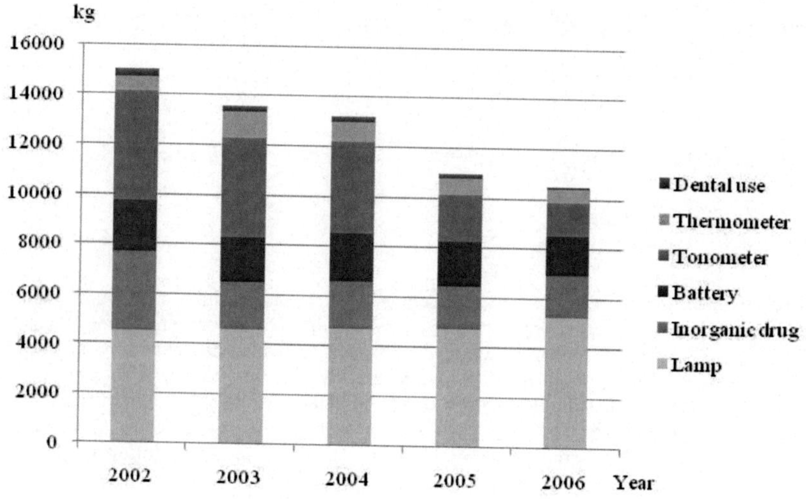


**Figure 1.** Annual usage of mercury in Japan

If the global mercury regulation comes into effect by intergovernmental agreement, the sale and production of traditional fluorescent lamps may be prohibited because one single fluorescent lamp made in Japan contains at least 7mg of mercury and the annual usage of mercury by fluorescent lamps has been slightly increasing and reached more than 5000kg in 2006 as shown in Figure 1 [2].

### 1.2 Revised Energy Conservation Law

Japanese government encourages community and individuals to save energy by energy conservation law and policy [3]. As the luminous efficiency and color rendering quality of LED has been improved for decades since 1960s, the application of LED has reached a level of lighting use in recent years. Besides, the recent LED lighting technology has been improved to consume only half amount of energy compared to the typical fluorescent lamp at the same luminous output. Therefore, LED has begun to receive attention as a next-generation lighting solution.

## 2 Market Demands

### 2.1 Quality Requirements

Quality requirements of LED lamps are represented as better luminous performances and lower energy consumption. The following major indexes describe customer requests regarding luminous outputs and power consumption.

- Illuminance (Brightness) : More than 500lx at 1m distance and more than 220lx at 2m distance
- Total luminous flux : More than 2000lm
- Color rendering index : More than Ra80
- Color temperature : 2700, 3500, 5000K
- Power consumption : Less than 22W
  (A fluorescent lamp generating the same brightness consumes about 40W.)

## 2.2 Cost Reduction

In order to meet the luminous requirements, hundreds of LED chips are used and aligned along a narrow linear printed circuit board in the case of a LED linear tube lamp, which is our major product. Most of the production cost is occupied by LED chips, and therefore, the unit price of a LED lamp is much more expensive than the price of a typical fluorescent lamp. This is the reason why the concurrent process is needed for product development of LED linear tube lamps in order to raise the efficiency of development and to reduce the production cost at the same time.

The following Table 1 shows that EVERLUCE has achieved outstanding improvement of luminous output at the same level of power consumption compared to competitors' products and also cost reduction to benefit to customers at much lower pricing than competitors through the implementation of concurrent engineering for the product development and improvement.

**Table 1.** Comparison of performance indexes and price with competitors

| Manufacturer | Illuminance (lx) | Power consumption (W) | Lifetime (Hs) | Unit price (USD) |
|---|---|---|---|---|
| A | 285 | 20 | 40000 | 300 |
| B | 267 | 20 | 50000 | 200 |
| C | 300 | 20 | 40000 | 200 |
| D | 213 | 14 | 40000 | 230 |
| E | 275 | 14~21 | 80000 | 300 |
| F | 249 | 18 | 40000 | 220 |
| EVERLUCE | 530 | 22 | 50000 | 160 |

## 2.3 Market Potential

According to the annual statistic data of lamp production in Japan from 2005 to 2009 published by Japan Electric Lamp Manufacturers Association (JELMA), the annual production quantity of linear fluorescent lamps is reported about 200 million units [4]. This means that there exists 200 million market potential for LED linear tube lamps to substitute for linear fluorescent lamps and the market scale is expected to amount to 24 billion USD (= 2 trillion JPY) on the assumption that the average price of LED linear tube lamps is 120 USD (= 10,000 JPY).

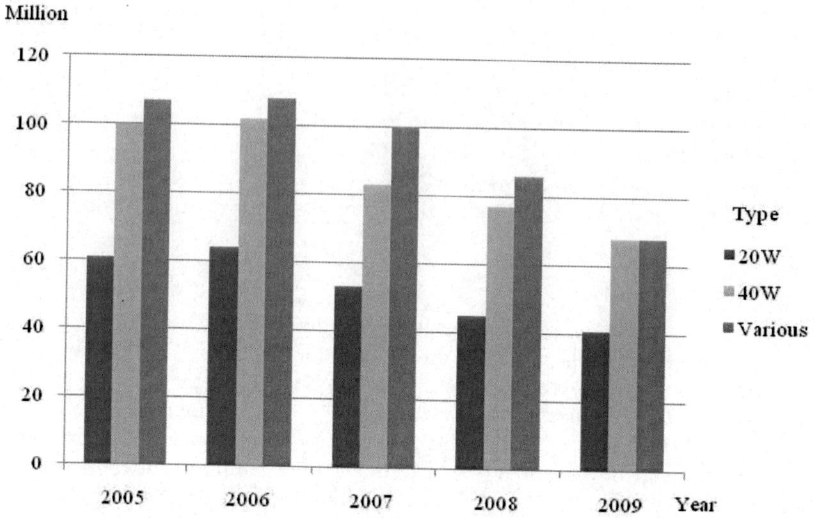


**Figure 2.** Annual production quantity of linear fluorescent lamps in Japan

# 3 Implementation of Concurrent Engineering

## 3.1 Supply Chain Management

EVERLUCE takes on main roles of strategy planning and execution, business management, marketing and sales, and product development. Additionally, one of the most important roles is to gather customer voices and to apply them to the product development process immediately and on a regular basis in order to keep providing customer satisfaction in the fast changing and growing LED lighting market. EVERLUCE focuses on managing the core values relating to the customer service and product development and collaborates with several partner corporations concurrently for stable supply of materials, development of component parts, and manufacturing under partnership contracts.

Stable supply of high quality LED chips with special price is essential to raise quality and price competitiveness because LED chips or packages are the most important components that occupy 50% of total bill of material cost and have dominant effects on luminous output and efficiency of LED lamps. Collaborative partnership with the Japanese best LED supplier helps to keep gaining competitive advantages on product quality and performance and also price over the competitors.

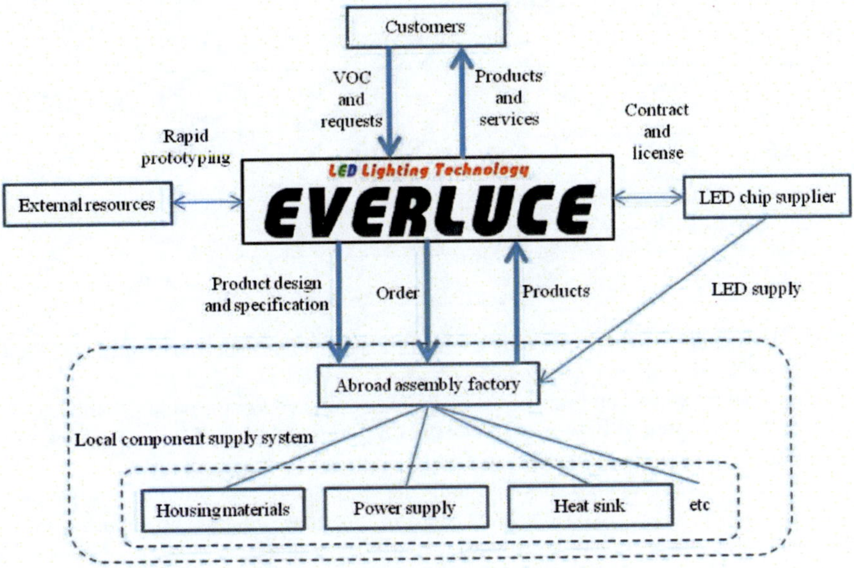


**Figure 3.** Structure of supply chain

All of the products are currently manufactured by an abroad original equipment manufacturer that has an abundance of knowhow and capabilities on LED lamp manufacturing and deserves to get plenty of grants from the local government. Production facilities capable of both stable mass production and mass customization are installed and working to manufacture our various types of LED linear tube lamps. For the stable supply of materials, best-quality local components have been selected and supplied. Quality evaluation regarding luminous performances and reliabilities are performed right after being produced at the factory site and second quality checks are performed in Japan after delivery before products are provided to customers.

Figure 3 describes the structure of supply chain to manage several independent processes relating to customer service, product development, component supply, and manufacturing in parallel.

### 3.2 Concurrent Product Development Process

As the implementation of concurrent engineering, the important component parts have been developed concurrently and in parallel with supply chain partners under collaboration relationships. Some examples about the concurrent development of major components, such as LED chip, power circuit module, and heat sink, are described below.

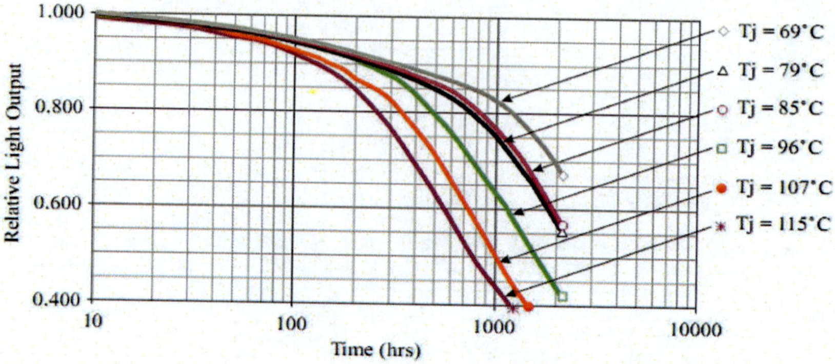


**Figure 4.** Example of light output variation by time and junction temperature of LED chips

There is no need to emphasize the importance of LED chip, which is the most important component to have a greater effect on the product performances such as luminous efficiency and power consumption and also on the production cost. LED chip supplier provides the newest and the best-quality components and we contributes to improve LED chips by sharing customer voices and requests that feeds back into improvement work with the LED chip supplier.

Power circuit module to driver LEDs generally contains AC-DC converters and several capacitors which cause power loss, in other words, power factor decreased. The optimized design of power circuit is required for reducing power consumption of the LED lamps. Optimizing and debugging works of the circuit design are performed with the assembly manufacturer in the product planning process.

Junction temperature of LED has a negative effect on the lifetime and luminous output directly according to Figure 4 [5]. If the junction temperature is higher, the light output of LED deceases more rapidly, and as a result, the lifetime is shortened. It is required to cool LED chips under 75°C in order to provide a 5- year warranty on the LED lamps. EVERLUCE and the assembly manufacturer make collaboration for development of material and structure design of heat sinks to improve thermal conductivity and cooling efficiency.

### 3.3 Rapid Prototyping Plan

We are planning to use external resources to accelerate the continuous quality improvement and new product development. Japanese enterprises have encouraged increase of the internal expenses for technology development for decades as shown in Figure 5 [6]. Small and medium enterprises in Japan owning unique and highly competitive core technologies have made continuous efforts to shift their capabilities to be compatible with mass customization from mass production during that time. And as a result, there are many Japanese small and medium enterprises creating core values by providing rapid prototyping with their unique and high-level technologies.

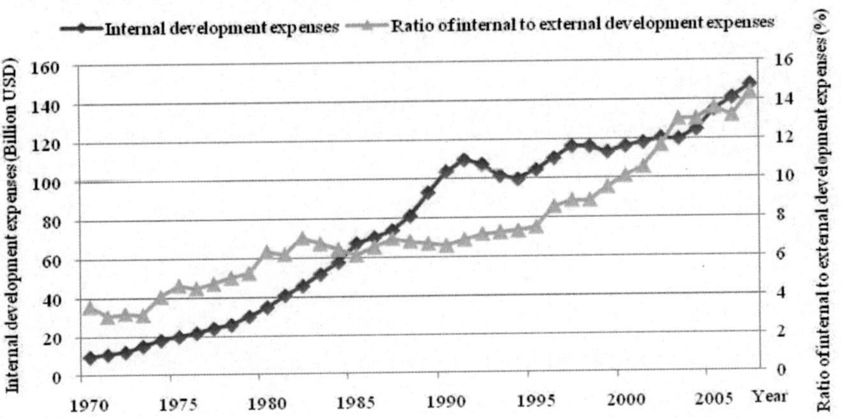


**Figure 5.** Development expenses of Japanese enterprises over the decades

## 4 Conclusion

The effectiveness of concurrent product development in the case of LED linear tube lamp is demonstrated with specific examples. Major component parts of LED lamp have been developed concurrently and in parallel with supply chain partners. As a result, we have achieved outstanding improvement of almost double luminous output at the same level of power consumption compared to competitors' products and also 40% cost reduction to benefit to customers at much lower pricing than competitors.

Furthermore, our next plan to use external resources in the new product development process is introduced. It is expected to be able to accelerate the continuous quality improvement and new product development through the rapid prototyping by the unique and high-level technologies of external resources in order to keep providing customer satisfaction in the fast changing and growing LED lighting market.

## References

[1] United Nations Environment Programme (UNEP). Available at: <http://www.unep.org/hazardoussubstances/Mercury/Negotiations/tabid/3320/Default.aspx>. Accessed on: Mar. 1st 2011.

[2] Statistic data of annual usage of mercury in Japan. Ministry of the Environment, Government of Japan. Available in Japanese at: <http://www.env.go.jp/chemi/tmms/seminar/jyoyaku/mat03.pdf>. Accessed on: Mar. 1st 2011.

[3] Revised energy conservation law. The Energy Conservation Center, Japan. Available in Japanese at: <http://www.eccj.or.jp/law/pamph/outline_revision/new_outline2010.pdf>. Accessed on: Mar. 1st 2011.

[4] Annual statistics of lamp production/sales in Japan [2005~2009]. Japan Electric Lamp Manufacturers Association. Available in Japanese at: <http://www.jelma.or.jp/03tokei/pdf/statistics01-07.pdf>. Accessed on: Mar. 1st 2011.

[5] Narendran N, Gu Y, Freyssinier JP, Yu H, Deng L. Solid-state lighting: failure analysis of white LEDs. J Crystal Growth 2004:268:449-456.

[6] Nikkei Monozukuri. Nikkei Business Publications, Inc. Tokyo, Japan. May 2010; 43-45.

# Concurrent Development of RIM Parts

Ricardo Torcato[a,b,1], Ricardo Santos[a] , Madalena Dias[a] , Elsa Olivetti[c] and Richard Roth[c]

[a]Faculdade de Engenharia da Universidade do Porto, Portugal.
[b]Escola Superior Aveiro Norte, Universidade de Aveiro, Portugal.
[c]Massachusetts Institute of Technology, USA.

**Abstract.** The aim of this paper is to present a research project which intends to identify and organize the knowledge required in the development process of Reaction Injection Molding (RIM) parts in the conceptual design phase, particularly as to material selection, mold design and the process planning for mold making and molding operations. The overall objective of the research is to verify if an Expert System, a computer program that uses knowledge and inference procedures to model the RIM development process, provides the required insight into metrics such as development lead time and manufacturing costs to deal with the decision makings required early on in order to reduce the subsequent redesigns and reworks. Based on the characterization of the industrial practice of RIM, it is our belief that the concurrent concept development is the appropriate approach to achieve this goal. This paper will focus on the structurization of the downstream processes and procedures of developing product design concepts for RIM parts; and on the dimensions of knowledge required in the concurrent concept development of RIM parts, aspects which are the basis of the knowledge base and production rules of an Expert System currently being developed.

**Keywords.** Concurrent engineering, conceptual design, process-based cost modeling, reaction injection molding.

## 1 Introduction

Early assessment of a material and its inherent production technology applicability in a particular product is fundamental for it to be considered in further product development stages. With the pressure to condense the product development cycle, it is essential to make the right choices early on, in order to avoid changes and reduce time-to-market.

Researchers have advocated that manufacturing companies must focus on how to improve the development process of a new product or, speaking in terms of time, how to reduce the time-to-market [1-5]. Competitive advantages derive from

[1] PhD student, LSRE - Laboratory of Separation and Reaction Engineering, Chemical Engineering Department, Faculdade de Engenharia da Universidade do Porto Rua Dr. Roberto Frias s/n, 4200-465 Porto, Portugal; Tel: +351 225081400; Fax: +351 225081440; Email: ricardo.torcato@fe.up.pt

D. D. Frey et al. (eds.), *Improving Complex Systems Today*,
Advanced Concurrent Engineering, DOI: 10.1007/ 978-0-85729-799-0_40,
© Springer-Verlag London Limited 2011

a fast product development cycle. If a product is launched early onto the market, the product's life cycle is usually longer, with extra revenues and profit. Furthermore, the market share and profit margin can increase, as there is less competition at the introduction stage. Emphasis on the time-to-market can be further justified as product life cycles are decreasing due to constant technological advancements [6]. The time-to-market strategy, while stressing time, does not imply that factors such as product performance, quality, reliability and cost are less relevant in product development. Time-to-market is regarded as one more objective, among others such as design for quality or low manufacturing cost, which is relevant for the product's competitive advantage [7].

As early as the 1990's, research has focused on concurrent engineering (CE) as a mean to reduce the time-to-market [8, 9]. CE has, since then, been recognized as "a viable approach in which simultaneous design of a product and all its related process in a manufacturing system are taken into consideration, ensuring required matching of the product's structural with functional requirements and the associated manufacturing implications" [10]. Moreover, early design decisions can have significant effect on the manufacturability, quality, product cost, product introduction time and, as a result, on the marketplace success of the product.

However, the systematic management of product knowledge is required to successfully complete and shorten the duration of the product development cycle. With technology, it is possible to use a computer-based approach to systematically acquire, represent, integrate, and coordinate the essential concurrent engineering knowledge. Expert Systems have proven to be valuable tools which allow the incorporation of knowledge of the different specialists involved in the decision making process [11-13]. Capturing, storing, processing and retrieving pertinent design, engineering and manufacturing knowledge in an integrated and artificially intelligent manner, using Expert System technology, can therefore, improve CE implementation.

## 2 Scope

The Reaction Injection Molding (RIM) process is visibly under the time-to-market paradigm. It is mostly applied in product development, prototyping and low to medium volume production, in markets such as automotive and medical equipment where time-to-market is highly valued.

RIM is a process to produce plastic parts from the injection of a reactive low viscosity mixture into a mold, allowing the production of parts with complex geometries, mainly polyurethanes. At the core of the RIM process is the rigorous mixing of two monomers introduced from two opposite jets into a semi-confined mixing chamber. The resulting reacting mixture is then discharged into a mold, where most polymerization occurs [14].

Studies reveal that the main limitation of the RIM industrial process, when applied to auto parts production occurs at the mixing stage, which results in the traditional RIM machines' lack of flexibility for formulation changes and the extreme dependence on operational conditions. As such, a series of studies were developed at the Laboratory of Separation and Reaction Engineering, LSRE, in

Faculdade de Engenharia da Universidade do Porto, FEUP, leading to two Ph.D. theses [15, 16], and a third one that is now being concluded. From these studies, RIMcop® – RIM with Control of Oscillation and Pulsation was developed, currently protected by patent PCT WO 2005/097477 [17]. The RIMcop® technology controls the mixing dynamics from pressure measurements of the monomer feeding lines to the mixing chamber, which allows a real time control of the process, a feature that is not yet available in state of the art RIM technology. Furthermore, the mixing is also ensured from a set of design changes to the mixing chamber and from the opposed jets forced pulsation [18-22].

Although initial research design addressed RIMcop® technology and how to position it competitively in the production of multifunctional auto parts, we verified that it is not ready for industrial implementation. Furthermore, literature review revealed that there is very little information on the RIM process, especially regarding the interaction between part design features and the cost of the downstream processes. As such, before considering RIMcop® as a technological development of the RIM process, we first acknowledged the need to identify and organize the knowledge required in the development process of RIM parts and processes. Only then will it be possible to understand how the RIMcop® process can influence design, manufacturing and cost.

In order to identify and organize the knowledge required in the development process of RIM parts and processes, field research and open-ended interviews to experts working at two US companies were carried out. Qualitative data was collected, synthesized and organized in datasheets in order to:

- Structure and formalize the downstream (from the concept development stage) processes and procedures of developing product design concepts for RIM parts, including the mold design and the process planning of mold manufacture and molding operation.
- Identify and define the dimensions of knowledge required in the concurrent product concept development process of RIM parts.

## 3 Results

### 3.1 Development of RIM parts

A visit to two US RIM companies (Armstrong Mold Corp. and RIM Manufacturing, LLC) and conversation with experts within each company allowed us to better define the scope of our study and the objectives of the research. The RIM process, including design and manufacture, was studied in two different industrial environments. Differences and commonalities were identified and conclusions were made as to the positive and negative attributes of RIM; the qualitative positioning of RIM when compared to the competing processes; RIM applications (such as medical equipment parts, automotive and trucks parts, and agricultural, construction and utility machinery parts); and the process flow.

During the visits we confirmed the general lack of explicit consideration of the downstream processes in the concept development stage. These interrelations are done in the minds of the different experts involved and consulted during the concept development stage. Often the expert makes the decisions without any explicit explanation as to the motives and reasoning behind them.

In this study we were able to acknowledge that for the development of a RIM part, the part design features are usually the dominant factor in mold design while the mold design characteristics, in turn, dominate the process plans of mold manufacture and molding operation. As a result, most of the costs of mold making and molding operation as well as the quality and reliability of molds and moldings are determined in the concept development stage of a new RIM part development project. Therefore, there is a need for a verified design that will provide the necessary insight into metrics such as development lead time and manufacturing costs to deal with the decision makings required in early stage design in order to reduce the subsequent redesigns and reworks. Based on the characterization of the industrial practice of RIM, we believe the concurrent concept development is the most suitable approach.

However, the concurrent development of RIM parts involves a substantial practical knowledge component (heuristic knowledge) on the relationship among part features, mold design requirements, mold-making processes characteristics and the selection of production molding equipment. The designs and process plans involved are predominantly based on the experience of designers. The processes rely heavily on engineers to define their designs in detail. Extensive mathematical analysis is often not used, as analytical models with sufficient accuracy and efficiency are not available. Calculations are limited to empirical rules. Hence, the designers of parts and molds are required to have a high standard of specific knowledge and judgment. Moreover, most decisions concerning the details of the design demand knowledge of the mutual influences between the various quantities. Changing one quantity in order to achieve better results, for example part design features, may have a negative effect on other influencing factors, for example the mold design and mold making process. This implies that knowledge and expertise of more than one specific area are required to have an optimum solution. Unfortunately, it is not easy to find such experienced engineers who possess all the required knowledge and expertise. The inherent complexity and intensive knowledge requirements of this concurrent development problem, as well as the scarcity of good human experts in the field are crucial aspects.

## 3.2 Concurrent concept development of RIM parts

Although CE entails that product design and process plans be developed simultaneously, the development process we propose for RIM parts is sequential, as each phase needs the output of the previous one, as well as the additional knowledge (see Figure 1). Two tasks can be performed simultaneously, mold making process planning and molding production plan. After the evaluation, the process can be repeated until an "optimal" result is achieved.

### *3.2.1 Material Selection*

The material selection phase involves the translation of part requirements (functional and manufacturing) into material properties. Then a screening of candidate materials is made based on the properties necessary to fulfill the requirements. In the material selection phase the following aspects need to be considered:

- Mechanical and physical properties
- Thermal properties
- Appearance properties
- Electrical properties
- Chemical properties
- Environmental performance
- Manufacturing requirements (molding, assembly and finishing)
- Economics (estimation of total cost, includes material and manufacturing costs)

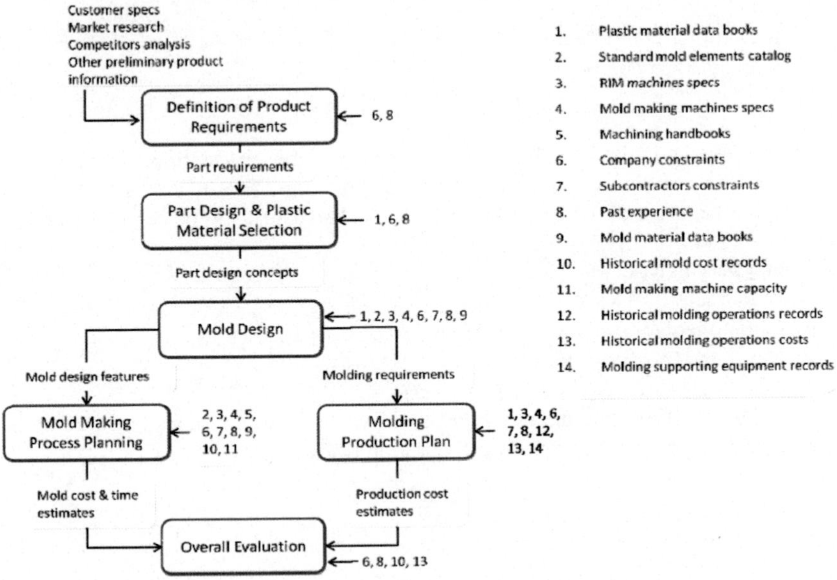


**Figure 1.** Concurrent concept development process of RIM parts

*3.2.2 Mold Design*

In the mold design phase, the part features (for each concept) are translated into mold features and molding requirements. This phase is critical in the concept development process, because it is the one that is worst to model (a large number of variables and of feasible outcomes) and is the one where decisions contribute most to the cost of the part. The aspects to be considered in this phase are:

- Type of mold structure
- Ejection system
- Feeding system
- Temperature control system
- Mold material selection
- Mold size determination

*3.2.3 Mold-Making Process Planning*

At the mold-making process planning phase, the mold features are converted into a plan for the manufacture of the mold; the major output is the mold cost and lead-time estimation. The aspects to be considered in this phase are:

a) Mold making processes

- Material removal processes (drilling, milling, CNC milling, EDM, etc.)
- Forming processes (casting, SLS, etc.)
- Heat treatment processes (hardening, nitriding, etc.)
- Surface finishing processes (polishing, blasting, etching, etc.)
- Assembly processes (tapping, fitting, etc.)

b) Mold components

- Cavity
- Core
- Remaining components
- Standardized components

*3.2.4 Molding-Production Planning*

The molding-production planning phase involves the translation of molding requirements into molding operations to produce the part. The major output of this stage is the estimation of the part production cost. The aspects to be considered are:

a) Molding cycle

b) Molding parameters

- Machine
- Components (system) preparation
- Molding temperatures
  - Control of components temperatures
  - Control of mold temperatures
- Injection pressure, flow rate and duration (time)
- Cure time

c) Finishing operations

- Trimming
- Cleaning
- Painting

*3.2.5 Overall Evaluation*

The final evaluation consists in adding up the costs and lead-times for mold making and molding operations. A decision is made whether one or more of the concepts are selected for further development. At this final stage it is also possible to change features of the concept(s) although this implies repeating the process from the beginning.

## 4 Conclusions

The research revealed the lack of literature on the RIM process. In order to identify and organize the knowledge required for the development process of RIM parts and processes, we visited two US RIM companies and we acknowledged that RIM is mostly applied in product development, prototyping and low to medium volume production, in automotive and medical equipment markets. Although literature refers to CE as the most suitable method to consider the downstream processes in the concept development stage [23-25], our study confirms, at least in these two companies, the general lack of explicit consideration of the downstream processes in the concept development stage.

We corroborate that concurrent concept development is crucial due to the need to incorporate expert knowledge in the part and process design, as well as the cost estimation techniques; and the need for a validated design that will provide insight into metrics such as development lead time and manufacturing costs to deal with the decision makings required in early stage design.

Even so, the complexity and intensive knowledge requirements, as well as the scarcity of human experts in the field, makes the need to develop an Expert System for the concurrent design of RIM parts evident.

## 5 Further Work

The next steps in the research will be to develop an Expert System architecture and framework, which we believe emphasizes the concurrent concept development process of RIM parts. The framework will then be validated by means of the prototype production and implementation in a selected company.

## Acknowledgements

Financial support for this work was in part provided by the MIT-Portugal Program and the national research grant PPTDC/CTM/72595/2006 and by LSRE financing by FEDER/POCI/2010, for which the authors are thankful. Ricardo Torcato acknowledges his Ph.D. scholarship by FCT (SFRH/BD/ 43053/ 2008) and Ricardo Santos acknowledges financial support from POCI/N010/2006.

## References

[1] Vesey JT. The new competitors: they think in terms of 'speed-to-market'. Academy of Management Executive 1991;5;2;23-33.

[2] Bayus BL. Speed-to-market and new product performance trade-offs. Journal of Product Innovation Management 1997;14;6;485-497.

[3] Menon A, Chowdhury J, Lukas BA. Antecedents and outcomes of new product development speed: An interdisciplinary conceptual framework. Industrial Marketing Management 2002;31;4;317-328.

[4] Afonso P, Nunes M, Paisana A, Braga A. The influence of time-to-market and target costing in the new product development success. International Journal of Production Economics 2008;115;2;559-568.

[5] Chen J, Damanpour F, Reilly RR. Understanding antecedents of new product development speed: A meta-analysis. Journal of Operations Management 2010;28;1;17-33.
[6] Minderhoud S, Fraser P. Shifting paradigms of product development in fast and dynamic markets. Reliability Engineering & System Safety 2005;88;2;127-135.
[7] Charney C. Time to market: reducing product lead time. SME, Dearborn, MI, 1991.
[8] Smith RP. The historical roots of concurrent engineering fundamentals. Ieee Transactions on Engineering Management 1997;44;1;67-78.
[9] Syan CS, Menon U. Concurrent engineering: concepts, implementation and practice. Chapman & Hall, London, 1994.
[10] Parsaei H, Sullivan WG. Concurrent engineering: Contemporary issues and modern design tools. Chapman & Hall, London, 1993.
[11] Rao SS, Nahm A, Shi ZZ, Deng XD, Syamil A. Artificial intelligence and expert systems applications in new product development - a survey. Journal of Intelligent Manufacturing 1999;10;3-4;231-244.
[12] Liao SH. Expert system methodologies and applications - a decade review from 1995 to 2004. Expert Systems with Applications 2005;28;1;93-103.
[13] Shukor SA. Axinte DA. Manufacturability analysis system: issues and future trends. International Journal of Production Research 2009;47;5;1369-1390.
[14] Macosko CW. RIM fundamentals of reaction injection molding. Hanser Publishers, Munich [etc.], 1989.
[15] Teixeira AM. Escoamento na cabeça de mistura de uma máquina de RIM Caracterização experimental por LDA e simulação dinâmica por CFD. FEUP, Porto, 2000.
[16] Santos RJ. Mixing mechanisms in reaction injection moulding - RIM An LDA/PIV experimental study and CFD simulation. FEUP, Porto, 2003.
[17] Lopes J, Santos RJ, Teixeira AM, Costa M. WO 2005/097477 Production Process of Plastic Parts by Reaction Injection Moulding, and Related Head Device. WIP Organization, 2005.
[18] Santos RJ, Teixeira AM, Costa M, Lopes J. Operational and design study of RIM machines. International Polymer Processing 2002;17;4;387-394.
[19] Santos RJ, Teixeira AM, Lopes J. Study of mixing and chemical reaction in RIM. Chemical Engineering Science 2005;60;8-9;2381-2398.
[20] Teixeira AM, Santos RJ, Costa M, Lopes J. Hydrodynamics of the mixing head in RIM: LDA flow-field characterization. AIChE Journal 2005;51;6;1608-1619.
[21] Erkoç E, Santos RJ, Nunes MI, Dias MM, Lopes J. Mixing dynamics control in RIM machines. Chemical Engineering Science 2007;62;18-20;5276-5281.
[22] Santos RJ, Erkoç E, Dias MM, Teixeira AM, Lopes J. Hydrodynamics of the mixing chamber in RIM: PIV flow-field characterization. AIChE Journal 2008;54;5;1153-1163.
[23] Carter DE, Baker BS. CE, concurrent engineering: the product development environment for the 1990s. Addison-Wesley Pub., Co., Massachusetts, 1992.
[24] Buyukozkan G, Dereli T, Baykasoglu A. A survey on the methods and tools of concurrent new product development and agile manufacturing. Journal of Intelligent Manufacturing 2004;15;6;731-751.
[25] Valle S, Vázquez-Bustelo D. Concurrent engineering performance: Incremental versus radical innovation. International Journal of Production Economics 2009;119;1;136-148.

# Preliminary design within CE using the MFF

Jože Duhovnik [a,1], Žiga Zadnik [b]

[a]Head of LECAD group, Faculty of Mechanical Engineering, University of Ljubljana, SI- Slovenia.
[b]Early stage researcher, Faculty of Mechanical Engineering (Laboratory LECAD), University of Ljubljana, SI- Slovenia.

**Abstract.** The objective of this manuscript is to present a concept of solutions of functional requirements in the concurrent development of a product by means of the descriptive matrix of function and functionality (MFF), based on the generative model and criteria for describing products, functions and functionalities [1]. The MFF model intends to improve the initial, preliminary design process, where only the most basic, sporadic information such as functions and functionalities are presented. Tasks are parallelized and integrated to reduce the time required. A developed MFF model was created as a tool in order to connect functional requirements with existing technical systems that partially or fully resolve functional requirements on the basis of a mathematical model and predetermined conditions, and they therefore do not depend solely on designer's intuition. Functional solutions within the MFF are based on mutual relations between the function and functionality, representing data definition. The concept of MFF improves the morphological matrix with the criteria and self-control by the preliminary design. The MFF model has been implemented into a prototype web application and confirmed on a concrete product for better and faster design management. It is based on the connection between processes, recognized in the nature, and searching for comparable or fulfilling technical processes at a certain level of knowledge development.

**Keywords.** Product development, preliminary design, parallelism of processes, functional matrix, functionality matrix, descriptive matrix.

## 1 Introduction

Market requirements are the basis for defining basic functional requirements, which in turn represent the initial information on a new potential product [2]. At the beginning of the design process, functional requirements are usually unarranged, incomplete and sporadically presented, which makes them necessary

[1] Head of LECAD group, Faculty of Mechanical Engineering, University of Ljubljana, Aškerčeva 6, 1000 Ljubljana, SI- Slovenia; Tel: +386 (1) 4771 416; Fax: +386 (1) 4771 156; Email: joze.duhovnik@fs.uni-lj.si; http://www.fs.uni-lj.si

D. D. Frey et al. (eds.), *Improving Complex Systems Today*,
Advanced Concurrent Engineering, DOI: 10.1007/ 978-0-85729-799-0_41,
© Springer-Verlag London Limited 2011

to be arranged, complemented and expanded. By means of structural enlargement of functions, product structure can be presented as a functional structure, which is at the same time the basis for defining the shape (physical structure) of the product [3]. In [1, 4, 5] matrix models were developed and presented. They enable generating functional structure of the product, described in matrices.

In order to reduce the time required for solving, arranging and improving functional requirements, there is an increasing demand for new forms of matrices that would enable concurrent and parallel solving of objectives.

The basic premise for concurrent engineering revolves around two concepts. The first is the idea that all the elements of a product preliminary design from function to functional requirements should be taken into careful consideration and based on the parallelization of tasks carefully managed. The second concept is that the proceeding design activities should all be occurring at the same time, or concurrently. The overall goal is that the concurrent nature of these processes significantly increases productivity and product quality [6, 7, 8].

The basic morphological matrix [9] is the basis for the development of the MFF model. With a small number of rows and columns, the model can yield a large number of solutions, which often makes them poor and unsuitable. The objective of the development of the MFF is – under the assumption of concurrence – to upgrade and update the deficiencies of the morphological matrix through the application of mathematically-based model for creating links between the function and functionality. Concurrent solving of functional requirements is widespread also in the areas of self-assessment, auto-solving or automated suggesting of solutions and the possibility of using modularity of individual sub-matrices.

The MFF is a synonym for tabular presentation of links between functions and functionalities. Functionalities are represented by technical systems [10] or shape models that in part or in whole fulfill the required functions. The matrix is used for the development of brand new products as well as for the development of variant design.

In [11], authors approach describing functions by defining terminology, related to the names of functions, while others describe functions of technical systems by means of physical laws [12]. With a view to unique identification, rules have been defined [1] by means of which functions, functionalities and products are described. Reference points for designing these rules are those presented in [13]. Functions are described by parameters, based on physical laws, which forms the basis for the development of a mathematical model through which the connection with functionalities is established.

Contemporary market requires ever shorter development time of a new product, which triggers the need for modular architecture of products. Modular architecture allows combining one or several functions in the functional structure with one element that solves them [14]. Such approach has several advantages; the main one being an increased number of product's variants [15].

Research and development activities within the product development process have their own characteristic and distinctive features, dominated by unpredictability, creativeness, mentality and abstraction. Due to these features it is difficult to thoroughly describe, develop and implement the design process in the initial phases of computer tools development [3].

## 2 The MFF model

Although the design of a product must often be done in a multidimensional world, engineers are often taught optimization techniques for the one-dimensional world. They do not know how to think in several dimensions because they have not been given the tools and techniques that can deal with that problem of a complex world.

The purpose of the MFF is to present a scientific approach to design and solve preliminary tasks in a faster, more concurrent way.

The matrix of function and functionality – the MFF represents a tabular presentation of links between functional requirements and functionalities. It can be devised if key elements are known, such as initial functional requirements and functionalities. The MFF is used when we want to improve the initial engineering process, where only the basic information is available and we need a tool with which it is possible in the concept phase to solve concurrently several functional requirements and generate new functional and design structures of a product. For a given functional requirement, the MFF is searching for a solving functionality or a materialized technical system, and vice versa for a functionality. Solving the matrix, more and more information is defined for a particular functional requirement or function. They are solved at the end of the process with suitable functionalities. For a function, which we do not know much about at the beginning of the design process, it is possible to determine a suitable solution by means of solving and describing, and to specify in more detail the type of the function and all corresponding parameters, winning parameters, intervals etc. The reason basically lies in the fact that during the initial engineering of a new product it is not possible to define precisely enough what function and what type of function it is all about as functions are initially described only with words, phrases and sentences. The objective is to find a suitable solution and a materialized technical system on the basis of initial data in the form of original lexical titles, which can only be done by means of the proposed MFF model.

Within the MFF, functional requirements are introduced into the relation on one side and functionalities on the other one, as shown in Figure 1. Functional requirements represent functions, while functionalities are represented via technical systems. Both functions and technical systems can be either simple or more complex, which depends on the initial description of individual systems. In order to present the MFF and for better understanding, it has been decided to use simplified and general descriptions of functions and technical systems. Looking at Figure 1, representing an MFF model, we can see that functions or functional requirements are generally marked with F$i$ and are placed in the first column, while individual technical systems are marked with TS$j$ and can be found in subsequent columns. Let's take a more detailed look at the nature of the function and the technical system within the MFF.

In the MFF model, technical systems are marked with general marks TS1, TS2,…,TS$j$ $j$=1,…,$n$, while in the case of implementations and concrete examples the marks are of course replaced by real names of technical systems.

Functions, defined in the MFF matrix, are described in the first column in the table, labelled *Function*. Each functions fills a new line. In order to present the model in a simple way, function names are marked with general marks, such as:

F1, F2,..., F*i*; *i*=1,...,*m*, while in concrete examples within an implementation, the marks are replaced by concrete and real names. Functions are defined on the basis of required functional requirements. For systematics and modularity reasons, they are described in input lists.

Defining and specifying the titles of functional requirements, functionalities and all other names, we should avoid as much as possible long and senseless titles, names and phrases since careless descriptions can make the concurrence of the engineering process significantly harder. Therefore, names should be carefully defined and at the same time describe the subject as clearly as possible. The two key questions are how to properly define the title and the meaning of names, and whether the title is suitable in the first place. Answers should be sought after in the areas of word formation, terminology, semantics, ontology and taxonomy.

Links between functions and functionalities that solve them are created by means of sub-matrices. Sub-matrices in the presented MFF model are colored gray and highlighted with a frame (Figure 1). Figure 1 shows five different sub-matrices within which we will explain concurrent solving and possibilities of multiple solutions. As a rule, sub-matrices are not logically distributed at the beginning as their internal distribution is determined by how the design process develops and by the presupposed number of functions and functionalities. The key feature of the matrix is its ability to follow the design process by arranging sub-matrices and their modularity. The whole matrix can be comprehensively edited and arranged during the design process. According to the given computer algorithm, the MFF presupposes a hierarchical order at the beginning of the process. It is based on matching percentage, which can be re-edited after the choice has been made. The MFF is added the possibility of two-level row- and single-level column sorting. During the engineering process and in line with engineering dynamics, the MFF is a dynamic matrix.

| **FUNCTION** | **FUNCTIONALITY / SOLUTION** | | | |
|---|---|---|---|---|
| | **TS1→** | **←TS2→** | **←TS3→** | **←TS*j*** |
| **Functional requirement - F1** [Suggested solution] ↓ | M 75 ↓<br>↑S 50 | | S 60 ↓<br>↑A 40 | |
| **Functional requirement – F2** [Suggested solution] ↑↓ | | M 100 ↓<br>↑S 100 ↓<br>↑S 87 ↓<br>↑B 56 | | |
| **Functional requirement – F3** [Suggested solution] ↑↓ | | | S 90 ↓<br>↑A 45 ↓<br>↑B 25 ↓<br>↑B 10 | |
| **Functional requirement – F4** ↑↓ | | | | |
| **Functional requirement - F*i*** [Suggested solution] ↑ | | | | B 100 |

**Figure 1.** Matrix of function and functionality

During the design process, more and more data are generated and they often prove to be incomplete and unclear. With this aim, the MFF consistently follows each piece of information, stores it and provides the designer with a comprehensive, global overview of product's sub-matrices in one place.

Sub-matrices involving at least one possible solution on at least one function within the presupposed building block or functionality are considered full and display partial and complete result. The result is displayed in the form of percentage values – numbers in the sub-matrix cell. The value or number is calculated on the basis of a verbal fraction. It crosses the value of functional requirements in the denominator and the value of function on functionality in the nominator. The fraction provides the acquired value of the solution, i.e. word matching.

Together with the number, an informative type of the current function is also displayed. In case there are no possible solutions within individual sub-matrices, results cannot be displayed and the sub-matrix remains empty. In the case of empty matrices, rules make it necessary to think about new possible functions and functionalities as the existing solutions cannot fulfill the presumed requirements. The number of functions within the sub-matrix is analogous to the number of possible solutions in the functionality column. Results-wise, only the functions with a specific possible solution are displayed. The functions not solving a given situation are not included in the display.

Functions in sub-matrices can be arranged into four types of functions, as follows: the main function, marked as M in the sub-matrix; the supplementary function, marked as S; the auxiliary function, marked as A and the binding function, marked as B in the sub-matrix. Each sub-matrix contains a functional description of a technical system, described in the MFF. The main function's name for each system is shown with a mark (M1). According to description rules, each building block can have only one function. It can happen that the MFF includes several technical systems with identical names of either the main, supplementary, auxiliary, or binding functions. In the prototype model implementation, real names of the existing status (description), are used for the main functions, as well as for the supplementary, auxiliary and binding functions of the technical system. Supplementary functions' names in the matrix model are shown in Figure 1 and marked with S1, S2,...,S$k$; $k$=1,...,$p$. A technical system can have more than one supplementary function. It is even possible that a technical system has no supplementary function if it was not planned in the actual descriptions. Auxiliary functions are shown with marks: A1, A2,...,Ak; k = 1,...,p. A technical system can have one or more auxiliary functions. Analogous to the explanation above, it can happen that a technical system has no auxiliary function. Binding functions are shown with marks: B1, B2,...; k = k=1,...,p. Contrary to supplementary and auxiliary functions, a binding function without a single binding function is not possible because it would make the description incomplete and the technical system would not fulfill the adequate criteria or rules of the pre-defined rules on describing functions, functionalities and technical systems. In other words, such technical system would not exist. Cumulative p value cannot be the same for the supplementary, auxiliary and binding function. Each technical system can have a different number of functions. For definition and uniqueness reasons, each function of a particular technical system in the MFF matrix is described by parameters, winning parameters and value intervals. However, it is not certain that it will be displayed: it has been mentioned above that it is displayed only when it solves a given functional requirement with a significant probability. Depending on the complexity of the function, it can be described by one or more parameters. In no case can it happen that a function would be left with no parameters since a function without parameters is no longer a function.

Automated solving and suggesting the final solution (the suggested solution in Figure 1 in each row of the first column) represents an upgrade of the MFF. It is presupposed that a possible solution is the one that most closely corresponds to a given functional requirement. The final solution is selected on the basis of individual percentage values, where the final solution will be the one with the highest calculated percentage value and the highest hierarchical type. On the other hand, it is not possible to claim that the automatically generated solution is always suitable and confirmed as final. It is only provides some sort of assistance in the decision-making phase. In any case, the end and final decision should always and in all cases be taken by the user, i.e. designer. The final solution is accepted only if the designer eventually decides for it. Let's point out at this point that in many cases it is even possible that a partial solution with a lower percentage value suits the final solution better because it has some parameters, intervals and winning parameters that in such case match the functional requirement better, even if the name of another function is more favorable and better in terms of the percentage value. Unfortunately, there is no sufficiently reliable decision-making tool available yet for such cases, leaving decision-making in the domain of designers, their experiences and personal views of a given issue.

## 3 Discussion and Conclusion

Design process represents the transformation and iterative process in which the first input information is transformed into the final design solutions. Design solutions as a result of the design process fully represent materialized technical systems. Within the design process, binding links between individual product functions and their shapes are created. The starting point for design process are functional requirements and needs, which may appear in a descriptive form through functions and functionalities or graphically over the shape model and known technical systems. The goal of the design process is to generate new conceptual product variations and new technical systems structures by binding of functional models with existing technical systems that can realize them. Morphological methods and tools based on the morphological matrix have a major role in creating the described goal.

The main contribution of the MFF method is the possibility of concurrent development of activities in all phases of the preliminary design and development process, allowed by a specific level of data. The MFF method therefore allows concurrent solving of several open functional requirements, which recognizes requirements for productivity, clear recognition of generators, binders and information users, and reduction of design and development times.

The MFF matrix is based on the interaction between function and functionality and is developed according to specific criteria, while we are striving for a solution that would be in relation to the morphological matrix built and determined on the basis of a mathematical model and predetermined conditions, therefore not only on designer's intuition.

The MFF aims to improve the main shortcomings of the morphological chart on the fields of mathematics-based modeling with implementation, automated suggesting and solving, ordering, modularity, sub-matrices, semantics, taxonomy and ontology.

The MFF model has been included and implemented into a prototype computer web application. By means of a developed central relational database it manages engineering data for the development of new conceptual product variants. Besides, the MFF model is currently presented and implemented on more than ten completely different, solved and described products in different areas. Figure 2 shows a concrete view of the MFF user interface for part of a product – a portable car vacuum cleaner, to be precise. The original implementation idea is also based on the fact that with the assistance of the MFF matrix, the designer is directed and led to new possible solutions.

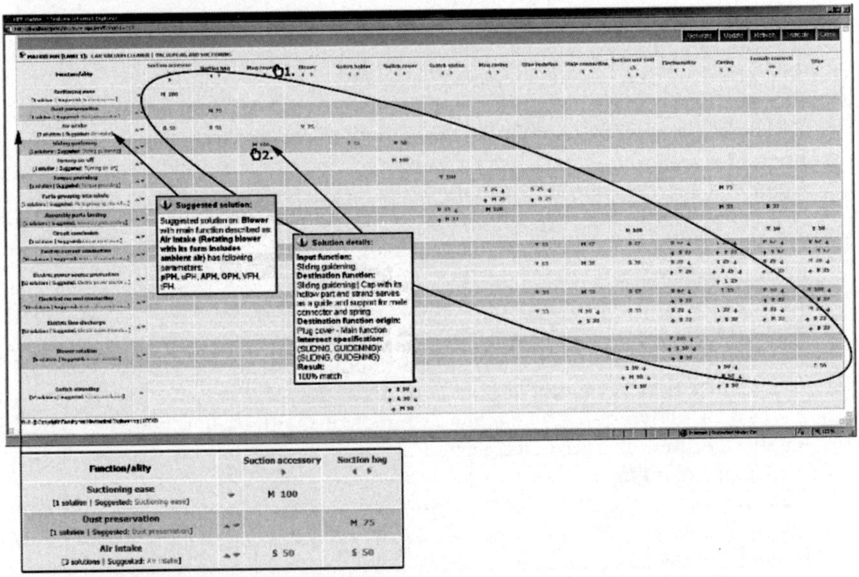


**Figure 2.** Implementation of the matrix of function and functionality

The presentation is aimed at direct users, developers and researchers of technical systems and recognized technical processes. It is based on the connection between processes, recognized in the nature, and searching for comparable or fulfilling technical processes at a certain level of knowledge development. A mission of the developed models is to contribute to and find within preliminary design processes adequate fundaments for better and faster design management.

## 4 References


[1] Zadnik Ž., Karakašić M., Kljajin M., Duhovnik J. Functional and Functionality in the Conceptual Design process. *Strojniški vestnik*, 2009, 55 (7-8), pp. 455-471.

[2] Kušar J., Duhovnik J., Tomaževič R., Starbek M. Finding and Evaluating Customers Needs in the Product-Development Process. *Strojniški vestnik*, 2007, 53 (2), pp. 78-104.

[3] Kurtoglu T. A Computational Approach to Innovative Conceptual Design. *PhD Thesis*, 2007 (The University of Texas at Austin).

[4] Karakašić M., Zadnik Ž., Kljajin M., Duhovnik J. Product Function Matrix and its Request Model. *Strojarstvo*, 2009, 51 (4), pp. 293-301.

[5] Karakašić M., Zadnik Ž., Kljajin M., Duhovnik J. Functional structure generation within multi-structured matrix forms. *Technical Gazette*, 2010, 17 (4), pp.465-473.

[6] Duhovnik J., Zargi U., Kušar J., Starbek M. Project-driven Concurrent Product Developement, Concurrent Engineering: Research and Applications, Vol. 17, Nr. 3, September 2009

[7] Duhovnik J., Kušar J., Tomaževič R., Starbek M. Developement Process with Regard to Customer Requirements, Concurrent Engineering: Research and Applications, Vol. 14, Nr. 1, September 2006

[8] Prasad B., *Concurrent Engineering Fundamentals*, 1996 (Prentice Hall PTR)

[9] Kljajin M., Ivandić Ž., Karakašić M., Galić Z. Conceptual Design in the Solid Fuel Oven Development. In *Proceedings the 4th DAAAM International Conference on Advanced Technologies for Developing Countries*, Slavonski Brod, September 2005, pp. 109-114.

[10] Hubka V., Eder W.E. *Theory of Technical Systems*, 1988 (Springer-Verlag Berlin, Heidelberg)

[11] Hirtz J., Stone R.B., McAdams D.A., Szykman S., Wood K.L. A Functional Basis for Engineering Design: Reconciling and Evolving Previous Efforts. *NIST Technical Note 1447,* 2002 (Department of Commerce United States of America, National Institute of Standards and Technology).

[12] Žavbi R., Duhovnik, J. Conceptual design chains with basic schematics based on an algorithm of conceptual design. *Journal of Engineering Design*, 2001, 12 (2), pp. 131-145.

[13] Duhovnik J., Tavčar J. Product Design Test using the Matrix of Functions and Functionality. In *Proceedings of AEDS 2005 Workshop*, Pilsen-Czech Republic, November 2005.

[14] Pavlić D. Sustav za konfiguriranje proizvoda modularne arhitekture. 2003 (Fakultet strojarstva i brodogradnje u Zagrebu).

[15] O´Grady P., Liang W.Y., Tseng T.L., Huang, C.C., Kusiak A. Remote Collaborative Design With Modules. *Technical Report 97-0,* 1997 (University of Iowa).

# Proposal of a NPD diagnostic method to identify improvement opportunities

Janaina Mascarenhas Hornos da Costa[a1], Creusa Sayuri Tahara Amaral[a] and Henrique Rozenfeld[a]

[a]University of São Paulo, São Carlos, SP, Brazil

**Abstract.** This paper proposes a cause-effect diagnostic method to aid in the identification of New Product Development (NPD) improvement opportunities, called the Diagile Method. The proposed method is based on the CRT method and incorporates best practices of project management and activities to identify relevant opportunities for improvement. Templates are provided to assist in each phase of the Diagile application and a tool is provided for conducting effective interviews with NPD personnel. The main Diagile outputs are a current reality tree and a personalized portfolio of NPD improvement projects. A controlled experiment involving two independent groups was performed to evaluate the method at a large multinational office supply manufacturer. The results of this research suggest several managerial implications for improving the NPD process. The first is that effective diagnosis plays an important role in improving the NPD process. A cause-effect diagnostic method enables a healthy discussion among multidisciplinary NPD teams seeking to improve NPD to make the right choice of NPD improvement projects. Finally, the company reported that Diagile has proved to be an excellent method to analyze the NPD process and identify its problems.

**Keywords:** new product development management, process diagnosis, business process management.

## 1 Introduction

Consumers have become increasingly selective and demanding in their choice of products, which must offer attractive and modern innovations in an ever briefer period. To meet these expectations, companies are obliged to implement rapid improvements in the management of the New Product Development (NPD) process, particularly with respect to reducing the product lifecycle [3,10], and hence, hastening the introduction of new innovations in their products. However, most companies do not find it easy to keep up with the requirements of this new scenario, a fact indicated by the high rate of products that fail after being launched, or even during the development process itself.

Companies should conduct Product Development Process improvement cycles aimed at identifying the main problems they need to overcome in order to achieve

---

[1] University of São Paulo. Av. Trabalhador São Carlense, 400, Production Engineering Department NUMA, 13566-590 São Carlos, SP, Brazil, Tel.: +55-16-3373.9433 / E-mail: andreapjubileu@gmail.com

D. D. Frey et al. (eds.), *Improving Complex Systems Today*,
Advanced Concurrent Engineering, DOI: 10.1007/ 978-0-85729-799-0_42,
© Springer-Verlag London Limited 2011

fully satisfactory results. Making a diagnosis of the process helps shed light on the internal and external factors that affect NPD, as well as its characteristics, such as the team's technical skills, how its activities are organized, available resources, customer demands or bargaining power with suppliers. Diagnostics contribute directly to the NPD improvement process, since organizations exist as entities that must be examined before a recommendation for intervention can be made [1]. It can therefore be safely stated that diagnostics is a highly desirable, if not essential, precursor of development, change and intervention in well informed and effective organizations [11].

Some examples of diagnostic methods are the Current Reality Tree [8,13], Cognitive Maps [2,8,13], Process Modeling [10,14] and Ishikawa Diagrams [7].

Diagnostic methods generally attempt to identify problems that occur or may occur in a process. Doggett [6] points out that "behind each problem is a cause for its occurrence," and that companies should therefore make an effort to identify the root causes of problems.

The Ishikawa Diagrams and Current Reality Tree (CRT) methods are frequently employed for this purpose. The CRT, in particular, has proved more appropriate for the identification of root causes, providing better quality results than the Ishikawa Diagrams [5,7].

The Current Reality Tree is a generic method to solve the restrictions or bottlenecks of a process or system using common sense, intuitive knowledge and logic [9]. One of the first steps to use the CRT may be a list of restrictions. These restrictions are called symptoms, undesirable effects, problems or dysfunctions. Then, for each restriction, a possible chain of cause-effect relationships responsible for its manifestation is examined. The objective is to trace back these restrictions to one or more roots of the problem, seeking to identify which central problems deserve attention [8].

Although the literature contains versions of the method of construction of the CRT originally proposed by Goldratt [8], an opportunity was identified to create a proposal of CRT construction focusing on NPD. Therefore, the purpose of this paper is to present a diagnostic method, called Diagile, aimed at identifying NPD improvement opportunities. This paper also reports on the results of the application of the method by means of a controlled experiment in a large multination company operating in the field of office supplies.

Based on the hypothetic-deductive approach, a survey and analysis of process diagnostic methods was made, particularly of cognitive methods.

The best practices of these methods, project management practices and practices of how to evaluate process improvement opportunities were then compiled to create the proposed method. To evaluate the proposed method, a controlled experiment was carried out at a large multinational company that develops office supplies. Two teams performed the diagnosis of the company's NPD: one team used the proposed method and the other one performed its diagnosis using the current reality tree construction method (Goldratt, 1994).

The next section describes the proposed method, while section 3 discusses the controlled experiment, and the last section of this paper presents the conclusions of this work.

## 2 Diagile

The proposed method for the diagnosis of NPD, called Diagile, is composed of nine phases (Figure 1), and is the result of an evaluation of the best practices of construction of current reality trees, and of project management activities, as well as the use of a scripted interview and activities aimed at and encouraging the search for improvement opportunities. A computational tool was developed, which is not described in detail here, to support the preparation of the interview script, the interviews, and the identification and association of undesirable effects.

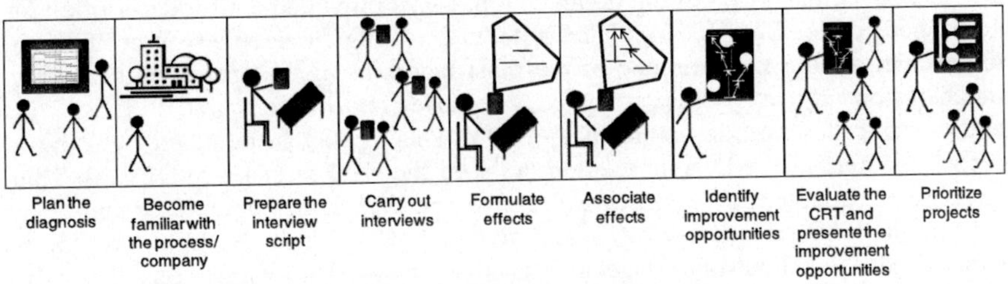


**Figure 1:** Phases of the Diagile method

The first phase of the Diagile method is planning the diagnosis. This phase consists of defining the team of interviewers, the people in the organization who will be interviewed, the dates for interviews, for meetings to be held during the project, and delivery due dates. The importance of this phase lies in the need to define a neutral team for the analyzed process in order to minimize bias of the team during the interviews and the construction of the diagnosis. In selecting the people to be interviewed, the team must keep in mind that NPD is a multifaceted and interdisciplinary process and it is therefore important to consider the opinions of other areas to develop a complete diagnosis.

In the second phase, the team must familiarize itself with the target process. This phase consists of making a preliminary analysis of the process to be diagnosed to give the team the necessary background for performing the diagnosis. This means collecting information about the process in question, its main activities, the systems it uses, interfaces with other areas, and data about its action in the market.

The third phase involves preparing an interview script. The script for the various interviews should be prepared aiming to obtain, from the interviewees of different areas, the information required for the construction of the CRT, i.e., the undesirable effects or problems that hinder the good performance of the activities of the process under study. The five dimensions of NPD to be taken into account are strategy, organization, activities/information, resources and knowledge. The standard NPD interview script, which is one of the elements of the Diagile method available at www.portaldeconhecimentos.org.br, can be used as an aid in this phase.

The fourth phase of the Diagile method consists of conducting interviews with the various interviewees, which can be carried out with the scripted interview. The interviews should not involve people of different hierarchical levels, since the objective is to identify the main restrictions and problems of the process. It is a well known fact that people do not feel comfortable about expressing their opinions about negative points in front of their superiors. Therefore, the interviewees should be informed that the final result of the diagnosis does not identify the names of the people that pointed out the problems, thus maintaining the neutrality of the information obtained. This phase of the method is fundamental, because this collection of information is what enables organizational problems to be identified.

The fifth phase consists of formulating the undesirable effects (UEs). The information collected in the interviews serves as the basis for identifying problems that have been occurring in the organization, based on which a brief description of each undesirable effect is formulated without exposing the people who brought up the problems. This phase is one of the most laborious of the project and requires much effort by everyone involved, because poorly formulated UEs will compromise the entire structure of the CRT. This phase can be aided by using a second element of the Diagile method: a list of the NPD's recurrent UEs [4]. This list is also available for free download at www.portaldeconhecimentos.org.br.

The sixth phase consists of associating the UEs. Once the UEs have been formulated, they should be associated through cause-effect connections. The UEs should be connected following the structures: "IF cause ... THEN effect". To facilitate this activity, the UEs should be grouped according to their area of knowledge, e.g., project management, people management and information and communication technology. A team should read the tree several times to check for the pertinence of the causal relationships that are established and seek new relationships in order to obtain a consistent CRT which effectively represents the real situation of the organization. The final result of this phase is the current reality tree itself, which is composed of the identified UEs and their causal relationships. This phase can be aided by using the NPD reference tree freely available at www.portaldeconhecimentos.org. The tree, which is available in the form of a matrix, associates all the recurrent UEs [4] with the UEs that are considered their causes. Thus, this matrix can be taken as the basis to build or analyze the associations created in this phase. For the graphic design of the tree, we suggest using the graph generation software yED, which is free (www.yed.com).

The seventh phase of the method consists of identifying opportunities for improvement. The CRT should be used to identify opportunities, seeking improvement projects that tackle primarily the root causes of the CRT, so that the largest possible number of problems can be minimized in a single proposal. The improvement projects should be inserted into the tree and associated with the UEs they aim to tackle. A portfolio of NPD improvement opportunities was compiled to support this phase. This portfolio contains 31 improvement opportunities, and a description is given of the objectives and minimum deliveries of each opportunity, as well as the undesirable effects it aims to minimize. The portfolio is the fourth element of the Diagile method and is also available for free download at www.portaldeconhecimentos.org.br.

The eighth phase involves evaluating the CRT and presenting the improvement opportunities to the NPD team and other people involved. The objective of this phase is to enable a discussion of the problems and improvement opportunities in the process. It should be noted that new effects and new causal relationships may be identified in the tree, and suggestions may be presented for new improvement projects.

The ninth and last phase is to prioritize improvement projects. The improvement opportunities should be prioritized by a multifunctional team. The purpose is to establish a systematic way to prioritize the projects so that decision-making is transparent to NPD personnel. To this end, several criteria should be defined and used to evaluate the improvement projects. To support this phase, a tool was developed for selecting improvement projects. This tool, which contains prioritization criteria and rules, is also freely available at www.portaldeconhecimentos.org.br.

The next section presents the results obtained through the application of the Diagile method in a controlled experiment.

## 3 Controlled Experiment

The controlled experiment was carried out to evaluate the performance of the Diagile method compared to the Current Reality Tree diagnostic method [8]. The experiment was conducted at a large multinational company that manufactures office supplies, which was dubbed Company A. The experiment was carried out with the voluntary participation of a group taking a postgraduate discipline, which involved nine students, who were divided into two groups, and the professor who teaches the discipline. The company consented to having the interviews conducted twice, enabling the two groups to interview the same people.

Two testing units were formed, called the control group and the experiment group. The control group performed the diagnosis using the traditional CRT construction method, while the experiment group used the Diagile method. The groups were not set up randomly but according to their educational level and their experience in performing diagnoses, in order to ensure their characteristics were homogeneous. The two groups carried out their diagnoses concomitantly. Durante and after the two diagnosis were completed, an analysis was made of the results achieved, the time spent, and the difficulties that were encountered.

As mentioned earlier, Company A permitted the groups to interview the same people twice, making a total of 32 interviews. The order of the interviews was decided randomly by the company, and the interviewees were not aware of which group was interviewing them. The activities of both the groups were supervised by the professor of the discipline, who stipulated due dates for performing the diagnoses and standard deliveries: one cause-and-effect tree diagram and one portfolio of improvement projects for the company.

Three general assessment criteria were defined for the experiment: (1) degree of difficulty in developing the diagnosis, (2) time spent on its development, and (3) quality of the results. To check the first criterion, the interviewees and members of the groups that performed the diagnosis answered the following questions:

- What was the degree of difficulty in identifying NPD problems?
- What was the degree of difficulty in identifying cause-effect relationships?

- What was the degree of work/effort involved in building the tree?
- What level of knowledge of NPD management is required to build the tree?

The result of the mean of the responses from both groups (Figure 3) confirms that the use of the Diagile method facilitates the construction of the cause-and-effect tree diagram, because the degree of difficulty involved in its construction is very low and because its use requires a very low level of knowledge.

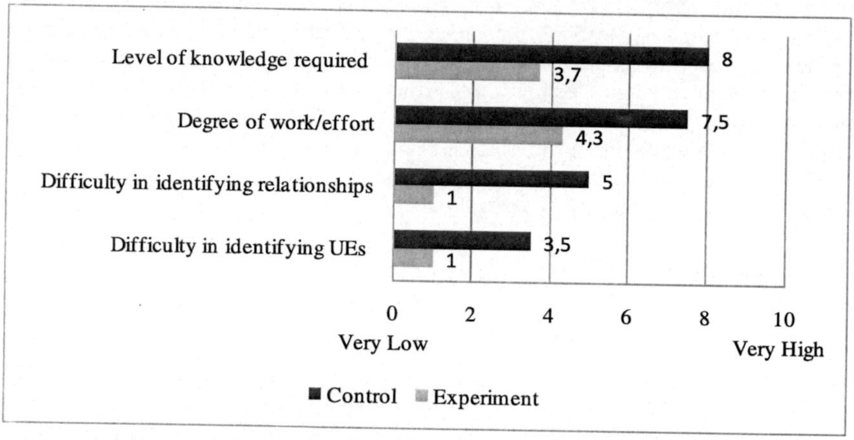


**Figure 2:** Analysis of the degree of difficulty in developing the diagnosis

The second criterion analyzed was the speed of the construction of the diagnosis that can be visualized. The experiment group, in addition to having one person less on its team, spent 31% less time than the control group to perform its diagnosis. This reduction was particularly visible in the activities of identification and association of the UEs, and in the identification of NPD improvement opportunities.

The last criterion analyzed was the quality of the constructed trees. The control group, which did not use the Diagile method, had to make 178% more corrections on the tree than the experiment group. Since both the teams of the experiment were composed of postgraduate students who are not considered NPD specialists, the smaller number of corrections of the diagnosis prepared by the experiment group demonstrates the better quality of the initial version of the tree created by using the Diagile method. Lastly, the higher efficiency and effectiveness of the proposed Diagile method was also perceived during the activities of identification and prioritization of improvement opportunities. The experiment group drew up a more relevant and coherent list of improvement projects than the control group, and also provided documentation for these projects in the form of project launch agreements.

Lastly, the group presented the CRT to the directorate of Company Alfa, which reported that the tree devised by the experiment group listed most of the problems the company encountered on a daily basis, and that, unlike the CRT prepared by the other group, the descriptions of the UEs were to a certain extent more sophisticated, containing current and technical terms.

## 4 Conclusions

In this paper, we described a diagnostic method that proved to be efficient in supporting the identification of NPD improvement opportunities. To evaluate the proposed method, a controlled experiment was conducted whose results confirmed that the use of the Diagile method increased the diagnostic efficiency. This finding was evident from the fewer hours spent in drawing up the diagnosis and by the fact that the experiment team had one member less than the control team. This can therefore be considered an increase in efficiency, since the diagnosis was completed in less time and with fewer resources.

The advantage of using the Diagile method was also revealed in the controlled experiment. First, the experiment demonstrated that the Diagile method increases the quality of the initial version of the diagnosis if one considers the construction of a CRT by a team who cannot be considered NPD specialists and who have no experience with this diagnostic method. The two teams of the experiment were composed of postgraduate students who are not NPD specialists and the end result demonstrated that the control group, which did not use the Diagile method, had to make far more corrections on the tree, in fact, 178% more than the experiment group. Another advantage of the Diagile method is that it encourages the search for improvement opportunities and provides a more relevant list of improvement projects for each diagnosis.

With regard to the use of controlled experiments, this approach was considered highly pertinent for the research area of this project, allowing to a comparative analysis of the quality of highly homogeneous data. Nevertheless, a few difficulties were encountered. It was found that the control group must employ methods, activities or tools that are already proven in the literature so that more reliable comparisons can be made of the groups' results.

This paper therefore demonstrated the usefulness of the proposed method and the greater efficiency and efficacy of the diagnosis using the Diagile method when compared with the traditional CRT construction method.

## Acknowledgments

The authors are indebted to their GEI2 colleagues for their helpful suggestions, and to the Brazilian research funding agencies CNPq and CAPES for their financial support of this work.

## References

[1] BUSHE, G. R., MARSHAK, R. J. Revisioning Organization Development: Diagnostic and Dialogic Premises and Patterns of Practice. The Journal of Applied Behavioral Science, 2009: 45(3), 348-368.

[2] CARBONARA, N., SCOZZI, B. Cognitive maps to analyze new product development processes: A case study. Technovation, 2006: 26(11), 1233-1243.

[3] CARRILLO, J. E., FRANZA, R. M. Investing in product development and production capabilities: The crucial linkage between time-to-market and ramp-up time. European Journal of Operational Research, 2006: 171, 536-556.

[4] COSTA, J. M.H., AMARAL, C. S. T., ROZENFELD, H. Systematization of Recurring NPD Problems: Opportunities and Challenges in Expediting Process Diagnosis. In: 15th International Product Development Management Conference, 2009, Twente.

[5] DOGGETT, A.M. A Statistical Comparison of Three Root Cause Analysis Tools. Jounal of Industrial Technology, 2004: 20 (2).

[6] DOGGETT, A.M. Root Cause Analysis: A Framework for Tool Selection. The Quality Management Journa. 2005: 12 (4).

[7] FREDENDALL, L.D., PATTERSON, J.W., LENHARTZ, C., MITCHELL, B.C. What should be changed? Quality Progress 2002: 35 (1),50-59.

[8] GOLDRATT, E. M. It`s not luck. Great Barrington: North River Press:1994.

[9] GOODRICH, D. F. The relationship of the theory of constraints implementation to change management integration in professional service organizations. Ph.D. thesis. School of Business and Entrepreneurship, New Southeastern University, 2008.

[10] JESTON, J., NELIS, J. Business Process Management Practical Guidelines to Succesful Implementations. Oxford: Elsevier: 2008 (2o edition.)

[11] MCCULLOCH, A., CRONSHAW, S.F. Reinstating the Lewinian vision: From force field analysis to organization field assessment. Organization Development Journal, 2008: 26(4), 89-103

[12] ORTT, J. R., DUIN, P. A. The evolution of innovation management towards contextual innovation. European Journal of Innovation Management, 2008: 11(4), 522-538.

[13] RAHMAN, S. The theory of constraints`thinking process approach to developing strategies in supply chains. International Journal of Physical Distribuition & Logistics Management, 2002: 3(10), 809-828.

[14] VERNADAT, F. B. (1996). Enterprise Modeling and Integration: principles and applications. Springer.

# Reference Planning Processes for Series Production

Nils Macke[a], Stefan Rulhoff[b,1] and Josip Stjepandic[b]

[a] ZF Friedrichshafen AG, Graf-von-Soden-Platz 1, 88046 Friedrichshafen, Germany.
[b] PROSTEP AG, Dolivostraße 11, 64289 Darmstadt, Germany.

**Abstract.** In the modern product emergence process production planning gets increasingly important and has to be executed in parallel to the product development. Different requirements and heterogeneous system architectures often result in inefficient planning processes. To standardize and optimize these planning processes, the ProSTEP iViP project group "Digital Manufacturing" has defined a reference process for production planning based on actual user experiences. The defined reference process supports the creation of planning processes tailored to the specific needs of many companies and locations and thus supports efficient production planning. Furthermore the reference process amends common understanding and transparency of terms and processes in order to enable efficient communication e.g. between planner and the IT department. This reference process represents also the foundation for the definition of necessary interfaces in supporting systems of the Digital Factory to enable integrated and continuous workflows. It allows easy reuse of existing planning documents and knowhow. The special importance of this approach is to build a "blueprint" for faster and easier implementation of production planning.

**Keywords.** Digital Factory, Product Emergence Process, Production Planning Processes, Reference Process

## 1 Introduction

Today global operating companies face growing demands in flexibility and shortened development cycles due to the increasing complexity of processes and products. This requires efficient workflows in the product emergence process more than ever. In parallel to the well organized product development processes, production planning gains increasingly more importance. But unlike the product development and production operations, which are supported by sophisticated software systems (CAx, PLM and PPC) the production planning phase in between (Figure 1 [1]) is affected by the frequent use of isolated Digital Factory solutions [2]. As a result, the production planning processes itself often seems to be uncoordinated and even department-centred. This leads to inefficient planning

[1] PROSTEP AG, Dolivostraße 11, 64289 Darmstadt; Germany; Tel. +49-6151-9287441;E-mails: stefan.rulhoff@prostep.com

D. D. Frey et al. (eds.), *Improving Complex Systems Today*,
Advanced Concurrent Engineering, DOI: 10.1007/ 978-0-85729-799-0_43,
© Springer-Verlag London Limited 2011

workflows with redundant activities, redundant work and transformation errors. Consequently, standardization of the necessary planning processes based upon established best practices in the realm of production planning makes it possible to unlock much potential [3].

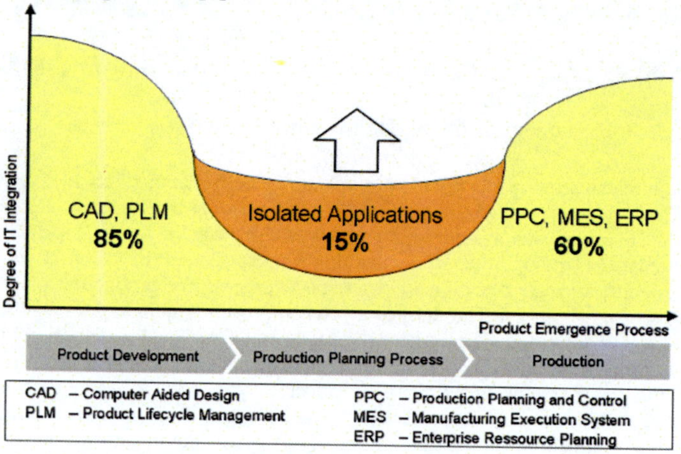


**Figure 1.** Isolated Applications for Production planning processes (adapted from Klauke [1])

In order to meet this challenge, ProSTEP iViP's "Digital Manufacturing" Project Group developed an uniformly defined, end-to-end reference planning process for series production. This reference process for production planning is defined in regard to interfaces with other business processes and in context to the product emergence process.

## 2 The Production planning process (PPP)

The PPP for a particular product typically begins with the product proposal or product concept that evolved during the "preliminary development" phase. Concurrently, product development activities also start within the engineering department. Figure 2 shows how the production planning process is incorporated in the product development process. Both the milestones and the interfaces between the processes involved in the different phases are included. The preliminary planning or industrialization subprocess begins with the milestone "Product concept released" and ends with the milestone "Work schedule for series production released". During the course of this subprocess, the work schedule is gradually drafted over a number of iterations, depending on the relevant milestones in product development. Initially, it will in general only be an extremely rudimentary version, as it is based on unreliable or incomplete data from the early design of the product under development. During the product development process, the schedule matures until finally an initial complete work sheet for subsequent production is released [4].

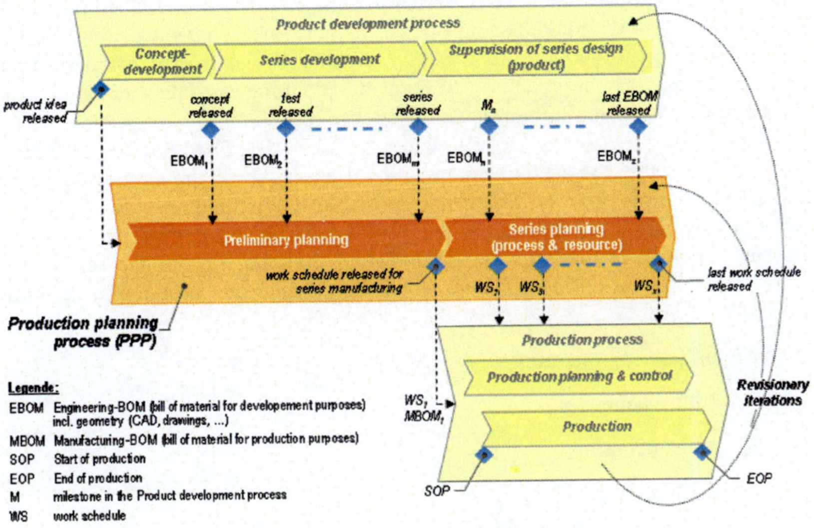


**Figure 2.** The production planning process (PPP)

The sub-process "Series planning", or "Adaptation and optimization/replacement", starts before the milestone "SOP" and ends with the release of the final work schedule. At the same time, the current version is continuously developed and adapted in order to account for changing conditions (CIP). On the one hand, this can involve design changes carried out during the product development process as part of series planning (e.g. model face-lifting). On the other hand, the changes may derive from production operations directly, for example because a machine has been replaced, a technology has been replaced by another, or the entire production process has been modified. The work schedule can also be a standard sheet which can be adapted for specific locations.

## 3 Reference planning processes for series production

The defined reference planning process is composed of several subprocesses structured in maturity phases and planning disciplines. There are maturity phases with gates where planning results of the different planning disciplines have to be consolidated. These phases can be processed iteratively. The preliminary planning phase (also referred to as industrialization) is broken down into three sub-phases [5]:

- Concept planning
- Rough planning
- Detailed planning

The final phase then consists of series planning (adaptation and optimization) whose primary task is to take account of all the necessary changes or optimization operations during production. In the way it considers the level of maturity of overall planning, this approach to structuring planning activities resembles the planning methods set out in the relevant literature [6].

Alongside the maturity level-related phases described above, planning disciplines constitute the second organizational criterion for the structuring of the production planning process. Since it is possible to distinguish between a large number of planning disciplines in the production planning field, it of utmost importance to restrict these to the fundamental planning disciplines that can be found in manufacturing companies when describing the production planning process. These are:

- Manufacturing planning
- Assembly planning
- Logistics planning
- Layout planning

Additional cross-departmental disciplines, such as quality management, are important throughout all the planning disciplines.

### 3.1 Manufacturing planning

Manufacturing planning comprises all the measures taken in order to design a manufacturing system as well as the selection of the necessary manufacturing resources and processes. Its scope of activity partially overlaps that of work preparation. When performing manufacturing planning, it is particularly important to take account of dependencies on and interactions with assembly, logistics and layout planning and integrate these in the planning process.

### 3.2 Assembly planning

Assembly planning, on the other hand, defines the steps involved in the assembly of various individual parts to create the final product and provides the necessary equipment (lifting cranes, robot arms, for example). This planning activity, which also includes the draft design of the assembly systems, is frequently performed by the department responsible for work preparation. Assembly in this context designates the process of putting together the modules, parts and amorphous materials to form a product.

### 3.3 Logistic planning

The aim of logistics planning is to ensure that the raw materials and semi-finished products, prefabricated components, assemblies or securing elements such as screws are available at the right place at the right time and in the correct, economically optimized quantities. During logistics planning, it is necessary, for example, to define the delivery and container concept. Optimized logistics planning permits production with low stock levels and at minimized cost while simultaneously ensuring responsiveness and versatility. The defined planning steps in the reference planning processes concentrates on the logistic processes inside the facilities.

## 3.4 Layout planning

As the last of the four focused planning disciplines, layout planning ensures that operating resources are located optimally in the production area so that, for example, the processes in an assembly line or assembly cell can run as efficiently as possible. To perform this task, it is very important that the knowledge and experience derived from the other planning disciplines is available to layout planning.

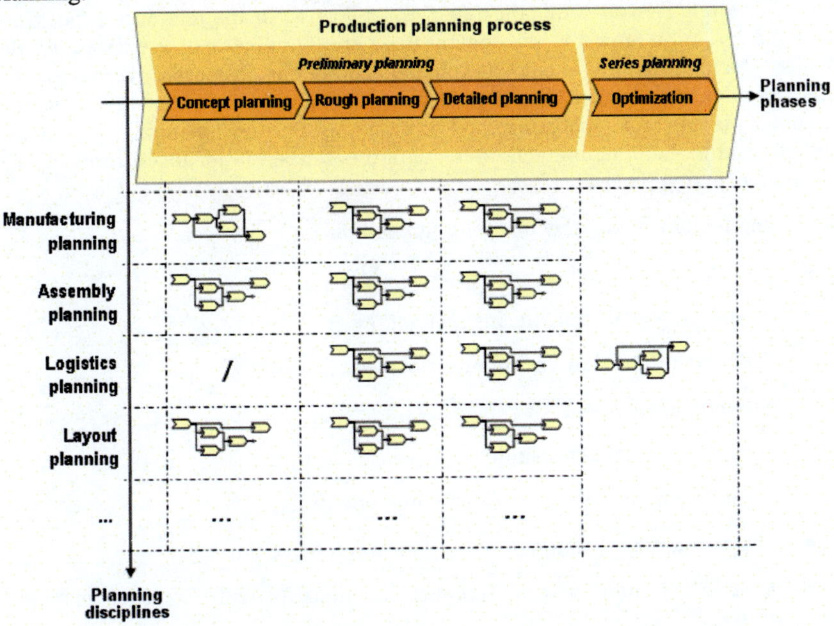


**Figure 3.** Production planning process composed of subprocesses

The level of detail presented in the planning processes developed up to this point, is nowhere near as adequate as needed for practical use for the harmonization or adaptation of a company's planning structures. For this reason, it is necessary to define and create detailed processes within the context of the structures determined thus far. Figure 3 illustrates the way these detailed processes of the production planning process are embedded within the matrix of corresponding planning disciplines and maturity levels. In the reference planning process each subprocess is described in detail with input, output and the necessary planning steps. There are two exceptions. In the maturity phase Concept planning there are no specific steps for the planning discipline logistic planning and in the maturity phase Series planning there is no differentiation between planning disciplines. The necessary planning steps there can affect parts of all planning disciplines depending on the use case.

This subdivision into subprocesses makes it clear that every module or subprocess in the end-to-end reference planning process may, under certain circumstances, be structured differently depending on the task that has to be resolved or the released data volume that is self dependent on the level of maturity achieved.

## 4 Use Cases

Several Use Cases demonstrate the application of the reference process in the company. With regard to the content there are two groups of use cases. While the first group provides exemplary planning activities that are covered by the reference planning process, the second group examines the use of the reference planning process for analyzing and optimizing existing planning processes, methods and tools. Planning activities that are described in idealized form by the reference planning process:

- Industrialization / new planning (green field)
- Adapt existing product mix (brown field)
- Introduce new manufacturing technology
- Relocate production, including machinery
- Procure replacement
- Increase level of automation

Analysis and optimization of planning processes:

- Check internal planning process against reference process
- Compare / align existing software functionality with reference planning process
- Examine penetration of IT tool using the reference planning process

Exemplarily the use cases "Industrialization /new planning (green field)" and Adapt existing product mix are described.

### 4.1 Use Case Industrialization

The use case Industrialization assumes the feasibility of a complete new production design in a "green field" (Figure 4). It is triggered by input from the business strategy such as the manufacture of new products, changed framework conditions such as modified or new laws, capacity bottlenecks or changed supply chains. The goal of this use case is the new planning of a shop floor. The result delivered by this use case is a manufacturing strategy that provides all the information required for industrialization.

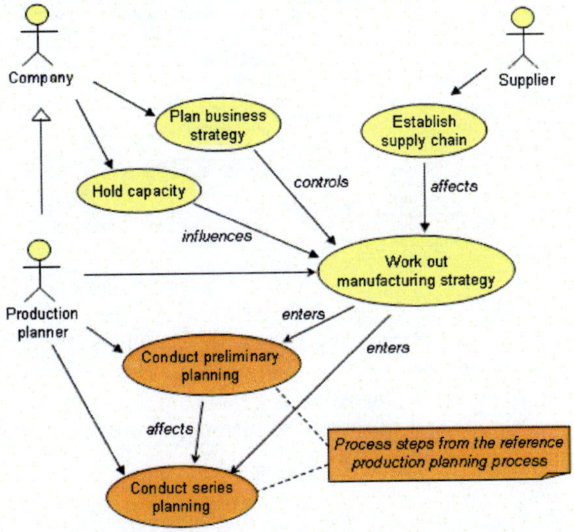


**Figure 4.** Use Case Industrialization / new planning (green field)

The three actors "Company", "Production planner" and "Supplier" are involved in this use case. The production planner works out the concept planning, rough planning, fine planning (included in the preliminary planning) and series planning based on planning processes of the reference planning process. This is incorporated, together with information relating to the business strategy, supply chain and capacity, into the manufacturing strategy.

## 4.2 Use Case Adapt existing product mix (brown field)

The use case "Adapt existing product mix (brown field)" is triggered when a new product is added to an existing product mix within an existing production infrastructure and is needed to ensure sufficient production capacity (Figure 5). The goal is to evaluate and, if necessary, improve the integration of a new product in the current shop floor situation. The use case ends with the successful expansion of the shop floor to include the new product.

The two actors "Company" and "Production planner" are involved in this use case. The production planner creates a manufacturing strategy within the framework of current capacities and the current product mix. Based on existing orders and existing capital, the production planner can ultimately adapt the manufacturing strategy in such a way that overall capacity on the shop floor is adapted to the new product mix. This ensures that there is sufficient manufacturing capacity for the new product.

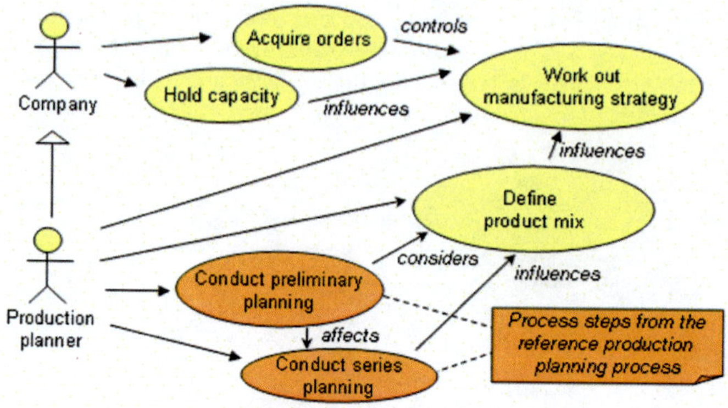


**Figure 5**. Use Case Adapt existing product mix (brown field)

These use cases shall demonstrate in which situations the reference planning process for production planning can be applied to. The reference process supports the planning of production processes in a standardized form and enables in this way the user to optimize and integrate the planning process in the product emergence process.

## 5 Outlook

Based on the defined reference planning process for series production and with regard to the requirements of the production planning, this year the ProSTEP iViP project group "Digital Manufacturing" will concentrate their efforts on the IT systems and supporting planning methods in order to enable efficient and integrated production planning processes.

## 6 References

[1] Klauke, S.: Methoden und Datenmodell der "Offenen Virtuellen Fabrik" zur Optimierung simultaner Produktionsprozesse (Methods and data model of the "open virtual factory" for optimizing simultaneous production processes). Düsseldorf: VDI-Verlag 2002

[2] VDI-Richtlinie 4499 (2009): Digitale Fabrik - Grundlagen, Blatt 1. Berlin: Beuth Verlag.

[3] Petzelt, D.; Rulhoff, S.; Deuse, J. (2010) Digitale Produktentwicklung und Digitale Fabrik – Getrennte Planungswelten. In: ProduktDatenJournal, Nr. 2, S.12.

[4] ProSTEP iViP White Paper (2010). Modern Production Planning Processes. Darmstadt: ProSTEP iViP e.V.

[5] ProSTEP iViP PSI Reommendation (2011). Modern Production Planning Processes. Darmstadt: ProSTEP iViP e.V.

[6] REFA Verband für Arbeitsstudien und Betriebsorganisation e.V. (1990). Methodenlehre der Betriebsorganisation - Planung und Gestaltung komplexer Produktionssysteme. München: Carl Hanser.

# Accuracy Evaluation System for Shipbuilding Blocks Using Design Data and Point Cloud Data

Kazuo Hieketa[a], Hiroyuki Yamato[a], Masakazu Enomoto[b] and Shogo Kimura[a,1]

[a] Graduate School of Frontier Sciences, The University of Tokyo, Japan.
[b] Graduate School of Engineering, The University of Tokyo, Japan.

**Abstract.** An accuracy evaluation system for shipbuilding blocks using design data and point cloud data by laser scanners is proposed in this paper. In this system, point cloud data obtained by laser scanner and design data are registered by extracted feature points and planes of longitudinal members. The gap between positions of feature points and surfaces of longitudinal members and shell plates extracted from point cloud data and design data are then shown quantitatively. At a measurement in a shipyard, the error in positions of longitudinal members due to inaccurate welding is evaluated by this system. The finding is a kind of knowledge for fabrication process.

**Keywords.** Shipbuilding Blocks, Laser Scanner, CAD, point cloud

## 1 Introduction

In the ship production process, ship building blocks are installed into the ship in the building docks. The efficiency of the building process depends on the accuracy of the blocks.

Accuracy of blocks is evaluated based on three-dimensional coordination of longitudinal members. When a block is measured by traditional methods, like Three-dimensional Coordinate Measuring System, reflectors must be properly located at the exact reference positions for high accuracy. It takes a long time to set up reflectors if there are many points to be measured. Laser scanners are sensors which measure shapes of objects by emitting laser and receiving its reflection. Measured results are formatted in point cloud data and the information includes three-dimensional coordinates, reflection intensity and color information, and these represent the locations the laser reaches.

[1] Student, Graduate School of Frontier Sciences, the University of Tokyo, Building of Environmental Studies, Room #274, 5-1-5, Kashiwanoha, Kashiwa-city, Chiba 277-8563, Japan; Tel:+81 (4) 7136 4626; Fax: +81 (4) 7136 4626; Email: kimura@is.k.u-tokyo.ac.jp; http://www.nakl.t.u-tokyo.ac.jp/index-en.html



D. D. Frey et al. (eds.), *Improving Complex Systems Today*,
Advanced Concurrent Engineering, DOI: 10.1007/ 978-0-85729-799-0_44,
© Springer-Verlag London Limited 2011

This paper proposes an accuracy evaluation system for shipbuilding blocks using laser scanners. The system compares the measurement data and design data by CAD system.

## 2 Proposed Accuracy Evaluation System

### 2.1 Overview

Figure 1 illustrates the overview of the developed system. First, point cloud and design data of shipbuilding blocks are obtained. Second, feature points are extracted from both point cloud and design data by plane fitting. Third, point cloud and design data are registered on the axes of feature points and the direction of the planes extracted by plane fitting. Finally, the difference between the measured point cloud and design data is evaluated. By calculating the distance between feature points extracted from point cloud and design data, the gaps between points in the desired shape and the corresponding point in the actual shape of longitudinal members are discovered. The system can apply to most of shipbuilding blocks, but the paper focuses on the parallel part without curved surfaces.

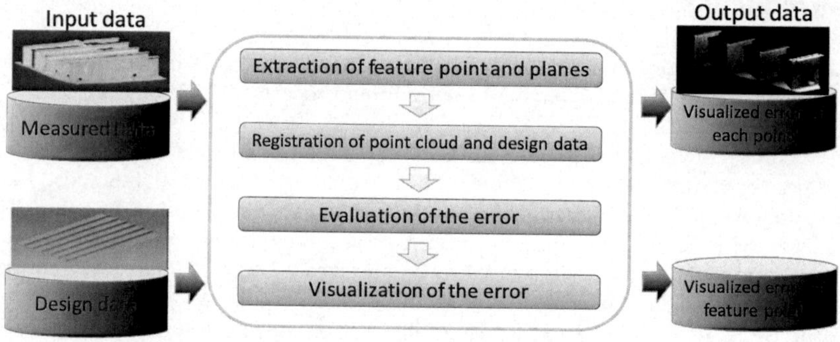


**Figure 1.** System Overview

### 2.2 Input Data

The point cloud used in this system has Cartesian coordinate system, RGB color system and text data format. The format of design data that this system uses is xgl format, which is used for rendering polygons drawn with OpenGL Architecture. Xgl format files include information about three-dimensional shapes of the objects. In xgl format files, surface of the object is expressed as an aggregate of triangle meshes. Xgl format uses xml 1.0 syntax.

### 2.3 Extraction of feature points

Fearture points are defined as intersections of planes extracted from longitudinal members. The positions of welded members are important at the install of blocks. Fearture points and planes of members are extracted as shown in Figure 2.

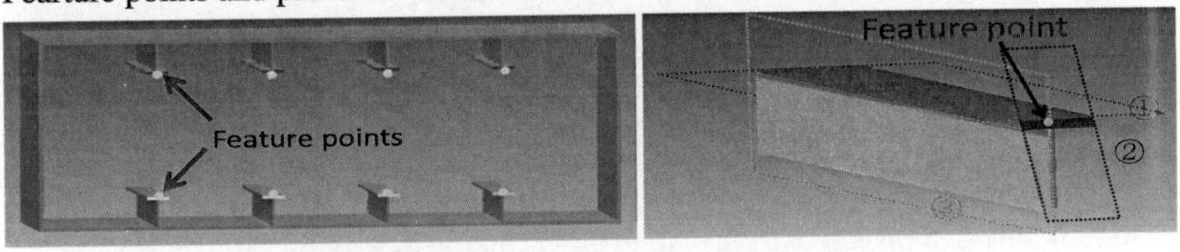


**Figure 2.** Extracted planes and intersections

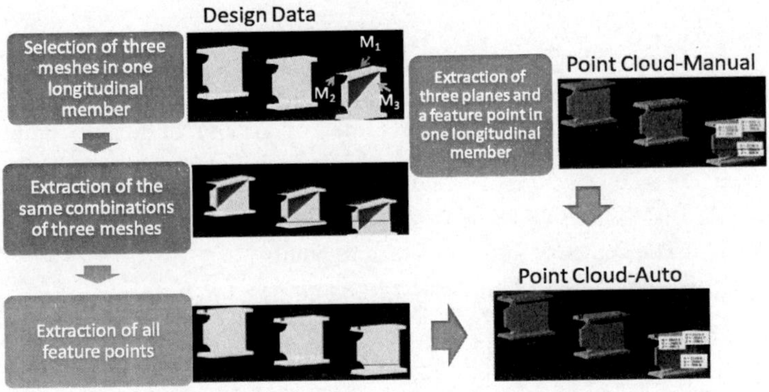


**Figure 3.** Overview of feature points extraction

Figure 3 illustrates the overview of feature point extraction. Feature points and planes on design data are obtained by three steps. First equation of three planes of one longitudinal member are obtained by picking three meshes existing on each plane. This combination of three mesh is called "Base-Meshes". Second "the same" combination of three meshes as Base-Meshes are searched. For every combination of three meshes on design data, a triangle composed of the center of gravity of each mesh are calculated. If the gravity triangle is congruent with that of Base-Meshes and all meshes are congruent with the meshes of Base-Meshes, the combination of three meshes is judged the same as Base-Meshes. Finally, equations and intersections of three planes of each the same combination is calculated. These intersections are feature points on design data.

Planes and intersections on point cloud data are extracted using feature points on the design data. First, three points existing on planes are picked up in one longitudinal member. This longitudinal member is called "Base-Longi." Equations of three planes and an intersection of planes are obtained by fitting planes in point cloud around picked points [1]. Through plane fitting, influence of ranging error of the laser scanner on the accuracy of intersections decreases. Second, positions of other intersections are presumed by applying positional relationships between intersections on the design data. Third, points around the intersection of Base-Longi are fitted to the points around the presumed intersection by using ICP algorithm [2] and the exact positions of other longitudinal members are recognized. Finally, three points picked on Base-Longi are transferred toward fitted longitudinal members and planes and intersections on other longitudinal members are calculated.

Feature points and planes can be extracted from point cloud manually with existing method for plane fitting shown in [1], by picking up points from all planes wanted to be extracted. But this method requires the user of the system to provide multiple selections of points when feature points are extracted from large shipbuilding blocks which have dozens of longitudinal members. The proposed method in this section can extract feature points automatically by using design data. This method requires selection of only 3 meshes and 3 points.

### 2.4 Registration

Feature points and planes are used as referrence points for registration. First, axes of feature points both on design data and point cloud data is fitted by linear fitting with principal component analysis. The center of gravity of feature points are calculated simultaneously.

$l_{pcd}$, $l_{cad}$ : Axes of point cloud data and design data

$g_{pcd}$, $g_{cad}$ : The center of gravity of feature points

Axes and the center of gravity of point cloud are registered in the following equation.

$$x_i' = \mathbf{R}(\mathbf{a})(x_i - g_{pcd}) + g_{cad} \tag{1}$$

$x_i$ : Point before transformation

$x_i'$ : Point after transformation

Rotation matrix $\mathbf{R}(\mathbf{a})$ is determined by quaternion $\mathbf{a}$ in the following definition and equation [3].

$$\mathbf{R}(\mathbf{a}) = \begin{pmatrix} a_0^2 + a_1^2 - a_2^2 - a_3^2 & 2(a_1a_2 - a_0a_3) & 2(a_1a_3 + a_0a_2) \\ 2(a_1a_2 + a_0a_3) & a_0^2 + a_2^2 - a_1^2 - a_3^2 & 2(a_2a_3 - a_0a_1) \\ 2(a_1a_3 - a_0a_2) & 2(a_2a_3 + a_0a_1) & a_0^2 + a_3^2 - a_1^2 - a_2^2 \end{pmatrix} \tag{2}$$

$$\mathbf{a} = \cos(\theta/2) + l_x \sin(\theta/2)\mathbf{i} + l_y \sin(\theta/2)\mathbf{j} + l_z \sin(\theta/2)\mathbf{k} \tag{3}$$

Rotation axis $l = (l_x, l_y, l_z)$ and rotation angle $\theta$ is determined in the following equation.

$$l = \frac{\mathbf{n}_{pcd} \times \mathbf{n}_{cad}}{|\mathbf{n}_{pcd}||\mathbf{n}_{cad}|} \tag{4}$$

$$\theta = \cos^{-1} \frac{\mathbf{n}_{pcd} \cdot \mathbf{n}_{cad}}{|\mathbf{n}_{pcd}||\mathbf{n}_{cad}|} \tag{5}$$

Finally, each point in the point cloud is superimposed on the design data by revision with the direction of extracted planes in section 2.3. Each plane extracted from point cloud and design data is registered so as to maximize following equation.

$$\max \sum_{i=1}^{n} \left\{ \frac{1}{w_i^2} (a_i \quad b_i \quad c_i) \cdot \mathbf{R}(\mathbf{a}') \begin{pmatrix} a_i' \\ b_i' \\ c_i' \end{pmatrix} \right\}^2 \tag{6}$$

$(a_i \quad b_i \quad c_i)$: Normal vector of plane extracted from design data

$(a_i' \quad b_i' \quad c_i')$: Normal vector of plane extracted from point cloud

Rotation matrix **R** is determined by quaternion **a'**. Rotation axis is $l_{cad}$.

Best rotation angle $\phi$ is calculated with Downhill Simplex Method [4]. Point cloud data is superimposed on the design data.

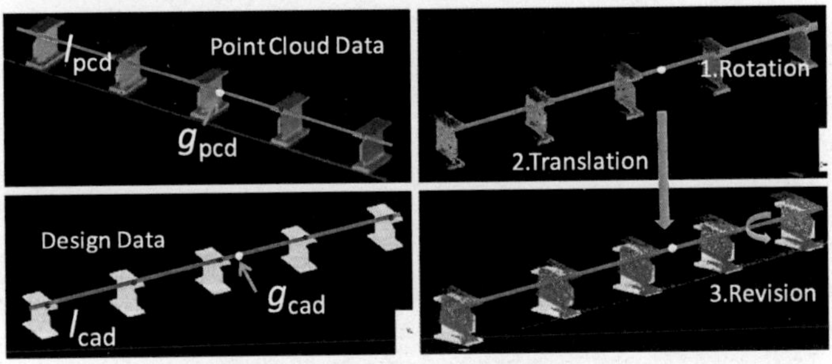


**Figure 4.** Overview of Registration

## 2.5 Evaluation and Visualization of the Error

### *2.5.1 Definition of the Error*

This system defines error in the following ways:

1) Signed distance from a feature point of point cloud to the corresponding feature point of design data decomposed into three directions (feature point error).

2) Absolute distance displacement from a point to the design data (each point error).

Feature point error is a metric of error in order to confirm quantitative accuracy of positions of longitudinal members. Each point error is a metric to confirm qualitative accuracy of positions and directions of longitudinal members.

### *2.5.2 Feature Point Error*

Accuracy of Longitudinal member influences welding cost. Welding cost of two longitudinal members depends on root gap and misalignment of members. So, distance between feature points of point cloud and design data is decomposed into directions of root gap, misalignment and welding seam. These three directions are determined by directions of normal vector of planes extracted from design data. Decomposed distance is displayed as text.

*2.5.3 Each Point Error*

Absolute distance from a point to the design data is defined as distance from a point to the nearest mesh [5]. The calculated errors are visualized in a point cloud with gradated color. Figure 5 illustrates point cloud with gradation and color bar. In this case, a point whose error is 0 mm is drawn as a black point. A point whose error is 8mm or more is drawn as a white point.

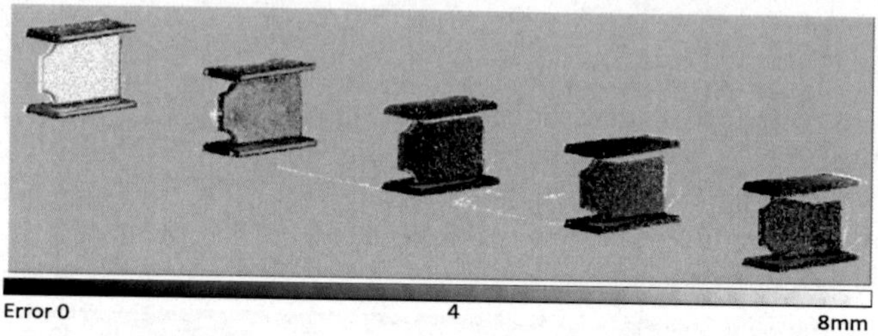


**Figure 5.** Point cloud with gradation color and color bar

## 3 Experiment

Two experiments were performed to validate this system. The objective of the experiment with measurement of test pieces is to verify registration and calculation of the error by this system. The expriment of sub panel measurement is performed to validate the whole of this proposed system in an actual shipyard with a real sample of shipbuilding blocks.

### 3.1 Test Pieces

Five test pieces on a measurement plate modeled in a part of a shipbuilding block were measured by laser scanner. Figure 6 illustrates dimensions and location of test pieces. Test pieces were located as shown in the left image of Figure 6, at the completely same position as design data. FARO Photon 120 was used in this experiment.

First, distance between feature points extracted from point cloud of test pieces are calculated ten times in two methods. One is the existing method mentioned in section 2.3, the other is automatically extraction method proposed in section 2.3 using point cloud and design data. The average error of gap between calculated value and measured value by a scale was 2.02mm by the proposed method. The result demonstrates the proposed method is sufficiently accurate although which has a little more error than the existing method, whose average error was 1.57mm.

Second, the error of feature points of point cloud and design data after registration was calculated. Distribution of the gap of feature points of each direction was plotted in the histogram shown in the right side of Figure 6. The average distribution of error is -0.03mm and standard deviation is 1.37mm. Two-sigma range is included from -2.77mm to +2.71mm. The accuracy of the error of feature points in each direction is less than 3mm.

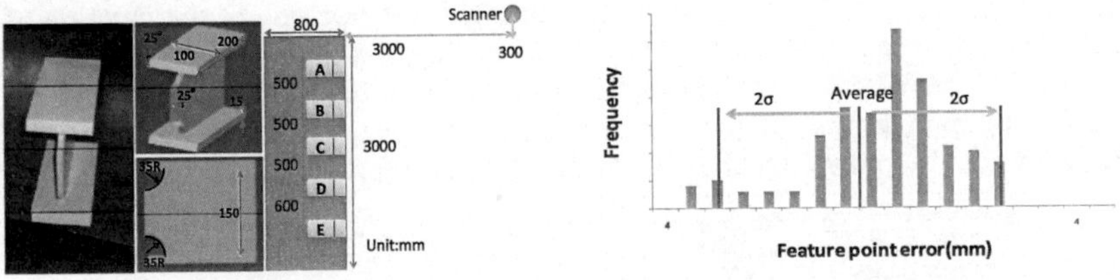


**Figure 6.** Dimensions and histogram of the gap of test pieces

### 3.2 Sub-Assembly Blocks

A sub panel of shipbuilding blocks was measured by laser scanner. In this case study, the point cloud of four longitudinal members, from the third to the sixth longitudinal members from the right in the top left image of Figure 7, was superimposed on the design data shown in the left below image of Figure 7.

Feature point error in Table 1 illustrates that Longitudinal member A (Longi A) has a large gap in a misaligned direction. Each point error shown in the right image of Figure 7 also illustrates the gap of Longi A, whose upper surface has a large gap compared with the design data although the gap of side surface is little.

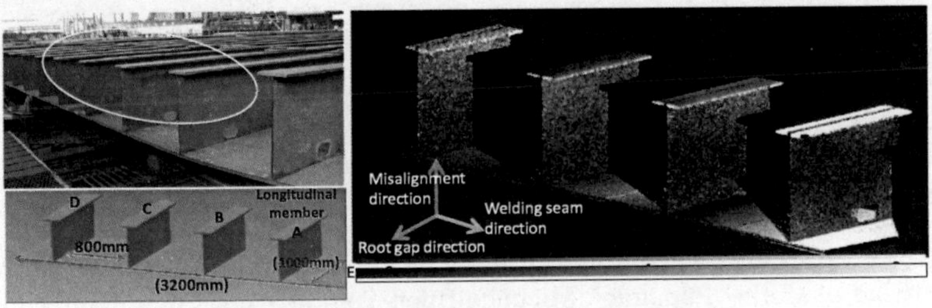


**Figure 7.** Overview of sub panel measurement

**Table 1.** Feature point error of the sub panel

| Longitudinal member | Root gap Direction | Welding seam direction | Misalignment Direction |
|---|---|---|---|
| A | -2.5mm | -2.1mm | -5.4mm |
| B | -0.1mm | -2.5mm | 2.4mm |
| C | 0.5mm | 2.6mm | 2.2mm |
| D | 2.1mm | 1.9mm | 0.8mm |

## 4 Discussion

Two shell plates of the sub panel were welded between longitudinal member B and C. The shell plate was divided into four parts as shown in the left image of Figure 8, and a plane was fitted to each part of the plate by moving least square method. As a result of comparison of normal vectors of the fitted planes, the angle between normal vectors of part 2 and 3, which are the right and left side of the junction of shell plate, was especially larger. The gap of misaligned direction in longitudinal member A occurred due to the inaccuracy in welding of the shell plate. The system identifies the problem of the process of joining steel plates based on the measured data and finds the knowledge to give suggestions for improving the process.

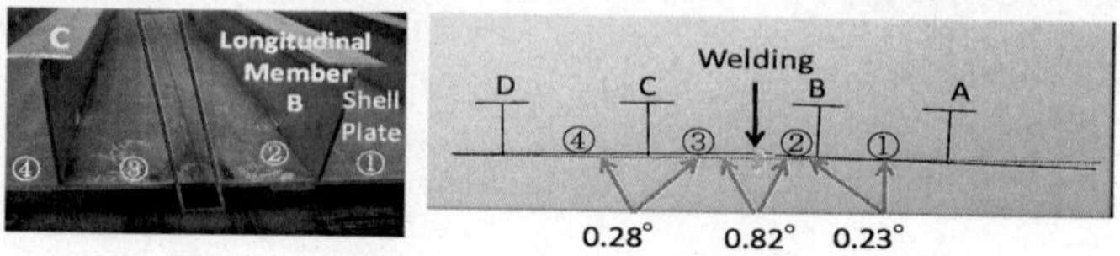


**Figure 8.** Overview of junction of shell plates

## 5 Conclusion

An accuracy evaluation system for shipbuilding blocks using design data and point cloud data was developed. This system showed the gap of positions of feature points of measured data and design data quantitatively, and visualized the gap of surfaces of longitudinal members and shell plates between design and measured data qualitatively. At the measurement of a sub panel in shipyard, the system identified the difference between design and measured data caused by welding process and the results gives suggestions for improving manufacturing process.

## 6 Acknowledgement

UNICUS Co., Ltd. gives us technical supports for employing point cloud data processing platform "Pupulpit". Michelle Buen Tumillba suggested improvements in language. The authors would like to thank all of them.

## 7 References

[1] Hiekata K, Yamato H, Enomoto M, Takahashi K. Development of Accuracy Measurement System for Shipbuilding Blocks. In: Conference Proceedings of Japan Society of Naval Architects and Ocean Engineers. 2010; 10: 375-376.

[2] Besl PJ, Mckay ND. A Method for Registration of 3-D Shapes. In: IEEE Transactions on Pattern Analysis and Machine Intelligence. 1992; 14: 239-256.

[3] Lohmann G. Volumetric Image Analysis. Born Digital, Inc, 2009; 213-215.

[4] Nelder JA, Mead R. A simplex method for function minimization. In: The Computer Journal. Oxford University Press, Oxford, 1965; 7: 308-313.

[5] Nonaka R, Kanai T, Makinouchi A, Suzuki T. Fitting Hexahedral Mesh to Scattered Point Data. The Report of Graphics and CAD, Information processing society of Japan, 2000; 98: 13-18(in Japanese).

# Design Support for "*Suriawase*": Japanese Way for Negotiation among Several Teams

Masato Inoue[a,1], Reona Mogi[b], Yoon-Eui Nahm[c], and Haruo Ishikawa[d]

[a]Assistant Professor, Department of Mcchanical Engineering and Intelligent Systems, The University of Electro-Communications (UEC Tokyo), Japan.
[b]Undergraduate Student, UEC Tokyo, Japan.
[c]Associate Professor, Hanbat National University, Korea.
[d]Professor, UEC Tokyo, Japan.

**Abstract.** One of the features of Japanese product development is product integrity development which can obtain a multi-objective satisfactory design solution by "*Suriawase*": Japanese way for negotiation among several teams of engineers including design teams, production divisions, and suppliers from the initial design phase to the detail design phase concurrently. This paper proposes a design support method for *Suriawase* which consists of two phases. Phase I can obtain feasible multi-objective design domain under the uncertain design environment while incorporating all designers' preference by applying Preference Set-based Design method. The designers can visualize and share the feasible design domain each other for *Suriawase* among several designers. Phase II can show the expanded design variables and possible design domain for function in case of the change of the design requirement domain from phase I by applying the quantitative dependency. We provide a suggestion of the direction of design modification for *Suriawase* among several teams. The designers can adapt the change of the design environment flexibly at the detail design phase.

**Keywords.** Negotiation, multi-objective design, set-based design method, quantitative dependency

## 1 Introduction

In general, it is said that a product development process of Japan is different from all the others. In Euro-American product development, the design phases are broadly divided definitely: the initial design phase and the detail design phase. Engineers' responsibilities are also divided definitely. In the initial design phase, the concept of a product can be allocated to the requirements of each part. In

[1] Assistant Professor, Department of Mechanical Engineering and Inelligent Systems, The University of Electro-Communications (UEC Tokyo), 1-5-1 Chofugaoka, Chofu-shi, Tokyo 182-8585, JAPAN; Tel: +81 (0) 42 443 5420; Fax: +81 (0) 42 484 3327; Email: inoue@mce.uec.ac.jp; http://www.ds.mce.uec.ac.jp/index-e.html

D. D. Frey et al. (eds.), *Improving Complex Systems Today*,
Advanced Concurrent Engineering, DOI: 10.1007/ 978-0-85729-799-0_45,
© Springer-Verlag London Limited 2011

addition, design environment does not too often change at the detail design phase. Therefore, designers can handle own parts under their own responsibilities at the detail design phase.

In contrast, the initial design phase and the detail design phase are not divided definitely in Japanese product development. One of the features of Japanese product development is a product integrity development which can obtain a multi-objective satisfactory design solution by "*Suriawase*" in Japanese: Japanese way for negotiation among several teams of engineers including design teams, production divisions, and parts manufacturers (suppliers) from the initial design phase to the detail design phase concurrently. In the initial design phase, uncertainties arise from many sources of variations [1, 2]. The initial design phase goes through the transition to the detail design phase with uncertain design concept for each part of a product. That is to say, the design requirements of each part cannot be defined definitely even at the detail design phase when the concept of the product is allocated to the requirement of each part. Therefore, the tentative design requirements of each part are defined by *Suriawase* among related teams, and then each team aims for the tentative design requirements at the detail design phase. This enables designers to go forward under the uncertainties at the detail design phase without a letup and shortens the product development time accordingly.

In addition, a design environment of Japan in general often changes even in the detail design phase for reflecting requirements of the customers and the marketplace in a product as much as possible before production process especially in an industry which has a heavy competition for development race such as automotive industry. For example, an unplanned design requirement sometimes must be considered as an additional requirement even though the design process proceeds to the detail design phase. The *Suriawase* development can delay the final decision by definition of the design requirements at the detail design phase. It is possible to adapt the change of the design environment flexibly at the detail design phase.

However, the *Suriawase* development needs to define the single-point value of the tentative design requirements of each part among related teams, consider the feasibility of the value as a whole, and repeat this procedure many times until the tentative design requirements involve feasible design domain. This repetitive operation depends on the contributions in personnel by communication and negotiation (*Suriawase*) among several teams because of the confliction among them, and they bear a heavy burden. They also have to redefine the tentative design requirements in case of the changes to the product specifications or the addition of the design requirements along the way.

Therefore, a design support method for *Suriawase* to obtain the flexible and robust design solution which can satisfy all requirements of several teams under the uncertain design environment and to provide a suggestion of the direction of design modification for *Suriawase* among several teams is required. This paper focuses on the uncertainties accompanied with the design process. We propose a design support method for *Suriawase* which can obtain multi-objective satisfactory design domain under the uncertain design environment and show the direction of design modification in case of the change of design environment.

## 2 Approach for Supporting *Suriawase* among Several Teams

### 2.1 Present Situation of Product Development Process Based on *Suriawase*

Japanese companies have had their engineers trained for multiple tasks and skills (called "*Tanoko*" in Japanese) based on long-term employment and long-term trading with suppliers because of the labor shortage after World War II. That is why Japanese engineers are good at works of many different teams and can find an optimum design solution faster than other countries through a trial and error processes among several teams. Therefore, the product development based on *Suriawase* takes root in Japanese culture.

The *Suriawase* development has an advantage in a product with integral architecture which individual parts of a structure are complexly intertwined with related functions as shown in the Fig. 1 such as automobiles, motorcycles, or light and nimble home electrical appliances [3]. It does not allow one-to-one correspondence between the structure (parts) and their functions, and does not enable, for example, the designer for part [$s_1$] to focus solely on function [$f_1$].

The present procedure of the product development based on *Suriawase* is as follows:

(1) The designers who take charge of the parts (e.g. [$s_1$] - [$s_3$] in Fig. 1) associated with a function (e.g. [$f_2$]) get together in a real large room.

(2) When they do not have the value of design requirement for the function which satisfy all requirements of the product functions (e.g. [$f_1$] and [$f_2$] for the part [$s_1$], [$f_1$] - [$f_3$] for the part [$s_2$], [$f_2$] - [$f_4$] for the part [$s_3$]), each designer negotiates with the other designers to understand the situation and background of the others each other to share the problem as shown in Fig. 2 (a).

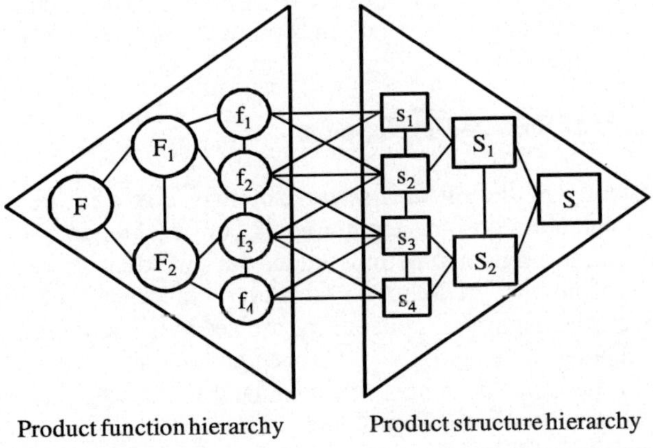


| F: Product function as a whole<br>$F_1$, $F_2$: Sub-functions of the product<br>$f_1$-$f_4$: Sub-sub-functions of the product | S: Product structure as a whole<br>$S_1$, $S_2$: Large modules<br>$s_1$-$s_4$: Small modules (parts) |
|---|---|

**Figure 1.** Product with integral architecture (modification of Fig. 1(1) of reference [3])

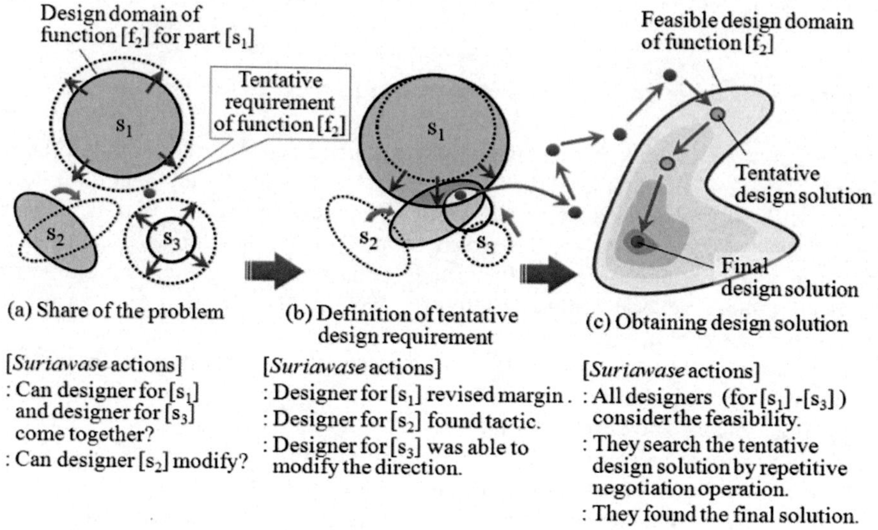


**Figure 2.** Present situation of product development process based on *Suriawase*

(3) They define the tentative design requirements of each part (e.g. [$f_2$]) which "are supposed to" satisfy all requirements of the product functions and aim for the tentative design requirements as shown in Fig. 2 (b).
(4) They consider the feasibility of the tentative design requirements and repeat the consideration and negotiation many times until the tentative design requirements involve feasible design domain as shown in Fig. 2 (c).
(5) After they found a tentative design solution which is located in the feasible design domain, they can find an optimum design solution (final design solution) as shown in Fig. 2 (c).

## 2.2 Problems of *Suriawase* Development

The first difficulty of the *Suriawase* development arises in the definition of the well-chosen tentative design requirements of each part which satisfy all requirements of the product functions under the uncertain design environment. If they do not define the well-chosen tentative requirement of the function, the numerous repetitive negotiation operations are needed. Therefore, the designers are required to capture own and the other design domains each other when they negotiate with the other designers to understand the situation and background of the others each other to share the problem.

The second difficulty of the *Suriawase* development arises in the variability in the design environment by the changes to the product specifications or the addition of the design requirements. Therefore, the designers are required to comprehend the impact on the other team (designer) coming from the own design modification and need to adapt the change of the design environment flexibly.

### 2.3 Approach for supporting *Suriawase*

In the present study, it is thought that we need to support before *Suriawase* among several teams. If the designers can capture the feasible design domain which satisfy all required functions under the uncertain design environment and can comprehend the impact on the other teams coming from the own design modification in case of the changes of the design environment, it enables thc designers to lead to more efficiency for *Suriawase*.

We propose a design support for *Suriawase* by following two phases.

(1) Phase I: Display of the feasible design domain

We propose a design support method to obtain the multi-objective satisfactory design domain. We display the feasible design domain to the designers by applying the set-based design method [4] which can obtain a set of design solution under the uncertain design environment.

(2) Phase II: Display of the impact of the design modification on the other teams

We propose a design support method to display the impact of the design modification on the other teams (the other parts and functions) to the designers by applying the quantitative dependency [5]. The designers can modify the own part of the product through a trial and error processes before *Suriawase* among several teams.

## 3 Phase I: Display of the Feasible Design Domain

### 3.1 From Point-Based Design to Set-Based Design

The traditional *Suriawase* development is the iterative process by point-based design. That is, it quickly develps a "single solution", evaluates the solution based on multi-objective criteria, and then iteratively moves to some other points until it reaches a satisfactory solution point. However, the precise value assignments do not include information about uncertainty, and a single-point solution provides limited information about the full range of possible designs under considerations.

At phase I, designers need to get a set of feasible design solutions instead of a single-point solution. Obtaining a set of design solution also provides design flexibility by allowing designs to be readily adapted to changing design environment. In addition, the designers' preference for the each design variables for each part and design requirement can be introduced for *Suriawase* among several teams. The previous series of our studies have proposed a preference set-based design (PSD) method that enables the flexible and robust design while incorporating designer's preference structure [6].

### 3.2 Preference Set-Based Design (PSD) [6]

Firstly, to capture the designer's preference structure on the continuous set, both an interval set and a preference function defined on this set, which is called the preference number (PN), are used. The PN is used to specify the design

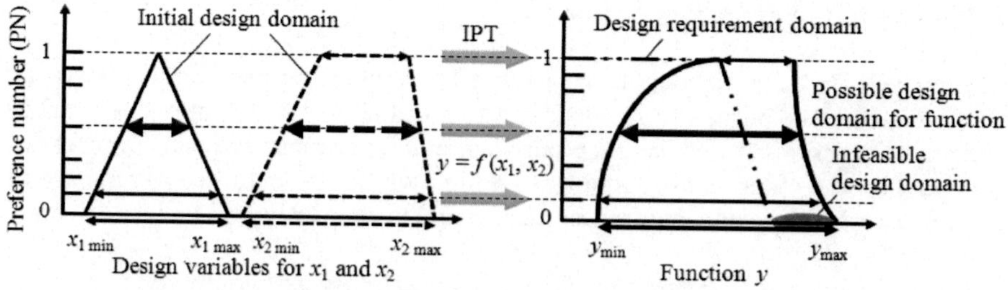


**Figure 3.** Procedure of set propagation

variables and design requirements for functions, where any shapes of PN are allowed to model the designer's preference structure, based on designer's knowledge, experience, or know-how. The interval set at the preference level of 0 is the allowable interval (not best but allowable), while the interval set at the preference level of 1 is the target interval that the designers would like to meet.

Secondly, the set propagation method that combines the decomposed fuzzy arithmetic with the extended interval arithmetic (*i.e.*, Interval Propagation Theorem, IPT [7]) was proposed to calculate the possible design domains for functions which are achievable by the given initial design domains for parts as shown in the Fig. 3. Then, if the overlapping domains between the design requirements domains and the possible design domains for functions exist, there are feasible design domains within the initial design domain. Otherwise, the initial design domain should be modified in set modification process. However, if the possible design domains for functions are not the sub-set of the design requirements domains, there also exist infeasible design domains in the initial design domain that produce functions outside the design requirement as shown in the Fig. 3.

Thirdly, the next step is to narrow the initial design domain to eliminate infeasible design domains, thus resulting in feasible design domains. The present method has been also used to define the possible design domain by capturing the designer's preference structure. In addition to the design robustness, we should take into account which one is preferred by the designer. The design preference and robustness are evaluated to eliminate infeasible design domains [6].

A set of design solution obtained by PSD method can be a feasible design domain under the uncertain design environment while incorporating all designers' preference. The designers can visualize and share the feasible design domain each other for *Suriawase* among several teams (designers).

## 4 Phase II: Display of the Impact of the Design Modification on the Other Teams

### 4.1 Modification of the Feasible Design Domain by Phase I

We have an assumption that a design environment changes by the changes to the product specifications or the addition of the design requirements after the phase I.

If the feasible design domain can adapt the changes of the design environment, the designers can select the final design solution from the design domain of phase I. Otherwise, the designers need to modify and expand the feasible design domain.

Therefore, we propose a design support method to display the impact of the design modification on the other teams to the designers and to provide a suggestion of the direction of design modification for *Suriawase* among several teams by applying the quantitative dependency [5].

### 4.2 Quantitative Dependency

To obtain the impact of the design modification on the functions, the quantitative dependency is applied. The quantitative dependency $\Psi_{x_i,y}$ between variable $x_i$ and $y$ can be estimated as shown ($y \propto x_i^n$):

$$\Psi_{x_i,y} = \left(\frac{y_{new} - y_0}{y_0}\right)\left(\frac{1}{r}\right) \tag{1}$$

where $y_0$ is the current value of the variable $y$ (function) and $y_{new}$ is the new value obtained by perturbing the value of variable $x_i$ (design variables) by $r$[%].

Moreover, we also consider the priorities of the design variables. The design variables which have a significant impact on the function should be changed significantly, and the high-priority design variables should be changed small. The values of change for $x_i$ and $y$ are obtained as shown:

$$\Delta x_i = \left(\frac{\psi_{x_i,y} / \omega_i}{\psi_{x_0,y} / \omega_0 + \hbar + \psi_{x_k,y} / \omega_k}\right)\Delta d \tag{2}$$

$$\Delta y = \sum_{i=1}^{k} f_{x_i,y}(\Delta x_i) \tag{3}$$

where $\Delta y$ is obtained by perturbing variable *xi* ($\Delta x_i$), ωi is the weight factor of $x_i$, and $\Delta d$ is basis variation. Equation (2) can be interpreted as: "the higher impact and the less consequential, the more the design variables are changed."

The Fig. 4 shows the expanding of the design variables and possible design domain for function in case of the change of the design requirement domain. The designers can modify the own part of the product through a trial and error processes before *Suriawase* among several teams. It is possible to adapt the change of the design environment flexibly at the detail design phase.

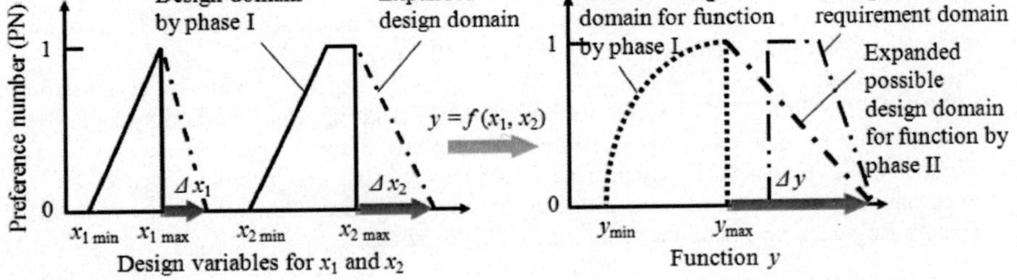


**Figure 4.** Expanding of the design variables and possible design domain for function

## 5 Conclusions

This paper proposes a design support method for *Suriawase*: Japanese way for negotiation among several teams. Proposed method consists of two phases. Phase I can obtain feasible multi-objective design domain under the uncertain design environment while incorporating all designers' preference by applying Preference Set-based Design method. The designers can visualize and share the feasible design domain each other for *Suriawase* among several designers. Phase II can show the expanded design variables and possible design domain for function in case of the change of the design requirement domain from phase I by applying the quantitative dependency. We provide a suggestion of the direction of design modification for *Suriawase* among several teams. The designers can adapt the change of the design environment flexibly at the detail design phase.

## Acknowledgements

The authors have investigated the difference of the product development between Japan and the others with the Structural Design and Fabrication Committee of JSAE (Society of Automotive Engineers of Japan) by hearing the design site. The authors gratefully acknowledge the members of the committee. This research was supported by the Japan Society for Promotion of Science, Grant-in-Aid for Young Scientists (B) No. 21760109.

## References

[1] Kogiso N, Ahn W, Nishiwaki S, Izui K, Yoshimura M. Robust topology optimization for compliant mechanisms considering uncertainty of applied loads, J. Advanced Mechanical Design, Systems, and Manufacturing 2008;2,1:96-107.

[2] Agarwal H, Renaud JE, Preston EL, Padmanabhan D. Uncertainty quantification using evidence theory in multidisciplinary design optimization, Reliability Engineering and System Safety 2004;85,1-3:281-294.

[3] Takeishi A, Fujimoto T. Modularisation in the Auto Industry: Interlinked multiple hierarchies of product, production and supplier systems, Int. J. Automotive Technology and Management 2001;1,4:379-396.

[4] Sobek II DK, Ward AC, Liker JK. Toyota's principles of set-based concurrent engineering, Sloan Manage Rev 1999;40,2:67-83.

[5] Kusiak A, Wang J. Dependency analysis in constraint negotiation, IEEE Transactions on Systems, Man, and Cybernetics 1995;25,9:1301-1313.

[6] Inoue M, Nahm YE, Okawa S, Ishikawa H. Design support system by combination of 3D-CAD and CAE with preference set-based design method, Concurrent Engineering: Research and Applications 2010;18,7:41-53.

[7] Finch W.W, Ward A.C. Quantified relations: a class of predicate logic design constraint among sets of manufacturing, Operating and Other Variations, ASME Design Engineering Technical Conference 1996; 18-22.

# Management Teamwork: Influence of Management vs. the Influence of Tools in Product Development Change

James Eoin Ryan[a,1], Michael Philipp Reik[b,1]

[a]Project Manager, Porsche Consulting GmbH, Bietigheim-Bissingen, Germany.
[b]Senior Consultant, Porsche Consulting GmbH, Bietigheim-Bissingen, Germany.

**Abstract.** In the Product Development Area, the influence of management on driving change has been explored considering the specific aspects and challenges in Product Development. However, the actual management activities in addition to commitment and patience, which lead to successful change are less well explored. In this paper the authors discuss the effects of the application of methods and tools on improving Product Development and the effects that management behaviour and activities has on effecting improvements. On the basis of case studies, they conclude that to drive truly successful change requires striking a balance between bottom-up and the right top-down activities and that management teamwork is an especially effective way of achieveing this balance.

**Keywords.** Lean Mangement, Lean Product Development Process, Product Creation Process, Change, Management Behaviour, Methods and Tools, Porsche

## 1 Introduction

In the past there has been much work done by researchers and professionals on the development of methods and tools for the improvement of Product Development [1,7,12] area. As with the area of Production and Logistics, the aim has been to provide specific approaches that support the generation of improvements.
In the Product Development Area, the influence of management on driving change has also been explored considering the specific aspects and challenges in Product Development [3,8]. However, the actual management activities in addition to commitment and patience, which lead to successful change, are less well explored.
In this paper the authors discuss the effects of the application of methods and tools on improving Product Development and the effects that management behaviour and activities has on effecting improvements. These are analysed and compared using the results of three case studies as references, in which Porsche Consulting implemented a large change project in Product Development at a client. Based on

[1] Porsche Consulting GmbH, Porschestr. 1, D74321 Bietigheim-Bissingen, Germany; Tel: +49 (0) 711 911 12100; Fax: +49 (0) 711 911 12203; Email: james.ryan@porsche.de; http://www.porscheconsulting.com

D. D. Frey et al. (eds.), *Improving Complex Systems Today*,
Advanced Concurrent Engineering, DOI: 10.1007/ 978-0-85729-799-0_46,
© Springer-Verlag London Limited 2011

these experiences the authors discuss the key learnings and how these can then be applied in practice.
In this paper the term 'Product Creation Process (PCP)' will be used. The PCP describes all the activities of an organization that are required to create new products up to the start of production [10]. There are many generally accepted definitions similar to the one above, which take this wide range into account [3,12].

## 2 Tools and Methods in the Product Creation Process

Much has been written about the use of various methods and tools in the PCP. Some of those mentioned are very specific to the PCP, i.e. PDS (Product Design Specification) [12] and some less so i.e. Kaizen Workshop [11]. While some of the methods are more recent in their use many have been in use for several years. A further point to note is that some authors have focused more on design methods while others have focused more on process design at system level.
A very comprehensive set of tools and methods is set out using the framework of the Design for Six Sigma frame work by Mollenhauer et. al. [7]. They also state that many of these approaches are well proven and have been in use for many years. Another in depth analysis of design approaches is provided by Huang et. al. [1]. Some of the Design Approaches, such as Design for the Environment, are specific to more recent demands on the PCP.
A methodical approach with many case studies in the use of methods and tools is well documented by Pugh [12]. Pugh also focuses on process design via the Total Design Process as a model. The question of viewing the PCP as a process and making the flow of information visible is discussed by Morgan & Liker [8]. PDVSM (Product Development Value Stream Mapping) is recommended as a tool to increase transparency regarding this flow.
Another interesting more recent development is the methods and tools being developed and implemented for software development [4] that are now finding other applications. The iterative approach to development combined with a variable design scope has found successful applications beyond software development.
At Porsche Consulting, to complement the theory of the Lean PCP [10,14], the methods and tools are organised in the framework of the PCP Systematic [10]. In the PCP Systematic, the methods and tools are grouped into three aspects; Process, Project Organisation and Reporting, which altogether provide a comprehensive approach to process design and implementation.

## 3 Management of the Product Creation Process

The topic of the attributes of a Lean PCP and what changes need to be made to achieve these attributes has been, in recent times, much written about and discussed. Earlier works focus more on the need for process design while later works tend to focus more on culture and leadership.

As mentioned in the Toyota Way [6] '*it is critical to keep in mind that product development has its own complex environment and fundamentally unique challenges'*. Similarly, Ryan & Reik [14] argue that the PCP is fundamentally different to other processes within the organization being one large dynamic system with very highly interrelated and interdependent elements producing very few finished products. They define a set of principles for steering change *(Value Orientation, Transparency, Synchronisation, Perfection)* in the PCP derived from company values based on this assumption.

Looking to earlier works, Imai [2] also makes references to indirect areas such as Product Development having great potential for productivity increases. He describes the need for a top-down approach focusing on Lean process design as well as bottom-up approach focusing on analytical methods. The humanistic aspect and the need for managers to understand the process is mentioned.

The process design aspect is focused on by Pugh. Pugh's [12] total design model, incorporating both static and fixed concepts, is a very good representation of the PCP as it should be understood. The full range of activities involved from market analysis to selling the product is shown. A thorough analysis of the management understanding required and organisational aspects is given by Pugh et. al. [13].

Leany & Marshall [5] draw a number of interesting conclusions, such as, that a strategic 'whole system' or holistic approach is required by management as opposed to the advocacy of any particular design or development methodology and that the bottom-up incremental implementations must fit within a strategic framework devised and supported by senior management.

That culture change is at the heart of changing Product Development is discussed by Morgan & Liker [8]. Many of the unique challenges involved in changing the PCP, such as the complexity, the imprecision of the process and the length of time needed are mentioned. Interestingly, they mention the need for leaders to be committed to and participate in change and that gaining this commitment is more challenging than in Manufacturing Areas. Furthermore, they mention the need for leaders to understand what is required for change, that it takes longer periods of time and that they themselves need to change the way management is done to promote the new culture.

Kennedy [3] focuses extensively on leadership aspects and on the need to change the system. He uses examples to show how through visionary, committed and confident leadership dramatic change can occur. Many of the aspects mentioned concur with those of Morgan & Liker [8] including, that management provide adequate resources for change and engage knowledgeable external help. He also states that the methods and tools are more a means for continuous improvement rather than radical change which occurs through system design and organisational change.

## 4 Case Study Examples

As was mentioned in Chapter 1, the conclusions in this paper are drawn from the work done with various clients. In order to put these into context, three representative case studies will now be discussed.

In each case a change project in the PCP is examined for a medium to large organization. In these cases, the processes and organizations are complex enough so that the effects of changes due to methods can be separated from that of the management approach at system level. The case studies are structured in three parts; part 1 relates to the application of methods & tools in the PCP, part 2 relates to the management approach to guiding change and part 3 deals with the overall result of the efforts undertaken by the organization in question.

### 4.1 Case Study 1. Large Semi-Conductor Manufacturer

The project being referred to relates to the implementation of Lean in the Product Development area as an extension to the implementation in the production and logistics area.
The company had introduced several methods and tools along with the introduction of Lean before the initiative began in the PCP. Some, such as Value Stream Analysis, were used along with the introduction of Lean in the Production and Logistics area. In the PCP area, several methods and tools specific to the PCP were also introduced. PCP Process Maps[2] were developed in Kaizen Workshops and used to plan and run development projects by the project management organization. Simultaneous Engineering teams (SE-Teams) were established along with a Chief Engineer organization, however team staffing and meeting attendance was heavily development biased. Kaizen workshops to improve specific aspects of the overall process were also conducted successfully using Business Process Mapping Techniques yielding improvements in the areas examined. The workshops were also used to promote the initiative and methodologies among employees. To support the change, a team of full time experts with specific focus on the PCP was established. However, the knowledge gained in the application of methods and tools came primarily from external sources and remained predominantly within the change team. Most of the transfer of knowledge to the employee base occurred through 'learning by doing' though there were not sufficient activities to make this sustainable and employee training was not sufficiently widespread. Some improvements did result from the implementation of methods and tools particularly through the early identification of problems during development projects.
While the change team management was active and engaged, only a small number of individuals from the development line management were active enough. Top management remained too distant from the change initiative and their activities were restricted a small number of short workshops and a number of initiative reviews. Individual training sessions were provided for senior managers to increase their understanding of Lean topics and the improvement potentials, however for many, their learning did not continue and understanding did not deepen enough. As a result, the combined effect meant that essential elements such as; goal setting, supervising the progress of change initiatives, business process reorganisation/system improvement, enabling employees through training etc. to

[2] The PCP Map is a type of process map specifiaclly tailored to adress the complex, system-type attributes of the PCP and the needs of those working in the PCP. It is used by Porsche Consulting and it's clients.

make the initiative truly successful not consequently pursued. Somehow, the sense of urgency due to their financial situation did not result in clear goals for change in the PCP that could be communicated by top management.
The initiative was, in part successful. Interestingly, there were positive examples where significant improvements had been made not only in the application of methods and tools but also through management practice in development areas in enabling change but, in the author's view, the initiative fell short of the true potential.

### 4.2 Case Study 2. Large Software Manufacturer

The project being referred to relates to the implementation of Lean in the area of Software Development. The objective was to implement new working methods in the development divisions and then implement a remodelled overall development process and organizational form that fitted with Lean principles and the new working methods.
The division in question placed a significant amount of effort into the implementation of methods and tools. A set of methods and tools was developed incorporating current best practices in software development. Kaizen workshops were used as a format in which to develop the methods and tools by experts and representatives of the employees impacted before piloting. Various external experts were consulted as to how best to implement the methods and tools and as to which other elements, i.e. process mapping, reporting transparency, training, needed to be addressed to ensure success. Extensive training and coaching support during roll out was made available to all teams. A good understanding and use of the new practices was quickly established, however significant variations between teams existed.
Upper management was very active in supporting the change spending a considerable amount of time trying to understand not just the effects of methods and tools but also how the PCP itself could be optimised. They conducted regular working sessions where, as a team, they discussed various topics and problems relating to the overall change. The optimisation of the end-to-end process was done considering the process as a system. The end-to-end process description was used as a guide for discussion, planning and division of responsibilities relating to change activities. These activities resulted in a good understanding of Lean topics and consistent communication from the management team although the team was reliant on external influences for focus, goal and aspiration setting for the overall change. The change activities were supervised through short but regular meetings where various operational aspects were discussed and decided by a wider management team. There was a good balance between top-down and bottom-up change and the management team approved the use of sufficient resources for bottom-up activities such as Kaizen workshops.
Overall, the initiative in the division in question was considered to be very successful both by employees and management and significant quantitative improvements were seen. The effects of quality improvements through better methods and tools and lead time reduction through an optimized end-to-end process combined to achieve this.

### 4.3 Case Study 3. Large Shipbuilder

The project being referred to relates to the implementation of Lean principles in the product creation process of large ships. The objective was to re-design the overall Product Creation Process to streamline the process and reduce labour and other costs. The project was undertaken in the company as part of a larger group-wide cost reduction programme.
A wide range of methods and tools some specific to the PCP, such as PCP-Maps and SE-Teams and some not specific to the PCP, such as Key Performance Indicators (KPIs), were introduced. Kaizen workshops were held with experts and impacted staff and experts to familiarise them with the tools and methods in question and to plan the concrete implementation steps necessary. Further to these activities, more complex methods such as a product based interface analysis were used as a basis for a Process-Orientated Reorganisation (PRO). Training was conducted intensively and considerable external resources were used for coaching both management and staff. On average, the resulting understanding of the methods and tools and their use was good and quantitative improvements were made despite a number of individuals resisting change.
The level of management activity was high though not all of the management team were equally engaged. They did participate in some management workshops where they received focused coaching and where various aspects of the initiative such as goal setting, the PRO and the resulting new roles were discussed. However, on average their understanding of Lean and the PCP remained methods and tools based rather that system orientated. There was significant external pressure placed on the management team by group management to make improvements and there were regular meetings held to supervise aspects to the change initiative. Management did ensure that the appropriate resources were made available for the implementation of methods and tools however, in sum the balance of activity was more bottom up than top-down and depended on key individuals.
Overall the initiative in question was very successful with a considerable increase in efficiency. The starting baseline was low in efficiency terms as the company had seen little change over the years and had very little exposure to Lean before the initiative began. From a qualitative perspective, on average the initiative was considered successful but there were individuals both among management and staff that did not consider the initiative successful.

### 4.4 Conclusions from Case Studies

When comparing all three cases it can be seen that all made improvements through the introduction of methods and tools, however there was a significant difference in the overall result between those where management was more active as in cases 4.2 and 4.3. They type of activities undertaken by management was also critical to the level of success. They provided the commitment and resources necessary for the staff to implement the methods and tools and make improvements over time. This is consistent with the behaviour prescribed by both Kennedy [3] and Morgan & Liker [8].

The differences between cases 4.2 and 4.3 are more subtle. In both cases they conducted management workshops and in both cases they re-organised according to the needs of their process while seeking advice from external experts.
In case 4.2, during regular workshops over time, they made a great effort to try and understand the Lean PCP theory and discuss their current state as compared to this as a group. Critically, they also considered their PCP as an end-to-end process and concerned themselves with how this could bc optimised at system level. The change activities were also planned and executed according to the vision of the optimised end-to-end process with team members assuming responsibility for sponsoring different change activities. They acted as a cross-functional team themselves creating one vision giving consistent leadership to staff. This type of management teamwork activity is, in the author's view, more likely to result in the type of visionary leadership described by Kennedy [3] and cultural change described by Morgan & Liker [8].

## 5. Putting the Leanings into Practice

Ryan and Reik [14] propose principles for change and recommend a parallel top-down and bottom up approach. As the case studies above show, the success of this approach is dependant on management conducting the right activities while encouraging employees to 'get on with it' and providing expert help. One of the principal difficulties encountered is to bring the different change activities together in one overall self-sustaining change initiative.
However, the expectations placed on management are often unrealistic. In a mature organization and culture, such as the one at Toyota described by Morgan & Liker [8] or at Porsche, managers have been trained in the use of tools and methods over many years and worked there way through various stations gaining experience and understanding. They have also been further trained for their management role and mentored by experienced senior managers who act as role models for the desired management behaviour.
In organisations wanting to change, this management experience is often missing. In the author's experience, staff or other stakeholders regularly ask for 'management commitment' whereby neither they, nor the managers being asked have an understanding of what is involved to make this commitment effective. In such cases, to ensure the success of the change initiative, training and coaching specific to the needs of management should be provided in the same way that the training needs of staff are addressed.

## 6 Conclusion

To drive successful change in the Product Development or PCP environment requires striking a balance between bottom-up and top-down activity. There are many well established methods and tools available for use by staff in bottom-up change. The need for commitment to change and patience by senior management has been much discussed, however it is often unclear as to what this entails, especially for the managers involved.

True success is achieved through the right top-down activities to compliment bottom-up efforts. This means more than supplying resources, getting external expertise and delegating tactical duties to competent staff. To achieve a solid balance, senior management must understand their situation, understand the theory, set a vision, optimise the end-to-end process, make a plan to implement change, be consistent in new behaviour and all while encourage and enable staff to 'get on with it' in terms of applying the methods and tools.

This obviously presents a great challenge however, the case studies show that the benefits that can be achieved through regular teamwork by a management team where the whole team commits to and promotes new behaviour are far greater that those achievable through the application of methods and tools alone.

## 7 References

[1] Huang GQ. Design for X - Concurrent Engineering Imperatives. Chapman & Hall, 1996
[2] Imai M. KAIZEN: The Key to Japan's Competitive Success, McGraw-Hill, 1986
[3] Kennedy MN. Product Development for the Lean Enterprise: Why Toyota's System Is Four Times More Productive and How You Can Implement It. Oaklea Press, 2003
[4] Larman C, Vodde B. Practices for Scaling Lean & Agile Development: Large, Multisite, and Offshore Product Development with Large-Scale Scrum, Addison-Wesley, 2010
[5] Leany PG, and Marshall R. An Introduction to Modern vehicle Design, Ed. Happian, Butterworth-Heinemann, 2001
[6] Liker JK. The Toyota Way: 14 Management Principles from the World's Greatest Manufacturer. McGraw-Hill 2002
[7] Mollenhauer JP et al. Design für Six Sigma+Lean Toolset: Innovationen erfolgreich umsetzen. Springer, Berlin. 2007
[8] Morgan JM, Liker JK. The Toyota product development system: integrating people, process, and technology. Productivity Press, 2006
[9] Porsche Consulting. JIT Basics, Lean Production, Sections 1-24. In: Internal Training Presentations
[10] Porsche Consulting. Training on Lean Product Creation Process, Training on PCP systematic. In: Internal presentations, 2009
[11] Porsche Consulting. Train the Trainer 1&2, In: Internal training presentations
[12] Pugh S. Total Design: Integrated Methods for Successful Product Engineering. Addison-Wesley, 1991
[13] Pugh S, Clausing D, Andrade R. Creating Innovative Products Using Total Design. Addison-Wesley, 1991
[14] Ryan JE, Reik MP. Applying the Core Elements of a Lean Enterprise to Product Development. In: Pokojski J (Ed.), Fukuda S (Ed.), Salwiski J (Ed.). New World Situation: New Directions in Concurrent Engineering: Proceedings of the 17th ISPE International Conference on Concurrent Engineering, Springer 2010

# Towards a cost modelling framework for outsourcing ERP systems

E. Shehab[a], M. Thomassin and M. Badawy

Decision Engineering Centre, Manufacturing Department, Cranfield University, Cranfield, Bedford, MK43 0AL, UK.

**Abstract.** ERP systems are costly, complex and require high-qualified people to manage them. In the current economic cliamt, when companies need to cut off costs, avoid expensive in-house skilled people and being focused on the core of their businesses, outsourcing ERP systems could be a solution to consider. The aim of this paper is to provide a framework that represents the first step towards the development of a cost modelling framework for outsourcing ERP systems which may help companies to select ERP providers and understand real outsourcing costs.

**Keywords.** ERP systems; outsourcing; implementation; agent based modelling

## 1 Introduction

The large expansion of the Internet during the last decade is one reason for the appearance of the Application Service Providers (ASPs). Using the Internet, companies can have access to achieving IT infrastructures which can support their ERP systems. The creation of this new business model was based on three elements: allowing companies to be focused on their core businesses, reducing costs of ERP adoption and avoiding spending money on in-house expertise [1].

ERP systems have become the heart of companies' information system since the last decade. According to Payne [2], ERP systems are "an approach to the provision of business support software that enables companies to combine the computer systems of different areas of the business – production, sales, marketing, finance, human resources, etc. – and run them off a single database". The concept of the single database is the core of the system. Thus, the departments of a company can share data and communicate in an easy way. Moreover, McAdam and Galloway [3] estimate that ERP systems permit "standardising business processes, ensuring integrity of data, and removing the number, complexity, and expense surrounding old independent legacy systems".

[a]Senior lecturer, Decision Engineering Centre, Manufacturing Department, Cranfield University, Cranfield, Bedford, MK43 0AL, UK. Tel: +44 (0) 1234 750 111 ext 5241. Fax: +44 (0) 1234 754 605; e.shehab@cranfield.ac.uk; www.cranfield.ac.uk/sas/aboutus/staff/shehabe.html

D. D. Frey et al. (eds.), *Improving Complex Systems Today*,
Advanced Concurrent Engineering, DOI: 10.1007/ 978-0-85729-799-0_47,
© Springer-Verlag London Limited 2011

Due to the complexity of outsourced ERP systems, many elements are at stake and make it difficult to evaluate its real cost. A framework to evaluate the real cost of outsourced ERP solutions will help companies to get a better understanding and will give them clues for decision-making in the case of adopting ERP systems.

Adoptaion of ERP outsourcing can avoid costs such as IT investments, staff recruiting and the need for in-house expertise making huge cost savings. However, outsourcing ERP systems also imply additional costs that do not exist in a more classical ERP adoption: knowledge transfer, communication systems, transition, etc. Therefore, a question arises. What is the real cost involved in the outsourcing of an ERP system? This research paper aims at understanding the main costs implicated during the outsourcing of the ERP system, especially the different activities implied during the implementation phase and at investigating how a cost model could be implemented using a modelling technique, such as agent based modelling or system dynamics.

## 2 Literature Review

Nowadays, the challenges of globalisation, with increasing competition, lead companies to find ways and means to structure their businesses so as to be flexible, reliable and meeting business demands. Typical ERP systems benefits include coordination, communication and organisational efficiency [4]. In 2008, companies have spent more than 20 billion Euros with ERP software firms [5].

However, installation of ERP systems can imply high risks, due to exceeding time and money investment. Hidden costs such as training for employees or maintenance make ERP systems installation risky. All these risks make ERP outsourcing through ASPs very interesting, as well as its lower costs and its increased flexibility.

Outsourcing is the best solution for companies which cannot afford paying for an in-house ERP adoption. The risks associated to the outsourcing are mainly about security issues; however companies avoid the risks associated to the installation and the maintenance of an in-house ERP system [6]. ERP Systems can be outsourced overseas (offshore). Even if outsourcing can save money, there is always the fact that ERP systems are too critical to be outsourced for many companies. The risks of downtime and loss of data are the biggest concerns for companies [6]. That is the reason why most of the companies which decide to outsource their ERP systems only outsource some modules which will not affect the company if a problem occurs. Outsourcing ERP systems is not only about cost, it is also a way for companies to be up-to-date concerning the latest technologies to be competitive. Thanks to outsourcing, they can have the latest software that other companies do not have yet in-house for the reason of extra cost or the need of expertise [7].

Outsourcing ERP system is complex and its implementation is quite close to a traditional ERP implementation. A contract is defined mixing customers and contractors inputs. Therefore, the final contract is composed on one hand of specification of services and of the delivery method. And on the other hand the customer defines the framework and all the legal phrases [8].

In order to understand ERP outsourcing, modelling is an essential step to get a good understanding of the AS-IS and TO-BE of the business processes. The modelling phase will avoid troubles such as lack of understanding of processes, or bugs and faults when the solution is implemented or tested. Besides, modelling allows optimising systems before their implementations.

Two kinds of models have been determined: the analytical model and the simulation model [9]. The analytical model depends on the number of parameters as inputs. The simulation model depends on a set of rules that will define how the model will evolve in the future according to its present state. The simulation model is commonly used when time dynamics is important. There are different approaches in simulation modelling, such as Agent Based Modelling (ABM) or System Dynamics Approach (SDA). The selection of a modelling tool is based on the level of abstraction and the way time is defined in the system that will be modelled.

The literature review has highlighted a lack of knowledge focused on outsourcing ERP systems based on modelling techniques. In order to fill this gap, a first analysis of reasons for outsourcing ERP systems will allow to get the key reasons for this new business model and then will allow to get the real cost drivers involved. Consequently, a better understanding of the cost drivers will also help to get a better understanding of the ERP activities during the implementation phase.

## 3 Analysis of ERP Outsourcing

Onshore, nearshore and offshore are three different types for outsourcing ERP systems. The classification is based on the service provider in relation to the receiving company. Companies have to benchmark the different opportunities for outsourcing their ERP system in the best way, which is extremely critical, to avoid troubles in the future that may influence their entire businesses.

According to Van Everdingen et al. [10], criteria for selecting providers who offer outsourced ERP system services are the product functionality, the implementation speed, the cost effectiveness, the interfacing with other products and the leadership in outsourced ERP system. Moreover, other criteria extracted from case studies [8] are that the scope is well defined, the duration of the engagement, the communication infrastructure, the cultural compatibility, the number of competent and skilled people available and finally the degree of implication of users needed.

ERP outsourcing has to be selected not for cost reduction only. The speed of implementation, skilled people, new businesses in other countries are some of the benefits of outsourcing that have to be seriously taken into account when companies contact ERP service providers. Then, offshore IT infrastructures can mean firing people within the company, which can be felt as a betrayal for employees. Offshore ERP systems can also create jobs (support for instance) in customers' countries. Besides, cost savings made by companies who have invested in other countries can be used to create new businesses, which means that new jobs will be created.

Culture is also a key challenge for companies who outsourced their ERP systems. In fact, customers have to match their policies and standards to meet providers' culture. If so, a comprehensive service level agreement is one tool which will avoid troubles and make the outsourcing in good conditions. Communication and implication are two of the several keys elements to successfully outsource. Here is one example extracted from a case study. A company has selected an ERP provider and decided to offshore its ERP system. Once the offshore model is in place, there is a huge loss of productivity from the offshore team (especially due to culture differences). The customer has to build a skilled internal team to communicate with the offshore team. All the ROI is lost in the loss of productivity and the creation of a new team internally.

Finally, companies are reluctant to use outsourcing techniques because of the risks. Data is too sensitive to be somewhere else, in another country. Companies are frightened by the fact that their data 'could' be in the middle of nowhere if the provider is not serious enough or if its security is not good enough.

### 3.1 Overview of Outsourcing ERP Cost Drivers

Identifying the cost drivers is important to get an overview of the main activities of outsourced ERP systems. Based on the literature review and real case studies [8] of companies who have outsourced their ERP systems, Figure 1 represents the nine cost drivers.

**Figure 1.** Cost drivers of outsourced ERP systems

### 3.2 ERP implementation phases

In order to get the real cost for an outsourced ERP adoption, it is important to know the main activities implied in the implementation phase.

An outsourced implementation phase is mainly different from a classical ERP implementation. The activities can be divided into two parts: activities which are executed onsite (it could be on the customer site directly or on a nearshore site) and activities which are executed offsite (offshore site).

The onsite team is onshore or nearshore in order to be close to the customer. Even if activities can be subdivided between the onsite and offsite teams, these two teams have to work hand in hand in order to lead the ERP adoption to success.

The onsite team aims basically at defining the customer's need (scope of the project, targets, etc.), the development of the specifications (functional, technical, processes); the testing of the system and finally the support to the users. Once the specifications done all along the ERP implementation process, they are sent to the

offsite team in order to transform the specifications into technical developments to build up the customized ERP system. The onsite team can prepare training within the company with the end users and the offsite team can prepare specific training that can be done online. The support is done by the onsite team whereas all the problems discovered by the users on the system are solved by the offsite team.
To summarize, the onsite team aims at designing the system with the customer whereas the offsite team aims at developing the specifications coming from the onsite team into a system. Figure 2 illustrates the activities involved in each team through the ERP implementation phases.

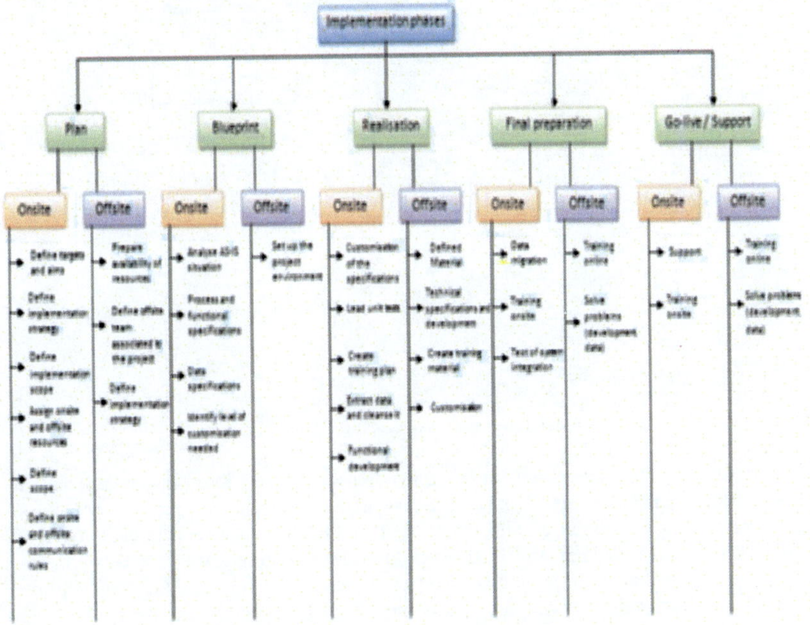


**Figure 2.** ERP Implementation Activities

## 4 Agent Based Modelling Analysis

An agent based modelling analysis has been employed to model the cost drivers and the activities of the ERP implementation. Agent based modelling technique requires [11] to: (i) Identify agents and their behaviour; (ii) Identify agents relationships; (iii) the location of the provider; (iv) Associate data to agents; (v) Validate the agent behaviour models and (vi) Run the model and analyse the outputs. It is common to use UML for representing models as object-oriented. Modelling agents is the first step to analyse a real world situation such as outsourcing ERP systems. It allows understanding the relationships between the agents and their parameters.

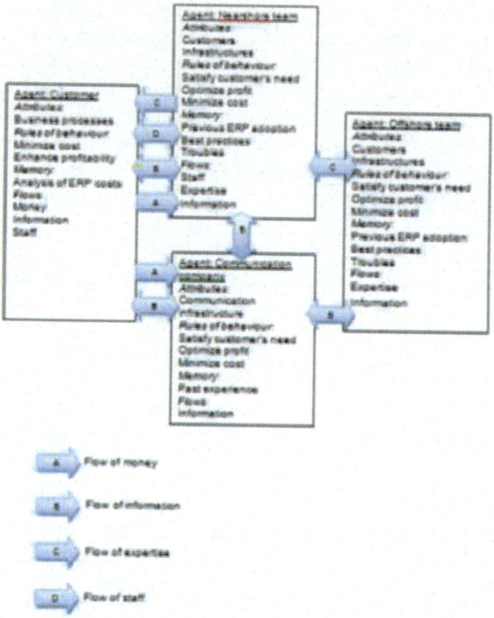


**Figure 3.** Agents in outsourcing ERP systems

Figure 3 illustrates the interactions between the four major agents in outsourcing ERP systems. It shows the case of a customer who wants to outsource offshore using a nearshore team which will be the link between the customer and the offshore team. Then the customer does not communicate directly with the offshore team, but with the nearshore team which relays the needs of the customer to the offshore team. At this level, four types of elements are exchanged between agents: money, information, expertise and staff. Money will be used to get the cost of outsourcing. Information is composed of the communication (such as needs from the customer), expertise is knowledge coming from the provider (nearshore and offshore teams) which will allow to develop the ERP system (implementation, etc.). Finally, the staffs are considered as a flow because of the training period which involves people to move (users to the nearshore team or vice versa).

Attributes are the main elements that agents own. For instance, customers have business processes, the provider has infrastructures and the communication company has communication infrastructure. Rules of behaviour are rules that will conduct agents' behaviours. For instance, all the agents have a rule which recommends minimizing costs. Finally, agents have to rely on their memory if they are facing a situation that had already happened in the past.

Figure 4 shows the environment in which agents will evolve and will affect their actions and interactions. The environment conditions the manner agents will interact. Agents in the same environment will be able to communicate easily. However, two agents in two different bubbles will have to use the communication tool (which will be the communication company agent) in order to be able to communicate. Nevertheless, if an agent from the customer wants to communicate and interact with an agent of the offshore team environment, this agent will not be able to communicate directly with it. The agent will have to go through the nearshore team environment and interact with a nearshore team agent that will relieve the request to the offshore team. This environment model is the closest to the reality (customers negotiate with the nearshore team and only the nearshore team has contacts with the offshore team).

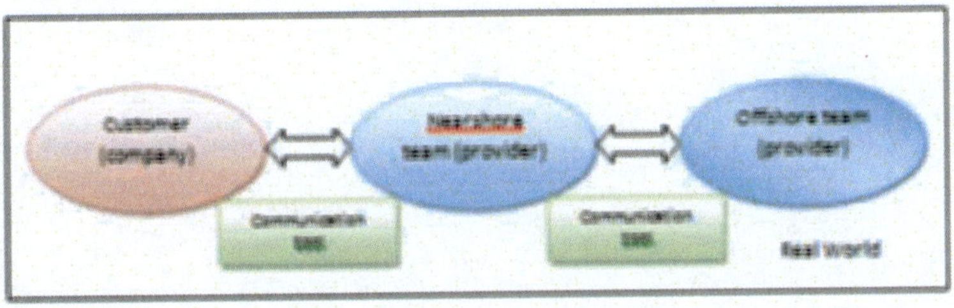


**Figure 4.** Environment for Agents

Based on the ERP implementation activities diagram (Figure 2), the UML class diagram illustrated in Figure 5 represents a general overview of agents, methods and attributes.

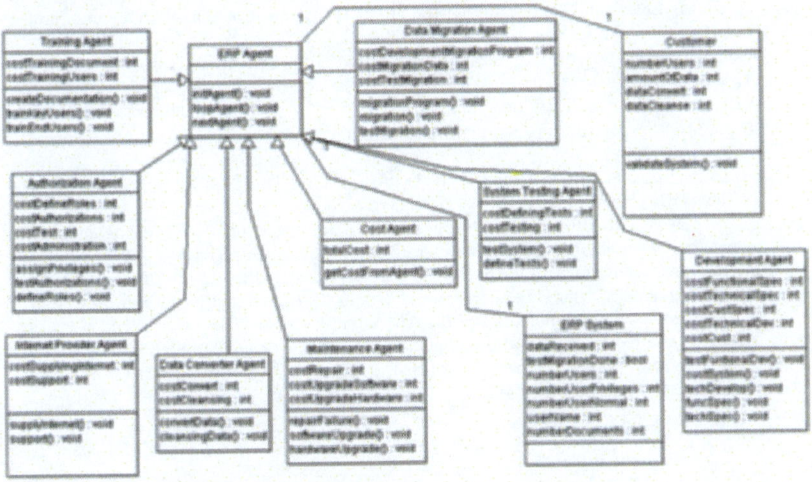


**Figure 5.** Implementation UML Class Diagram

## 5 Conclusions

The paper presents a framework which will assist in developing cost modelling of outsourcing ERP. ERP Cost Drivers and ERP implementation activities have been developed. An Agent Based Modelling analysis and UML have been created. Outsourcing ERP systems will be given a fully comprehensive literature review to surround the problem. Moreover, the costs involved are explained, as well as the activities for the implementation. In the case people would like to know the real cost of outsourcing, the model is presented and ready to be implemented and customized according to the need. Finally, outsourcing is a real alternative to classical in-house ERP adoption. The selection of the provider is crucial and can lead to a successful ERP implementation or to a total reverse with huge money losses.

## 8 REFERENCES

[1] Ekanayaka Y, Currie WL, and Seltsikas P. Delivering enterprise resource planning systems through application service provider. Logistics Information Management 2002; 15(3); 193-203.
[2] Payne W. The time for ERP?. Work Study 2002; 51 (2); 91-93.
[3] McAdam R. and Galloway A. Enterprise resource planning and organisational innovation: a management perspective. Industrial Management & Data Systems. 2005; 105 (3); 280-290.
[4] Stevens C. Enterprise resource planning: a trio of resources. Information Systems Management. 2003; 20 (4); 61-67.
[5] Shehab EM, Sharp MW, Supramaniam L and Spedding TA. (2004) "Enterprise resource planning: An Integrative Review. Business Process Management Journal. 2004; 10 (4); 359-386.
[6] Newell M. Leveraging opportunities for an ERP implementation to driver enterprise transformation. An Everest White Group Whitepaper. 2008.
[7] Olson DL. Evaluation of ERP outsourcing. Computers & Operations Research. 2007; 34 (12); 3715-3724.
[8] ERPwire (2009), 'Is ERP outsourcing necessary for all organizations?', available: http://www.erpwire.com/erp-articles/erp-outsourcing-necessity.htm (Accessed on August 1,2009).
[9] Hendel A, Messner W and Thun F. (2008), 'Rightshore! Successfully Industrialize SAP Projects Offshore', Springer.
[10] Borshchev A and Fillipov A. From system dynamics and discrete event to practical agent based modeling: Reasons, techniques, tools. Proceedings of the 22nd International Conference of the System Dynamics Society, July 25–29, 2004.
[11] Van Everdingen Y, van Hillegersberg and Waarts E. ERP adoption by European mid-size companies. Communications of the ACM. 2000; 43 (4); 27-31.
[12] Macal C. and North M. Tutorial on agent-based modelling and simulation part 2: how to model with agents. Proceedings of the 2006 Winter Simulation Conference, Karlsruhe, Germany; 2006; 73-82.

# Activity-based costing model for MGPD

**Xianming Cai**[†], **Satish Kumar Tyagi**[‡], **Kai Yang**[‡1]

[†]Siemens Energy Corporation, Orlando, FL-32817, USA

[‡]Department of Industrial and Systems Engineering, Wayne State University, MI-48202, USA

**Abstract.** Many product development processes are multi-generational in nature and may require redesign of the product at each generation. This is due to the fact that an optimized product development strategy for a single generation may not be the best option for multi-generation scenarios. To provide measurements for deployment of multi-generational product development (MGPD), an integrated project cost model has been introduced in this paper. This model posits that integrated product development cost consists of three factors, namely: product development cost, service cost and associated risks costs. Further, the risks cost includes three main factors: technical risk, market risk and operation risk. This specialized product development framework addresses the basic needs of MGPD, such as design for future reuse and design for modularity, and also serves as a decision tool for project management. For the purpose of validating the proposed model, an empirical case study is illustrated by considering three candidate concepts. Therefore, the proposed strategic cost model measures the on-going MGPD status, as well as provides a cost tools to support the management decision-making.

**Keywords.** MGPD,concept design, associated risk cost, service cost

## 1 Introduction

According to recent market trends, product development (PD) is not confined to implement innovative technologies and realize the benefits from single generation of a product (Kim et al., 1999). The success of such multi-generational product development (MGPD) is explicitly related with the project cost accrued from different phases of produce life-cycle such as development, marketing, service, and associated risks until the product is disposed. Further, the associated risk parameter has most importantly three types of cost which includes marketing risk transformed cost, technical risk transformed cost, and operation risk transformed cost. The researchers have also realized that almost 70% of the product cost is committed at the early design stage thereby preliminary design decisions affect cost the most (Boothroyd *et al.,* 1994). Thus, one of the motivations of this research is to provide

---

[1] Professor, Department of Industrial and Systems Engineering, Wayne State University, MI-48202, USA. Tel: 313-577-3858. Email:ac4505@wayne.edu

D. D. Frey et al. (eds.), *Improving Complex Systems Today,*
Advanced Concurrent Engineering, DOI: 10.1007/ 978-0-85729-799-0_48,
© Springer-Verlag London Limited 2011

engineers a tool that is easy to use and can exploit cost-saving potential from the multi-generational perspective at the first stage (see Fig. 1) of MGPD.

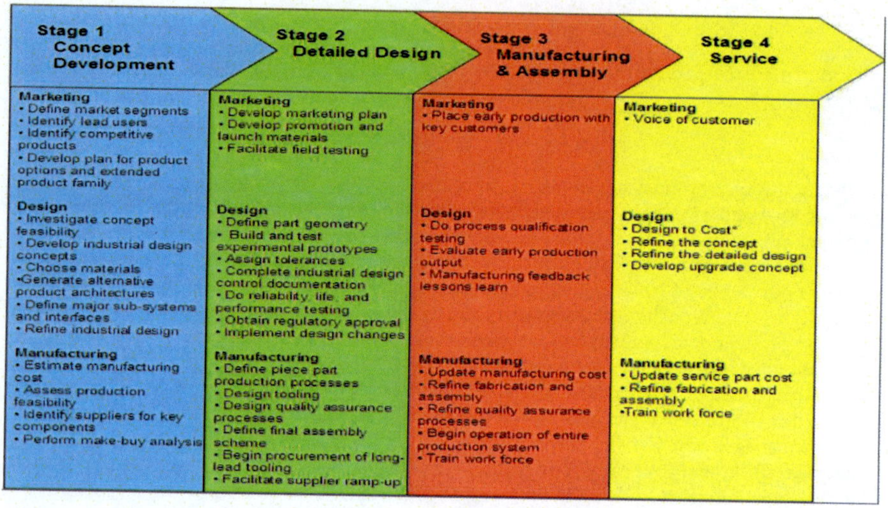


**Figure 1.** Extended Product Development Process

The current paper presents a novel framework to introduce a revised cost model for MGPD environment. This framework benefits to the PD teams in making decisions compatible with future generations at a cheaper cost. Moreover, this framework will assist design teams to forecast a project that is easy to manage and can be completed in a relatively small window of time. Therefore, this paper aims to provide a new direction and approach to maximize the business profit, the ultimate goal of an organization, gained from MGPD.

The study in this paper focuses on a product with general characteristics such as a complex module structure, a long development cycle time, a long lead time in production, and a high cost in parts. These processes are characterized as highly complex to organize and manage. This research work is primarily considering the manufacturing-based NPD process, which generally involves the activities that are used to (1) Determine that a new product is required to serve some need (2) Conceive of a concept for the product based on the wants and needs of customers (3) Develop all the technical specifications for the product (4) Devise a production process (5)Validate both the design and the production (Yang 2007)

## 2 Multi-generational product development (MGPD)

Comparing with Ulrich and Eppinger's (2003) generic PD process, inclusion of an extended concept development stage is proposed. This stage includes certain

activities at "System-Level Design" for a single generation of PD as shown in Fig. 1. The scope of the extended PD process covers the initial concept generation to the product service, including disposal of product at the end of the life cycle.
The basic processes of developing a MGPD are similar to the development of a single generation product. The objective of MGPD is to optimize the overall resource planning and to maximize the profits over multiple product generations. The starting point of the development of a next generation product is the updating or continuous design from the first generation design. Therefore, more effort should be involved in the development of a first generation product. However, in the real world, the development of a next generation product sometimes starts while the previous generation is still in under development. Moreover, connections among the four stages of MGPD are parallel instead of sequential. Fig. 3 shows a sample of the timeline concurrence between several MGPD processes in 3D.

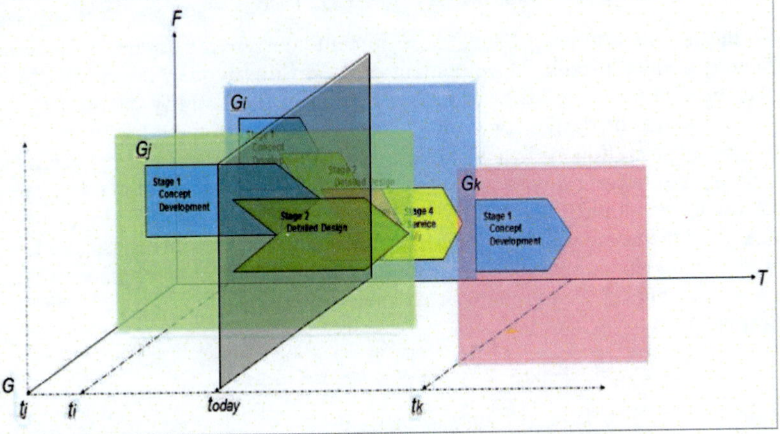


**Figure 2.** 3D Model of Timeline Concurrence among MGPDs.

## 3 Activity-based Costing for Integrated Product Development

Various cost estimation practices (Gutowski 1994, Esawi *et al.,* 2003) have evolved over a period of time, all having the fundamental aim of assigning cost in a more accurate and effective manner which largely depends on the legitimacy and appropriateness of data. No "one right way" of performing cost estimation task in the best way has been found in the exhaustive literature search. An advanced management estimation technique developed in the recent past called '*Activity Based Costing* (ABC)' (Innes et al., 2000) is one way to eliminate the differences. ABC inherently generates a lot of cost relevant information that can be used for early cost estimation. In this practic for this research, the project cost ($C^{PT}$) includes product development costs ($C^{PD}$), product service costs ($C^{service}$), marketing risk transformed costs ($C^{MR}$), technical risk transformed costs ($C^{TR}$), and operation risk transformed costs ($C^{SR}$) as shown in Eq. (1).

$$C^{PT} = C^{PD} + C^{Service} + C^{MR} + C^{TR} + C^{SR} \tag{1}$$

3.1 Product development cost

$C^{PT}$ is considered to be the combination of material costs ($C^{Matl}$) from product manufacturing, labor costs ($C^{Hr}$) from product design and production, and processing costs ($C^{Mfg}$) from manufacturing and assembly. Cost of materials is considered as a deterministic factor and has been studied during the MGPD stage 1. The processing cost from manufacturing and assembly is also considered as a deterministic factor of each concept, which is the result of the preliminary manufacturing capacity study during MGPD stage 1. Labor cost is the summation of all the manpower required in the MGPD stages 1 to 3. The manpower related cost spent is considered as a part of service cost during the MGPD stage 4.

$$C^{PD} = C^{Matl} + \left(C^{Hr} + C^{Mfg}\right) \times R \tag{2}$$

During multi-generational product development, design for future reuse could significantly reduce product development costs. In this model, design for future reuse has been represented as the Reusable Index (R). To simplify the problem, we consider the scope of design for future reuse only between the current generation and the next generation. When a module or part meets requirements from two generations, then a certain portion of the labor and processing-related costs can be considered to be amortized into the next generation product development. To calculate this, the number of total recognized design requirements is assumed as n, and m represents the number of design requirements that target both generations. k is the number of design requirements that have been fulfilled during each candidate concepts. Thus,

$$R = \left(1 - \frac{k}{n}\right) + \frac{k}{2n} = 1 - \frac{k}{2n} \qquad 0 \le k \le m \le n \tag{3}$$

$$C_i^{PD} = C_i^{Matl} + \left(C_i^{Hr} + C_i^{Mfg}\right) \times R_i.$$

3.2 Service cost

All the costs and expenses occurring during stage 4 have been considered as service cost. To generalize and simplify the study, we consider service cost to be the combination of material/ hardware related to cost of the adopted design concept. Let $T^{Warranty}$ represents the duration of the warranty contract, which is pre-determined based on the company policy, and $T^{Life}$ represents the average lifetime of the product, which serve as key design parameters and can be estimated at the end of MGPD stage 1. Service cost for each candidate concept can be estimated by

$$C^{Service} = C^{Matl} \times \left(T^{Warranty} / T^{Life}\right) \tag{4}$$

3.3 Associated risk cost

3.3.1 Technical risk transformed cost

Technical-related risk occurs when the product has been introduced to the market too early, and the techniques adopted in the concept have not been studied and tested adequately. Logically, we consider this type of risk as increasing when the concept development time is reduced. During this study, the technical-related risk is represented as a certain portion of the PD cost. Thus,

$C^{TR} = C^{PD} \times \mu(t_\alpha)$ ,for each candidate concept $C_i^{TR} = C_i^{PD} \times \mu(i,t_\alpha)$ (5)

$\mu(t_\alpha)$ is defined as the technical risk transforming index. Let tα represent the duration of MGPD stage 1, which is the length of concept development, and tβ represents the duration of MGPD stages 1, 2, and 3, which is the length of traditional product development. As mentioned, TWarranty represents the duration of the warranty contract, which is just the duration of MGPD stage 4. Thus we have twt= tβ +TWarranty covering the duration of MGPD. Therefore, the technical risk transforming index is defined as:

$$\mu(i,t_\alpha) = \begin{cases} 0 & t_\alpha = t^{wt} \\ \mu_i^s & t_\alpha = t^s \\ +\infty & t_\alpha = 0 \end{cases} \tag{6}$$

$t^s, \mu_i^s$ are pre-determined values. It could be an input from an expert system or historical data, which is also an "interface" of the integrated cost model. Based on the nature of $\mu(t_\alpha)$ , and the study of logarithm functions, the technical risk transforming index is defined as:

$$\mu(i,t_\alpha) = -\frac{\ln t_\alpha}{\ln \sqrt[\mu_i^s]{t^{wt}/t^s}} + \frac{\ln t^{wt}}{\ln \sqrt[\mu_i^s]{t^{wt}/t^s}} \text{ and its derivative is } \frac{d\mu(i,t_\alpha)}{dt_\alpha} = -\frac{1}{\ln \sqrt[\mu_i^s]{t^{wt}/t^s}} \times \frac{1}{t_\alpha} \tag{7}$$

3.3.2 Market risk coefficient

Market-related risk occurs when the product has been introduced to the market too late, and the potential market share has been taken by a competitors' product. Logically, we consider this type of risk as increasing when the design time and manufacturing time increase. During this study, market-related risk is also represented as a certain portion of the PD cost. Thus,

$C^{MR} = C^{PD} \times \lambda(t_\beta)$, for each candidate concept $C_i^{MR} = C_i^{PD} \times \lambda(i,t_\beta)$ (8)

$\lambda(t_\beta)$ is defined as the technical risk transforming index. Let tb represents the best case for marketing, which is the earliest time of product market introduction, and tw represents the worst case for marketing, which is the longest time of a similar product being established in the market. And ta represents the average time that the studied company needs to introduce its product. Obviously, $t^b \le t^a \le t^w$, and technical risk transforming index is defined as:

$$\lambda(i,t_\beta) = \begin{cases} 0 & t_\beta = t^b \\ \lambda_i^a & t_\beta = t^a \\ +\infty & t_\beta = t^w \end{cases} \tag{9}$$

$t^b, t^a, t^w, \lambda_t^a$ are pre-determined values. They could be inputs from an expert system or historical data, which is also an "interface" of the integrated cost model. Based on the nature of $\lambda(t_\beta)$, and the study of logarithm functions, the technical risk transforming index is defined as:

$$\lambda(i,t_\beta) = -\frac{\ln(t^w - t_\beta)}{\ln \sqrt[\lambda_i^a]{\frac{t^w - t^b}{t^w - t^a}}} + \frac{\ln(t^w - t^b)}{\ln \sqrt[\lambda_i^a]{\frac{t^w - t^b}{t^w - t^a}}} \text{ and its derivative is } \frac{d\lambda(i,t_\beta)}{dt_\beta} = \frac{1}{\ln \sqrt[\lambda_i^a]{\frac{t^w - t^b}{t^w - t^a}}} \times \frac{1}{(t^w - t_\beta)} \tag{10}$$

3.3.3 Operation risk coefficient

Operation-related risk refers to the risk and uncertainty during project management, especially the schedule risk. Logically, under the assumption that the resources of the project is limited, we consider this type of risk as increasing when time left for the detailed design and manufacturing / assembly is reduced. During this study, operation-related risk is also represented as a certain portion of the product development cost. We have

$C^{SR} = C^{PD} \times \sigma(t_\beta - t_\alpha)$, for each candidate concept $C_i^{SR} = C_i^{PD} \times \sigma\left[i,(t_\beta - t_\alpha)\right]$ (11)

$\sigma(t_\beta - t_\alpha)$ is defined as the technical operation transforming index. As discussed, tα represents the duration of MGPD stage 1, which is the length of concept development, and tβ represents the duration of MGPD stages 1, 2, and 3, which is the length of traditional product development. The operation risk transforming index is defined as:

$$\sigma\left[i,(t_\beta - t_\alpha)\right] = \begin{cases} 0 & t_\beta - t_\alpha = t_\beta \\ \sigma_i^m & t_\beta - t_\alpha = t_i^m \\ +\infty & t_\beta - t_\alpha = 0 \end{cases} \quad (12)$$

$t_i^m, \sigma_i^m$ are pre-determined. They could be inputs from an expert system or historical data, which is also an "interface" of the integrated cost model. Based on the nature of $\sigma(t_\beta - t_\alpha)$, and the study of Logarithm functions, the technical risk transforming index is defined as

$$\frac{\sigma\left[i,(t_\beta - t_\alpha)\right]}{d(t_\beta - t_\alpha)} = -\frac{1}{\ln \sqrt[\sigma_i^m]{t_\beta / t_i^m}} \times \frac{1}{(t_\beta - t_\alpha)} \quad (13)$$

3.4 Integrated model

Therefore, for each candidate concept, the integrated cost model with consideration of technical risk, marketing risk and operation risk, as well as the cost during the product service could be defined as:

$$C_i^{PT} = C_i^{PD} + C_i^{Service} + C_i^{PD} \times \mu(i,t_\alpha) + C_i^{PD} \times \lambda(i,t_\beta) + C_i^{PD} \times \sigma\left[i,(t_\beta - t_\alpha)\right]$$

$$C_i^{PT} = C_i^{PD} \times \left\{1 + \mu(i,t_\alpha) + \lambda(i,t_\beta) + \sigma\left[i,(t_\beta - t_\alpha)\right]\right\} + C_i^{Service}$$

Finally, we have the total project cost for each concept, which can be estimated as:

$$C_i^{PT} = \left[C_i^{Matl} + (C_i^{Hr} + C_i^{Mfg}) \times R_i\right] \times \left\{1 + \mu(i,t_\alpha) + \lambda(i,t_\beta) + \sigma\left[i,(t_\beta - t_\alpha)\right]\right\} + C_i^{Service}$$

## 4 Illustrative Example

The approach adopted in this research is simulating the product development environment using historical data. First, two already present mature product generations are selected since they have sufficient and accurate project data such as, part cost, project schedule, product module structure etc. to support the simulation. The second step is to select a target project, which had carried on multiple concepts at the concept development stage. Moreover, the most important criterion is that the final selected concept must have been reused again during the following generation product development. Hence, a similar product development environment is created to demonstrate integrated cost model, and MGPD theory. Unfortunately, the original data cannot be directly used due to the sensitivity and security of the information. In this case, the original figures have been transformed via certain rules, but the interrelationships still remain identical to the original. Thus, the transformed data can still be used to illustrate how to use this intergraded cost model as a tool for concept selection, as well as project management. In this case, three candidate concepts have been identified and developed during the MGPD stage 1. At the end of stage 1, the design engineers have selected one of these concepts to pursue a further detailed design. Thus, this integrated cost model can generate a figure, which provides a summarized overview from all aspects of product development to determine the concept with the lowest cost.

Generally, the concept-I represents the existing mature design with less reusability. It has low material cost, low labor cost, short concept development time, short product lead-time, and short product life time. The concept-II represents the new design targeting future generations. It has high level future reuse, and also scores high in various other costs. The concept-III is considered a "compromise" option. The results suggest that concept-II is the obvious best choice, when the project has sufficient resources. The concept-I would only be considered when the project is under high pressure to introduce the new product as fast as possible. As discussed, $t_\beta$ represents the total time of MGPD stages 1, 2 and 3, which is also considered as the representation of project resources. Based on a determined $t_\beta$, we can choose the candidate concept that has the lowest project cost.

However in real world, one of the concepts could be selected by management decision earlier in the process, in favor of new technology, customer special requirements or many other reasons. In that case, the project team has to figure out the detailed project plan, which has to minimize the risk and cost. One of the important decisions has to be made is the duration of the concept development, or how much time the design engineers have to develop a mature concept, which could result in fewer engineering changes during the later manufacturing and assembly stage, as well as meet most of the requirements and have a lower overall cost. Fig. 3 shows the possible project costs for concept-I, while $t_\beta$ is from 13 months to 35 months, which is also the boundary of best and worst cases identified for the target project. Alternatively, $t_\alpha$ is from 0 months to 34 months.

## 5. Conclusions

The research conducted in the paper targets multigenerational products, which have higher parts costs and longer development time. Abstract of The literature review suggests that existing research has already reached the boundary of this 3-D world and the only way left is the fourth dimension, time. The fundamental purpose of this study is to achieve the maximum profits considering the scope for certain time duration in the future. The reference model has been specially introduced into our study. This three-layer model combined with the four-stage MGPD model creates a roadmap that is how to conduct MGPD step by step.

Detailed discussions of practice have occurred at each one of three-layer at each one of four-stages. Thus, more efforts have been put in to develop an integrated cost model, which can serve as the ultimate tool at the objective layer. An integrated cost model has been proposed to address the needs from MGPD, as well as incorporating the different aspects of MGPD. A case study has been presented to illustrate the performance of the proposed cost model.

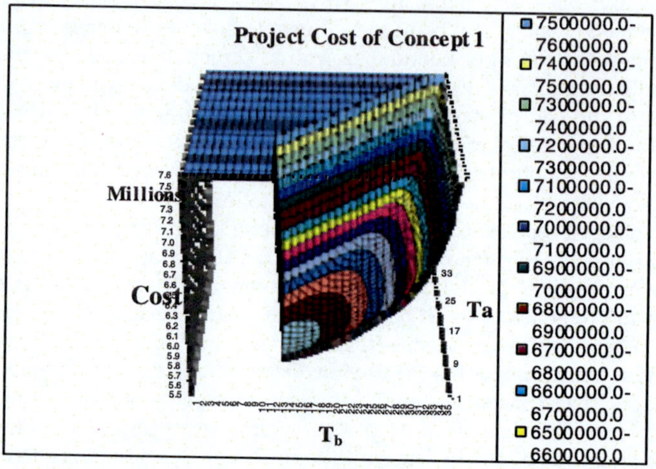


**Figure 3.** Project Cost of Concept-I

## References


[1]. Boothroyd, G., Dewhurst, P., and Knight, W., 1994. Product Design for Manufacturing and Assembly. New York.

[2]. Esawi, A.M.K. and Ashby, M.F., 2003. Cost estimates to guide pre-selection of processes. Materials & Design, 24(8), 605-616.

[3]. Gutowski, T., 1994. Development of a theoretical cost model for advanced composite fabrication. Composites Manufacturing, 5(4), 231-239.

[4]. Innes, J., Mitchell, F., and Sinclair, D., (2000). Activity-based costing in the U.K.'s largest companies: a comparison of 1994 and 1999 survey results. Management Accounting Research, 2000, 11, 349–362

[5]. Kim, N., Bridges, E., and Srivastava, R.K., (1999). A simultaneous model for innovative product category sales diffusion and competitive dynamics. International Journal of Research in Marketing, 16(2), 95–111.

[6]. Ulrich, K.T., and Eppinger, S.D., Product Design and Development. 2nd edition 2000 (McGraw Hill, Boston).

[7]. Yang, K., 2007. Voice of the Customer Capture and Analysis. New York, McGraw-Hill Professional.

# Planning Manufacturing in a Concurrent Engineering Environment: A Case Study

Malachy Maginness[a], Essam Shehab[b,1] and Chris Beadle[c]

[a, b] Decision Engineering Centre, Manufacturing Department, Cranfield University, College Road, Cranfield, Bedfordshire, MK43 0AL, United Kingdom
[c] Rolls-Royce plc., PO Box 31, Moor Lane, Derby, DE24 8BJ, United Kingdom

**Abstract.** This paper presents an investigation of current state practice of the Manufacturing Engineering discipline for concurrent manufacturing planning. The research adopted a case study approach and has been conducted at a globally operating manufacturer of aerospace products. The investigation establishes how information systems and the cross-functional teaming enable integrated processes for planning the manufacturing method to progress simultaneously with design in a lean and efficient manner. It applies value stream analysis to understand where value and non-value is added in these transactional processes.

**Keywords.** New Product Introduction, Manufacturing Engineering, Aerospace

## 1 Introduction

Robust New Product Introduction (NPI) business processes are crucial for conducting product development in complex engineering projects. The product development challenge of the aerospace industry lies in delivering high complexity product systems at the highest levels of quality. Legislative, environmental and economic factors compel product solutions to satisfy performance requirements in areas of fuel efficiency, emissions control and operating costs. In turn aerospace manufacturers are driven toward complex NPI projects with considerable technical challenges, development period and financial investment. The ability of manufacturers to deliver quality and cost effective products in a short time to market is a basic competitive requirement. As such, NPI processes must be suited

[1] E.Shehab, Manufacturing Department, Cranfield University, College Road, Cranfield, Bedfordshire, MK43 0AL, United Kingdom; Tel: +44 (0)1234 750111 x524; Email: e.shehab@scranfield.ac.uk



D. D. Frey et al. (eds.), *Improving Complex Systems Today*,
Advanced Concurrent Engineering, DOI: 10.1007/ 978-0-85729-799-0_49,
© Springer-Verlag London Limited 2011

to the requirements of Concurrent Engineering and be efficiently and effectively integrated across the product development supply chain. This research is conducted with the support of a globally operating manufacturer of aerospace products based in the United Kingdom. Here, NPI processes are administered across a matrix of business units and collaborating functions. The Manufacturing Engineering function is responsible for planning manufacturing and assembly methods that delivery product designs at levels of cost, quality and lead time that are consistent with customer requirements. The manufacturing planning challenge is delivering capable, multi-process manufacturing methods simultaneously with the development of the product solution by the Design Engineering function. The sponsoring company has adopted Concurrent Engineering in product development projects and has invested in integration practices and information system technologies to support the approach. These include examples of PLM (Product Lifecycle Management) and CAD/CAM (Compute Aided Design/Manufacture) software; cross-functional teams to manage component level NPI; and design for manufacture (DfM) dialogue. This study is motivated by a need to better understand the Manufacturing Engineering role and manufacturing method planning within Concurrent Engineering. Issues impacting Concurrent Engineering effectiveness include a 'hostile' downstream attitude to receiving unfinished design work that is likely to be changed, and the willingness of upstream and downstream parties to make agreement when determining a design [4]. By applying value stream mapping and analysis, this study seeks to understand the practices that enable such unfinished work to be used in NPI processes for planning the manufacturing method in the context of a large aerospace project.

## 2 Related Literature

A growing area of literature applies lean principles to transactional processes in product development [1, 3, 5-10]. Lean principles represent an extension of best practice in product development and add the efficiency concepts of value management to the principles of integrated problem solving. Existing integrated problem solving practice (Concurrent Engineering, cross-functional integration *etc.*) are congruous with application of lean in the product introduction. A criterion for identifying activities within a transactional process by value type (Directly Value Adding, Enablers of Value Adding activity, or Non-Value Adding wastes) is established for determining where value is added by activities within a value stream [9]. Furthermore, a methodology for value stream analysis of transactional engineering processes has been successfully demonstrated in the context of aerospace Design Engineering [5]. This research seeks to contribute to this field of inquiry and investigate current state processes for planning manufacturing method.

## 3 Method

The case study was defined in three elements: the NPI processes of interest, the NPI project for which these were conducted, and the responsible business unit within the company. The work breakdown structure for NPI projects identifies distinct standard processes for planning the method of manufacture which are expected to interact with one another. The true nature of process interactions was revealed from the collected data. Case Study 1 examined processes dedicated to the development of tooling for a multi-process manufacturing method. Case Study 2 examined the design and manufacture of special work holding fixtures (tools required in the manufacturing method) from the point of receiving component definition through to tooling design and ultimate handover to shop floor operations.

The second and third elements of the case defined the context and reflect the complexity of the product development supply chain. Large product projects at the sponsor company are disseminated into individual component level NPI projects and are carried out in individual business units. Each case investigated, for a certain component from a contemporary project, how the NPI processes were carried out by the respective business. The same product project was selected for both cases. As the most immediately recent project it represents the state-of-the-art practice in the company. Data collection was assisted by the ready availability of project documentation and the continued presence of engineers involved.

The same data collection and analysis method was followed in each case. Group workshops with representatives from across the engineering disciplines were held at the start of each case. Post-it® Notes were the flexible and interactive means of capturing an initial high-level process description. A semi-structured interview was used for first-hand data collection and complemented a review of available project documentation. Interview candidates were selected from the management level of the relevant business unit through to component-level project managers and specialist Manufacturing Engineering resources that participated in the component NPI project in question. The question set was designed for one hour's interview length and was structured in accordance with two major concepts that are established in literature for understanding transactional processes. Opening questions follow the Supplier-Input-Process-Output-Customer (SIPOC) model and captured information inputs into activities, output information and the dependencies of other activities on these [2]. Secondary questions are developed from an existing value steam analysis question set [5]. This captured time metrics, governance concepts (disciplines involved) and tools used (software systems).

Process mapping and value stream concepts served as the analysis method. In each case, a map was created to depict how process activities were executed at component level. Activity boxes linked by arrows recorded the information flow of and the role-activity format allowed the association of a specialist discipline with activities to be seen throughout the process. Corroboration of interviewee accounts was made by examination of associated project documentation (including project plans), and validation of the findings in a concluding group workshop.

## 4 Findings

Practices that enable manufacturing method planning to progress simultaneously with those of Design Engineering (model creation and analysis processes that determine the component form) are identified. This includes the integration support given by the information technology system tools and cross-functional teaming.

### 4.1 Progressive Definition Release

Quality manufacturing plans rely matching the design intent. Mechanisms of progressive definition and the application of 'surrogate' product information resolve the paradox of beginning manufacturing planning when final intent is determined later, at the end of the design process. 'Progressive definition release', describes how component information created by Design Engineering is integrated into Manufacturing Engineering processes. Both cases demonstrate how progressive information is integrated into the planning work stream. This enables planning to be executed concurrently with a continuing design effort. 'Buy-off' (formal agreement that a method matching design intent can be delivered) becomes a progressive process based on partial or incomplete definitions.

Progressive product information definitions that are useful inputs into buy-off and planning processes are identified. These allow manufacturing decision-making and resource planning (tools or materials ordering, especially those with long lead times) to advance to the extent that production trials are commenced ahead of final designs. The structure of the progressive definition modifies the medium (ranging from new sketches, sketches that modify past component drawings, CAD models and finally engineering drawings) in which product information (material, geometry or tolerances) is stated. This structure reflects the increasing maturity of the design. Accepting preliminary design information removes waiting from the project critical path. For example, certain upfront planning decisions (securing capital funds or long lead time orders) are based on sketches or 'design envelopes' (geometry limits known early). Later, geometry contained in CAD models is used to plan manufacturing methods ahead of detailed Engineering Drawings.

'Surrogate' information is used alongside incomplete product information to support buy-off and commence planning. This is based on past projects with comparable component geometry. Templates of information include a generic Method of Manufacture or 'router'. In its final form the Method of Manufacture documents the sequence of operations (including relevant processes and equipment) that transform raw material into a final shape. Design 'standards' is a mechanism to enable buy-off of preliminary definitions (sketches and models) with reduced risk. Geometry and tolerances that represent the best potential process capability are agreed upfront between Design and Manufacturing Engineering. Planning work progresses on the assumption that design geometry and tolerances will be defined consistent with agreed standards and will thus be compatible with planning definitions when completed. Cross-functional teaming is an important function of managing progressive release. In this team the Manufacturing Engineering representative (a Component Owner) sets out the requirements for and negotiates the schedule and structure of progressive definition release.

### 4.2 Planning Definition Driven from CAD Models

Planning definitions are generated in a manner that is consistent with the Method of Manufacture sequence of manufacturing operations. Each stage of operation is defined in a Stage Model and Stage Drawing. Stage Models describe the component shape at the end of each operation. Stage Drawings define that model's geometry and specify details including work holding areas and datum points. The planning process and the definition of Stage Models is driven from the CAD model of the component design and are associated with that model's geometry. A single design model propagates numerous Stage Models and Stage Drawings. A range of downstream planning definitions derive from these. These include the design and manufacture of machine tools, production tools, work-holding fixtures, creation of NC (Numerical Control) machining and CMM (Coordinate Measurement Machine) sequences and creation of Manufacturing Instructions (technical information required by the operator to complete manufacturing operations). As such the process of creating the Stage Models and Stage Drawings interacts with the processes that define these downstream items. Using CAD models, rather than drawings, to generate planning definitions is crucial for integrating the planning processes and sharing product information. The PLM (Product Lifecycle Management) system employs consistent CAD software (UG-NX). Design and manufacturing information is stored and directly accessed on a shared environment for progressive release. Design and process planning models (Stage Models and downstream derivatives) are created in the same CAD application (Unified Graphics or UG/NX software). By using models within a consistent information system any revisions to the master design model can cascade through to the Stage Models and derivative planning definitions automatically with minimal rework or change management effort. In this way planning a quality manufacturing method, which is dependent on matching the design intent is supported. Integration within the PLM environment facilitates this. Process structure plays a role in managing the risk of rework or revisions to planning definitions. Limits to the downstream use of preliminary design model data are set within the process. Committing resources or investment in planning (e.g. to design and procure fixture tools) is suspended until the cross-functional team confirms certain geometry is fixed. Accordingly certain Stage Model and derivative definitions are protected from further revision in later updates and resources are thus invested with reduced risk.

### 4.3 Appraisal of the Value Added in the Planning Processes

Value stream analysis principles were applied to the mapped manufacturing planning processes [5, 9]. Direct Value Add activities are identified as those that define the product or process. Planning definition activities correspond with this; *i.e.* creating Stage Models, Fixture Models, NC Programming and manufacture/machining of tooling. Activities that reduce risk and uncertainty are also classed as value adding. Good quality simulation and method trials satisfy this. Enabler activities help form the value adding output. Stage Drawings, Tooling Drawings correspond to this and describe CAD model geometry in a manner that enables downstream planning, such as tooling fabrication to proceed.

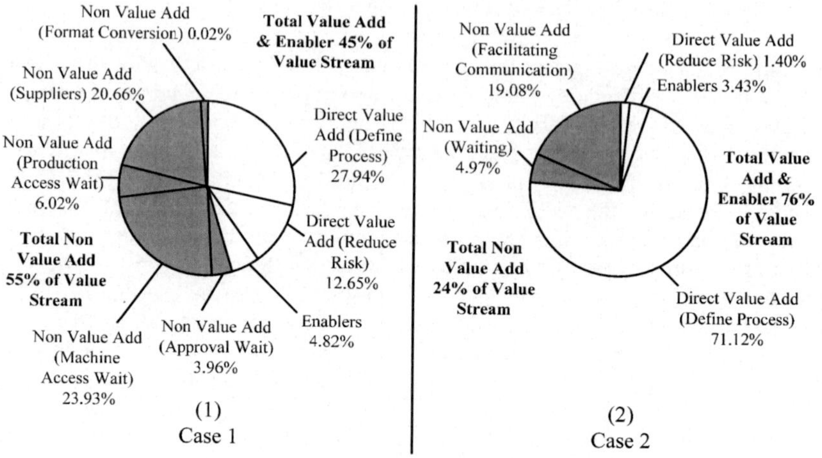


**Figure 1.** Value Stream Contributions in Cases 1 & 2

The forgoing analysis validates the activities of the current state processes that support the Manufacturing Engineering value proposition- the delivery of a quality and cost effective method of manufacture that matches the design intent. These are transactional processes aimed at making a physical shop floor process. It is possible to identify where non-value add, or waste exists. In Case 1 it is calculated that the direct value add and enabling activities contribute only 45% of the total value stream lead time (Figure 1-1). In Case 2 this is 76% (Figure 1-2). Analysis of the waste types helps to understand proportions of the waste within the total lead time of the Manufacturing Engineering NPI processes.

Waiting is the significant feature of the value stream in both cases. The convention followed in analysis is that lead time for an individual activity is the sum of cycle time (pure working time) and waiting time (delays and interruptions). Waiting is identifiable with iterations of physical trials of the manufacturing method (Case 1), and the purchasing and tendering procedures (Case 2). In Case 1, the method planning is dependent on batch trials and altering the production tools with additional machining. Waiting time is encountered in trials when accessing plant equipment shared with other production areas. Machining of tooling used in the method encounters a similar wait. Waits are compounded by the iterations of trials required. Despite risk reduction activities being classified as 'value add' it is notable that the non-value add waiting component associated with trials accounts for 30% of the total lead time for delivering a quality method. Reducing dependency on physical iterations themselves with enhanced virtual simulation capability is an improvement opportunity. Additionally, the majority of the lead time for laboratory evaluation of parts produced by the trials is described as waiting. The laboratory requires upfront involvement in the project to ensure resources are available when required for such evaluation. The fixture tool design and delivery process investigated in Case 2 reveals additional wait aspects including interaction with purchasing and tendering processes. A significant contribution is made to lead time where Manufacturing Engineering processes interact with those of the purchasing function. Activities associated with purchasing and tendering are approximately 20% of the total lead time and lie on the critical path. These correspond to the non value add concept of facilitating communication. This 'non-value add' category endorses the view that communication should be seamless in lean processes [3]. Further waste is evident in the value streams. Transportation type waste is identified in the transfer of model data between information systems. These actions are 'breaks' in the integrated information system. Breaks are observed where alternative software and/or isolated PC systems are used outside of the PLM environment (*i.e.* non-PLM model data). This insulates downstream planning definitions from the design model. The cascade of updates is prevented which necessitates additional change management. The definitions propagated in this manner require manual intervention and updating in order to reflect changes to design intent.

In Case 2, the Manufacturing Engineering team made extensive and consistent use of the PLM system for creating and distributing planning definitions. Motion waste remains evident where physical transfer by hard drive media is necessary between computers running the standard PLM systems and separate stand-alone machines upon which preferred or legacy software exists. Similarly inconsistent file formats require conversion from an original model format (when created in the standard CAD software) and/or transferred to an alternate system.

## 5 Conclusions and Future Work

This paper reports two case study investigations of Manufacturing Engineering for New Product Introduction in the context of large aerospace product development. The practices enabling a quality manufacturing method to be planned and delivered within a Concurrent Engineering project approach are identified. The investigation into this area of product development decision making and the application of value stream analysis to the transactional processes, is the contribution of this paper.

The timing and structure of progressive release supports the conduct of planning processes such that ramp-up can occur in a manner that is simultaneous to the completion of design. This permits a greater window of opportunity for design to define the innovative product solutions that drive the competitive strategy in this industry. The decision-making observed here balances the factors of Quality, Cost and Delivery by using incomplete product information progressively received from design sources, and manufacturing knowledge from past projects. Furthermore, the work of planning definition is understood to propagate a range of specialist processes. Matching design intent throughout all these is a key aspect of delivering a quality manufacturing method. Driving planning processes from the design CAD model, through an integrated information system efficiently ensures quality.

Applying lead time metrics to the analysis of the product development value stream is a relevant approach. Reducing the time between decision-making events is critical for achieving a rapid and flexible NPI capability. Lean principles help the process more quickly arrive at decision-making events with lower uncertainly. The case studies reveal areas in current NPI processes where wasteful activity lengthens the lead time for the processes. The value analysis is quantitative data (lead time metrics) upon which improvement actions can be justified as necessary. These case studies have investigated discreet, component level facets of a large aerospace NPI project. The project complexity is manifested in numerous components and system families, all subject to NPI processes in a Concurrent Engineering manner. Further study shall seek to understand the complexity of managing numerous simultaneous NPI processes, and the application of information systems and the cross-functional approach to support Manufacturing Engineering and the quality of manufacturing method planning.

## 6 Acknowledgements

The authors would like to thank the sponsors, Rolls-Royce plc, the Decision Engineering Centre in the School of Applied Sciences at Cranfield University, and the Engineering and Physical Sciences Research Council for supporting this work.

## 7 References

[1] Baines T, Lightfoot H, Williams GM, Greenough R. State-of-the-art in lean design engineering: A literature review on white collar lean. Proceedings of the Institution of Mechanical Engineers, Part B: Journal of Engineering Manufacture 2006; vol. 220, no. 9, pp. 1538-1547,

[2] Browning TR, Fricke E, Negele H. Key concepts in modeling product development processes. Systems Engineering 2005; vol. 9, no. 2, pp. 104-128

[3] Chase J. Value Creation in the Product Development Process. Masters thesis, Massachusetts Institute of Technology, Cambridge, 2001.

[4] Clark KB, Fujimoto T. Product Development Performance: Strategy, Organization and Management in World Auto Industry. Harvard Business School Press 1991.

[5] Darwish M, Haque B, Shehab E, Al-Ashaab A. 2010 Value Stream Mapping and Analysis of Product Development (Engineering) Processes. Proceedings of The 8th International Conference on Manufacturing Research (ICMR 2010), University, Durham, UK, 14-16 September 2010.

[6] Haque B. Lean engineering in the aerospace industry. Proceedings of the Institution of Mechanical Engineers, Part B: Journal of Engineering Manufacture 2003; vol. 217, no. 10, pp. 1409-1420.

[7] Haque B, Moore MJ. Measures of performance for lean product introduction in the aerospace industry. Proceedings of the Institution of Mechanical Engineers, Part B: Journal of Engineering Manufacture 2004; vol. 218, no. 10, pp. 1387-1398.

[8] Liker JK, Morgan JM. The Toyota Way in Services: The Case of Lean Product Development. Academy of Management Perspectives 2006; vol. 20, no. 2, pp. 5-20.

[9] McManus H. Product Development Value Stream Mapping (PDVSM) Manual. MIT Lean Aerospace Initiative, Cambridge, 2005.

[10] Reinertsen DG. Let it flow. Industrial Engineering (American Institute of Industrial Engineers) 2005; vol. 37, no. 6, pp. 40-45.

[11] Reinertsen DG. The Principles of Product Development Flow: Second Generation Lean Product Development. Celeritas Publishing 2009.

# Concurrent engineering yesterday, today and tomorrow

Alemu Moges Belay[a], Petri Helo[b,1] and Fentahun Moges Kasie[b]

[a]PhD student, University of Vaasa, Dept. of production, Vaasa, Finland.
[b,1]Professor, University of Vaasa, Dept. of production, Vaasa, Finland
[b]Institute of technology ,Dept of Mech and Industrial Eng.,Hawassa, Ethiopia

**Abstract.** The purpose of this paper is to give an in-depth insight for the development and improvement of concurrent engineering in several applications. In addition to that the conceptual model that includes the new application areas is developed and presented in simplified way so that the new practitioner in the field of concurrent engineering can understand easily. The methodology that is followed in this paper is review of concurrent engineering since the philosophy has started (80's). The paper will start with historical background of CE, explains its general characteristics, different types of processes execution while doing overlapping and parallelism. Another area that will be covered is CE in relationship with other improvement principles and philosophies (like…supply chain, BPR, TQM) and etc. The challenges of CE are presented in this article to give a glance for managerial decision making purposes. Different perspectives towards CE and application areas is presented in a way that reader can understand in simplified but with clear objective in the implementation of CE. From the findings, though there are some challenges, still there are untapped application areas (like services) that could exploit the benefits of CE if it is considered with cautions and analyzed in advance using system dynamics.

**Keywords.** Concurrent Engineering, Product development, Supply chain

## 1 Historical Background of Concurrent Engineering

Both manufacturing and services have made several changes to be competent in global markets in order to sustain their competitiveness in dynamic and turbulent global markets. One of the adopted management and manufacturing philosophies is CE that companies use to withstand the challenges, use their resources efficiently and wisely. Companies can achieve this through a CE approach of production [8]. The main objective of CE is to minimize production life cycle of a product by making processes more efficient [7].
Different researches have carried out by different stakeholders on implementation of CE to reduce product life cycle. For instance, Shorts and Boeing – have adopted

---

[1] PhD student, University of Vaasa, Department of production, Vaasa, Finland, Tel: +358443126822; Email: albel@uwasa.fi

D. D. Frey et al. (eds.), *Improving Complex Systems Today*,
Advanced Concurrent Engineering, DOI: 10.1007/ 978-0-85729-799-0_50,
© Springer-Verlag London Limited 2011

CE techniques for the design and build of a relatively short production run aircraft [5]. Consumer electronics; CD players or PC printers (6-9 months), automotive industry reduced from 5-8 years to 36 months and less [17]. Xerox is one of a good example that began a significant effort in the early 1980s to improve quality, cost, and cycle time. Improvement were on product cost (10%); reducing of rejection (93%); reduction of lead time from 52 to 15 weeks [24]. A 1990 report on CE by Business Week, heralding the concept as "promising to create the most wrenching upheaval in manufacturing in a half-century," enumerated the following benefits; Development time: 30-70% less; Engineering changes: 65-95% fewer; Time to market : 20-90% less; Quality: 200-600% higher; productivity: 20-110% higher; sales: 5-50% , Return on assets: 20-120% higher [24].Hence, CE has a significant role for industries to overcome the challenges to this dynamic global competition.

**General Characteristics of Concurrent Engineering**

Concurrent engineering has been defined in many ways by different authors, some of the definitions of Concurrent Engineering CE are:

1. CE is "an engineering management philosophy and a set of operating principles that guide a product development process through an accelerated successful completion" [30].
2. CE is "an approach to the integrated, simultaneous design of products and related processes, like manufacture and support" [14].
3. CE is "a process which can integrate all the steps in the process of product development including the design stages and manufacturing process and it can put them in a form in which we can observe and consider them concurrently" [19]. All the above definitions focus on integration and parallel engineering activities. According to [16] the key components of CE are understanding of customer needs; stability in the product specification; a structured, systematic approach to product development; ability to build effective teams; a realistic & defined product development process; availability of resources; early involvement of all stakeholders support the parallel design of product and process; and appropriate technological support.

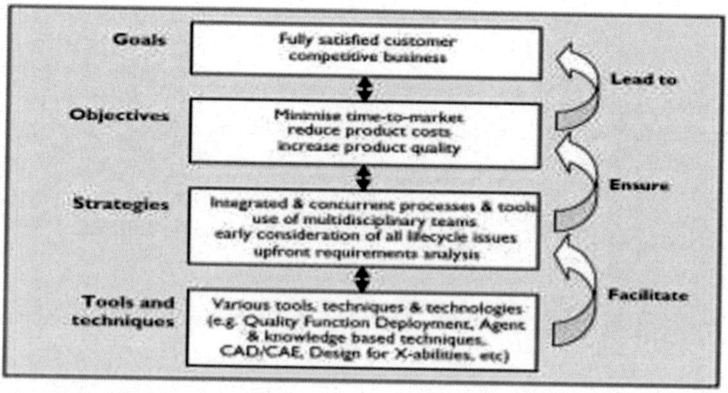


Figure1 A framework for understanding CE [28]

## 2 CE and Overlapping Processes in New Product Development

The integrated activities of marketing, marketing research, R&D, industrial design, engineering, and various involved disciplines are known as overlapping. NPD is a system that its activities encompass the dynamic interaction between internal and external factors such as customers and suppliers [9]. CE needs the co-ordination of the whole processes and simultaneously designs of a product, development, and preparation for quantity of production [3]. CE bring a product to market, design engineering, and manufacturing are jointly managed to work in parallel, in sharp contrast with the traditional approach [11].

## 3 Concurrent Engineering and Supply Chain

According to [33] the vital dimensions of CE are attention to customer, organization of the company and supplier. It has been also mentioned that enhancing the involvement of manufacturing, suppliers and customers has long-term benefits rather than perceived customer satisfaction [36]. Direct, logical link between supply chain and the design of roles are indicated origination [3].

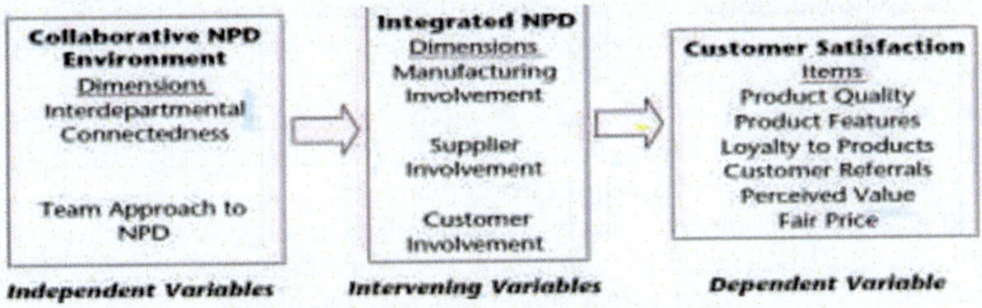


Figure 2 Explanatory/Theoretical Models [31]

Many researches are focused on the significance of customer-supplier relationships i.e. supply chain during NPD [18,25].

# 4 Application Areas of CE

**Education**: CE techniques have been to develop a comprehensive model for designing of a new department in the university [35].

**Intelligent Design Planner**: the research performed by Jiao, Khoo & Chen presented a prototype intelligent concurrent design task planner, which combines the strength of genetic algorithms and an iterative design analyser for the scheduling of a complex design process of a manufacturing system [13].

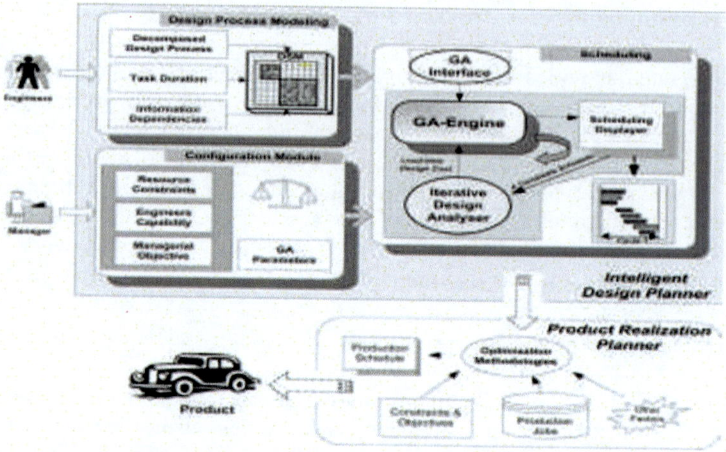


Figure 3 Intelligent concurrent design planner [17]

**Chemical Process Design:** According to [10] CE concept particularly improves the way a chemical design engineer deals with "external factors" that influence a process design. Researchers introduced a framework that easily identify and classify these external factors.

**Project Scheduling:** CE is necessary to model the engineering process and to develop techniques that can schedule activities concurrently by allowing an optimal degree of overlap of activities under due consideration of uncertainty [27]. Nicoletti & Nicold developed a decision support model to decide which activities in a project primarily need to be concurrently scheduled, enhance requirements re-configurability and minimize errors and unplanned evolution of the activities [18]. Besides CE is designed to facilitate the simultaneous consideration of all project related issues and processes from the conception stage [3]

**Operation Testing:** an analytical model for the scheduling of tests in overlapped design process was developed by [25], where a downstream stage starts before the completion of upstream testing.

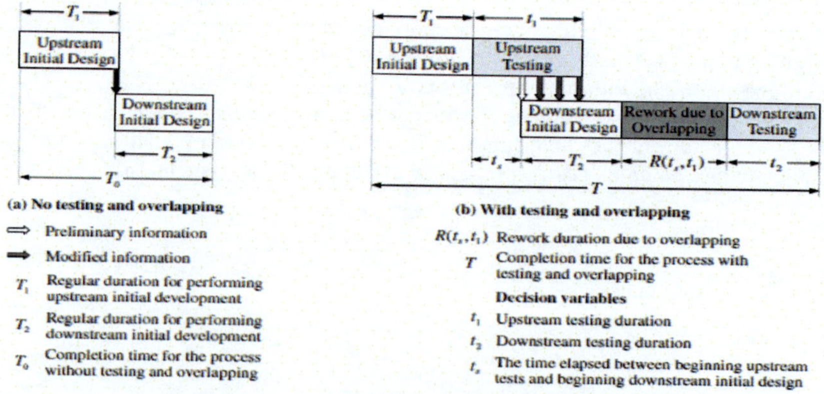


Figure 4 Product development processes [30]

**Construction:** the adoption of CE in construction is shown by literatures [2, 14]. If we consider construction, operation and maintenance phases at early stage of a project would undoubtedly lead to an overall improvement in the project performance [2].

## 5 Relation with Other Concepts

Several researches indicate the positive relationships between CE and other improvement philosophies( its compatibility with TQM, BPR, IT and etc.)
**Total Quality Management TQM and CE:** the effect of quality management on the speed of NPD and compared CE and TQM that leads to several common principles [4,28,29]. It is indicated that the possibility of CE characteristics to be incorporated in TQM approach (ISO9000:2000 standard) [22]. QFD fits ideally as a "front-end" process to CE [12]. [34] Reveals that TQM, teamwork, value analysis (VA) and QFD are all positively correlated with the speed of NPD.
**BPR and CE:** Bovey described that BPR in a CE environment covers all dimensions from all of a business's functional groups need to be brought together. According to [6] if correctly applied, CE and BPR can be very effective in improving the performance of a company.
**IT & CE:** The main step while implementing CE is effective cross-functional teams, which integrate the development process using both organisational and information management methods [1]. A PDM helps engineers and others manage both data and the PD processes, and hence support a CE framework in a company [16]. Kong et al. also developed mathematical model [15].
**Automation and CE**: CIM and concurrent engineering (CE) are multidisciplinary subjects concerned with providing computer assistance, control and high level integrated automation; at all levels of manufacturing (and other) industries, by linking islands of automation into a distributed processing system[21]. The design DFA concept is also widely used [32]. CE nowadays focused on those tools which facilitate it, CAD/CAE/CAM and MRP products [17].

## 6 Challenges/Limitations of CE

The challenge of developing successful products, to increase the user's experience, needs an interrelation approach across all the key functions involved in NPD. [11] Argued on major limitations of CE. For instance, in the design of integrated circuits, subdividing the work into modules smaller than individual components can be impractical, counterproductive, resource limitations, product technology, and when subcontracted to suppliers. [20] Focused on four critical problems that challenge management while implementing CE in complex product development projects. These problems referred as iteration, parallelism, decomposition and stability. Another limitation of CE is its human resource implication. When humans are added to CE, it is logical solution but it can become messy as Filipczak indicated. Tucker & Hackney underlined that the main reasons for the failure of CE projects are the lack of formal methodologies to assist organizations with the processes required to move from sequential to concurrent product development phases [22, 32].

## 7 The conceptual model and summary

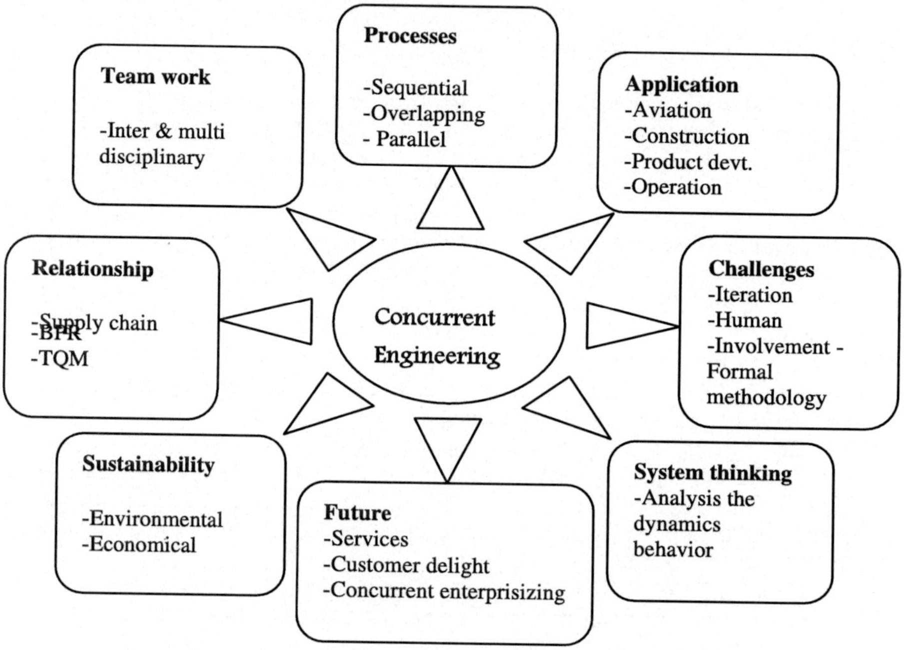


Figure 5 Conceptual model based on different perspectives in CE

As figure 5 shows, there are many stakeholders and perspectives involved in concurrent engineering so that before implementation it is important to analyze systematicaly and see the dynamic behaviour in advance. This will help for managerial decision to see the payoff of adopting and implementation of concurrent engineering. The other point that it shouldn't ignored is sustainability issue and the future consideration in customer involvement and delight. This is because CE in 1980's (design practices) has extended to new applications with various perspectives.

## 8 References

(1) Abdalla HS. Concurrent engineering for global manufacturing. Int. J. Production Economics 1999;60-61: 251-60

(2) Anumba CJ, Baugh C, Khalfan, MA. Organizational structures to support concurrent engineering. Industrial management & data systems 2002;102:260-70

(3) Anumba CJ, Siemieniuch CE, Sinclair MA. Supply chain implications of concurrent engineering. International Journal of Physical Distribution & Logistics Management 2000; 30: 566-97
(4) Belay AM, Helo P, Takala J, Kasie FM. Effects of Quality Management Practices and Concurrent Engineering in Business Performance. International Journal of Business and Management 2011;6:45-62
(5) Bradley D. Concurrent engineering for bespoke products. Assembly Automation 1995;15:. 35-37
(6) Brooks B.Realising the benefits of concurrent engineering through BPR. Computing & Control Engineering Journal 1996
(7) Chou S, Trappey A, Pokojski J, Smith S. Global Perspective for Competitive Enterprise, Economy and Ecology: Proceedings of the 16th ISPE International Conference on Concurrent Engineering 2009
(8) Curran R, Chou SY, Trappey A. Collaborative Product and Service Life Cycle Management for a Sustainable World. Proceedings of the 15th ISPE International Conference on Concurrent Engineering. Verlag London Limited 2008: Springer
(9) Harmancioglu N, McNally RC, Calantone RJ, Durmusoglu SS.Your new product development [NPD) is only as good as your process: an exploratory analysis of new NPD process design and implementation. R&D Management 2007; 37:339-424
(10) Herder PM, Weijnen PC. A concurrent engineering approach to chemical process design Int. J. Production Economics 2000;64: 311-18
(11) Hoedemaker GM, Blackburn JD, Van Wassenhove LN. Limits of concurrency. Decision sciences 1999;30:1-18
(12) Jarvis M. Concurrent engineering. Work Study 1999;48:88–91
(13) Jiao LM, Khoo LP, Chen CH. An intelligent concurrent design task lanner for manufacturing systems. Int J Adv Manuf Technol 2004; 23: 672–81
(14) Kamara JM, Anumba CJ, Evbuomwan NFO. Developments in the Implementation of Concurrent Engineering in Construction. International Journal of Computer-Integrated Design and Construction 2000; 2: 68–78.
(15) Kong L, Bhuiyan N, Thomson V. The Value of Organizational Structures. Concurrent Engineering 2009; 17: 61-72
(16) Kumar R, Midha PS. A QFD based methodology for evaluating a company's PDM requirements for collaborative product development. Industrial management and data systems 2001;101:126-31
(17) Landeghem RV. Experiences with a concurrent engineering self-assessment tool. Int. J. Production Economics 2000; 64:295-309
(18) Li Y, Jian J, Yan R, Liao W. Aircraft Tooling Collaborative Design Based on Multi-agent and PDM. Concurrent Engineering 2009; 17: 139-46
(19) Lindquist A, Berglund F, Johannesson H. Supplier Integration and Communication Strategies in Collaborative Platform Development. Concurrent Engineering 2008; 16: 23-35

(20) Luh D, Ko Y, Ma C. A Dynamic Planning Approach for New Product Development. Concurrent Engineering 2009; 17: 43-59

(21) Maffin D, Braiden P. Manufacturing and supplier roles in product development. Int. J. Production Economics 2001;69: 205-13

(22) Melin T, Isikveren AT, Rizzi A, Stamblewski C, Anders HV. How industry concepts of concurrent engineering enhance aircraft design education. Proc. Imeche Vol. 221 Part G: J. Aerospace Engineering JAERO127 imeche 2007

(23) Najmi M, Ip-Shing F. Concurrent Engineering in Total Quality Management Environment", 8th International Conference on Concurrent Enterprising 2002, Rome, Italy.

(24) Nicoletti S, Nicolo F. A concurrent engineering decision model: Management of the project activities information flows. Int. J. Production Economics 19981;54: 115-27

(25) O'Neal C. Concurrent Engineering with Early Supplier Involvement: A Cross- Functional Challenge. International Journal of Purchasing and Materials Management, Spring 1993; 1-9

(26) Qian Y , Xie M, Goh TN, Lin J. Decision Support Optimal testing strategies in overlapped design process. European Journal of Operational Research 2010; 206: 131–43

(27) Ranky PG. Features Concurrent Engineering and Enterprise Modeling. Assembly Automation 1994;14:14-21.

(28) Rolstadas A. Planning and control of concurrent engineering projects. Int. J. Production Economics 1995; 38: 3-13

(29) Sun H, Zhao Y, and Hon Keung Yau HK. The relationship between quality management and the speed of new product development. The TQM Journal 2009; 21:576-88.

(30) Sun H, Zhao Y. The empirical relationship between quality management and the speed of new product development Total Quality Management 2010;21:351–61

(31) Tan CL, Tracey M. Collaborative New Product Development Environments: Implications for Supply Chain Management. The Journal of Supply Chain Management Summer 2007;1-15

(32) Tucker D, Hackney R. Towards the integration of concurrent engineering environment within organizational strategy. Journal of management development 2000;19:179-89

(33) Veryzer RW, de Mozota BB. The Impact of User-Oriented Design on New Product Development: An Examination of Fundamental Relationships. J Prod Innov Manag 2005;22:128–43

(34) Willaert SA, de Graaf R, Minderhoud S. concurrent engineering in a wider context. J. Eng. Manage. 1998;15:87-109

(35) Yang CC, Chen SH, Jiun-Yan Shiau JY. A DFX and concurrent engineering model for the establishment of a new department in a university. Int. J. Production Economics 2007;107: 179–89

(36) Yassine A, Braha D. Complex Concurrent Engineering and the Design Structure Matrix Method. Concurrent Engineering 2003; 11: 165-76

# Set-Based Concurrent Engineering process within the LeanPPD environment

Muhammad Khan[a], Ahmed Al-Ashaab[a, 1], Athanasia Doultsinou[a], Essam Shehab[a], Paul Ewers[b], Robert Sulowski[c]

[a] Manufacturing Department, School of Applied Sciences, Cranfield University, MK43 0AL, United Kingdom

[b] Visteon Engineering Services Ltd, United Kingdom

[c] Sitech Sp. z o.o, Poland

**Abstract.** This paper presents a newly defined set-based concurrent engineering process, which the authors believe addresses some of the key challenges faced by engineering enterprises in the 21st century. The main principles of Set-Based Concurrent Engineering (SBCE) have been identified via an extensive literature review. Based on these principles the SBCE baseline model was developed. The baseline model defines the stages and activities which represent the product development process to be employed in the LeanPPD (lean product and process development) project. The LeanPPD project is addressing the needs of European manufacturing companies for a new model that extends beyond lean manufacturing, and incorporates lean thinking in the product design development process.

**Keywords.** Set-based concurrent engineering, lean thinking, LeanPPD, product development

## 1 Introduction

Lean thinking is an improvement philosophy which focuses on the creation of customer-defined value and the elimination of waste. Lean thinking has been a subject of research for nearly three decades, the focus of which has been on improving manufacturing processes, administration, management and the supply chain. However, new engineering products continue to under-perform in their lead times, cost, and quality. There has been comparatively less research done to improve product design and development (PDD): the design process, from the concept stage, to the detailed development of products and their related manufacturing processes. The reasons for this are many; however PDD has the greatest influence on the profitability of any product [3]. Research undertaken to improve PDD with lean thinking may prove instrumental in the progress of engineering.

[1] Manufacturing Department, School of Applied Sciences, Cranfield University, MK43 0AL, United Kingdom, Tel: +441234750111, ext.5622, E-mail: a.al-ashaab@cranfield.ac.uk

D. D. Frey et al. (eds.), *Improving Complex Systems Today*
Advanced Concurrent Engineering, DOI: 10.1007/ 978-0-85729-799-0_51,
© Springer-Verlag London Limited 2011

Most product development models have been developed in order to meet the challenges and situation at the time that they were created. However, the market and environment for engineering products is almost as turbulent as white water rapids. In order to develop a product development model that is fit to consistently perform in a rapidly changing market and environment, a changeless core is required. While comparing various engineering companies a number of US researchers realised that the Toyota motor co. had established a changeless core that was based on principles, focus and discipline [7]. Toyota product development focuses on three central elements: value, knowledge (or learning) and improvement. The authors believe that this focus has enabled them to please customers through optimal designs, minimise design rework, and achieve high profit levels. In order to reach the aforementioned achievements, they developed a process that is now known as set-based concurrent engineering (SBCE); however there is no publication that describes the methodology of this process in detail that would help organisations to introduce and implement it into their product development process.

This paper presents a newly defined SBCE process for the LeanPPD project and is structured into five sections, namely a review of SBCE, SBCE principles, SBCE baseline model, SBCE process and finally conclusions and future work.

## 2. A Review of Set-Based Concurrent Engineering

[7] advocate that set-based concurrent engineering (SBCE) is potentially the underlying cause for Toyota's various successes. They looked for evidence of a scientific product development approach in the Japanese and US automotive industries, and found it being practiced at the Toyota Motor Co. This work provided a case study of Toyota PD, but does not present a detailed process or methodology for SBCE. [5] built on this case study and provided more explanation for the SBCE process. The authors describe SBCE through an organised group of principles and a number of additional mechanisms that have been briefly described. The authors described the process as follows:"Design participants practice SBCE by reasoning, developing, and communicating about sets of solutions in parallel. As the design progresses, they gradually narrow their respective sets of solutions based on the knowledge gained. As they narrow, they commit to staying within the sets so that others can rely on their communication."

The above definition means that traditional product development approaches including point-based CE select one conceptual design as early as possible in the development process. This causes costly re-work as well as some of the resources are not going to be available at the re-work stage. In SBCE, the selection of the design is delayed as the design set is gradually narrowed based on the knowledge available to support decision taking. This will reduce or eliminate the re-work. [6] also compiled a textbook that described a number of Toyota PD mechanisms in detail with convincing arguments and simple rationale. The author states that the secret to lean product development is "learning fast how to make good products" and maintains this focus on learning, creating 'usable' knowledge and producing consistently profitable operational value streams throughout. Operational value streams are described as "the output of development and run from suppliers through plants into product features and out to customers" [6]. Ward emphasises on SBCE, supported by trade-off curves as the key elements of lean product development. Trade-off curves are graphs that evaluate one design attribute against another for a number of alternatives. In this approach the team breaks the system down into sub-systems and sub-sub-systems, identify broad targets at each level, and create multiple concepts for each component and whole system. They then filter concepts by aggressive evaluation, while capturing information in the form of trade-off curves and finally filter and converge based on the knowledge acquired.

Some of the typical challenges faced in product development are addressed by SBCE. These are explained in Table 1.

**Table 1: SBCE and Challenges faced in product development**

| Challenge | How SBCE Addresses the Challenge |
|---|---|
| Rework | Problematic design options are ruled out by developing and evaluating multiple alternatives in parallel |
| Sub-optimal Designs | Customer value is internalised and communicated holistically to all designers |
| Knowledge Crisis | An effective and coherent knowledge life-cycle facilitates the capture, representation and provision of the right knowledge to the right people at the right time |
| Lack of Innovation | Specific time and resources are scheduled for innovation, and multiple options must be considered as part of the process |
| High Unit Cost | By reducing rework, focusing on customer value, and improving communication and the process of PD, unit cost is reduced |

## 3. Set-Based Concurrent Engineering Principles

This section presents the main principles of Set-Based Concurrent Engineering as identified in several literature sources. The principles have been classified in five categories (Table 2), namely strategic value research and alignment, map the design space, create and explore multiple concepts in parallel, integrate by intersection and establish feasibility before commitment.

**Table 2: SBCE categories and principles**

| Category | Principle |
|---|---|
| 1. Strategic value research and alignment | • Classify projects into a project portfolio [4,6]<br>• Explore customer value for project X<br>• Align each project with the company value strategy<br>• Translate customer value (product vision) to designers (via concept paper) [4,5] |
| 2. Map the design Space | • Break the system down into subsystems [6]<br>• Identify essential characteristics for the system [6] |

| | |
|---|---|
| | • Decide on what subsystems/components improvements should be made and to what level (selective innovation) [6]<br>• Define feasible regions based on knowledge, past experience and the Chief engineer, and consider the different perspectives/functional groups [5] |
| 3. Create and explore multiple concepts in parallel | • Pull innovative concepts from R&D departments [6]<br>• Explore trade-offs by designing multiple alternatives for subsystems/components [5]<br>• Schedule time for innovation and problem solving while the set of alternatives is broad [4,6]<br>• Ensure many possible subsystem combinations to reduce the risk of failure [6]<br>• Extensive prototyping (physical and parametrical) of alternatives to test for cost, quality, and performance [4,5,6,7]<br>• Perform aggressive evaluation of design alternatives to increase knowledge and rule out weak alternatives [5,6]<br>• Information goes into a trade-off knowledge base that guides the design [6]<br>• Communicate sets of possibilities [4,5,7] |
| 4. Integrate by intersection | • Look for intersections of feasible sets, including compatibility and interdependencies between components [4,5,6]<br>• Impose minimum constraints: deliberate use of ranges in specification and initial dimensions should be nominal without tolerances unless necessary [5]<br>• Seek conceptual robustness against physical, market, and design variations [5,6]<br>• Concurrent consideration of lean product design and lean manufacturing |
| 5. Establish feasibility before commitment | • Narrow sets gradually while increasing detail: functions narrow their respective sets based on knowledge gained from analysis<br>• Delay decisions so that they are not made too early or with insufficient knowledge [5,6]<br>• Design decisions should be valid for the different sets and should not be effected by other subsystems [5]<br>• Stay within sets once committed and avoid changes that expand the set [5]<br>• Control by managing uncertainty at process gates [5]<br>• Manufacturing evaluates the final sets and dictates part tolerances [5]<br>• Manufacturing begins process planning before a final concepts has been chosen and thus act on incomplete information [5]<br>• Delay releasing the final hard specification to major suppliers until late in the design process [6] |

## 4. Set-Based Concurrent Engineering Baseline Model

After a critical analysis of the SBCE principles The captured SBCE principles have been analysed against the traditional product development approaches [2] and the various descriptions of the SBCE process, a number of phases were defined in order to represent the SBCE process. Although the phases may appear similar to some traditional PD models, the activities within them are unique and thus the phase names are intentionally unusual. Figure 1 illustrates graphically the SBCE baseline model that has been developed based on the captured principles. Customers and suppliers are involved throughout the product development process. During the first phase: value research, the initial product concept definition is developed based on a strategic and thorough internalisation and analysis of value. In phase 2: map the design space, design participants or subsystem teams define the scope of the design work required as well as the feasible design options/regions. In the third phase: concept set development, each participant or subsystem team develops and tests a set of possible conceptual sub-system design solutions; based on the knowledge produced in this phase some weak alternatives will be eliminated. In phase 4: concept convergence, sub-system intersections are explored and integrated systems are tested; based on the knowledge produced in this phase the weaker system alternatives will be purged allowing a final optimum product design solution to enter phase 5: detailed design.

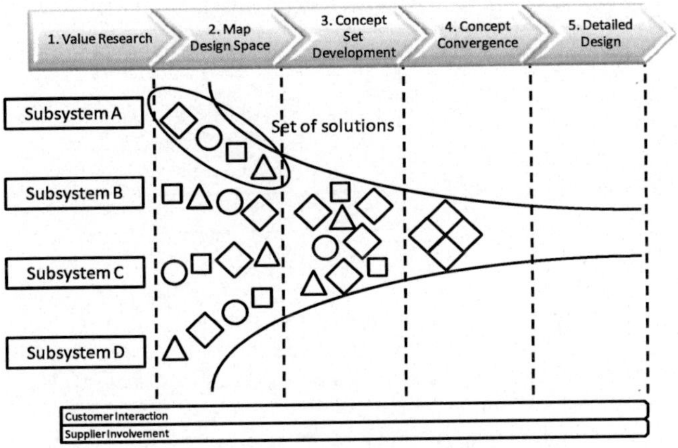


**Figure 1: SBCE baseline model**

## 5. Set-Based Concurrent Engineering Process

The SBCE process consists of several key phases, as shown in Figure 2.

Each phase is divided into activities, which are described as follows:

1. Value Research:
   1.1 Classify project type: the project will be classified and defined according to the level of innovation that will be incorporated (e.g. minor/ major modifications)
   1.2 Identify customer value: Customer value should be thoroughly understood in order to determine system targets and will be used throughout product development to measure the leanness of the alternative product designs

1.3 Align with company strategy: The project will be aligned with the company strategy to assess how the company can take strategic advantages from the project, such as increased process value
1.4 Translate customer value to product designers via product concept definition

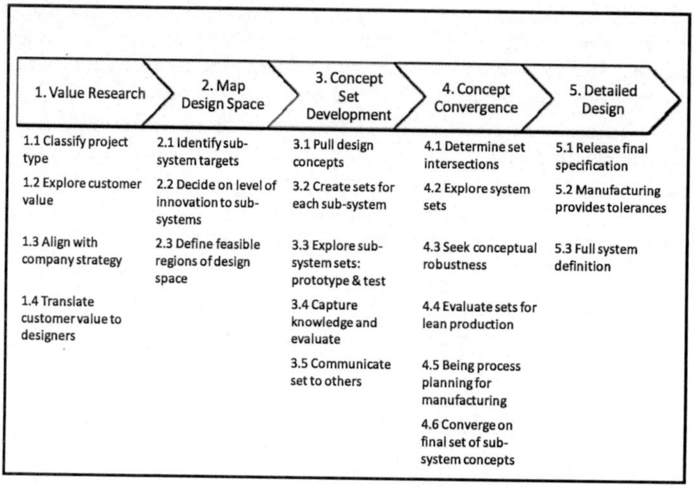


**Figure 2: SBCE process and activities**

2. Map Design Space:
   2.1 Decide on the level of innovation for system/ subsystems/ components (as appropriate): Each team will decide on which system/ subsystems/ components improvements should be made and to what level
   2.2 Identify system/ subsystem/ component targets: Each team will analyse their architecture and identify their own lower-level targets (essential characteristics) based on the system targets and product concept template
   2.3 Define feasible regions of design space: These are defined based on knowledge and past experience while considering the views/ constraints of different functional groups
3. Concept Set Development:
   3.1 Extract design concepts: Innovative concepts can be extracted from previous projects, R&D departments and competitor products
   3.2 Create tests for each subsystem: Design teams brainstorm so that a set of design solutions is created
   3.3 Explore subsystem sets: simulate/ prototype and test; Alternative solutions must be simulated/ prototyped and tested for cost, quality and performance
   3.4 Dynamic knowledge capture and evaluation: Knowledge that has been created must be captured (in a quantitative and qualitative manner) in order to evaluate the sets

3.5 Communicate sets to others: Each team will present their set to others in order to receive feedback and understand constraints

4. Concept Convergence:
   4.1 Determine set intersections: Potential systems will be integrated by the intersection of feasible sets, including compatibility and interdependencies between components
   4.2 Explore system sets: Potential systems should be simulated/ prototyped (parametric and physical) and tested for cost, quality and performance
   4.3 Seek conceptual robustness: Conceptual robustness will be sought against physical, market and design variations in order to reduce risk and improve quality
   4.4 Evaluate sets for lean production: Once the potential systems have been explored, they will be evaluated for lean production to assess the costs, efficiency, problems, etc.
   4.5 Begin process planning for manufacturing: Based on the evaluations, manufacturing can begin process planning for the possible sets that have been agreed to be feasible
   4.6 Converge on the final set of subsystem concepts: Based on the evaluations and knowledge captured, sub-optimal system designs have to be eliminated and the proven optimal design from the system set is finalised
5. Detailed Design:
   5.1 Release final specification: The final specifications will be released once the final set is concluded
   5.2 Tolerances' provision: Manufacturing will provide part tolerances to the design teams
   5.3 Full system definition: Further detailed design work will follow.

## 6. Conclusions and Future Work

The research presented in this paper provides an overview of the Set-Based Concurrent Engineering (SBCE) process that is being developed as part of the LeanPPD project [1]. This research extends the work of [5] and others who identified the merits of SBCE by providing a detailed and structured SBCE process. In order to develop a product development model that is fit to consistently perform in a rapidly changing market and environment a changeless core is required. The authors are developing a product development model that has a changeless core based on principles, which focuses on three central elements: value, knowledge (or learning) and improvement. The authors believe that this focus will enable companies to please customers through optimal designs, minimise design rework, and achieve high profit levels. The presented set-based concurrent engineering (SBCE) process addresses challenges that are faced by engineering companies in the 21st century and provides significant benefits over traditional approaches with regards to challenges such as rework, knowledge provision, and lack of innovation.

The research presented in this paper is work in progress. The activities defined in the SBCE baseline model and currently being developed so that they embody state-of-the-art product development methods while maintaining a focus on the principles upon which the process is based. Longitudinal studies are in progress within automotive, aerospace and home appliance sectors to understand the specific organisations needs to which the process will be applied and tested.


## Acknowledgements

The research presented in this paper has been conducted as part of a European project titled 'Lean Product and Process Development (LeanPPD)'. The project involves multiple research and industrial partners from the UK, Spain, Germany, Italy and Poland. The project is supported by the Commission of European Community, contract number NMP-2008-214090) under the NMP Programme (Nanosciences, Nanotechnologies, Materials and new Production Technologies). The authors wish to acknowledge the European Commission for its support as well as the other partners in the consortium (http://www.leanppd.eu).


## References


[1] Al-Ashaab, A., Shehab, E., Alam, R., Sopelana, A., Sorli, M., Flores, M., Taisch, M., Dragan Stokic, D. and James-Moore, M., (2010). The Conceptual LeanPPD Model. The 17th ISPE International Conference on Concurrent Engineering September, Cracow, Poland

[2] Atkinson, D and Al-Ashaab, A. (2008) A review and critical analysis of global New Product Introduction and Development, Int J of Product Development, vol. 6, no.2, pp. 118-141.

[3] Duverlie, P. and Castelain, J. (1999), "Cost estimation during design step: parametric method versus case based reasoning method", The international journal of advanced manufacturing technology, vol. 15, no. 12, pp. 895-906.

[4] Morgan, J.M. and Liker, J.K. (2006) The Toyota Product Development System, New York: Productivity Press.

[5] Sobek, D. K., Ward, A. C. and Liker, J. K. (1999), "Toyota's principles of set-based concurrent engineering", Sloan management review, vol. 40, no. 2, pp. 67-84.

[6] Ward, A.C. (2007) Lean Product and Process Development, Cambridge, MA, USA: Lean Enterprise Institute.

[7] Ward, A.C., Liker, J. K., Cristiano, J. J. and Sobek, D. K. (1995), "The second Toyota paradox: How delaying decisions can make better cars faster", Sloan management review, vol. 36, pp. 43-43.

# Capacity fade model of Lithium-ion batteries for Practical use

Wataru Imamura[a,1], Noriataka Eda[a], Kenji Tanaka[a], Hideaki Horie[a], Hiromichi Akimoto[a]

[a]School of Engineering, The University of Tokyo, Japan

**Abstract.** Lithium-ion (Li-ion) batteries are essential components for electric vehicles and smart grid systems that will become increasingly widespread in the future. However, the complexity of the capacity fade mechanism of Li-ion batteries has been a technical barrier to their practical use. This study proposes a capacity fade model composed of three factors: capacity loss, increase in internal resistance, and diffusion effect. The model derives sufficient information from battery data and conditions of use for evaluating and predicting capacity fade. The battery data includes discharge curves and electrode materials. The conditions of use are elements correlated to the three factors listed above and include the number of cycles and storage time. This paper discusses the availability of our proposed model by applying it to Li-ion battery data obtained from reference papers.

**Keywords.** Lithium-ion battery, capacity fade, lifetime assessment

## 1. Introduction

An appropriate method for evaluating the capacity fade of lithium-ion (Li-ion) batteries is required for an eco-friendly society in future. The capacity fade mechanism is so complex that it is now impossible to predict the battery lifetime for practical use, such as electrical vehicles and smart grid systems. Therefore, we need to elucidate the capacity fade mechanism of Li-ion batteries and develop an evaluation model that would help our society achieve sound and reasonable development by eliminating defects from the market, opening the door to secondary-use markets and promoting investment.

There are mainly two approaches to studying the capacity fade mechanism. One is to directly observe the causes of capacity fade through electrochemical analysis such as X-ray diffraction, X-ray photoemission spectroscopy and scanning electron microscopy [1][2]. The second approach is to develop mathematical models for predicting the service life [3] [4]. However, with former approach the level of capacity fade could not be quantified and second approach models do not identify the cause of capacity fade and lack versatility since they were applied to a small number of cases using specific cells and conditions.

Ramadass et al. [5][6] proposed a capacity fade model composed of three factors: secondary active material, primary active material and rate capacity loss, and experimentally quantified the level of capacity fade. They also developed a mathematical model using the three factors for predicting capacity fade. However, the approach is unsuitable for practical use since the proposed factors were obtained under specific experimental circumstances.

[1]Graduate student, School of Engineering, the University of Tokyo, 7-3-1, Hongo, Bunkyo-ku, Tokyo, 113-8656, Japan; Tel: +81 3 5841 6522; Fax: +81 3 5802 3374; Email: Imamura@triton.naoe.t.u-tokyo.ac.jp

D. D. Frey et al. (eds.), *Improving Complex Systems Today*,
Advanced Concurrent Engineering, DOI: 10.1007/ 978-0-85729-799-0_52,
© Springer-Verlag London Limited 2011

The objective of the present study is to propose a capacity fade model of Li-ion batteries practical for evaluating and predicting residual performance. The proposed model defines three factors of capacity fade that are applicable to all types of cells and conditions, and that can be obtained through a simple process. The evaluation process and the prediction process using the three factors are introduced in this paper. This paper also discusses the availability of our proposed model by applying it to Li-ion battery data obtained from reference papers.

## 2. Proposed Capacity Fade Model

### 2.1. Modeling Approach

Li-ion battery performance is reduced after use or storage. In Fig. 1, the progress of capacity fade can be seen by the change in the discharge curves. With our proposed model, the capacity fade is attributed to three factors: capacity loss, increase in internal resistance and effect of diffusion. The capacity fade due to these factors is expressed as characteristic shifts in the respective discharge curves and the linear combination of these three shifts represents the change in discharge curves by capacity fade, as illustrated in Fig. 1. This fact indicates that monitoring the respective changes in the three factors is sufficient for evaluating the capacity fade of Li-ion batteries. The changes in factors depend on several conditions such as the electrode materials and temperature. Thus, the capacity fade can be predicted by using the data on the three factors and correlated conditions.

### 2.2. Capacity Loss

Capacity loss is caused by the decrease in number of movable Li ions. Li ions move between the positive and negative electrodes to charge and discharge. However, some Li ions become fixed in the negative electrode or in the reactant on the electrode surface, resulting in the Li-ion battery lowering the limit on discharging capacity. Thus, the discharge curve shifts in the direction of the capacity axis, as shown in Fig. 2.

### 2.3. Increase in Internal Resistance

An increase in internal resistance is caused by film growth on the surface of the negative electrode. During the charge/discharge cycle, the electrolyte reacts with the negative electrode and reactants form on the surface of the electrode. This reactant film interferes with the movement of Li ions necessary for charging or discharging. This results in the Li-ion battery requiring more voltage to charge and discharge. Thus, the discharge curve shifts in the direction of the voltage axis, as shown in Fig. 3.

### 2.4. Diffusion Effect

The diffusion effect is caused by the lag in the movement of Li ions described by the diffusion equation. The lag is equivalent to the internal resistance. Thus, the discharge curve shifts in the direction of the voltage axis, and the extent of the shift increases with the increase of voltage, as shown in Fig. 4. This type of capacity fade can be estimated by the diffusion equation if the diffusion coefficient is known.

### 2.5. Conditions Correlated to Capacity Fade Factors

Capacity fade of Li-ion batteries occurs under various conditions. Elucidating the correlation between the conditions and the three capacity fade factors allows us to predict the capacity fade. The conditions are classified into two categories: electrode materials and status of use. Different combinations of positive and negative electrode materials influence the battery performance. Capacity fade also varies with different combinations of electrode materials. Based on the status of use, the capacity fade processes are divided into two types: cycle degradation and storage degradation. Cycle degradation depends on the number of charge/discharge cycles, storage degradation depends on the storage time, and both types depend on the temperature, current density, state of charge and depth of discharge.

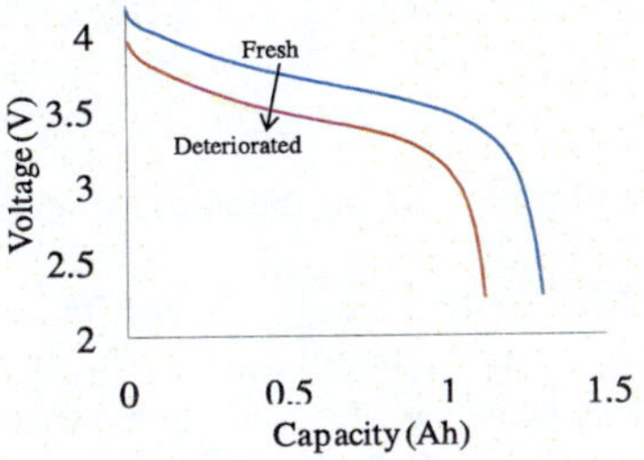


Fig. 1 Change in discharge curve by capacity fade

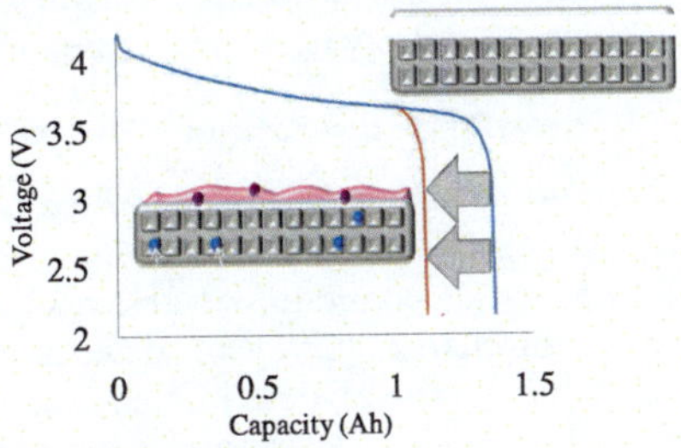


Fig. 2 Capacity loss

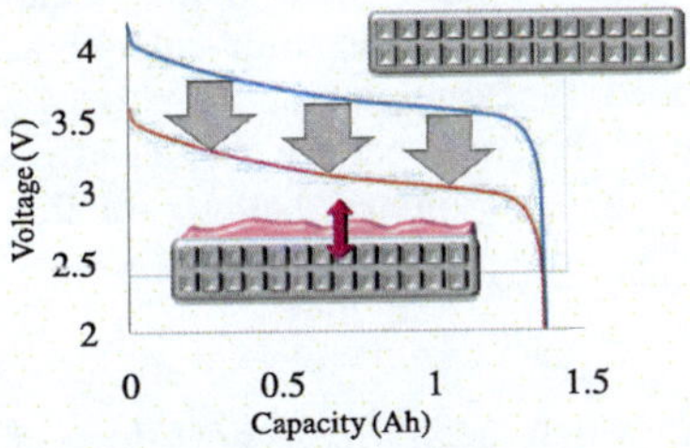


Fig. 3 Increase in internal resistance

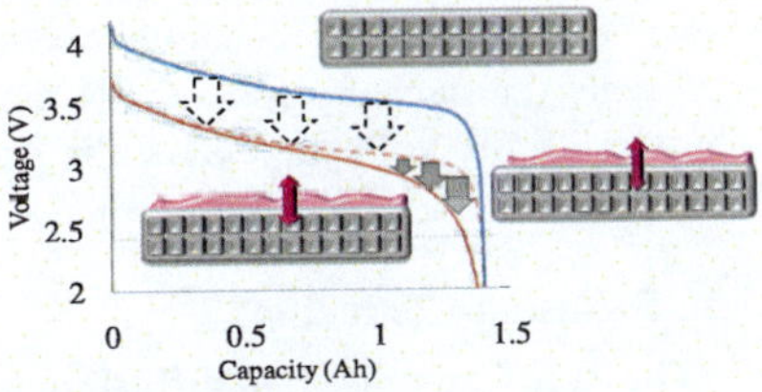


Fig. 4 Diffusion effect

## 3. Evaluation Method for Capacity Fade

### 3.1. Flow

Fig. 5 shows the capacity-fade evaluation flow for Li-ion batteries. In order to begin the evaluation, the proposed capacity fade factors are obtained from the battery database. The database includes the discharge curves of both fresh and deteriorated Li-ion batteries, the capacity fade of which is needed for evaluation under the different conditions. The three capacity fade factors are quantified based on the manipulation of data on the fresh and deteriorated battery discharge curves.

### 3.2. Digitization

The evaluation flow starts with digitizing the discharge curves in order to obtain the value of the increase in internal resistance.

In the applied digitization method, numerical data on the voltage and capacity is obtained at regular intervals along the voltage axis from the upper to lower operation limit voltage. The increase in internal resistance is obtained by comparing the capacity data between fresh and deteriorated batteries. The increase is calculated as expressed in Eq.(1). $\Delta R_I$ represents the increase in internal resistance. $I$ is the discharge rate used to describe the discharge curves. $\Delta V$ is the correction voltage that represents the absolute value of the decreased voltage resulting from the increase in internal resistance. The correction voltage is calculated as the difference between the voltages of the fresh battery and the deteriorated battery near the voltage axis.

### 3.3. Resistance Correction

In the second step, the resistance is corrected and the value of the capacity loss is obtained.

Resistance correction neutralizes the effect of the increase in internal resistance in the discharge curves of deteriorated batteries. This manipulation shifts the discharge curves of deteriorated batteries in the direction of the voltage axis for correction voltage, as shown in Fig. 6. After the resistance correction, the digitized voltage values are converted as expressed in Eq.(2) where $V_{RC}$ is the resistance-corrected voltage and $V_D$ is the original deteriorated voltage.

Capacity loss is obtained by comparing the deteriorated and resistance-corrected battery capacity data on the lower operation limit voltage, as shown in Fig. 7. The capacity loss is calculated as expressed in Eq. (3). $\Delta C$ represents the capacity loss. $C_F$ is the capacity of the fresh battery on the lower operation limit voltage. $C_D$ is the capacity of the deteriorated battery on that voltage.

### 3.4. Diffusion Correction

The third step is the diffusion correction, which neutralizes the effect of diffusion in the discharge curves of the resistance-corrected batteries. This manipulation reduces the size of the fresh battery discharge curve in the direction of the capacity axis for capacity fade, as shown in Fig. 8. After the diffusion correction, the digitized voltage values are converted as expressed in Eq. (4) where $C_{DC}$ is the diffusion-corrected capacity and $C_F$ is the fresh capacity.

The diffusion effect is shown in the figure as the area bounded by the diffusion- and resistance-corrected discharge curves. Thus, the resistance increase caused by diffusion is presented as the gap between the voltage of the diffusion- and resistance-corrected batteries, as shown in Fig. 9. $V_{DCj}$ is the diffusion-corrected battery voltage and $V_{RCj}$ is the internal-resistance-corrected battery voltage under the condition that one corresponding capacity is equivalent to the other corresponding capacity. The resistance increase due to the diffusion effect at a certain capacity is calculated as in Eq. (5). Using the data, the diffusion coefficient is obtained by discrete simulation based on the diffusion equation.

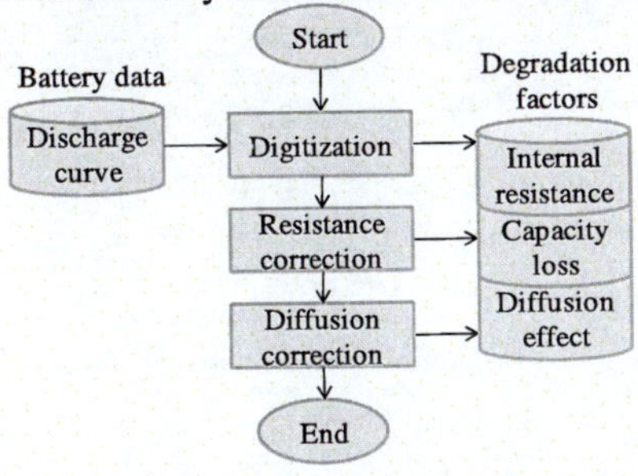


Fig. 5 Evaluation flow of capacity fade

$$\Delta R_I = \frac{\Delta V}{I} \quad (1)$$

$$V_{RCi} = V_{Di} + \Delta V \quad (2)$$

$$\Delta C = C_F - C_R \quad (3)$$

$$\Delta R_{Dj} = \frac{V_{DCj} - V_{RCj}}{I} \quad (4)$$

$$C_{DCi} = C_{Fi} \times \frac{C_{DN}}{C_{FN}} \quad (5)$$

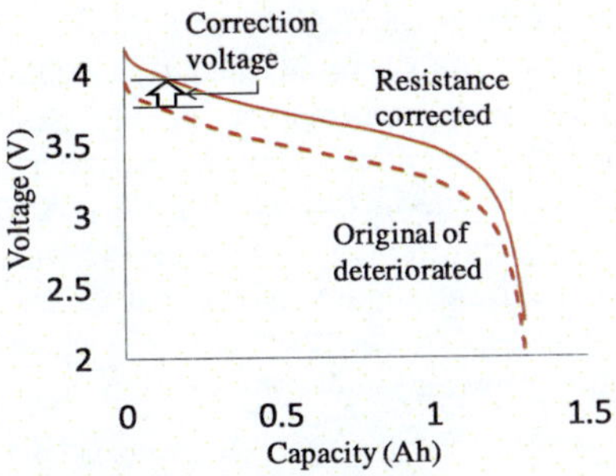


**Fig. 6** Resistance correction

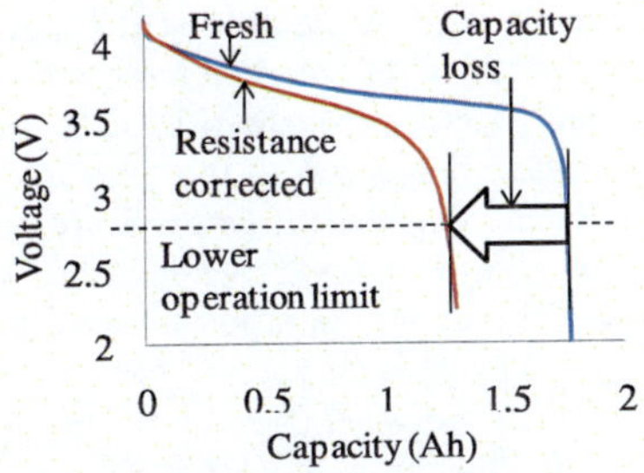


Fig. 7 Obtaining the capacity loss

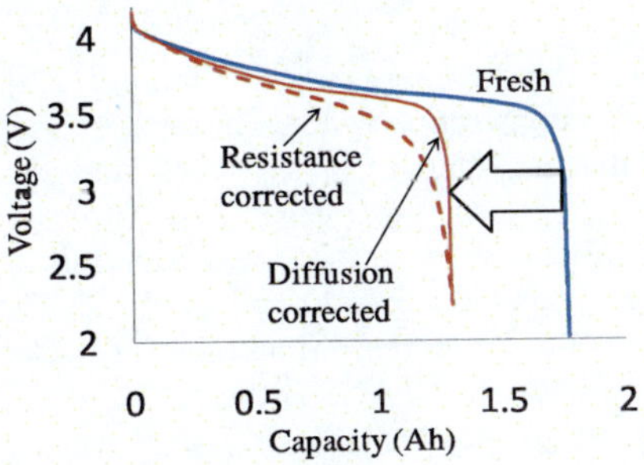


Fig. 8 Diffusion correction

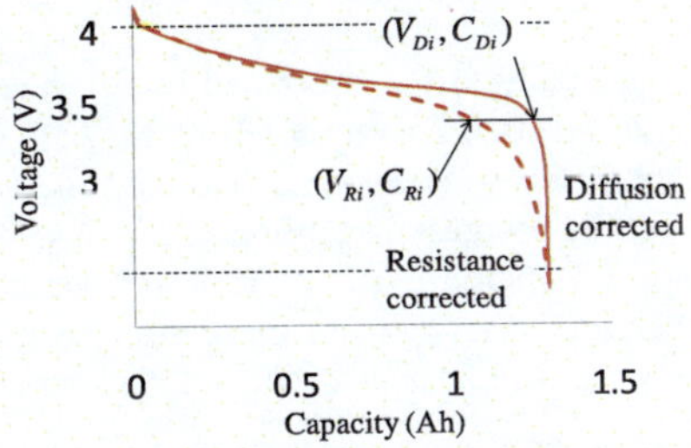


Fig. 9 Choice of voltage and capacity sets for calculating diffusion coefficient

# 4. Prediction Method for Capacity Fade

Fig. 10 shows the capacity-fade prediction flow for Li-ion batteries. This prediction system uses two databases, one for battery data and the other for capacity fade factors data. The battery data includes the discharge curves and conditions. The discharge curves are used as the basis for prediction and the conditions are used to determine the capacity fade factors for prediction. Examples of conditions are shown in the section 2.5. The prediction of capacity fade is achieved through three calculations using the estimated capacity fade factors. After the predicted state is realized, the estimated factors are revaluated through comparison between the predicted and actual state.

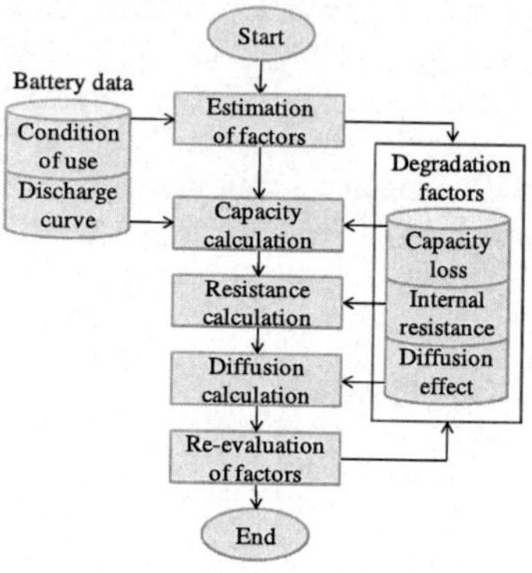


Fig. 10 Prediction flow of capacity fade

# 5. Case Study

## 5.1. Conditions of Verification

The validity of the proposed capacity fade model is verified by applying it to Li-ion battery data obtained from reference papers. In this section, the battery data obtained from one paper[5] that indicates cycling performance at elevated temperature is applied to the proposed model for verification. The Li-ion battery used for the experiment was a Sony 18650 cell with rated capacity of 1800 mAh. The material of the positive electrode was $LICoO_2$ and the negative electrode was carbon. The upper operation limit voltage was 4.2 V and the lower was 2.0 V. The cells were cycled at 120, 300, 500 and 800 cycles at 45°C. The discharge rate was a constant current of 1 A. For charging the cells, conventional constant current – constant voltage (CC-CV) was adapted.

## 5.2. Capacity Fade Evaluation

Capacity loss is obtained by digitization of the discharge curves in the paper[5], as an original curve shown in Fig. 11. For the digitization result, the capacity loss at each cycle number is obtained as presented in Table 1.Capacity loss increases with the increase in number of cycles.

The increase in internal resistance is obtained by resistance correction as described earlier. The correction results are shown in Fig. 11 and Table 2. The increase in internal resistance also increases with the increase in number of cycles

[1]Graduate student, School of Engineering, the University of Tokyo, 7-3-1, Hongo, Bunkyo-ku, Tokyo, 113-8656, Japan; Tel: +81 3 5841 6522; Fax: +81 3 5802 3374; Email: Imamura@triton.naoe.t.u-tokyo.ac.jp

from 150 to 500, but decreases slightly from 500 to 800. The decrease could be due to digitization errors caused by overlapping of the original graphs, but the decrease in resistance increment is reasonably induced.

The diffusion effect is also obtained by diffusion correction as described earlier. Fig. 12 shows the results of diffusion correction. The figure indicates that the resistance increases due to the effect of diffusion at every cycle number.

## 5.3. Capacity Fade Prediction

Based on the discharge curves at fresh, 150 and 300 cycles, the capacity fade and discharge curves of the battery at 500 and 800 cycles are predicted using the proposed capacity fade factors. Here, to simplify the prediction process, the capacity loss of the diffusion effect is excluded and the capacity fade factors at 500 and 800 cycles are estimated by linear projection of capacity loss and increase in internal resistance at 150 and 300 cycles.

Fig. 13 shows the results of prediction using the estimated factors and original data at 500 and 800 cycles. The predicted discharge curves are illustrated with dotted lines and the original curves are illustrated with continuous lines. The difference in the lines represents prediction errors.

Table 3 shows the evaluation of the prediction errors of the battery data. At 500 cycles, the error in battery capacity is around 8% and the error in battery resistance is less than 1%. At 800 cycles, the error in battery capacity is around 6% but the error in battery resistance is almost 50%. These results indicate that simple linear projection of the increase in internal resistance from 150 and 300 cycles to 800 cycles is not suitable for prediction.

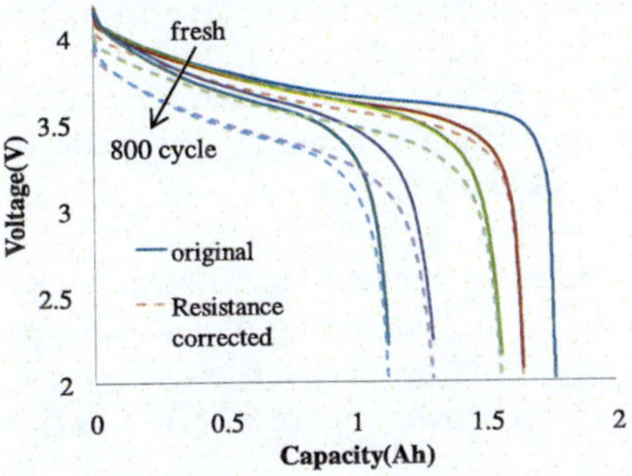


Fig. 11 Original and resistance-corrected curves

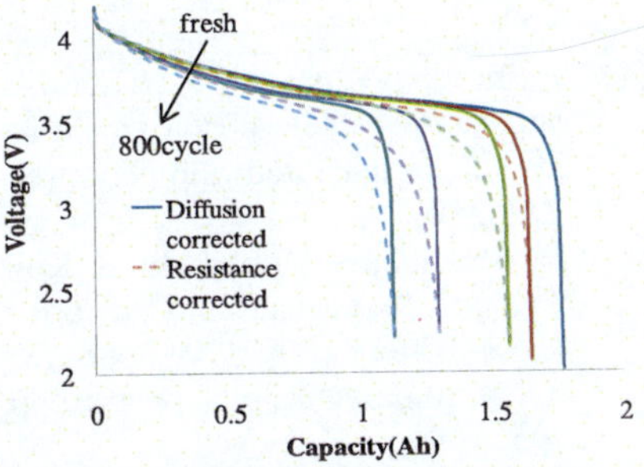


Fig. 12 Diffusion-corrected curves

Table 1 Value of capacity loss

| Cycle | 150 | 300 | 500 | 800 |
|---|---|---|---|---|
| Capacity Fade[Ah] | 0.13 | 0.21 | 0.48 | 0.65 |

Table 2 Value of increase in internal resistance

| Cycle | 150 | 300 | 500 | 800 |
|---|---|---|---|---|
| Resistance Increase [Ω] | 0.05 | 0.14 | 0.22 | 0.20 |

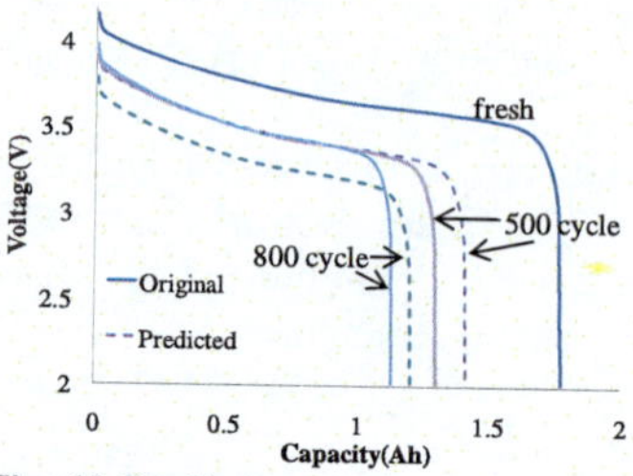


**Fig. 13** Prediction results

Table 3 Prediction errors

| | CN=500 | CN=800 |
|---|---|---|
| Capacity | 8.16% | 6.42% |
| Resistance | 0.33% | 46.82% |

## 6. Conclusion

This study proposes a capacity fade model for Li-ion batteries composed of three factors. The model allowed us to quantitatively evaluate and predict the capacity fade of Li-ion batteries using discharge curves and other conditions easily obtained for practical use.

Using battery data from reference papers, the model was applied to the estimation and prediction of capacity fade. In this particular case, the prediction was not very precise due to the small amount of available battery data. For further study, we are now collecting battery data through experiments to improve the accuracy of the model and its development.

## 7. References


[1] J. Shim, R. Kostecki, T. Richardson, X. Song, K.A. Striebel: Electrochemical analysis for cycle performance and capacity fading of a lithium-ion battery cycled at elevated temperature, Journal of Power Sources 112 (2002) 222–230.

[2] M. Shikano, H. Kobayashi, S. Koike, H. Sakabe, E. Ikenaga, K. Kobayashi, K. Tatsumi: Investigation of positive electrodes after cycle testing of high-power Li-ion battery cells II An approach to the power fading mechanism using hard X-ray photoemission spectroscopy, Journal of Power Sources 174 (2007) 795–799.

[3] A.P. Schmidt, M. Bitzer, Á.W. Imre, L. Guzzella: Model-based distinction and quantification of capacity loss and rate capability fade in Li-ion batteries, Journal of Power Sources 195 (2010) 7634–7638.

[4] N.A. Chaturvedi, R. Klein, J. Christensen, J. Ahmed, A. Kojic: Modeling, estimation, and control challenges for lithium-ion batteries, American Control Conference (2010) 1997–2002.

[5] P. Ramadass, B. Haran, R. White, B.N. Popov: Capacity fade of Sony 18650 cells cycled at elevated temperatures Part II. Capacity fade analysis, Journal of Power Sources 112 (2002) 614–622.

[6] P. Ramadass, B. Haran, R. White, B.N. Popov: Mathematical modeling of the capacity fade of Li-ion cells, Journal of Power Sources 123 (2003) 230–240.

# Part VII
# Emotional Engineering

# Color and illumination in the project of interior aircraft

Viviane Gaspar Ribas[a,1], Fernando Molinari Reda[b]

[a]University of Federal of Paraná
[b]Embraer Brazilian Company

**Abstract.** The purpose of the article is to present a project developed in partnership with the University Federal do Paraná and EMBRAER the periods from 2003 to 2006 whose focus was the development of a roadmap chromatic in accordance with the profile of flight of commercial jets. Will be an overview of the process of design that was adopted for the development of the project steps involving this project for its effective conclusion, in order to demonstrate that the design must be prepared to assume greater role in the development of products differentiated value of technology. The project included not only with the participation of professors and students in the Department of Design of University Federal do Paraná but also with engineers, architects, designers, pilots, psychologists flight of the company among other partners in the project.

**Keywords.** Aircraft Project; Design Process; Boarding of the Manufacturers.

## 1 Introduction

The article aims to present information about the effect of color on human behavior in the interiors of environments, whether architectural or transportation. The color, by their physical characteristics, affects human beings in their attitudes, choices, feelings, impulses, desires and rejections. The human mind and body react consciously and unconsciously to the colors and thus cause effects on respiration, circulation, blood pressure, mood, thermal sensations, and other sensations. Many studies claim that the applications of color in environments affect the individual's performance in their activities, whether business or leisure. Alter behaviors, such as mood, satisfaction, motivation or even the performance of their daily activities and also considering the complexity of the tasks in the approach to care for their achievement.

It is intended to show the reader how the colors, when applied to environments and added to the factor light (illumination), can be powerful tools in creating a sense of space. When the tones gradually change from dark to pale, creating subtle gradations of brightness that accompanies the gaze towards the distance suggested.

[1] Professor of Design of UFPR (Designer Bal. Msc and Dr. Engineering), Paraná. Rua General Carneiro 460, 8° andar, Curitiba PR Brasil, Tel/ Fax: +55 (41) 33605360 Email: viviane.gasparibas@ufpr.br; http://www.design.ufpr.br

D. D. Frey et al. (eds.), *Improving Complex Systems Today*
Advanced Concurrent Engineering, DOI: 10.1007/ 978-0-85729-799-0_53,
© Springer-Verlag London Limited 2011

Exploring the nature of color is key. The red and yellow (warm colors in general) tend to move while the blues and other cool colors tend to recede. Thus, the present study as a contribution to presentation of the product development process technology Wash Light LED Strip RGB (Red Green Lightings Emition Diodes Blue - provided by Avionics Services) developed for the interior of commercial aircraft.

The research result is presented below; the text is structured so as to highlight the initial design process adopted for the development of a new integrated component of the flight profile of commercial aircraft, followed by closing remarks.

## 2 Color in design

"The color, with its unlimited possibilities aesthetic and psychological is not intended to create environments exhibitionists, but make life more pleasant." [1] the color is as important as the whole history of civilization. For accompanied man since prehistoric times to the present, as well as the pursuit of their senses.

Like the color, the physical environment to have always been present in human life and its origin stems from the need to have adequate space. Not just for shelter and protection from the weather, but also to develop their general activities, leisure and home, furthering their social and existential needs. [2].

For example, red has been associated with vigor, anger, stress [3], excitement, stimulation and happiness [4]. But the blue and blue-green has been related to the relaxation [3], comfort, safety, peace and quiet [3]. The shades of blue nuances of blue are related to less distress, anxiety, drowsiness, and nicer than red.

"The man and the environment can`t be considered as independent elements, instead there is an interaction between them, and considering the feelings and human perceptions, then every action, reaction, and ways of acting are interfered with by the environment. In the design of environments, the color appears as a powerful element which can cause feelings or even promote emotional well-being. However, only the color of the spaces is not sufficient to achieve the goals determined in the conceptual phase of a project, one must consider the functional aspects of space, the characteristics of tasks and users, though, the psychological and even physiological in individuals living in this space. As with architectural spaces, the colors in the composition of transportation undergo psychological parameters, and can be used as a metaphorical suggestion of cultural meanings indexed, thus contributing both to create strong attraction and impact as for relaxation and comfort.

The colors are elements of human behavior modifiers. It is worth mentioning that through the stimulation of the senses, the more sensations are possible to be stimulated and aroused, more efficient are the objects. [5]. Add new concepts and technologies to design and generate user satisfaction, is understood as a prerequisite that must be considered in product design. So the design professional to venture activity, for example, light design should be prepared to work with several interfaces for knowledge, addressing issues from the area of design, to management, technology, psychology or even engineering and architecture.

## 3 Air transport X passenger comfort

Currently, air transport, because of its speed and security, is responsible for millions of people move each year. The size of this market makes aircraft manufacturers and airlines to invest in new equipment and staff training programs. However, usually this type of investment does not focus on solving problems that only now beginning to be considered and that are directly related to passenger comfort.

For example, the fact that an airplane has a very modern cockpit, the passenger does not mean that you are feeling comfortable, especially if he suffers from some kind of phobia related to air travel. A highly significant difference for the passenger is his physical and psychological comfort. [6].

On the issue of passenger comfort, there are several works to enlarge it, ranging from the in-flight service, cooling the cabin space, seating, lighting, entertainment, leisure, among others. And as the airline industry has several suppliers of parts and aircraft parts, in general these are responsible for the project, which can only meet the design requirements and passed by the industries or customers or develop new projects that meet with the trends the aeronautical market. [6]. The "Roadmap to the Integrated chromatic flight profile.

In this part of the article was organized descriptive design development above named product of a partnership between the Federal University of Parana, Curitiba-based firm Embraer and headquartered in Sao Jose dos Campos. The dynamics that occurred came from the project management scope which consisted of several steps that have been developed to ensure that the project includes all activities necessary to achieve pre-established goal and desired success. The project scope was defined by an agreement called the Project Charter (formalized by contract to provide service between the parties). That agreement called Project Roadmap Interactive Chromatic Lighting System built into the Agreement and the Envelope of Flight of Commercial Aircraft, aimed to associate the color lighting system inside the cabin of passengers. Just create an innovative environment as the colors associated with desirable psychological and physiological states of the passengers. The result should allow the user to experience new sensations air transport, for example, those that allow him to enter a contemplative state where phobias eases arising from fear of flying. In this context, the color applied to the Lighting System in Aircraft Interior, provided that associated with human activities undertaken in each moment of flight, sent on ergonomic issues. The direction of the result was to promote the safety, health, comfort and welfare of passengers. For example, generate an environment conducive to reassure the passengers during the most stressful moments, such as take-offs and landings; facilitate the process of loading and unloading; highlight the flavors' of food during the on-board service, creating a feeling of comfort and wellbeing during times of rest, or even induce passengers to sleep on long journeys.

The color applied to the lighting system should lead to perceiving new passenger information, such as spatial sensations, thermal and aesthetic of the interior, which alters your mood and makes him feel more secure and comfortable. Besides being the color modifier element of human behaviour, it was believed that the development of this project would enable ease phobias, from the state of tension that the tunnel leads to the interior of aircraft. So the facts the user feels more comfortable and safe, and, consequently, happier, could be a decisive factor to win their loyalty. It is noteworthy that the comfort factor is targeted by the users of airlines as a deciding factor in choosing the company, and this factor is above price, speed and security [7].

All these considerations not only show the difference in innovation of this project, but also indicate that, above all, your goal is to add more values to the subject aircraft. The planning and project execution in the initial phase the project team prepared a document covering all phases of the project scope, project flow chart, detailing the stages and steps to be performed. Each stage contained a description of the problem for analysis, methods, results, assumptions and restrictions. This document also guide the implementation of the stages of research served as quality control, shift in decision making and facilitate the search for information.

Detailed planning has guided the team in making decisions or changes to the route design, and even lends a better project management and provides the knowledge areas of PMI. Each stage of the project was the responsibility of both teams (sometimes company employees, hours of university teachers and students), however, the dedication (hours) varied depending on the expertise found at the same time or in discussions held between the project coordination. At each step, it is stated the executor and supervisor of the steps.

Regular meetings were held throughout the duration of the project (2003-2006), with no regular intervals between some of the meetings which dealt with administrative matters hours of research process and project development. The tools of time management, information exchange and documentation used by the team helped to ensure agility in decisions or to improve group interaction. This factor has become relevant in view of the members of the project, being geographically located in different states, or even in different workplaces.

Another factor which determined the exchange of confidential information was the question of the project, formally subject matter between the company Embraer and the Federal University of Parana in the confidentiality agreement. Whose contents is UFPR get Embraer and Embraer UFPR receive documentation and information involving sensitive and confidential technical and commercial aspects, among other information, which may even be considered sensitive to national security.

As the project progressed, the team for being innovative handle design, performed other activities throughout the development must be done. Soon some of the activities contained in the execution of steps or delivery of results documented and considered unsatisfactory resulted in changes in deadlines and deliverables for next steps. On three of the project if they did change the schedule originally proposed. But in fact occurred and, without changing the project costs, as several steps demanded only of information from experts in the areas that supported the project. All results and documents generated by both teams were fully disclosed to all members of the project team. A summary of the design phases can be observed in their execution order in the TABLE 1, but in summary form.

**Step 1 - Recognize the procedures of the agreement of the flight envelope**

In this step was identified as a major contributor to the project test pilot of the company. This was submitted to an open interview, and being all the team members gathered together and armed with the document called internally by Typical Profile of Flight 190, for the pilot to clarify any technical procedures adopted in the various phases of aircraft flight. With the activity performed was possible to identify all the data needed to prepare the document that summarizes the information arranged in table, which describes the approximate times of each activity, the name of the flight phase, synthesis phase and operating procedures that occurred the aircraft during flight. The document generated by the team was again subjected to a review by the pilot, with that ensured the reliability of the information collected.

**Step 2 - Recognize the activities undertaken during the flight in the cabin by passengers**

As in the previous step was identified group of employees as both commissioner and head of the commissariat commissariat as all participated in the interview closed, which was conducted by Typical Flight Profile document, and a document from Step (1) and guidelines for developing interviews. This interview aimed to obtain more detailed information about human activities which are carried by passengers during flight procedure, and information about the behavior of passengers identified, reasons for these behaviors, and reason for travel.
With this procedure, sought to show at what time the flight, behaviors can interfere with normal routines such as phobias fear of flying in several phases of flight. Another activity, extra deemed important by the team to better understand human behavior of the passengers was extracted by means of an internal survey conducted with employees of Embraer held for a period of five (5) working days using the intranet. Data from approximately 850 employees. The results were compiled statistically collected and stored in a separate document and thus confirm which flight phases in which people show more Joy, Satisfaction, Anxiety, Fear or rabies.

**Step 3 - Analyze the activities of the special situation of human flight.**

In this step, first, was identified as employees of the psychologists of the project company and the pilot group and the Commissioner attended the new interview. This activity was driven by Typical Flight Profile document, as well as the Document generated in Steps 1 and 2, to facilitate the interaction of time with such experts in the field of psychology. Besides this activity, was sent a spreadsheet with the intention of detailing the psychological profile of the passengers and thus completing the stage of collecting information about the profile of Flying under the prism of the Pilot, the Commissioner, Passenger and Psychologists.

**Step 4 - Co-relate the parameters of the sheets 1, 2 and 3**

At this stage members of the Federal University of Parana in Curitiba gathered within the university to compile all the documents generated in the previous steps (1, 2 and 3). The documentation was assessed and correlated the most relevant information from the technical aspects of the operational procedure of flight, achieved by the human passengers as well as information about the behavior and psychological profile of passengers. After the generation of these documents, the information was available for consultation in one document found in the documentation project that was developed in order to facilitate the search for information later in the meeting which was held among all participants and this in the next step.

**Step 5 - Analyze the influence of colors that affect passengers**

At this stage, a presentation was organized, which provide all information about colors and their influences on humans. Next was shown the list of information contained in this presentation. Although the activity was developed correlation of documents 4:05 already defined and, therefore giving the continuity of the project.

**Step 6 - Set the color interactive script**

Performed to generate a scheduled activity (1) Screenplay Interactive Chromatic in the RGB (RedGreenBlue) which describes in detail all RGB, drew up a meeting between the team members for the presentation of results. With use of this document was presented the proposed colors. The result was asked what drove the team to prepare a review of all information. The results of this review led to a new path in order to streamline the number of flight phases as presented in the document profile Typical Flight 190. This occurred because the time of each flight procedures to be short enough to not affect too much the use of colors. The result of this reduction in the number of colors / sensations in phases, this summary of the number of colors based on human feelings you want. After time afforded by the new theoretical orientation as a result of the problem was made a second proposal. In the activity schedule of the Roadmap Interactive RGBs defined in the Agreement Chromatic Flight Envelope, the activity of test scheduling RGBs defined in the roadmap we used a 3D simulation program which generates all of the proposals set for chromatic and simulation study. The activity of a test (1) time in the Roadmap Interactive Chromatic Standard Existing Embraer in 3D Visualization System team at the time of the flight was conducted with members of the group to check the image quality or even chromatic light interference in the computing environment. The following activities were carried out in meetings between the members that there is a big difference in quality and perception of the color space between the real space and computing power. Soon a review or even change the roadmap at this stage would be performed in real space.

**Step 7 - Validate the script interactive color in real environment**

At this stage all activities were planned since the programming of RGBs within a 1:1 mock-up of the aircraft used as the standard model EMB190 (see FIGURE 1). The project documents was made available to all members for a long time to planning delegating the activities to be fulfilled for the success of tests in real environment.

FIGURE 1: Picture of the interior mock-up of the EMB 190 1:1 at the time the tests.

| |
|---|
| The analysis of the results were discussed among the group members at the time of simulation, comparing effects of colors on objects, space, skin types and food usually served. RGB adjustments were made in different situations, with lights core mock-up 1:1 totally erased, light or half full access. All colors are registered in the database of images and can be viewed on the project documentation. Were administered questionnaires and forms general. Before applying the questionnaire was presented to the volunteers an overview of the project, with only the number of colors simulated to check if the desired effect of the colors had been struck. The tests were based on several authors. The results are recorded in digital photography and in digital cinema. The activity of this set the stage Roadmap Interactive Chromatic Final. At the end of the project all work was recorded in documentation sent to the administrator of the project which includes the area of project management. Documents from the project plan (with the scope statement) document the steps, minutes of meetings generated in the period, pictures, forms, materials generated by all members, are available at the company in digital media and printed, and serves as a basis for consultation for other projects later. |

TABLE 1: Resume of the steps the project

## 4 Concluding

In the process of developing a new product, as well directed and managed, it creates a better product, more competitive, increasing also the level of technological expertise and manufacturing processes used. From the volume of information circulating, to generate new knowledge, of a collective, every step of the process, turns into individual and organizational growth. Particularly in this project, where design orientation was in the process of product development technology Roadmap Interactive Chromatic, "and which was related to the various areas of the company as well as all stakeholders of the system as user, manufacturer, members of the team project and the industrial product. It is noticed that on this basis, the process of developing this new product involved a commitment that pervades the designer of course, the communicative and functional level, responding to the physiological, psychological and social use, as well as psychological aspects involved in designing the new product. Products with "known needs" are constantly evolving to meet market expectations. In an unending process of invention and reinvention of solutions to problems, where the practice is the search for new realities, potentially creative and innovative, involving diverse contexts and subject to specific desires, emotions, or conflicting ideologies. But the creation and design professionals should always seek to overcome market expectations, seeking always a need, a feeling, a yearning not yet surfaced. Within this context, one has sought to project "Roadmap Interactive Chromatic", develop a different proposal from the currently used in commercial jets. It was through the interface 'Light and Colour' that the design team, sought to develop a proposal aimed at increasing the comfort of the user / passenger.

Minimizing phobias about indoors, fear of flying that is dormant for a flight and also giving the user a pleasant environment for the time they spend flying, as pleasant and smooth as possible. The project not only served as an apprenticeship, and took the first steps in the Company for the University regarding the development area of Interior Design, where the search for the user with the product, focused on the cognitive field, has given grants to future projects. The agreement signed with the Federal University of Parana, according to statements the company Embraer has brought the academic knowledge, combined with demand and that experience can reach a satisfactory and feasible regarding their deployment and operation, creating a technological edge in the product that will reflect the well-being of the user (this product was launched in December 2005).

Managing Scope understands the phases and steps taken to insure that the project includes all activities necessary and only needed to reach the pre-established goal successfully. The project scope was defined by an agreement called the Project Charter (formalized process for the contract to provide service between the Federal University of Parana and EMBRAER).

All changes were recorded in clear and easy to understand, in order to avoid discrepancies due to the desires and expectations misunderstood. Some changes as inevitable and necessary, as provided for Project Management such as: the inclusion of stage 4.1 (described above) a review of Step 6 (already described above). So there were changes of schedule, without damage to the estimated cost of the project or even man-hours.

Communications requirements are referenced to the team need to keep up with the data for the project. The higher the level of complexity of information increased the need for exchange of information to be made decisions concerning the project. They were developed for the exchange of information already consolidated in the market, namely the Internet, to distribute e-mail communications with information, schedules of meetings, exchange of material produced between team members, documents undergoing review.

## 5 Reference List

[1] GOLDMANN, Simão. **Psicodinâmica das cores** – 1966.

[2] GREENE, T. C., BELL, P. A. & BOYER, W. N. **Coloring the environment; Hue, arousal, and boredon.** Bulletin of the Psyconomic Society, 21, p. 253-254, 1983.

[3] LEVY, B. I. **Research into the psychological meaning of color**. American Jornal of Art Therapy. 23, 58-62, 1984.

[4] PLACK, J. J. & SHICK, J. **The effects of color on human behavior**. Journal of the Association for the Study of Perception, 9, p. 4-16, 1974.

[5] NORMAN, Donald A. **Boas razões para as máquinas sentirem medo**. SP: Scientific American, ano 02, nº21, 2004, pág.18-19.

[6] FIOROTI dos Santos, Leonardo. **Apreciação ergonômica da influencia da iluminação e cores nos passageiros de aeronaves, caso Embraer 170.** Dissertação de Mestrado em Engenharia Aeronáutica e Mecânica, ITA. Orientador: Ezio Castejon Garcia. Coorientadora: Viviane Gaspar Ribas El Marghani, São José dos Campos, SP, 2004.

[7] IRWIN, Douglas A. e PAVCNIK, Nina. **Airbus versus Boeing revisited: international competition in the aircraft market**. Journal of International Economics, 2004, nº64, pág. 223–245.

# KANSEI ENGINEERING: METHODOLY TO THE PROJECT ORIENTED FOR THE COSTUMERS

Viviane Gaspar Ribas El Marghani[a, 1], Natália Sgarbi Salata[a], Ana Claudia Gabardo[a]

[a]University of Federal of Paraná

**Abstract.** Thus, this work presents an original contribution to discussion on the topic. Initially, he discusses what the literature shows about the subject and then focuses on what types of KE exist to be exploited. The survey sought information from the major bases of research papers available, and using it to bibliographic sources and database platforms available in academic pursuit (CAPES Periodicals) on the subject. This paper intends to make a characterization of the theme KE and through it provide an understanding of the breadth of the subject. Despite the rather ambitious goal, the aim is to bring the subject up for discussion, so that academics and practitioners together to reflect on this important subject, find better alternatives for interaction and look for appropriate ways of spreading this topic in the country. The research result is presented below, the text is structured so the initial highlight the context of KE, their definitions and basic guidelines, the existing types of KE.

**Keywords.** Kansei Engineering; Method of design; Customer in Process.

## 1 Introduction

O *KANSEI ENGINEERING* (KE) is a methodology for product development, which translates the impressions, feelings and demands of users and solution-specific design parameters [1]. Despite KE exist for a long time, about 30 years, little is known about the design methodology in the teaching of design. With the improvement of communication processes, consumers had become more demanding and more competitive industry, so that functionality by itself, no longer meets the requirements of the market. Therefore, current design methodologies for design have been seeking new methods that allow the generation of innovative ideas. In this step, creativity is considered the heart of design, and it is the main ingredient in the process of product development at every stage design specification, and should, therefore, stimulate it. Furthermore, although the use of engineering design methods provide satisfactory results to the designer, they turn out to be quite inadequate, considering that today's consumers seek more and more,

[1] Professor of Design of UFPR (Designer Bal. Msc and Dr. Engineering), Paraná. Rua General Carneiro 460, 8° andar, Curitiba PR Brasil, Tel/ Fax: +55 (41) 33605360 Email: viviane.gasparibas@ufpr.br; http://www.design.ufpr.br

D. D. Frey et al. (eds.), *Improving Complex Systems Today*,
Advanced Concurrent Engineering, DOI: 10.1007/ 978-0-85729-799-0_54,
© Springer-Verlag London Limited 2011

products that are "able to touch the strings Heart, playing around with your emotions, capture the culture and reason, (...) can also arouse the active participation and at the same time, move him [2].

With respect to KE which is presented in Brazilian literature is greatly simplified in terms of detail compared to the literature in other countries, detailing the KE and apply broadly.

Initially, the article discusses what the literature shows about the issue, and then the article focuses on what types of existing and KE that can be explored by researchers with a brief explanation of each type, and some examples of work done. For this, the survey sought information from the major bases of research papers available, and using it to bibliographic sources and database platforms available in academic pursuit (CAPES journals), digital libraries and Science Direct IEEExplore with the keywords to search Kansei Engineering.

This paper intends to make a characterization of the theme KE and bring the subject up for discussion, so that academics and practitioners together to reflect on this important subject, find better alternatives for interaction and look for appropriate ways of spreading this topic in the country.

## 2 Literature

The Kansei Engineering (KE) was founded at the University of Hiroshima 30 years ago [3]. His focus was to implement the feeling of the customer / user and their requirements depending on the product and design. This feeling is called in Japanese "Kansei". According to Nagamachi (2001), there are three focal points of KE: how to accurately understand the consumer, how to reflect and translate the understanding into product design and how to create an organization focused on Kansei design. The methodology is essentially a mechanism for the systematic development of new and innovative products, but can also be used as a tool for improving product concepts [1]. As input data for KE is measured, conditioned and then processed by a Kansei Engineering System (KES). The resulting information tells how the psychological feeling is directly related to the product that is projected to be either material or immaterial.

### 2.1 Different types of KE

Kansei Engineering (KE) can be achieved in different ways. There are six (6) types of KE-developed, proven and tested. (See Table 1).

Table 1 - Types of KE

| | |
|---|---|
| **KE tipo I** | This is the simplest and quickest way to do an analysis KE. A strategy for specific product and market segment is identified and developed in a tree structure presents similarities with the Ishikawa diagram and QFD (see Figure 2). The decision in favor of a particular market segment involves properties of a product that should be known by the project team and considered when designing the new product. The structure starts with a zero (0) which is divided into several sub-concepts. These sub-concepts can be evaluated separately at various levels until the design parameters of the product can be easily determined. An example is the car Mazda Miata (known in Europe as MX5). In this project the goal was to build a low price sports car for young male drivers. KE applied the result determined by users who focused 'cheap car is tight. "After the tree was chosen at two levels by a car with a body length of 3.98 and 2 seats [1]. |
| **KE tipo II Sistema de engenharia kansei (KES)** | This type of KE is a computer-assisted manner where the feelings of users and the product properties are stored in a database, so correlated. To build a program, among other knowledge is the need to obtain data on lifestyle and habits of the target group. [4]. applied this technique to a program of designing a kitchen, where the alleged clients described the dream kitchen in your own words. The system picks a cuisine based on lifestyle and habits of the customer. |
| **KE tipo III Hibrido** | Hybrid Kansei Engineering, according to [5] is a combination of support and consumer support of the designer. This type of Kansei consists of KES (shown in the previous section) and the Reverse Engineering Kansei.Kansei Engineering is well known reverse, because it uses a database of products that can be used, and the user is from a drawing or a concept of the possible feelings that might be stimulated with these "drawings" or "concepts. "The database is specially developed by a designer, which feeds the system with their ideas through the user interface, which analyzes the product parameters and compares it with stored data. These data in turn are connected to a database of words related to feelings of users that feed back the designer to generate new ideas. |
| **KE tipo IV Matemático** | This type uses more than just intelligent systems. In this particular type using a mathematical model that is built on the rule-base to get the result of ergonomic Kansei words. In this procedure, a mathematical model involves a kind of logic that plays a role similar to the rule-base [5].Sanyo applies this kind of success with KE for color copies that reproduce more faithfully the skin tone of the Japanese people. The first step is the "human sensory perception of image processing [6]. We surveyed 50 people to rate 24 images of human faces with slightly different skin tones, and offered a five (5) points of analysis. The values of the resulting scores were used to define fuzzy sets that represent the degree of want of skin tones. With fuzzy logic was possible to determine the three factors of color that is processed in a RGB system obtained the type of skin color more desirable [7]. |
| **KE tipo V Virtual** | Virtual Kansei Engineering (Vike) uses of virtual reality, putting the user in a 3D virtual environment, which can be manipulated directly. It is a combination of a computer system with virtual reality systems to help the user's selection of a product using his experience as a resource to the virtual space [7]. It has the example of applying this methodology in the development of kitchens and living rooms Matsushita Electric by dinner in partnership with the University of Hiroshima. First the consumer responds to questions regarding your lifestyle and inserts your words Kansas. The system proposes the kitchen that fits the user to Kansas, using a database based on the feelings of 10,000 women and kitchen design imagined for them. If consumers feel satisfied in virtual space, the kitchen project decided by Vike system is transferred to the factory and delivered in a matter of two weeks. The new kitchen is assembled in the presence of CONSUMERS. [8]. |
| **KE tipo VI Colaborativo** | It is a type which uses software for collaborative work in groups or the internet. In this case offers the opportunity to bring the views of both clients and designers. By doing so, the initial stages of development are being reduced and simplicity. Using the World Wide Web, builds a structure of group project, which has a clever system and database of Kansas. There are many benefits of this type of KE, among them: greater commitment to the project, collaborative work among participants, efficiency, speed product development, greater dialogue between producer and consumer, or the effective participation of many people, offering a diversity ideas for the project [9]. |

## 3 A recent case of application of KE

As a way to illustrate how the KE has been applied, and within the group studied from 2001 to 2010, they selected an article that demonstrates the process of development and application of KE. The case to be presented was published recently in the International Journal of Industrial Ergonomics 40 (2010) pages 237-246, by the authors [10] Department of Industrial Design, National Cheng Kung University, located in Taiwan, with financial support in number of A023 contract, 2007.

The work used the method of genetic Several algorithms have been applied in many fields. In the field of design [10] developed an automatic system design for a quick way to obtain a product and its corresponding image using fuzzy neural networks and genetic algorithms, among others. In this study, Kansei Engineering was used to quantify the responses of consumers with product styles. The characteristic data of each product style were evaluated, and the style of product that are closer than imagined by the customer was selected using genetic algorithms. We detailed the steps to build the project template (see Table 2).

Table 2 – Steps of project

| | |
|---|---|
| Step 1 | To determine the importance of a product in human life and to identify the habits of the user operating from different perspectives. Collection of various styles of products and linguistic variables establishing an association between the product and the questionnaire.. |
| Step 2 | Build a database of encryption products with style, basic items and categories of the theory of Morphological Analysis. |
| Step 3 | Apply Type I quantification analysis for the values the contribution of styles of products for each component. |
| Step 4 | Consolidating the linguistic variables and evaluations of individual assessments and the weights of each linguistic variable through an analytic hierarchy process. |
| Step 5 | Building a program of genetic algorithm, the contribution of values, first make a consolidated assessment, and then using it to get the combinations that conform to the categorical value of fitness. |
| Step 6 | Decoding combinations of these categories used 3D models to confirm the efficacy of the genetic algorithm. |
| Step 7 | Using 3D imaging and rapid prototyping models to inspect the finished products |

The product was analyzed in two sub-systems: body and jar. The categories of target products were classified according to a scale morphology based on three rules of morphological characteristics: 1) size limits, 2) Proportion of the set and 3) Graphic Design. To evaluate the images of products and semantics of linguistic variables was used a questionnaire design firms and experienced designers. The first linguistic variables that describe the appliances were selected from the universe of 100 files in a survey of twenty students. Thus, 20 variables were selected the most appropriate language to describe the coffee product. The study developed a genetic algorithm using the Matlab software, and followed the procedures shown in Figure 1.

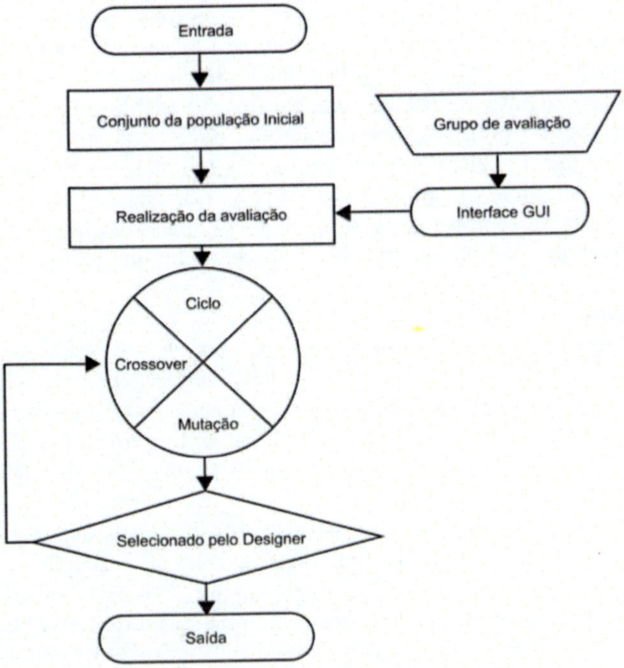


Figure 1 - Proceedings of the genetic algorithm. Source: [10]

Obtained a value of the demand of the seven linguistic variables, with the best solutions obtained by comparison with data from the correct code. In the genetic algorithm program was included in the four sub-systems in the coffeemaker, the pair-wise comparison matrix of linguistic variables, the individual scores for the linguistic variables, the scores of contribution of each linguistic variable, and the output area of the best solutions. Finally, a fitness value was calculated as the standard chosen for the sum of variable scores linguistic variables and the best fitness value. Thus, twelve optimal solutions were obtained from data collected from customers.

After obtaining the best solutions using a genetic algorithm, the CAD software Pro Engineer Wildfire was used to construct three-dimensional models and rapid prototyping of the selected model (see Figures 2 and 3).

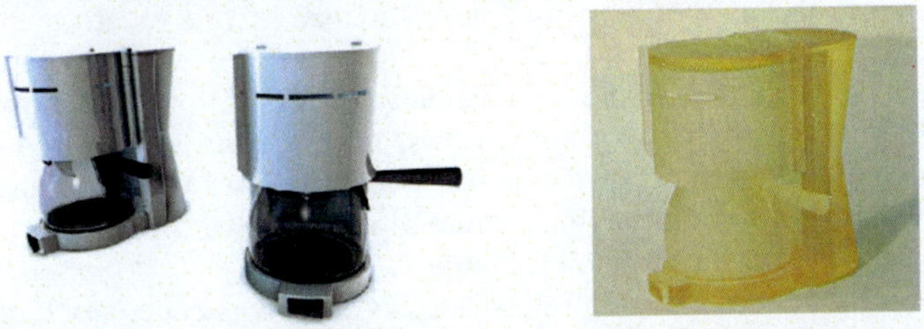

Figure 2 and 3 - 3D modeling and prototyping of the final model. Source: [10]

This case demonstrated the use of genetic algorithms to assess the feasibility of a new product and demonstrated that the best solutions based on linguistic data based on the hierarchy of the process can increase the efficiency of the process of designing and producing products can best meet the needs of client. The concept of using a model that is based on morphological theory, enables rapid development of forms of products and programs for accelerated development of the product. The deviation of cognition between virtual and real models is reduced in the 3D rendering, virtual models of optimal solutions can be quickly converted into real solutions using a rapid prototyping system. The application of this model of genetic algorithm theories proved that it can achieve faster and more accurate results.

## 4 Conclusions

Many are the products developed in Japan which has used the engineering Kansas. Recently, it has also been applied to construction products, as well as urban design. The list of organizations, not just the Japanese who introduced KANSAS engineering / ergonomics by the year 2002 according to [8] include different sectors: Automotive, construction machinery, appliances, office machines, civil, textile, cosmetics, etc.

Since most companies that fall into these sectors and introduce the Kansas are of Asian origin which agrees with the fact that most research and scientific publications are concentrated in China (especially Taiwan), Japan and Korea. All products developed by KE good sales on the market until now, because it aims to incorporate consumer sentiment and images into new products. The most important point in the KE is conducting a survey of customers to understand their feelings, since the beginning of product development. At this stage, the project team to investigate the psychophysiological aspects of users, using a type of KE then the project team would have better conditions to take design decisions with a view to meet the needs of customers of products.

Are several factors that should be considered during a project including consumer preferences [11], the manufacturing processes [12], the color element [13], the texture of the material [14] interfaces involving the use product [15], among others. In summary, all these factors involve substantial expenditures of time and money to optimize a design, in general these factors are manipulated in the conceptual design phase.

In the Conceptual Design, handled and processed information are scattered, many coming from the designer's own reasoning, and therefore not formalized [16]. Because of this, there is the need for extensive use of creativity by the designer, which, in conclusion, means that there is great influence of this creative ability in the final quality of the project.

At that stage, he is starting hand of abstraction - or putting the problem in general terms, represented by one or more verbs, representing the desired action, and a noun representing the object of the action - which allows us to glimpse other views and other solutions, besides having a more global view of the problem, it also allows to identify other solutions to the problem.

[17] put this theme in terms of generalization, saying it was necessary to express the problem to solve in the form of a neutral solution. According to [18] are the advantages of systematization: ability to obtain and examine more chains of partial

solutions, able to analyze more variants of solutions and obtain more models and prototypes. In practice, all the steps involved in the conceptual design phase are interconnected, and the process is highly iterative and interactive. Finally, it should be stressed that great importance is the early stage of project cost and success of the final product. Decisions taken at this stage have great difficulty and high cost in proportion, to be altered in later stages of a product [19].

Thus, KE is a highly positive approach because corroborates with the decisions of the project team which uses more precise information from the manipulation of data that departed from users. With this design process is increased efficiency not only in its conceptual stage, but at the production stage, since the products are in better accordance with the needs of customers / users. The methodology helps KE theories of decision making as it allows the project team visualize more accurately the needs and desires of customers / users, so the project becomes more rapid, precise and similar results with human aspirations.

## 5 Reference List

[1] SCHÜTTE, Simon. Designing Feelings into Products. Integrating Kansei Engineering Methodology in Product Development. Linköping Studies in Science and Technology. Thesis No. 946, ISBN: 91-7373-347-4 ISSN: 0280-7971, 2002.

[2] SALOCCHI, Claudio. *Gemma Gioielli*. 2004. [Online] Available from URL: <http://www.gemmagioielli.com/inglese/index.htm.>. [Accessed 2004 May 26th].

[3] NAGAMACHI, M., 1999. Kansei engineering. In: Karwowski, W., Morris, W.S. (Eds.), The Occupational Ergonomics Handbook. CRC Press, New York, pp. 0–2 (Chapter 102).

[4] IMAMURA, K., J. NOMURA, et al. (1997). An Application of Virtual Kansei Engineering to Kitchen Design. Kansei Engineering 1. M. Nagamachi. Kure, Kaibundo Publishing co.,LTD: 63-68.

[5] NAGAMACHI, M., 1995a. Kansei engineering: a new ergonomic consumer-oriented technology for product development. Int. J. Indus. Ergonomics 15 (1), 13–24.

[6] FUKUSHIMA, K., Kawata, Y., Genno, H., 1995. Human sensory perception oriented image processing in a color copy system. Int. J. Indus. Ergonomics 15 (1), 63–74.

[7] NAGAMACHI, M., 1989. Kansei Engineering. Kaibundo Publishing,Tokyo.

[8] NAGAMACHI, M., 2002, Application of kansei engineering and concurrent engineering to a cosmetic product. Proceedings of the ERGON-AXIA.

[9] NISHINO, T., 1999. HousMall: an application of kansei engineering to house design consultation. Proceedings of the International Conference of TQM and Human Factors, pp. 349–354.

[10] HSIAO, Shih-Wen; CHIU, Fu-Yuan Chiu; LU, Shu-Hong. Product-form design model based on genetic algorithms. nternational Journal of Industrial Ergonomics 40 (2010) páginas 237-246. Taiwan.

[11] HONG, S.W., HAN, S.H., KIM, K.J., 2008. Optimal balancing of multiple affective satisfaction dimensions: a case study on mobile phones. International Journal of Industrial Ergonomics 38 (3e4), 272e279.

[12] PINE, B.J., 1993. Mass Customization: The New Frontier in Business Competition.Harvard Business School Press, Boston.(Pine, 1993),

[13] HSIAO, S.W., CHIU, F.Y., CHEN, C.S., 2008. Applying aesthetics measurement to product design. International Journal of Industrial Ergonomics 38 (11), 910e920.(Hsiao et al. 2008),

[14] CHANG, C.-C., 2008. Factors influencing visual comfort appreciation of the product form of digital cameras. International Journal of Industrial Ergonomics 38 (11), 1007e1016.

[15] RAMIREZ, Artacho; MAS, Diego MARZAL, Alcaide., 2008. Influence of the mode of graphical representation on the perception of product aesthetic and emotional features: an exploratory study. International Journal of Industrial Ergonomics 38 (11), 942e952.(Artacho-Ramirez et al., 2008)

[16] DURKIN, J. & DURKIN, J. *Expert Systems* – design and development. New York, Prentice Hall, 1998.

[17] PAHL, G. & BEITZ, W. *Engineering Design* – a systematic approach. Translated by Ken Wallace and Lucienne Blessing. Berlin, Springer Verlag, 1996.

[18] VDI-GKE (Verein Deutscher Ingenieure – professional engineers' body). *VDI Guideline 2221*, systematic approach to the design of technical systems and products. Dusserdorf, VDI Verlag, 1987.

[19] ULLMAN, D. 1997. *The Mechanical Design Process*. McGraw-Hill, New York.

# Effect of Tonal Harmonic Feature in Product Noise on Emotional Quality

Hideyoshi Yanagisawa[a, 1], Shunsuke Yokoo[a], Tamotsu Murakami[a]

Koichi Ohtomi[b] and Rika Hosaka[b]

[a]The University of Tokyo, JAPAN.
[b]Toshiba Corporation, JAPAN.

**Abstract.** The product sound is an important factor that affects the product emotional quality. In the design of product sound quality, a designer needs to find the design factors that affect the emotional quality and determine the characteristics of their effects. The authors previously proposed a method for extraction of potential emotional factors by analyzing human sensitivity towards unexplored design and applied the method for designing product sound quality. From the result using vacuum cleaners as a case study, the authors found that the existence of prominent peak tones in sound has the potential to improve sound quality. However, prominent peak tones are usually regarded as a factor of annoyance. In this paper, we propose an indicator for adjusting the frequency and level of peak tones to improve a product sound quality. We have assumed that the harmonic features of peak tones in noise can be used as the indicator. We created vacuum cleaner sounds having three peak tones whose harmonic features such as tonal consonance and modality are different. To evaluate the effectiveness of the harmonic features, we conducted a pairwise comparison-based sensory evaluation with two groups of participants, one consisting of those who play some musical instrument and the other of those who do not. From the experiment, we found that the peak tone harmonic features can be perceived by both groups of participants and significantly decrease their annoyance at vacuum cleaner sounds.

**Keywords.** emotional quality, design for sound quality, harmonic feature.

## 1 Introduction

In the design of mature products, an *emotional quality* is becoming increasingly important due to diversifying customer's needs [1]. An emotional quality is a quality of a product or service that evokes the user's specific impressions, feelings or emotions towards that product (e.g., comfort, luxury, delight). The product's

[1] Assistant Professor, Department of Mechanical Engineering, The University of Tokyo, 7-3-1, Hongo, Bunkyo-ku, Tokyo, JAPAN; Tel: +81 (3) 5841 6330; Fax: +81 (3) 5841 6330; Email: hide@mech.t.u-tokyo.ac.jp

D. D. Frey et al. (eds.), *Improving Complex Systems Today*,
Advanced Concurrent Engineering, DOI: 10.1007/ 978-0-85729-799-0_55,
© Springer-Verlag London Limited 2011

sound has been regarded as an important factor that affects the emotional quality [3]. For example, most people are annoyed by loud noises but become relaxed when hearing the sound of gentle ocean waves. The task of the sound designer is to design a product sound that evokes a target emotion (impression) in the user.

The most important issue in the design of emotional quality is extracting quantitative criteria for evaluating such a quality. Without such quantitative criteria, the designer cannot set a clear goal for design because he/she is not sure what kind of evaluation criteria the customers have, or how to design a product to increase a target emotional quality. The designer has to rely on his/her assumptions based on sensitivity and tacit knowledge. If there is a gap between the designer and customers in terms of their sensitivities, the designer will misinterpret how a customer will evaluate his/her design.

To date, several studies have been carried out to extract such quantitative criteria of a product's emotional quality using sensory evaluations and statistical methods[4]. Most conventional approaches formalize emotional qualities expressed by adjectives using sensory evaluation with existing products. For example, researchers often construct evaluation criteria for emotional qualities using sound samples recorded from existing products and their emotional evaluations scored by human subjects. However, the variety of existing products is limited. The obtained evaluation criteria may not cover areas of a design space where future designs would appear. To address this problem, the authors previously proposed a method to cover such untouched areas using composite design samples[1]. We applied the method to extract potential factors for the future design of a vacuum cleaner's sound quality and found that the existence of a perceivable peak tone around 500Hz improves the emotional quality. However, a perceivable peak tone is generally regarded as a factor of annoyance which the sound designer aims to reduce. This finding implies the possibility of using peak tones as a new design factor that improves an emotional quality.

Tone consists of its frequency and power. A musical instrument such as a piano employs combinations of tones. In music, the theory of harmonics provides ways of combining tones that affect human emotions such as consonance and key. Quantitative features of tonal harmony have been developed such as tonal consonance[5]-[7] and modality[8]. In this study, we assume that such harmonic features can also be applied to peak tones in a product sound.

In this paper, we discuss the potential of the peak tones' harmonic features as a new design parameter for product sound quality. We synthesize vacuum cleaner sounds with three peak tones having different harmonic features. We conduct a pairwise comparison-based sensory evaluation of the synthesized sounds using adjectives that express the sound's emotional quality. With the result of the experiment, we discuss the effectiveness of the harmonic features of peak tones in product sound design.

## 2 Product Sound Quality and Tonal Harmonic Feature

The sound made by a product is one of the important factors that affect the product's emotional quality. For quite a long time, sound engineering mainly dealt with the reduction of the overall sound pressure level (SPL) emitted by a product. Within the last decade, however, the focus started to switch more towards aspects related to the quality of the sound. The biggest change is that the design goal switched from objective values, such as the "decibel" levels that can be physically measured, to subjective ones such as emotional qualities. To design such emotional qualities of product sound, it is necessary to develop metrics to quantitatively evaluate such subjective qualities. Zwicker et al. developed sound quality metrics (SQM) as an evaluation metric of the product sound quality. SQM provides values for simple perceptions of sound such as loudness, sharpness, roughness and fluctuation strength [9].

However, a product's emotional qualities include more complex affective perceptions, such as pleasant, annoying, luxurious, etc. To deal with such complex sensitivity in sound design, most conventional approaches conduct sensory evaluations using affect-laden words to score target emotional qualities. Statistical methods are used to compose a map between SQM and complex emotional qualities[10]. Several applications have been studied based on the approach[11]-[14]. Most research so far, however, has not considered the diversity and potential of human sensitivity.

Human sensitivities for emotional qualities differ from person to person. The range of individual differences depends on the particular emotional quality. Averaging the results of sensory evaluations to measure emotional quality is appropriate only when individual differences are negligible and most subjects share a similar sense. To address this issue of diversity, Yanagisawa et al. proposed a new emotional quantification method which pays special attention to the diversity of customers' sensitivities[15]. Based on personal differences regarding sensitivity, the method extracts multiple emotional scales from SD (semantic differential) scales, which are used in a sensory evaluation called the semantic differential (SD) method[16]. An SD scale consists of a pair of adjectives representing an emotional quality, such as "powerful-weak". The method was applied to construct evaluation criteria for the design of a vacuum cleaner sound using SQM as design features.

To construct the evaluation criteria for emotional qualities, we often use sound samples recorded from existing products and their sensory evaluations scored by human subjects. However, the variety of existing products is limited. For example, the sounds of existing products may not be exhaustive enough in the design space to construct general evaluation criteria that can be used for future design. People may evoke different sensitivities towards an unknown sound. To cover feature areas untouched by existing products in the design space, Yanagisawa *et al.* proposed a method to create composite sounds synthesized from the original sounds of existing products[1]. In the method, the strategy for creating such sounds is to disperse them on a target emotional quality and on untouched areas of the design feature space. From the result of applying the method to vacuum cleaner sounds, it was found that the existence of a perceivable peak tone around 500Hz has the potential to improve product sound quality.

According to musical theory, harmonic features such as harmony and key affect human emotions. For example, we generally perceive a harmony when the frequency ratio between two tones is an integral number. Plomp and Levelt found that "tonal consonance" between two tones correlates to a frequency difference that corresponds to their critical bandwidth[5]. For example, two tones around 500Hz are harmonized when the frequency difference is more than 100Hz because the critical bandwidth around 500Hz is about 100 Hz. Kameoka & Kuriyagawa [6] defined "dissonance" based on the tonal consonance of Plomp and Levelt. They found that the dissonance among more than two tones corresponds to the sum of dissonances of all combinations between the tones. Based on the above ideas, Sethares [7] defined the quantitative dissonance.

The key between tones such as major and minor evokes the listener's emotions such as happy and sad in general. Cook & Fujisawa[8] defined the degree of major and minor, which they called "modality," based on the equality of tonal intervals in three tones. The following formula gives the modality.

Although the above harmonic features are formalized in music theory, we do not find experimental investigations in terms of their effects on product sound quality. In product sounds such as for vacuum cleaners, prominent peak tones in the noise are regarded as annoying sounds. Sound designers have aimed to reduce such peaks so far.

## 3 Sensory Evaluation Experiment of Peak Tones' Harmonic Feature in Product Noise

### 3.1 Objective

We conduct a sensory evaluation experiment to investigate how the harmonic features of peak tones in the noise of a product sound affect the listener's perception and feeling. We synthesize vacuum cleaner sounds having three perceivable tones whose frequencies are adjusted based on different harmonic features. We ask the participants to evaluate the synthesized sound samples using adjectives in terms of their perception and feelings. From the result of the sensory evaluation, we analyse the quantitative relation between the harmonic features of the tones and the sensory responses given by participants.

### 3.2 Sound Samples, Evaluation Method and Evaluation Words

We recorded in an anechoic room existing vacuum cleaner sounds whose peak tones are not prominent. We synthesized three perceivable tones into the recorded sound. We created eight patterns of the three tones that cover all combinations of harmonic features, i.e. consonance – dissonance and minor – major, and two sound pressure levels (53 and 58 dBA). The level of the three tones is the same. We selected a 523Hz (C4) tone as a base frequency and varied the frequencies of the remaining tones between C4 and G4. 523Hz (C4) corresponds to the frequency of the vacuum cleaner's motor. From the authors' previous work[1], we found that the prominent peak tones caused by the motor's frequency affected the emotional quality of sound. The loudness and sharpness of the eight sounds are respectively on the same level.

To evaluate the synthesized sound samples, we used Scheffe's paired comparison with Nakaya's modification. In the method, we randomly selected from samples and asked participants to compare all combinations of the samples in terms of the evaluation criteria. We selected five adjectives used as evaluation criteria in the paired comparison. We selected "pleasant" as a total impression of the synthesized product sound. We assumed that the words "consonant" and "sad" correspond to the harmonic features. To check the effect of harmonic features on loudness and sharpness, we selected "loud" and "sharp". Since the loudness and sharpness of the samples are respectively on the same level, we assumed that the respective senses of "loud" and "sharp" of sound would not vary significantly.

The sensory evaluation consists of two phases. In the first phase, we asked the participants to evaluate the pleasantness of eight synthesized sounds and the original vacuum cleaner sound using paired comparison with "pleasant" as an evaluation word. In the second phase, participants evaluated the eight synthesized sounds using paired comparison in terms of the senses of "consonant", "sad", "loud" and "sharp". In all steps, the participants listened to the sound samples using a monitor headphone. To reduce the order effect of samples, we inserted white noise between the samples.

### 3.3 Participants

Having some playing experience with a musical instrument may affect the cognition of harmonic features. To check personal differences based on such experiences, we invited five students from among members of the university orchestra club and ten students who do not play any musical instrument. We denote the group of musically experienced students by G1 and the group of inexperienced students by G2. All participants are students from the University of Tokyo in their twenties.

## 4 Effect of Peak Tones' Harmonic Features on Sound Quality

### 4.1 Method of Analysis

Based on Scheffe's paired comparison, we calculate the evaluation score for each sample by summing the comparison scores. We conduct three-way ANOVA with the evaluation score for each word as the response, and the harmonic features and type of participant group as factors so as to determine the factor's main effect and their interactions. We conducted ANOVA for all combinations of the two participant groups and two tonal levels, i.e. four conditions for each evaluation word.

### 4.2 Effect of Peak Tones' Harmonic Features on Perception of "Consonant", "Sad", "Loud" and "Sharp"of Product Sound

Regardless of the participant group or tonal level, tonal consonance significantly affects the sense of "consonant" from the result of ANOVA ($p$<0.01). The contribution ratio of tonal consonance is dominant (G1-low: 84.8%, G1-high: 71.6%, G2-low: 65.2% and G2-high: 71.6%). This result suggests that the tonal consonance of peak tones in noise can be recognized regardless of the listeners' experience with a musical instrument. We found that an interaction effect exists between tonal consonance and modality ($p$<0.01). For all four conditions, the sound having minor tones is "sadder" than the major tones if the tones are consonant. In sounds with dissonant tones, the difference between major and minor is not significant. Thus, the modality of tones in a product sound is effective to evoke certain emotions only if the tones are consonant.

The perception of "loud" and "sharp" should correspond to only loudness and sharpness, respectively. We assumed that the feeling of "loud" and "sharp" among the samples should not differ because all synthesized samples are on the same levels of loudness and sharpness. Although we found that tonal consonance has a significant effect in the "G1 and low tone" condition ($p$<0.05, contribution ratio is 31.2%), the effect of the participants' difference is significant ($p$<0.05, contribution ratio is 41.4%) from the result of ANOVA towards "loud". We did not find any significant effect in the other conditions. On the other hand, we found that tonal consonance has a significant effect on "sharp" in all conditions ($p$<0.01). The participants perceive sounds having consonant tones as "sharper" than the dissonant tones. Thus, the tonal consonance of peak tones has the potential to affect the sharpness of a product sound independently from SQM's sharpness.

### 4.3 Effect of Peak Tones' Harmonic Features on Pleasantness of Product Sound

In all conditions, tonal consonance significantly affects the pleasantness of sound, especially in the high level tones($p$<0.01). In the low level tone with G1, the effect of the interaction between consonance and participant type is significantly high($p$<0.01). Thus the level of peak tone is important if one wishes to minimize personal differences in the effect of consonance on pleasantness.

Figure 1 shows the score of "pleasant" for each condition for the original sound and the synthesized sounds with harmonic features. Each line represents one of four different combinations of participant group and tonal level. The synthesized sounds having consonant tones are evaluated as more pleasant than the original sound. Conversely, synthesized sounds having dissonant tones are evaluated as unpleasant compared to the original one. This result suggests that tonal consonance is an essential factor in determining whether prominent peak tones have a positive or negative effect on the pleasantness of a product sound. In other words, tonal consonance has the potential to be used as a design guideline for product sound quality.

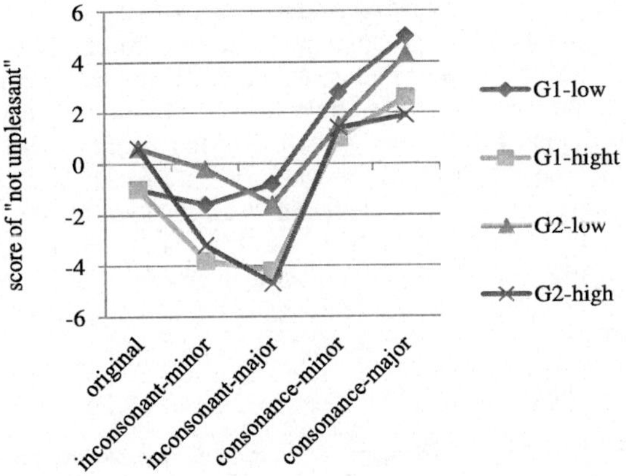


**Figure 1.** Factorial effects of harmonic features of tones synthesized in vacuum cleaner sound on feeling of "pleasant" for each condition of participant's experience and tonal level.

## 5 Conclusion

By analysing human sensitivity towards unexplored design space areas, we found, in a previous work[1], that the existence of a perceivable peak tone around 500Hz, which corresponds to the frequency of the inside motor, in noise has the potential to improve the emotional quality of a machine sound such as in a vacuum cleaner. In this work, we assumed that the theory of harmonics in music can be applied to adjust the intervals between such peak tones to design the total sound quality. We introduced quantitative harmonic features such as tonal consonance and modality to design the intervals of three prominent peak tones in a vacuum cleaner sound. To evaluate the effect of the harmonic features on the product sound quality, we synthesized the vacuum cleaner sound with three peak tones around 500Hz with different harmonic features and conducted a paired comparison using the synthesized sounds with two kinds of participants, i.e. those who play music and those who do not. From the experiments, we found the following results regardless of the participant's experience with a musical instrument.

- The participants perceived the tonal consonance of peak tones in machine sound.
- A minor key setting of the peak tones evoked a "sad" feeling only if the tones are consonant.
- The consonance of peak tones significantly affected the perception of "sharp" independently from SQM's sharpness.
- Consonant peak tones significantly increased pleasantness as compared to the original vacuum cleaner sound without peak tones. In contrast, dissonant peak tones decreased pleasantness compared to the original sound.
- Tonal level is important to accentuate the effect of tonal consonance on the sound's pleasantness.

The above results suggest that tonal consonance is a key factor when designing the intervals between peak tones in product sound quality. Modality has the potential to adjust the emotional factor of a product sound, i.e. happiness and sadness.
It must be noted, however, that the above results are limited to frequencies of around 500Hz. The tonal level seems sensitive to the effect of harmonic features. For future work, we need to investigate the harmonic features' effect at different frequencies and the effect of tonal level in detail. Although we use a vacuum cleaner sound as a case study, we believe that the results can be applied to other product sounds consisting of stationary noise, such as a fan, air conditioner, etc.

## References


[1] Yanagisawa H, Kataoka A, Murakami T, Ohtomi K, Hosaka R. Extraction of latent emotional factors by analyzing human sensitivity towards unexplored design: application to product sound design. Int Conference on Engineering Design, 2009.

[2] Yanagisawa H, Fukuda S. Interactive reduct evolutionary computation for aesthetic design, Trans AMSE, J Computing and Information Sci in Eng 2005; 5(1):1-7.

[3] Lyon, RH. Product sound quality from perception to design, Sound and Vibration 2000; 37(33): 18-23.

[4] Yanagisawa H. Kansei quality in product design, In Fukuda S, editor. Emotional engineering - Service development: Springer, 2010; 289-310.

[5] Plomp R, Levelt WJM. Tonal consonance and critical bandwidth, J Acoustical Society of America, 1965; 38: 548-560.

[6] Kameoka A, Kuriyagawa M. Consonance theory part 1: Consonance of dyads, J Acoustical Society of America, 1965; 45: 1451-1459.

[7] Sethares WA. Local consonance and the relationship between timbre and scale. J Acoustical Society of America, 1993; 94(3): 1218-1228.

[8] Cook ND, Fujisawa TX. The Psychophysics of harmony perception: Harmony is a three-tone phenomenon. Empirical Musicology Review, 2006;1(2): 106-126.

[9] Zwicker E, Fastl H. Psychoacoustics - Facts and models, third edition. Springer, 2006.

[10] Otto N, Amman S, Eaton C, Lake S. Guidelines for jury evaluations of automotive sounds. In Proc. the 1999 SAE Noise and Vibration Conference, 1999.

[11] Chatterley JJ, *et al.* Sound quality testing and analysis of multiple brands of sewing machines, In Proc. IDETC/CIE 2005, DETC2005-85636.

[12] Vastfjall D, Gulbpl M, Kleiner M Garling T. Affective evaluation of and reactions to exterior and interior vehicle auditory quality. J Sound and Vibration 2002; 255(3): 501-518.

[13] Fastl H. Neutralizing the meaning of sound for sound quality evaluations, Internoise, 2004.

[14] Ellermeier W, Zeitler A, Fastl H. Predicting annoyance judgments from psychoacoustic metrics: Identifiable versus neutralized sounds, Internoise, 2004.

[15] Yanagisawa H, Murakami T, Noguchi S, Ohtomi K, Hosaka R. Quantification method of diverse kansei quality for emotional design - Application of product sound design, In Proc DETC'07/ASME Int 27th CIE, 2007, DETC2007-34627.

[16] Osgood CG, Suci GJ, Tannenbaum P. The Measurement of meaning. University of Illinois Press, 1957.

# Pragmatic Concurrent Engineering: Emotion, Experience and Engineering

Shuichi Fukuda

Department of Mechanical Engineering, Stanford University, Stanford, CA. 94305, USA

**Abstract.** Our products have been developed on a feed forward linear model. This could roughly be described as a differential approach. We looked at local region and optimized. The early concurrent engineering emphasized to bring downstream information as much upstream as possible. But this is not the optimization of the whole product lifecycle. It is the optimization of production. Increasing diversification, complexity and globalization call for a much broader perspective. Optimization throughout a whole product life cycle including product use is now called for. In order to achieve such a goal, feedbacks from user experience must be considered and the system must be adaptive to changing situations. Today, our path may be continuous, but the changes are sharp. i.e. , changes yesterday were differentiable but they are not differentiable today. Our world was closed in the 20th century but now it is an expanding open world. We have to get prepared for the unexpected in order to explore such a New World. We have to look more globally. This calls for an integral approach. Emotion and experience play important roles in such an approach which complement the factors in our traditional differential approach.

**Keywords.** Emotion, Experience, Differential Approach, Integral Approach, Pragmatism

---

Consulting Professor, Department of Mechanical Engineering, Stanford University Stanford, CA, 94305, USA Email: shufukuda@cdr.stanford.edu

D. D. Frey et al. (eds.), *Improving Complex Systems Today*,
Advanced Concurrent Engineering, DOI: 10.1007/ 978-0-85729-799-0_56,
© Springer-Verlag London Limited 2011

## 1 Concurrent Engineering Yesterday and Today

The primary goal of the initial concurrent engineering (CE) was to achieve shorter time to market with the currently available resources. To achieve this goal, how downstream information can be brought upstream was emphasized. This is because current engineering substantially started with DARPA's project "DICE: *D*arpa's *I*nitiative in *C*oncurrent *E*ngineering". Presumably, their goal was to establish a methodology to catch up with their enemy when they find out that their enemy is developing a new weapon with functions better than theirs. Thus, how time to product realization can be reduced so that they can realize their weapon in time or before their enemy with functions at least comparable to their enemy's was their primary concern. The product realization processes were linear so that it was necessary to bring downstream information as much upstream as possible to reduce development time. Thus, information had to be processed in parallel (Figure 1).

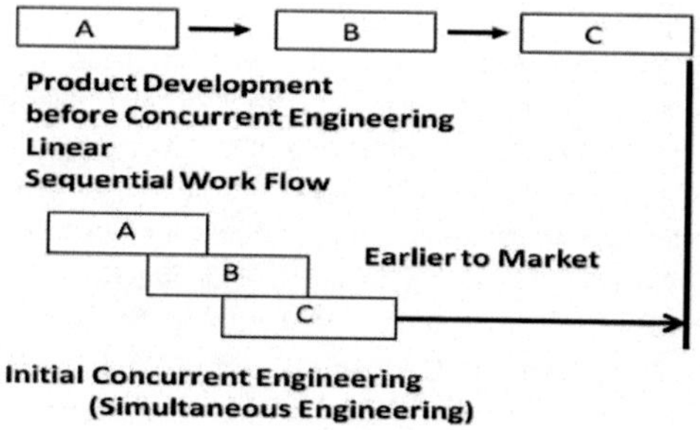


Figure 1. Concept of initial concurrent engineering

Therefore, "concurrent" in this initial version of CE meant processing information in parallel or at the same time. Soon it became clear that to achieve this, engineers have to collaborate. Thus, the second version of CE was called Collaborative Engineering. "Concurrent" in this 2nd version meant cooperating.

No matter how the interpretation of CE may change, CE has been producer-centric up to now. CE has been a tool to optimize the product realization for the benefit of the producer. Traditional product realization looked at issues at each process and optimized them locally, and then linked these pieces of information from process to process. Namely, our traditional product realization is a linear feed forward model. It is a differential way of thinking.

It is easily understood that the current product realization is based upon such a differential way of thinking, if we observe how inspections are carried out.

They are carried out from process to process in order to check whether a part meets the design specifications. In other words, the current product realization is verification-based.

## 2. Concurrent Engineering Tomorrow

### 2.1 Value

Such verification-based product realization was effective because in the 20th century changes are small and if any, they are gently curved. But in the 21st century, changes become sharp (Figure 2). Thus, such a differential way of thinking will not be effective anymore. We must develop another approach which is more flexible and adaptive.

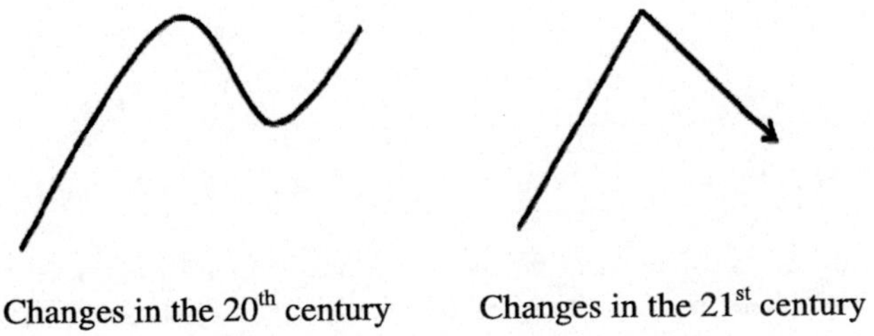


Figure 2. Changes of 20th century and 21st century

When we look for another approach, we have to get back to the basics. Why do we develop products? It is because it provides value. Value is defined by

Value= Performance/Cost

In the 20th century when product development was producer-centric, performance meant functions, because customers would like to have products with better functions. But globalization and increasing amount of information brought about diversification in requirements. As the word "customer" means, they would like to customize our products to their own needs and to their own tastes. They are no more satisfied with products with better functions at the time of delivery. They are more interested in how products can be customized and adapted to their own needs and preferences. Thus, the meaning of performance has changed. It now means emotional satisfaction [1].

## 2.2 Quality

In the 20th century, quality of a product has been considered in such a framework as shown in Figure 3. Producers made efforts to improve quality along each axis (each quality element) independently. But diversification calls for personalization. Customers would like to have a quality profile that meets their preferences, environments and situations
characterized profile to satisfy customers emotionally (Figure 4).

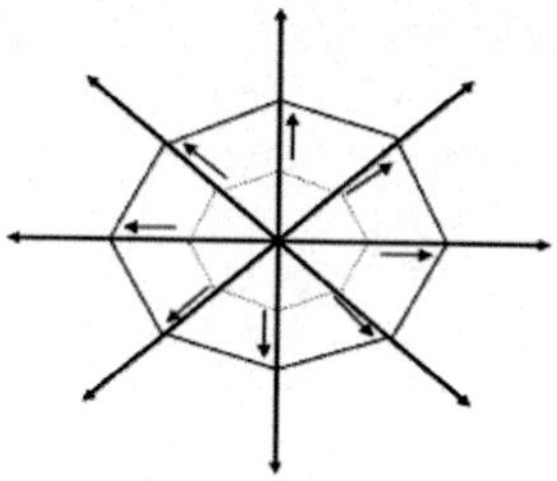

Figure 3. Traditional Quality Chart

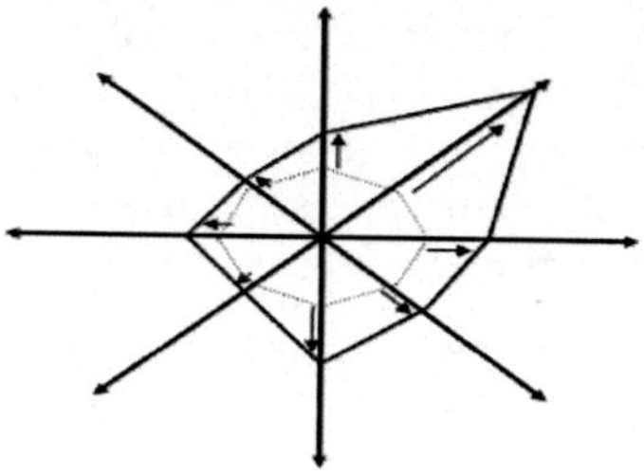

Figure 4. Personalized Quality Profile

In short, the focus of product development has changed from verification to validation. Producers have to look at quality in a much broader and longer perspective, not at each process or at each quality element (quality axis), but across all quality elements and throughout the whole product lifecycle including use. In fact, customers feel really happy and emotionally satisfied when products meet their own specific needs and preferences. Thus producers have to look at quality not as one time value, but as lifetime value (Figure 5). In other words, we should consider quality or product development not in a differential way, but in an integral way.

## 2.3 Experience

Behavior economics emphasizes the importance of experience in creating value. But what they are discussing is user experience. We should remember that customers not only would like to customize our products, but they also would like to be creative. Engineers emphasize the importance of creativity to meet the quickly increasing diversification of customer's requirements. But we should remember that customers themselves would like to be creative. If we can provide customers with opportunities to enjoy their creativity to the fullest extent, their emotional satisfaction would increase considerably. Then, why do we hesitate to get them involved in our product development? We can change our design and manufacturing to get them involved. Then, performance in the definition of value will be real performance and quality will no longer be evaluated at the point of delivery but will be evaluated as the integral of all these pieces of experience (Figure 5).

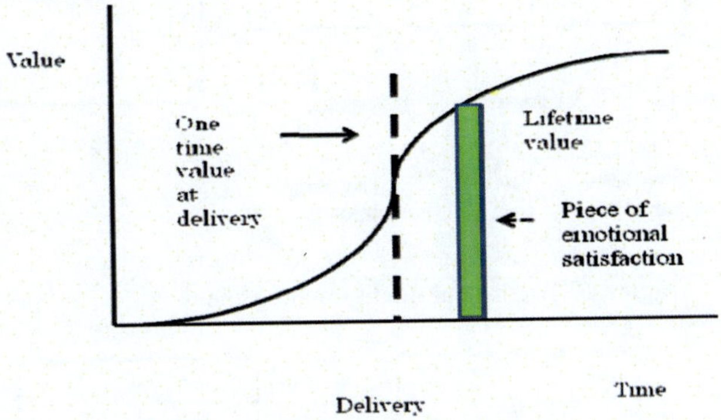


Figure 5. One time and lifetime value

## 2.4 Working Together: Value Co-creation

Prahalad and Ramaswamy [2] proposed the concept of value co-creation where the producer and the customer work together to create value. They call this value "unique value", because the customer and the producer's value will be identical. In the past, the producer developed products from their own perspective. Thus, value meant nothing other than profit to the producer and was not identical with that of the customer. If the customer and the producer work together, then they can identify the common goal so that they can create value common to both of them. This unique value means profit to the producer and emotional satisfaction to the customer at the same time. But their perspective is still producer-centric. Their emphasis is on how the producer and the customer can coordinate in identifying the common goal and their discussion is based on the current differential framework,

## 2.5 Looking from the Other Side

But if we look at production from the other side, i.e., from the customer's point of view, design, manufacturing, inspection and maintenance will be revolutionized.

Let us take inspection for example. Current inspection procedures are sequential. Inspection is carried out from process to process to remove poor quality parts, poor in quality from the standpoint of the producer (Fig.6). What matter more is that only flaws which the producer can expect are removed. Poor quality elements which the producer cannot predict might get through inspection undetected and impair the satisfaction of the customer (Fig.7).

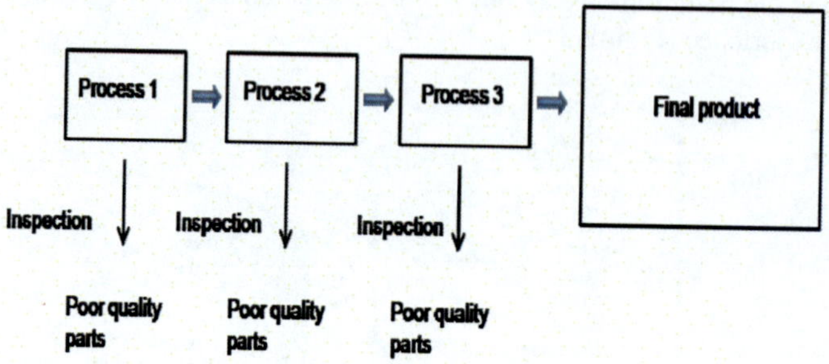


Figure 6. Traditional quality management

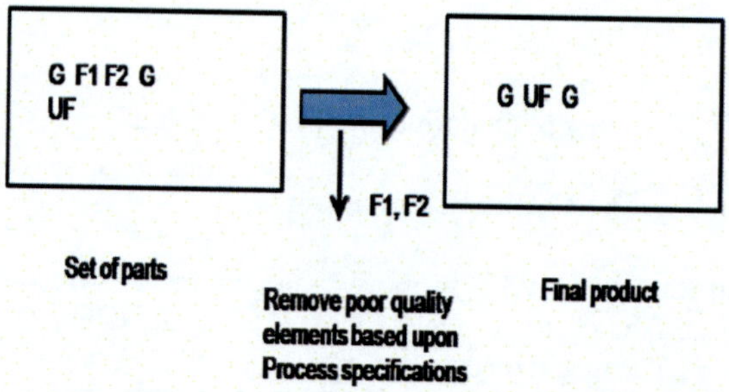


Figure 7. Shortcoming of traditional inspection procedures
(Predictable flaws F1, F2, -- can be removed.
But unpredictable flaw UF could get through undetected.
G stands for good quality element.)

But if we look at quality from the customer's side based on a personalized quality profile, then we can remove quality elements which are poor in the eyes of our customers. The remaining elements in the final product may be not good from the standpoint of the producer, but we should remember that our feeling of health is not identical with medical doctor's definition of health. If we live a happy life, we feel healthy, no matter what a doctor may say. He or she may not be healthy by doctor's definition. He or she may have some disorder. But if he or she is living a happy life, he or she "feels" healthy. We, engineers, like medical doctors, have been focusing too much on our definition of health and forgot to regard health (quality) from the perspective of daily life. If we look at quality from such user perspective or user emotion, inspection will be revolutionized and could be changed to monitoring-based as shown in Fig.8 and we can reduce a considerable amount of inspection [3]

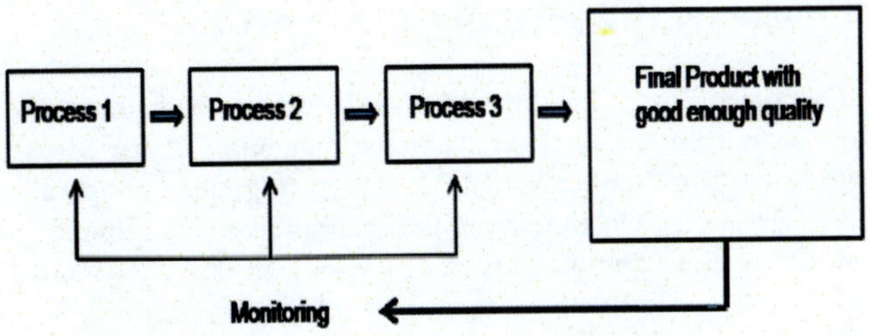


Figure 8. Monitoring based quality management
(Only quality elements good enough for customers are monitored.)

## 2.6 DIY Adds Value

If we manufacture our products ourselves, we attach greater value to them than those we just receive. If you prepare your dishes yourself, then they taste better than dishes delivered to you. This is what economics teaches us. What is more important is that it satisfies the human need of self actualization. Maslow [4] proposed the hierarchy of human needs (Figure 9).

The more we go up, the more important emotional satisfaction becomes. At the highest level, human beings would like to actualize themselves. If we can let our customers enjoy their creativity in product development, then we can satisfy their highest human needs.

Maintenance will also be greatly changed. It is no more an activity to restore the degrading functions back to the design level, but it becomes a challenge how we can customize our product. It is no more cost increasing chores, but it will create a great value. It will be truly performance on the part of our customers [5]

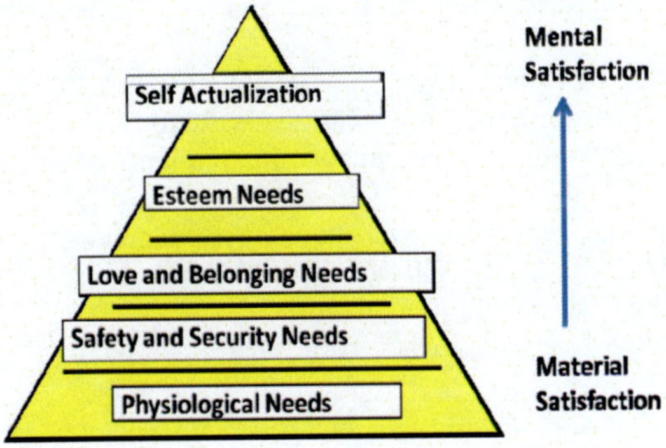


Figure 9. Maslow's hierarchy of human needs

## 3. What "Concurrent" Means in CE Tomorrow

Therefore, "concurrent" in CE tomorrow will come to include the whole definitions of "concurrent", i.e., 1. Occurring or existing at the same time. 2. Meeting in the same point; converging. 3. Acting together; cooperating. 4. In agreement. In other words, CE tomorrow will become very pragmatic [6] ,and it will be a methodology based upon an idea of "All's well that ends well" [7].

## 4. Summary

Traditional CE was based upon a linear feed forward model. In the $20^{th}$ century, changes were small and gentle. Therefore, point-wise optimization was effective. And value/quality was evaluated based upon the functions of a final product. This may be called a differential approach.

But in the $21^{st}$ century, changes are sharp so that such a differential approach is not effective anymore. Globalization and increasing amount of information brought about diversification of customer's requirements. Customers come to look for emotional satisfaction. They would like to be creative and to customize our products to their own needs and to their own tastes. Thus, lifetime value/ quality become important. To achieve this goal, our design and manufacturing must be changed to allow for customer involvement. This is an integral approach, based upon pragmatism.

## References

[1] Fukuda S. ed. Emotional engineering: service development. Springer, London, 2010.
[2] Prahalad CK, Ramaswamy V. The future of competition: co-creating unique value with customers. Harvard Business School Press, Boston. 2004.
[3] Fukuda S. Regrognostics: reverse quality management. DETC2010-28497, ASME IDETC/CIE 2010, Montreal, Canada. 2010
[4]Maslow AH. Motivation and personality. Intercollins College, New York, 1987.
[5] Fukuda, S, Virtual reality for reverse quality management. WINVR2011-5580, ASME WINVR 2011, Milan, Italy, 2011.
[6] http://en.wikipedia.org/wiki/Pragmatism
[7] http://en.wikipedia.org/wiki/All's_Well_That_Ends_Well

# Author Index